AF893740

Smith-Purcell Effect

Smith-Purcell Effect

V.P. Shestopalov

NOVA SCIENCE PUBLISHERS, INC.
Commack, NY

Assistant Vice President/Art Director: Maria Ester Hawrys
Office Manager: Annette Hellinger
Graphics: Frank Grucci
Acquisitions Editor: Tatiana Shohov
Book Production: Ludmila Kwartiroff, Christine Mathosian, and Tammy Sauter
Editorial Production: Susan Boriotti
Circulation: Cathy DeGregory and Maryanne Schmidt

Library of Congress Cataloging-in-Publication Data available upon request

ISBN 1-56072-446-3

Copyright © 1998 by Nova Science Publishers, Inc.
6080 Jericho Turnpike, Suite 207
Commack, New York 11725
Tele. 516-499-3103 Fax 516-499-3146
E-Mail: Novascience@earthlink.net
Web Site: http://www.nexusworld.com/nova

All rights reserved. No part of this book may be reproduced, stored in a retrieval system or transmitted in any form or by any means: electronic, electrostatic, magnetic, tape, mechanical, photocopying, recording or otherwise without permission from the publishers.

The authors and publisher haven taken care in preparation of this book, but make no expressed or implied warranty of any kind and assume no responsibility for any errors or omissions. No liability is assumed for incidental or consequential damages in connection with or arising out of information contained in this book.

This publication is designed to provide accurate and authoritative information with regard to the subject matter covered herein. It is sold with the clear understanding that the publisher is not engaged in rendering legal or any other professional services. If legal or any other expert assistance is required, the services of a competent person should be sought. FROM A DECLARATION OF PARTICIPANTS JOINTLY ADOPTED BY A COMMITTEE OF THE AMERICAN BAR ASSOCIATION AND A COMMITTEE OF PUBLISHERS.

Printed in the United States of America

CONTENTS

Preface ix

Introduction xiii

Part I THEORY OF SMITH-PURCELL EFFECT

CHAPTER 1. Plane waves and quasi-optical wave pencils of Smith-Purcell radiation 5

S 1. Unbounded electron flow over strip grating 5

S 2. Smith-Purcell radiation for the gratings made of perfectly conducting rectangular bars 15

S 3. Quasi-optical Smith-Purcell wave pencils 35

S 4. Smith-Purcell radiation of the point charge, single clot and finite sequence of charged clots 47

S 5. Influence of the finite size of gratings on the Smith-Purcell radiation 58

CHAPTER 2. Smith-Purcell and Cherenkov radiation. Electron-wave model of Smith-Purcell effect 109

S 6. Theory of Smith-Purcell and Cherenkov radiations 110

S 7. Electron-wave excitation of Smith-Purcell radiation. 123

CHAPTER 3. Wave modeling of Smith-Purcell radiation 155

S 8. Wave model of Smith-Purcell effect (scattering of waves of dielectric waveguide by strip diffractional gratings placed perpendicular to the waveguide) 159

S 9. Theoretical study of the wave Smith-Purcell radiation on the dielecticwaveguide and beveled strip grating 183

Part 2. EXPERIMENTAL STUDY OF SMITH-PURCELL EFFECT

CHAPTER 4. "Hot" experiments with ``pure" Smith-Purcell effects 201

S 10. Smith-Purcell experiments and their analysis 201

S 11. Experiments of Ishigaro and Tako 205

S 12. Experimental methods for the study of Smith-Purcell radiation in the millimeter range and the results of experiments. 206

S 13. Interaction of Smith-Purcell radiation with the field scattered by a grating (inverse Smith-Purcell effect). 215

CHAPTER 5. Excitation of open resonators by Smith-Purcell radiation 235

S 14. Open resonators with local reflection diffractional grating 235

S 15. Interaction of electron flow with the field of Smith-Purcell radiation in open resonators 244

S 16. Polarization effects of Smith-Purcell radiation in open resonators 248

CHAPTER 6. Experimental wave modeling of Smith-Purcell effect 269

S 17. Justification of the method of wave modeling of transformation of surface waves to volume waves by diffractional gratings 269

S 18. Experimental modeling of Smith-Purcell and Cherenkov radiations 281

S 19. Antenna properties of Smith-Purcell radiation for beveled gratings 288

Part 3. APPLICATIONS OF SMITH-PURCELL EFFECT

CHAPTER 7. Generators of diffractional radiation 315

S 20. Basic generator of diffractional radiation (continuous flight regime) 317

S 21.	Reflection, impulse and autodyne operation regimes of the generators of diffractional radiation	328
S 22.	Package generators of diffractional radiation	359
CHAPTER 8.	Application of Smith-Purcell effect in other areas of science and technology	393
S 23.	Measurements of the dynamical parameters of electron flow (remote sensing method based on Smith-Purcell radiation)	393
S 24.	Antenna properties of Smith-Purcell radiation	401
S 25.	Phase measurements for millimeter and submillimeter waves on the basis of Smith-Purcell effect	411
S 26.	Interferometer for the integrated schemes of millimeter range	416
S 27.	The device for excitation of planar dielectric waveguide and other elements of millimeter technology based on Smith-Purcell effect	419
S 28.	Diffractional input of energy to dielectric waveguide	421
S 29.	Diffractional logical elements of optical computer	423
S 30.	Formation of holograms with the help of Smith-Purcell radiation	426
S 31.	Application of Smith-Purcell effect for generation and amplifying of millimeter and submillimeter waves in semi-conductors	428
References		451
Index		469

PREFACE

It was published more than forty years ago in the "Physics Review" Journal the work [1] by S. J. Smith and E. M. Purcell where it had been described for the first time the results of experimental investigation of the radiational process occurring when electrons move close to a diffractional grating (DG). This remarkable natural phenomena was later called Smith-Purcell effect (SPE) or Smith-Purcell radiation (SPR). We have introduced together with these terms the completely equal definition of this effect: diffractional radiation (DR).

In accord with Einstein relativity principle the charged particle, electron, cannot radiate if it moves uniformly and straightforwardly in vacuum with the speed less than the light velocity (c). The case is somewhat different if electron moves in a medium or in a presence of an obstacle. Then the Cherenkov, transfer or diffractional radiation may take place.

Cherenkov radiation occurs when electron moves with the speed (v_e) greater than the light velocity in the medium. Figuratively speaking electron "jumps out" from the static field, which surrounds it, and the field "tail" remaining after the electron is the radiation field. Phase velocity of light in unbounded medium with dielectric and magnetic permittivities ϵ and μ is determined by the formula $v_\Phi = \frac{c}{\sqrt{\epsilon\mu}}$. Electromagnetic radiation forms the cone with the angle $\cos\theta = \frac{c}{v_e\sqrt{\epsilon\mu}}$. Since $\cos\theta$ is always less than unit, Cherenkov radiation is possible only when $v_e > v_\Phi$.

Transfer radiation is formed when electron crosses the boundary between two media with different dielectric permittivity. The variation of light velocity in a medium has a break and a part of particle's field is radiated.

SPR occurs when electron flies close to an object. Physically SPR is connected with polarization of obstacle's material. Electron, being close to an object, generates the static charge inside an obstacle. If electron moves straightforwardly and uniformly, the generated charge moves in the same way and hence constant current appears in an obstacle. Finally the charge reaches obstacle's boundary and disperses, and the current vanishes; the current varies in time and alternating current radiates alternating electromagnetic field. Such a radiation is known as SPR and it is especially intensive if there are many obstacles and they are placed periodically. Usually SPR is called diffractional if it occurs when electron moves close to infinitely conducting bodies. Sometimes it is called radiation of blinking dipole [2].

In the general case any scattering of eigenfield of moving charged particles on inhomogeneities may be related to DR. It is confirmed by applicability of diffraction laws in the study of DR. As well as in the process of diffraction of usual electromagnetic waves the eigenfield of electron flow is transformed (scattered) by obstacles to electromagnetic waves of scattering (diffractional radiation) which may spread in various directions depending on geometrical size of an obstacle, wavelength and speed of a charged particle.

The given analysis shows that Cherenkov radiation is the zero harmonic of diffractional radiation in a medium.

One should also pay attention to some distinctive features of DR:

radiation in vacuum is possible only at negative harmonics;

there exists pumping of energy from one radiating harmonic to another, i. e., Wood anomalies take place;

amplitude of DR harmonic decreases with growing number.

SPR or DR by its nature is more complicated and richer in features and characteristics in comparison with other radiational effects occurring as a consequence of uniform and straightforward movement of particles. It is widely used for remote sensing of electron flows and electromagnetic fields, elaborating radio and electron devices. Diffractional electronics [3] has been created on its basis. DR is applied in semiconductor, quantum and acoustic electronics, in super-gratings, plasma physics, antenna technology.

Theoretical and experimental studies of SPE have been carried out for more than thirty years by the Kharkov diffraction school. The results are presented in a number of monographs [3–5]. This book, which is specially prepared for the western reader, covers mainly the data obtained by my disciples and myself at the Kharkov Institute of Radiophysics and Electronics of the Ukrainian Academy of Sciences (IRE UAS) during the last two decades. It consists of three parts. In the first part theoretical investigations of SPE are systematically stated. In the second part experimental results are presented. The third part is devoted to application of SPE in various branches of science and technology.

Before the early sixties the studies of SPE were episodically organized by several groups in the USA (Brooklyne Institute of Technology), in Italy (Florence University), in Japan (Tokyo and Tohoku Universities) and also in Russia (Institute of Physical Problems, Institute of Physics, Institute of Radioengineering and Electronics of the Russian Academy of Sciences). The most valuable contribution to the solution of this problem was made by the scientists of the Kharkov Institute of Radiophysics and Electronics and Kharkov State University: V. A. Abdulkadirov, S. D. Andrenko, I. M. Balaklitskij, A. A. Vertij, G. S. Vorobiev G. P. Ermak, A, Ya. Kirichenko, G. I. Komar', A. P. Evdokimov, V. K. Korneenkov, V. V. Kryzhanovskij, V. G. Kurin, K. A. Lukin, V. S. Miroshnichenko, A. A. Petrushin, N. A. Popenko, S. A. Provalov, Yu. B. Sidorenko, Yu. K. Sirenko, B. K. Skrynnik, N. N. Suslov, A. S. Sysoev, O. A. Tretyakov, G. I. Khlopov, A. I. Tsvik, A. A. Shmat'ko, V. P. Shestopalov. They are the authors of most of the papers.

Creating the monograph has become possible owing to the kind agreement of the Nova Science Publishers to take upon itself the troubles concerning its publication. The author wants to express his deep gratitude to Dr. Columbus for the interest to his works and energy in getting things done. I am also thankful to Youri V. Shestopalov for translating and editing the monograph and for his comprehensive assistance.

INTRODUCTION

1. At the first stage SPE was studied using the blinking dipole approximation [2] and later using the approximation of given current within the frame of the idealized model of monochromatic electron flow (under the assumption that one longitudinal electron wave of space charge current is formed in a flow).

The blinking dipole model constructed on the basis of Huygens principle in order to explain the results of Smith and Purcell experiments [1,2] enabled to establish the wavelength and total energy losses of a dipole for SPR (for known electrons' velocities, angle of radiation's observation and grating's period). Calculations carried out by the formulas for λ and energy losses and their comparison with Smith-Purcell experiments [1] make possible to study some characteristics of SPR, in particular, to conclude that the light was strongly polarized. At the same time one cannot correctly evaluate radiation's intensity for arbitrary λ by the formula for SPR power.

Application of the method of given current for the study of SPE is a considerable step forward in the development of the theory. This method enabled to apply all that enormous theoretical background which had been well tested for the problems of volume waves diffraction by various objects including gratings.

In this monograph it is widely used rigorous numerical-analytical methods of the diffraction theory which turn out to be effective for solving the problems of surface waves scattering by gratings (including scattering of eigenwaves of the electron flow). Although in such statement it is not possible to take into account inverse influence of SPR on particles' movement, one can cover practically all situations of energy transformation of moving charged particles to SPR choosing the form of the charge (current) of electron flow.

One of the most impressive results of this theory is an opportunity of generating comparatively big intensities of SPR in the wave ranges differing from optical (in particular, in millimeter and submillimeter wave ranges); radiation of waves at the minus first harmonics; excitation of diffractional-Cherenkov radiation of the electron flow moving close to the strip grating placed on dielectric's surface (in this case zero diffractional harmonic of SPR may radiate whose excitation condition coincides with classical excitation condition of Cherenkov radiation).

2. However, it follows from the general classical analysis of electron-wave processes of energy exchange that under the influence of modulating

super-high frequency (SHF) field it is formed superposition of longitudinal and transversal electron waves propagating with different phase velocities. Slow (SWSC) and fast (FWSC) waves of space charge are related to the longitudinal waves; slow (SCW) and fast (FCW) cyclotron electron waves, which appear in an electron flow as a result of its focusing by magnetic field, are related to the transversal waves. Unlike electron waves of space charge (WSC), transversal waves have polarization properties caused by the movement (and rotation) of electrons along spiral trajectories. In particular, FCW and synchronous electron waves (SEW) are right-polarized while SCW and SEW are left-polarized [4]. These changes in dynamical characteristics of electron flow will undoubtedly lead to variation of "thin" structure of SPR which may be determined after the solution of corresponding problem.

The presence of "thin"structure of SPR excited by SWSC and FWSC is predicted in [6]. Here it is shown by the perturbation technique that forming of SWSC and FWSC in a flow leads to excitation of SPR in various directions. Angle variation of these waves is negligible (about $1^?$? hence one can hardly discover such effects experimentally. Unlike [6] it is established within the frame of given current approximation that the difference between radiation angles of not only longitudinal but also transversal electron waves may be essential and for real flow's densities under the certain conditions it may reach several degrees [7]. These theoretical results stimulated the development of experimental studies.

Thin structure of DR in the short-wave part of millimeter wave range has been discovered and investigated for the first time in the works [8–18]. The special DR analyzer has been created for experimental studies which enabled not only to separate DR of longitudinal and transversal electron waves but also to establish new effects: variation of polarization characteristics of excited electromagnetic field.

3. Discovery of SPR thin structure is especially important for developing physical foundations of diffractional electronics and creating of principle new devices for electronics of millimeter and submillimeter wave ranges: generators, amplifiers, electron parameter devices of diffractional radiation together with elaborating remote sensing methods for measurements of dynamical characteristics of electron flow (or other charged particles) in the devices of indicated wave ranges, since the ways of flow's diagnostics known from classical electronics are not applicable in the given case [19].

4. Further development of electron-wave mechanism of DR excitation

is reflected in theoretical and experimental studies of energy exchange and radiation of electron flow interacting with electromagnetic field scattered by periodical structure [12-16]. Here the flow most effectively interacts with the field of the space harmonic whose phase velocity coincides with phase velocity which has been formed in a pencil of electron wave: if a flow is synchronized with SWSC it gives back (amplifies) field's energy of a surface wave, and if a flow is synchronized with FWSC it takes this energy away (absorbs it).

An important characteristic of interaction of electron flow with scattered field is the possibility to control the energy exchange of electrons with space diffractional grating by varying the angle (phase) of incident field (here synchronizing parameter depends on this angle). These features of the process of energy exchange make the devices of diffractional electronics considerably different from the classical surface waves devices.

5. Operation principle of the devices of diffractional electronics is based on the effect of radiation of coherent clots of charge's densities of electrons forming longitudinal (transversal or superposition of longitudinal and transversal) electron wave of convectional current, which propagates along the flow close to a diffractional grating with constant (phase) velocity satisfying the condition of DR excitation. Such clots of charge's densities (or electron current's wave) are excited in a flow of electrons when they interact with the field of slow space diffractional harmonic, which is formed as a result of electromagnetic wave's diffraction on a periodical structure. Electrons' flow excites optimal energy density of DR when SWSC is "purely" synchronized with the field of slow diffractional harmonic. Then one can amplify electromagnetic waves when they interact with incident field scattered by a grating. If there is inverse communication of DR with electrons' flow, it is also possible to realize autogeneration of oscillations.

In a generator of diffractional radiation (GDR) electron-wave processes of energy exchange occur at eigenoscillations of an open resonator (OR) where total field is the sum of volume field in OR and surface waves of space diffractional harmonics close to DG. The condition of auto-oscillations' excitation here is exceeding of the power, which is given back by a high frequency electron flow to the field, over the power stored in OR during the period of high frequency oscillation.

6. The considered processes of DR excitation are confirmed by experimental studies [14,18]. Effects of amplifying and absorption of scattered field by an electron flow have been discovered and investigated together with the pos-

sibilities to control energy exchange when the phase of incident field varied. It was proposed and realized the remote way of measurement of dynamical characteristics of an electron flow interacting with an electromagnetic field scattered by a periodical structure [18]. It was determined experimentally by this method in the millimeter wave range phase velocities of the propagation of electron waves of space charge, synchronization parameters, reduction coefficients, effective plasma frequencies and other pencil's characteristics. The forming of effective plasma frequencies in an electron flow was discovered differing from ω_R of the electron flow with preliminary modulation taking into account dynamical correction which may be rather big under certain conditions ($\omega_R = R\omega_p$, where R is reduction coefficient, ω_p is the plasma frequency of unbounded flow). The possibility to measure dynamical characteristics of an electron flow in the millimeter and submillimeter devices considerably simplify solution to theoretical problems of optimization of their output data since one can use experimentally measured parameters of the given flow [4].

7. Observation of radiation phenomena for real particles in natural conditions or in a laboratory is connected with the difficulties caused by organizing the experiment, where one should provide the presence of particles' source and indication of their radiation field. Purely wave modeling of SPR (Cherenkov and transfer radiation) is possible [3] since some characteristics are determined kinematically by the properties of the medium and inhomogeneities taking into account charge's value and its velocity. One treats radiation processes as transformation of particle's eigenfield to radiation field representing a moving particle as a source of surface electromagnetic field. Particle's eigenfield is formed artificially by various electromagnetic structures, which can form the required surface fields.

Experiments of DR modeling are carried out with the help of plants which are used in radio engineering for antenna measurements. After the study of radiation far field pattern and redistribution of input and radiated power it is possible to get qualitative data which enable one to determine radiation conditions, radiation angle (its dependence on target distance for the minus first harmonic), radiation power (concentrated in the petals of far field pattern and strongly dependent on target distance and grating's properties). The results of SPR wave modeling essentially complement the data obtained after the studies of real flows of electrons, i.e., in the so-called hot experiments.

8. Since the most impressive SPE applications are related to the vacuum electronics of millimeter and submillimeter wave ranges, we will treat at

greater length GDR created on the basis of this effect.

There are various regimes of excitation of diffractional radiation in OR depending on purpose and functions of GDR: continuous generation regime with single flight of electron flow through the interaction space (flight regime of GDR), continuous generation regime with multiple interaction (reflection regime), impulse generation regime, autodyne generation regime and also a number of special regimes: heterodyne (reflection GDR with double interaction, regimes of amplifying of electromagnetic waves in open electrodynamical structures et al. The basis for the study of generation regimes is experimental data which has been obtained for the first time at the specially made operating models of the generators of dismountable constructions. They stimulated the development of GDR theory, creation of specific devices and statements of new problems determining the perspectives of diffractional electronics. Some of these results are reflected in [22-34] and presented in the given monograph.

The flight regime of GDR is characterized by continuous excitation of oscillations by an electron flow moving close to a diffractional grating, which is placed on one of OR mirrors. Operation principle of the generator is based on SPE occurring when electrons have single flight through the interaction space. This regime has been realized for the first time at the GDR model described in the works [22]. It was established by experimental data that DR excitation regime is perspective for creating highly effective sources of continuous oscillations. However for the solution to this problem one had to carry out systematic theoretical and experimental studies of linear and non-linear processes of energy exchange and output generator's characteristics in different stages of oscillations' excitation: in before-generation, starting, non-stationary stage of oscillations' establishment and in the stage of establishing stationary oscillations. It is important also to study spectral and fluctuation characteristics of excited oscillations, the possibilities to reduce starting currents, increasing generation's power, broadening the ranges of frequency adjustment, miniaturization of GDR, etc. Solution to these problems enables one to verify new effects and appropriateness of characteristics' variation in a flight GDR (FGDR) which are used nowadays for constructing specific devices.

In order to broad functional capabilities and applications of GDR it is proposed reflection regime of oscillations' excitation (RGDR) based on physical properties of DR and multiple interaction of an electron flow with the field of OR [28]. Combination of phase control of an electron flow which are used in FGDR and non-power control of energy exchange in a reflection clystron enables to provide in the considered device non-typical for FGDR excitation regimes: oscillations' excitation with essentially reduced starting current, effective non-power control of frequency generation and power, high modulation characteristics et al [27].

Impulse generation regime of diffractional radiation in OR of GDR is realized by letting current's impulses through the space situated close to a diffractional grating [31-33]. This generation regime is used for creating high-coherent impulse sources of electromagnetic oscillations, impulse GDR (IGDR) having high monochromatic index and stability of the bearing frequency, increased power of HF signal in generation impulse and other positive features which are non-typical for known impulse devices.

Autodyne operating regime of GDR is connected with variation of output or internal generator's characteristics under the influence of coherent eigenradiation reflected by the considered object (when autodyne signal is being registered the autodyne effect with internal detecting is observed in the feeding chain of GDR; the effect with external detecting is observed in the load chain; in the last case GDR operates simultaneously as a receiver and amplifier). Since GDR is a high-coherent source of millimeter and submillimeter waves, the study of autodyne regime is of essential interest. High q-factor of the oscillation system of GDR and considerable partition of SHF energy output from the electron flow (by means of diffractional radiation) determine high qualities of autodyne signal. Besides, the low steepness of GDR electron readjustment leads to its minimal sensibility relative to non-stabilities and pulsations of a feeding source. All these factors enable to obtain the highest operating data of GDR in autodyne regime in comparison with analogous autodyne devices made on the basis of magnetrons and clystrons.

It has been established in the process of investigating the energy exchange between an electron flow and electromagnetic field scattered by a grating that under the certain conditions electron flow amplifies the field reflected by the grating. It is observed in electron-wave approximation when SWSC is synchronous with surface diffractional harmonic. In this case the power of diffractional radiation "adds up" with the field of primary (input) signal of the reflected field since SWSC radiation is directed at the same angle as the reflected field. Amplifying coefficient grows due to the inverse communication of diffractional radiation with an electron flow which is realized with the help of reflecting mirror in OR or other type of open electrodynamical system.

At present it is studied experimentally the amplifying regime of FGDR (when currents are less than starting current), in two-cascade GDR and also in wide-band open waveguide-type electrodynamical system [21,34]. Results point out to the practical possibility of creating amplifiers of diffractional radiation in millimeter and submillimeter wave range.

9. After the comprehensive study of SPE and operating regimes of GDR one can directly start investigations of the problems of creation of package GDR in millimeter and submillimeter wave ranges. The basic attention should be paid here to working out generators operating in flight regime. Package GDR working in other regimes are necessarily considered for the construction of specific devices. Creation of package GDR requires solution of the following problems: construction of magnetic focusing systems (MFS) having minimal mass and given value of the induction of magnetic field in the chosen air gap; determination of optimal values of the induction of magnetic field in the given air gap; establishing optimal air gap of MFS [35-37]. One has to point out that the main contribution to the amplitude of total pulsations Δz of an electron flow in GDR of the indicated wave ranges are made by pulsations caused by transversal components of the tensity of electric field, space charge of the dense electron flow and transversal component of SHF field close to reflection diffractional grating. This conclusion well corresponds to the results of working out package GDR whose optimal induction of focusing magnetic field is approximately determined by the expression $B \approx (2\lambda)^{-12}$, where B, Tl, λ, mm.

If the main contribution is made by the pulsations of anode lense, one must provide $B \approx \lambda^{-1}$ in order to keep $\frac{\Delta z}{\lambda} = const$ (in place of $\sim \lambda^{-12}$).

For the wavelengths more than 2 mm the value of optimal starting current grows up to $J_S \approx 40 - 50\mu$ A for the working current $J_w \geq 0.1$ A owing to the application of electron flows with the thickness $z \approx 0.10$ mm, what decreases effective length of interaction from 14 mm to 10 mm and drives aperture of OR mirrors to 20 mm in the direction of flow's movement. As a result of application of such resonators GDR have been created for which MFS with air gap of about 32 mm were suitable [38].

In submillimeter wave range one must decrease the thickness of electron flow up to $z_0 \leq 0.04$ mm since its optimal thickness providing minimal value of the product of starting current and starting current's density is determined by the expression $z_{opt} \approx 0.19\beta_e\lambda_0$. However creation of electron-optical systems with thin electron flows meets a number of technological and of principle difficulties: the limiting value of current's density falls with decreasing flow's thickness in metallic-porous cathodes. The use of matrix cathodes [39] is perspective for increasing current's density of thin electron flows. Generalizing the results of working out package GDR let us point out that nowadays it is created package sources continuously covering all millimeter wave range with working current 0.13 A and accelerating voltage 2.4–4.0 kV on the basis of common construction which occupies the gap of about 40 mm in the magnetic field.

Owing to optimization of interaction space and OR parameters one can reduce the gap occupied by a device up to 32 mm without noticeable worsening of output data of GDR but with decreasing in two times the mass of package devices. Such sources have already been created in the whole millimeter range and they work in continuous and impulse regimes. Besides, GDR operating in autodyne regime have been worked out. The sensibility $2 \cdot 10^{-14}$ Wt$\cdot$Hz^{-12} has been reached.The possibility to register displacements $\Delta L = 10^{-3}\lambda$ has been shown (equal to units of micrometer); RGDR have been created which considerably reduced (up to tens of microamperes) starting current; small submillimeter GDR have been experimentally investigated.

A number of significant problems have been solved during the last decade in the Kharkov Institute of Radiophysics and Electronics in order to improve output data of GDR and to broad their functional capabilities: influence of vibrations have been reduced and temperature frequency coefficient have been lowered; efficiency and output power of continuously working GDR have been increased. In particular, the efficiency of interaction of electron flow has been raised almost in two times owing to the use of double gratings; the efficiency is approximately 15 per cent in the long-wave part of millimeter range and the level of output power is more than 100 Wt. Such gratings may be used in the short-wave part of millimeter wave range when pencil's thickness is 0.01–0.02 mm and current's density $i_0 \geq 30$ A cm^{-2}. Recuperation of energy of electron flow enabled to raise the technical efficiency in 1.5–2 times what was more than 20 per cent. It has been considerably improved the amplitude and frequency generator's control; mechanical frequency readjustment in the frequency range 500 MHz for the power's variation within the limits 3 dB has been provided.

In the constructed GDR the short-time frequency instability is 10^{-10} for tens of microseconds and it is determined by the fluctuations of accelerating voltage [40, 41]. The prolonged frequency instability is 10^{-5} when instability of accelerating voltage is about 10^{-3}. The level of amplitude noises of GDR with the same feeding blocks (for the level of pulsations of accelerating voltage $\approx 10^{-3}$) is 10 – 20 dB lower than in the devices of the type of a lamp of running wave, magnetrons and clystrons and it reaches 140 – 150 dB Hz^{-1} in the frequency band having the deviation $\Delta f \sim 5$ kHz from the bearing frequency [42].

The most essential and characteristic features of GDR are practically invariable width of its spectrum, steepness of electron and current frequency readjustments for constant ratio of the period of diffractional structure to the length of generated wave in the whole indicated range of wavelengths. GDR has the wide range of combined frequency readjustment what is provided by the coordinated variation of distance between the mirrors and anode voltage. It exceeds 20 per cent in package GDR when output power grows in five times. In case of the strict focusing of electron flow the range of combined readjustment of one device reaches one octave.

Maximal magnitudes of output power from one device to another in the whole millimeter range varies in GDR proportional to the squared value of optimal wavelength as well as in other electron devices of this wave range. With shortening the optimal wavelength effective thickness of electron flow and maximal value of synchronization parameter describing coordination between flow's velocity and phase velocity of "synchronous" harmonic decrease in proportion.

As the range of combined frequency readjustment of each device broads proportional to optimal frequency with shortening the wavelength, i.e., its relative magnitude does not change, the range of electron frequency readjustment narrows almost proportional to the wavelength, i.e., its relative magnitude narrows proportional to the squared value of wavelength. This phenomenon reduces the level of noises close to bearing frequency.

With variation of wavelength one has to change the shape and size of OR and DR for providing necessary communication of an electron flow with the field of OR. For example, plotting diffractional grating on a part of a plane mirror of semi-symmetrical OR [25] increases the feeding coefficient of OR by diffractional radiation and SHF field's intensity in the domain of the flight of an electron flow if to compare with the devices of the same class. The technology of gratings' manufacturing has been considerably improved, what has essentially raised the power of a generator of millimeter range in comparison with orothron [43].

Further increasing of the coefficient of transformation of DR to eigenoscillations of OR and raising field's intensity in the domain of flow's movement are carried out by pressing SHF field of OR in the direction perpendicular to the movement of electron flow. To provide this effect one must increase mirrors' curvature in this direction by replacing a cylindrical mirror by a plane one whose axis is parallel to the direction of electrons' movement [43]. For stronger pressing of the field's spot in this direction (for $\lambda < 2$ mm) and keeping its invariable length astigmatic mirrors are used. Their curvature radius along the flow's movement is in several times greater than in transversal direction. The composed astigmatic mirrors find applications in GDR of submillimeter wave range.

Since the losses in OR of GDR caused by DG as a rule exceed other losses and DG itself is a necessary element connecting the field of OR with an electron flow, then the choice of grating's parameters considerably influences efficiency of GDR excitation. Optimal value of the grating's filling coefficient is determined not only by the ratio of ohmical losses in a grating to the squared value of the amplitude of synchronous harmonic [44] but depends also on the losses in a load. If communication coefficient of a generator with a load $\sigma = 0.2$–0.5, the filling coefficient (d) of a diffractional structure should be equal to $d \approx (0.4 - 0.45)l$.

In the long-wave part of millimeter wave range one can improve interaction by means of reflection diffractional structures with elementary cells in the form of cylinders with narrow slots [45]. In addition to reducing the losses in such structures the tensity of electrical field considerably grows in the domain of the slot what leads to increasing of pencil's thickness, which effectively interacts with the field of OR.

Maximal interaction efficiency in GDR is achieved when the depth of reflection DG is close to the quarter of wavelength, $h \approx 0.25\lambda$. However, as it follows from experimental studies, one must reduce the grating's depth up to $h < 0.20\lambda$ in order to eliminate accompanying regime of LRW and to broad the readjustment range of GDR (it especially concerns submillimeter GDR).

Package GDR–autodyne have been worked out operating in the regime of external detecting. They essentially exceed other types of autodyne SHF devices by the coefficient of autodyne amplifying. One must point out that external detecting does not completely reveal capabilities of autodyne GDR and it can characterize only semi-conductor diodes used as detectors of autodyne signals. Conditions of switching on detectors in the case of external detecting require their strong communication with autodyne generator what does not allow to increase autodyne power over the limits determined by the detector [46].

Disadvantages of semi-conductor detectors become especially clear in reflectometrical plants for diagnostics of high-temperature plasma heated by powerful SHF generators of submillimeter wave range, which require reliable receiving devices being able to work in strong electromagnetic and radiational fields. For the solution of these problems autodyne GDR with electron signals' detecting have been worked out. They differ from usual GDR by the presence of two-electrode detectors placed at the exit of a pencil from the space of its interaction with the field of OR. By reacting on changes of energy of an electron pencil electron detector registers negligible variation of the pencil's current caused by the receiving signal. Sensibility of such detector does not yield to semiconductor one and by reliability and dynamical range it considerably exceeds the latter.

Owing to the combination of amplifying properties of autodyne GDR with high operation characteristics of electron detectors the transmitting-receiving devices of the entire millimeter wave range have been worked out having high phase accuracy and sensibility, broad dynamical range and regulated electrical parameters. Such devices combining the functions of a transmitter, an amplifier of input signal, a heterodyne device and a mixer have a number of universal capabilities. They can register quasi-stationary as well as fast processes having the frequencies up to several tens of MHz. Together with active autodyne regimes they may be applied as detectors and mixers with high sensibility in the millimeter and submillimeter wave ranges.

10. The variety of modifications of package GDR enables to widely use them in different areas of science and technology. Since information about the surrounding nature obtained as a result of interaction of electromagnetic radiation with crystals, plasma formations, earth's and ocean's surface, biological objects and the rest considerably depends on the quality of radiation itself, the "clearer" the initial signal the "thinner" effects may be discovered and registered.

In the spectroscopical experiments one needs first of all high frequency stability, low noise level, constant character of amplitude and phase parameters of the radiation, i. e., it must have high space and time coherent factor. There were no sources of such radiation in optics before the invention of lasers, hence the latter led to revolution in the most of optical disciplines. In SHF range as well as in the considered region of millimeter and submillimeter wave ranges the sources of coherent radiation appeared much earlier than lasers. However, the development of modern science and technology requires further improvement of the quality of electromagnetic radiation.

One of examples of the progress on this way is the creation of GDR. Having high-quality oscillation system this generator has much higher radiation's characteristics than the LRW constructed earlier. Let us notice that LRW was applied in the so-called LRW-spectroscopy which enabled to study in the wide frequency range spectral properties of the matter in the millimeter and submillimeter wave ranges. Its main advantage in comparison with Fourier-spectroscopy is rather high coherent factor of LRW what provides essential increasing of the measurement accuracy of LRW spectrometers. LRW also has comparatively wide range of electron frequency readjustment. However, further improvement of the characteristics of spectrometers is connected with raising the quality of radiation. In this way GDR is the most perspective source of electromagnetic oscillations [47]. Active inculcation of microprocessor technology may considerably simplify the creation of wide-range high-frequency GDR spectrometers.

The situation becomes in principle different in those spectroscopical experiments where not the frequency but some other characteristic varies, for example, the magnetic field. Then the decisive factor for registering the spectrum of electron paramagnetic resonance (EPR), which provides high spectrometer's accuracy, is the stability of slowly scanned magnetic field and frequency stability of a generator. As a rule, stabilization of magnetic field is not difficult. In particular, in the spectrometers of millimeter range at strong magnetic fields they bring superconductor solenoid in the regime of "frozen" current with very high stability and field's stretch is made by means of additional windings with low currents. Hence, the main contribution in raising the accuracy of the EPR spectrometer in the indicated wave range is made by generator's stability what is effectively provided in GDR.

The demands on stability of GDR become even more strict if one has to considerably increase the system's sensibility. With this purpose high q-factor resonance cell are used. Such situation appears either when very small concentrations of paramagnetic centers are applied or in the case of wide lines of absorption which can be registered with big difficulties. One of the most important parameters influencing the sensibility of EPR spectrometer or he generator's level of noises. In accord with the general theory of EPR spectrometer the minimal quantity of discovered spins is proportional to the half-degree noise-factor. Studies of GDR in the millimeter wave range show that its noise level is 20 dB lower than in LRW what outline the perspective of the use of GDR for solving the problems of spectroscopy of magnetic resonance. The presence of sufficiently strong autodyne effect in GDR enables to considerably simplify the scheme of radiospectrometer by eliminating frequency auto-readjustment. Besides, the autodyne regime for the registering of EPR spectrometers with the help of autodyne GDR may essentially improve the quality of information.

The problems connected with relaxometry in paramagnetics and pumping of dynamically polarized nuclei [48] are very close to the spectroscopical ones considered above. The main factor providing the success of experimental work is the long-time stability (in nuclear experiments the duration of measurements reaches tens and hundreds of hours) and comparatively high power: 1–2 Wt at the frequency 150 GHz. Here such experiments are assumed where powerful cryogenic plants are applied working at low and super-low temperatures. All these factors together with strong scattered fields of superconductor magnets require considerable moving off GDR from the working matter of a nuclear target. Here the main part of power is lost in a feeding transmission line. The latter may exclude relaxometrical measurements or provide effective nuclear polarization since here paramagnetic works in nonlinear regime, what requires a certain minimal threshold power. Situation aggravates especially in those cases when it is necessary to create nuclear targets of considerable volume operating at the temperature not lower than 1 K. Here the more substantial raising of power is required due to small times of spin-lattice relaxation (in comparison with working temperatures −0.5 K).

The experience of applying polarized nuclear targets shows that in a number of cases short-time stability (minutes) of GDR-pumping is also important. It depends on the mechanism of dynamical polarization and the type of nuclei being polarized in a target (for example, for deutrons' polarization it is necessary to use more stable generator). Practically all home nuclear targets use GDR for deutrons' polarization.

It is studied nowadays within the frame of the development of the program on polarization of nuclear targets the possibility to create the target of new generation on the basis of a stream of atom polarized hydrogen. For this one must use a high-stable generator which can work at the frequencies $180-220$ GHz. GDR is the most perspective type of a generator for solution of this problem. It provides sufficiently high power and frequency stability in the indicated frequency interval.

One of the ways of diagnostics are radio-locating measurements. When the frequency of sounding SHF signal is much greater than the frequency of electrons' collisions, one can investigate the behaviour of plasma domains with given concentration and determine the velocity of its movement by the signals reflected from plasma.

Development of the methods of SHF plasma heating and appearance in the future of pilot thermonuclear plants with strong levels of neutron and gamma radiation require working out diagnostic apparatus stable in regard to such factors. One can use GDR-autodyne with electron signals' detecting as a receiving-transmissing device for the control of behaviour of plasma boundary [49,50]. Such device enables to measure the velocity and displacement of plasma between the moments of its emergence and disintegration, to radiate and receive signals in a one guiding channel without junctions and commutators, to adjust from harmonics of powerful generators of plasma pumping by the high q-factor resonance system. The internal electron detector with high sensibility and broad dynamical range may provide reliability of autodyne GDR at arbitrary levels of external SHF signals and other radiation influences. It gives reason to consider the use of such receiving-transmitting device to be more perspective for radio-locating diagnostics of plasma than homodyne systems.

The radiowave method for the control of functional state of biological objects by means of autodyne GDR has been created, which enabled to carry out remote measurements of their electrophysical parameters. High phase sensibility and accuracy together with low internal noises enables autodyne GDR to obtain information about changes occurring in biological objects at low levels of radiated signal as when they are influenced by external factor and when these factors are absent. Owing to this method it has been registered the variation of reflection coefficient of the retina of isolated frog's goblet under the influence of a light impulse and relaxation processes in the intervals between the influencing impulses.

It has been measured by autodyne GDR the linear velocity of moving objects and vibration frequencies of immovable bodies. The accuracy of such device in the eight-millimeter range is equal to units of micrometer at the distance of several hundreds wavelengths from an object.

GDR radio-location of millimeter and submillimeter ranges is determined as location of near and super-near objects with the signal's delay of about 10^{-4} sec when it propagates to and from the located object. For these purposes GDR is the only type of a generator which can carry out forming and coherent processing of millimeter and submillimeter radio signals without complicated amplifying chains directly at the bearing frequency. The achieved level of the power radiated by GDR in the short-wave part of the millimeter wave range (about 1–3 Wt) provides the highest possible value of RLS energetic potential (the ratio of radiated power to the limiting sensibility of a receiver, 10^{16}–10^{17}) and essentially exceeds analogous parameters for other types of radiation sources (10^{11}–10^{12}). The models of high-coherent RLS on the basis of GDR has been created for applications in radiophysical remote sensing of the environment: it has been obtained for the first time with the help of these RLS the spectral characteristics of coherent signals scattered by hydrometeors, inhomogeneities of "clean" sky, plants et al; experiments were made up to the range of infralow Doppler frequencies.

11. SPE is widely used in antenna technology. As it is indicated above, SPE is developed in transformation of surface waves to volume waves. This physical phenomena may be used for creation of directed antenna system, where the sharply directed radiation far field pattern is formed due to the diffraction of inhomogeneous plane (surface) wave of a guiding system on the elements of DG connected with waveguide's field. Transmitting-receiving antenna of the millimeter wave range consists of DG with two similar dielectric waveguides placed on it [51]. Waveguides are oriented related to the grating in such a way that their axes are perpendicular to strokes' lines. If a surface wave propagates along the waveguide, then under the fulfillment of radiation condition the energy of a surface wave is transformed to volume waves which fly off the grating forming the far field pattern in free space. The incident angle of a part of radiation reflected back to DG is equal to the one at which the volume wave left the grating. In accord with reciprocity theorem the coming wave excites another waveguide placed on the grating. Hence the considered system (DG with two identical waveguides of surface waves) is an effective receiving-transmitting antenna with disjunction factor not worse than -70 dB.

It has been created on the basis of SPE a family of various antenna systems which were widely applied for construction of aerospace complexes, multi-beam radiometers of the millimeter wave range [52].

12. Many other applications of SPE are related to vacuum electronics, plasma physics and solid state physics, creation of element base, units and radiosystems of the millimeter and submillimeter wave ranges. Let us consider shortly some of them.

Let us point first of all to the ways of remote measurements of the dynamical parameters of plasma and electron flow which are based on the analysis of SPR excited by the waves of space charge. It enables to determine reduction coefficient of the plasma frequency of an electron flow and of a flow of cold plasma. It has been established in the process of analysis of SPR that modulation of electron flow by SHF field led to excitation of space and time spectrum of electron waves propagating with different phase velocities. If to let such a flow go close to DG, SPR is excited in the directions connected with these phase velocities and grating's parameters. This connection is determined from the conditions of SPR excitation, which give the reduction coefficient of plasma frequency with respect to the angle of inclination of maximal SPR intensity related to the initial angle. The analyzers of electron flow in the short-wave part of of the millimeter and centimeter wave ranges have been created with the help of indicated method. Analogous measurement devices may be created also for semi-conductor active media [8, 19, 20].

SPR finds broad application in working out the devices of the millimeter and submillimeter technology. Communication devices between linear and planar dielectric waveguides in the form of a slanted comb-type grating are used as the element base.

Interferometers in integral implementation have been worked out acting in the regime of surface waves under the fulfillment of the conditions of the diffraction of the second order Bragg waves. Both ends of a dielectric waveguide serve as channels of interfering signals and results of interference are registered in free space [4].

One of actual problems is effective input (and output) of energy to (or from) the dielectric waveguide [53]. This problem may be solved with the help of SPR when surface waves of a dielectric waveguide are transformed to volume waves in the presence of DG. Here dielectric waveguide remains regular since DG occupies only small external part of it.

Frequency shifters on the basis of SPR have been created consisting of a dielectric waveguide and DG. One can investigate automatically phase characteristics of electromagnetic fields by means of these devices (in particular, over the aperture of big mirror antennas [4]).

It is of considerable interest to study logical devices based on SPR for the future optical computers [54].

Part 1.

THEORY OF SMITH-PURCELL EFFECT

The simplest ideas about the radiation of volume waves by electrons moving uniformly and straightforwardly over diffractional gratings based on the model of blinking dipole [55-57] enabled to determine only frequency and polarization properties of the radiation as well as its orientation. The simplified models of blinking dipole led to mistakes in calculation of energetic characteristics. However, these data have been called in question for a long time. It was explained by the fact that very small radiation power was registered in the optical range and the blinking dipole theory predicted its strong fall with growing wavelength. Hence until the early sixties the Smith-Purcell effect was treated episodically without essential results [58-60].

During the last decade the conception of given current has been widely disseminated in the theory of the effect of diffractional radiation. Its success has been predetermined by the development of new methods in the diffraction theory [61,62]. Solution to the problems of radiation of unbounded plane monochromatic electron flow moving over infinite gratings gave the possibility to obtain the results essentially different from those followed from the blinking dipole theory. It has been proved already in the first works [63-65] of the author and his disciples that the power of diffractional radiation remained constant with shortening the wavelength λ if the current's density in an electron flow grew proportional to λ^{-1}. Hence it has been established theoretically that the practical use of SPE was not only possible but perspective far from the optical frequencies, in the millimeter and submillimter wave ranges.

A certain contribution to the construction of the modern theory of diffractional radiation has been made by the works [66] where the broad class of problems of this theory has been solved for the structures admitting factorization technique. Other methods of diffraction theory (the method of Riemann-Hilbert problem [61], semi-inversion method and others [62]) enabled to solve in rigorous statement the problems of diffractional radiation not only describing the movement of unbounded electron flows close to various infinite gratings but also the cases of bounded flows and periodical structures.

A number of new theoretical results [5] has been recently obtained in the studies of SPE, which were of great interest for further development of this scientific direction. It has been investigated the properties of excitation of diffractional, Cherenkov and diffractional-Cherenkov radiations by an electron flow moving close to metallic grating with dielectric substrata. The possibilities of raising the intensity of SPR by varying the gratings' parameters have been analyzed and "cold" modeling of the processes of excitation of SPR and Cherenkov radiation have been realized.

In order to ascertain the thin structure of SPR the electron-wave mechanism of excitation of diffractional radiation by a flow of electrons interacting with the field scattered by a grating has been studied.

As it has been pointed out above, at first the theory of SPE was based on the blinking dipole assumption [1,56,57,67]. In accord with this theory electron e^- and its reflection e^+ forms the dipole with varying dipole moment when flying close to the surface of a metallic grating with constant velocity $v_e = \beta_e c$ (see Fig. 1). Coherent radiation is excited due to periodical variation of the dipole moment $p = el$. Closer an electron moves to a grating, stronger the given effect . It is usually assumed within the frame of this model that the distance a (sight or aimed parameter) is considerably smaller than the grating's period l ($a << l$).

The condition of coherent radiation is easily found from Huygens construction (Fig. 1). The line BC is the front of all radiated waves; the difference between electrons' flight time t_1 on one grating's period and the time t_2 of field's wave propagation on the segment AB is divisible by the oscillations' period T with the wavelength λ, i. e., $t_1 - t_2 = nT$ where $t_1 = \frac{l}{\beta_e c}$; $t_2 = \frac{l}{c}\cos\theta$; then the radiation wavelength

$$\lambda = \frac{l}{n}(\beta_e^{-1} - \cos\theta), \quad n = 1, 2, 3, \ldots. \tag{1.1}$$

It follows from this formula that radiation of electrons moving close to a grating depends on structure's period and SPR is observed only at a certain angle θ for the fixed velocity $\beta_e = \frac{v_e}{c}$ of electrons and given wavelength λ.

If electrons move at a small distance from a grating ($b = 0.1l$), total energy losses of the dipole with the moment $p = el$ spent for radiation of the principal space harmonic is determined by the formula [1,2]

$$W = \frac{16\pi^4 l^2}{300\beta_e\lambda^2}L. \tag{1.2}$$

Hence one may conclude that SPR grows with increasing grating's length L, decreases with growing β_e (proportional to β_e^{-1}) and essentially falls with growing λ ($W \sim \lambda^{-2}$). Besides, it is not possible to study by (1.2) the radiation intensity via the angle θ.

As it will be shown in what follows, the use of the conception of given current for the creation of SPR enables one to solve almost completely the problem of determining the main characteristics of diffractional radiation by applying the rigorous mathematical approach which is well-tested in the diffraction theory.

CHAPTER 1

Plane waves and quasi-optical wave pencils of Smith-Purcell radiation

The simplest problem in the study of radiation of volume waves by charged particles is the problem on diffractional radiation, SPR, for the plane unbounded monochromatic electron flow moving uniformly and straightforwardly over unbounded periodical grating. This model, which is in general adequate to the physical process, in many cases may be rigorously analyzed by the mathematical methods when the approximation of given current is used. The essence of the considered problem is in the following. If all the values determining the movement of an electron flow are known, one can find its own electromagnetic field and then solve the problem of diffraction of the field of moving flow on a grating lying close to its trajectory. The scattered field is a superposition of inhomogeneous plane waves forming the localized field close to the grating and homogeneous plane waves dispersing to the surrounding space. As it has been indicated above, the dispersing plane waves were the diffractional radiation. The main feature of this effect is in transformation of eigenfield of uniformly moving charges localized in the space to homogeneous plane waves.

This model of SPE enables not only to expose qualitatively its physical essence but also points to the way of construction of the quantitative theory of the phenomena. When creating the theory of the effect of diffractional radiation one comes to the already studied diffraction problems. The problem of diffractional radiation for a periodical strip grating may be solved rigorously. The physical results, which follow from this solution, have the basic character. They give the possibility not only to compare theoretical results obtained in the modeling of SPE in more complicated situations but also to reasonably organize the experiment confronting its data with exact calculation characteristics.

It is also of great interest to study SPE within the frame of the model of given current for 1) plane monochromatic electron flow bounded in its width and length, 2) point charge, 3) single charged clot, 4) finite sequence of charged clots flying along an infinite grating. The influence of the finite size of a grating on SPR has been also studied.

§1. Unbounded electron flow over strip grating

1. Let the monochromatic electron flow with instantaneous variable charge's component

$$\rho = \rho_0 \delta(z-a) e^{i(k\alpha y - \omega t)} \tag{1.3}$$

move with constant velocity perpendicular to infinitely thin perfectly conducting strips of an infinite periodical grating (with period l and slots' width d, Fig. 6). Here ρ_0 is flow's modulation amplitude, ω is modulation frequency, c is velocity of light in vacuum, $k = \frac{\omega}{c}$, $\alpha^{-1} = \beta = \frac{v}{c}$ is relative flow's velocity; $\vec{v} = \vec{y}_0 c\alpha^{-1}$, $\delta(z-a)$ is Dirac delta-function, $\vec{x}_0$, $\vec{y}_0$, $\vec{z}_0$ are unit Cartesian vectors, a is aimed distance.

Electron flow with the charge's density (1.3) is similar to the waves of space charge arising due to inverse influence of the radiation on real initial non-modulated flow. Hence the field's source in the form (1.3) is the real model for the study of the generated diffractional radiation. Let us note also that representation of the current's density in this form considerably simplify mathematical statement and solution of the problem as a whole.

Within the frame of given current approximation the problem of finding SPR is reduced to the determination of electromagnetic field, which is as a superposition of the eigenfield of charges moving in unbounded free space and the field scattered by a diffractional grating. Eigenfield is the solution to inhomogeneous Maxwell equations whose right-hand side contains the given charge's density (1.3) and current's density $\vec{j} = \vec{y}_0 \rho\beta c$. The field of diffractional radiation must satisfy homogeneous Maxwell equations, radiation and boundary value conditions together with the condition of finiteness of energy in the given volume of space. This yields the following representation for the field in the domain over and under the grating:

$$\vec{H}^+ = \vec{x}_0 2\pi\rho_0 \text{sign} z e^{-k\gamma|z|+ik\alpha y} + \vec{x}_0 A \sum_{n=-\infty}^{\infty} \frac{a_n}{\alpha_n} e^{ik[\gamma_n(z+a)+\alpha_n y]};$$

$$\vec{H}^- = \vec{x}_0 2\pi\rho_0 \text{sign} z e^{-k\gamma|z|+ik\alpha y} + \vec{x}_0 A \sum_{n=-\infty}^{\infty} \frac{b_n}{\alpha_n} e^{ik[-\gamma_n(z+a)+\alpha_n y]}. \tag{1.4}$$

Here (and in what follows) the factor $e^{-i\omega t}$ is omitted and the constant

$$A = -i2\pi\rho_0\sqrt{1-\beta^2} e^{-2\pi\frac{a}{\beta\lambda}\sqrt{1-\beta^2}} \tag{1.5}$$

is introduced for the convenience of calculations. In addition to this,

$$\gamma_n = \sqrt{1-\alpha_n}; \ \alpha_n = \frac{\kappa\alpha+n}{\kappa}; \ \kappa = \frac{kl}{2\pi} = \frac{l}{\lambda}. \tag{1.6}$$

a_n, b_n are the sought for unknown coefficients of the Fourier-field (amplitudes of space harmonics) which are determined from the solution to the boundary value problem for total electromagnetic field (1.4). It is chosen the branch with $\phi < \pi$ of the root $\sqrt{B+iD} = re^{i\phi}$ in the calculations of γ_n. Demanding for (1.4) the fulfillment of boundary value conditions on strips and slots of a grating one obtains the system of functional equations of the first kind with respect to $b_n = -a_n$ which turns out to be a coupled summation-type equations with trigonometric kernel. This equations are of little use for the analysis of diffractional radiation. Their regularization or, in other words, separation of completely continuous and reversible (by the method of Riemann-Hilbert problem [61]) parts of the linear operator, leads to infinite Fredholm-type system of linear algebraic equations. The latter is solved numerically by the method of reduction in the broad range of wavelengths. Explicit representations for the sought for fields are obtained by successive iterations for some values of grating's parameters and wavelength of diffractional radiation. Application of the method of Riemann-Hilbert problem for the studies of SPE is considered in Chapter 4.

2. The very form (1.4) of the field is evidence of the fact that it is the countable set of plane H-polarized waves, where their finite number are homogeneous plane waves and the rest are inhomogeneous waves. The latter have imaginary values γ_n and hence they exponentially decrease with growing z in the direction off the plane $z = -a$ where the flow is situated. Homogeneous plane waves correspond to real γ_n which go off the grating without attenuation: it is the field of SPR.The condition of its existence is determined by inequality $(\kappa\alpha + n)^2 \le \kappa^2$:

$$\frac{|n|\beta}{1+\beta} \le \kappa \le \frac{|n|\beta}{1-\beta}; \quad \frac{\kappa}{|n|+\kappa} \le \beta \le \frac{\kappa}{|n|-\kappa}; \quad (n = -1, -2, -3, \ldots). \tag{1.7}$$

(1.4) enables us also to determine the direction of diffractional radiation

$$\psi_n = \arccos \frac{\kappa\alpha + n}{\kappa} \quad (n = -1, -2, -3, \ldots), \tag{1.8}$$

where ψ_n is the angle between wave vector of the n-th space harmonic and the axis Oy in the yOz-plane. It follows from (1.7) that

$$\lambda_n = \frac{1}{-n}(\beta^{-1} - \cos\psi_n) \ (n = -1, -2, -3, \ldots), \tag{1.9}$$

where λ_n is the wavelength of diffractional radiation.

Domains of diffractional radiation may be visually represented on the so-called Brilluin diagram [5] (Fig. 5). Its constructions determined by the choice of coordinate system $(\kappa, \kappa\alpha)$ and establishing the set of points where $\gamma_n = 0$; they lie on the beams starting from points $\kappa\alpha = n$ on the axis of abscissae and inclined to it at the angles $\frac{\pi}{4}$ and $-\frac{3\pi}{4}$. Radiation domain of the given harmonic with the index n (determined by inequality (1.7) is inside the sector bounded by two beams. The beams starting on the diagram from the apex of the sector $\kappa\alpha = n$ and lying inside it are the lines of constant radiation angle for the harmonics with longitudinal index n. The beam with inclination angle $\frac{\pi}{4}$ corresponds to zero radiation angle $\psi_n = 0$, $\frac{3\pi}{4}$ – to the inverse angle $\psi_n = \pi$; the beam $\kappa\alpha = n$ corresponds to diffractional radiation at the angle $\psi_n = \frac{\pi}{2}$. There is no radiation in the domain $\alpha = \beta^{-1} < 1$ since the relative flow's velocity with $\beta > 1$ is impossible. Shaded triangles are out of radiation domains of all harmonics, hence, they form "prohibited zones" for SPR. For those κ and α, which correspond to internal points of the triangles, the scattered field is the superposition of surface waves and diffractional radiation is not present at all.

For the given κ and β the working point on the diagram is determined as intersection of the line $\kappa = const$ parallel to the axis of abscissae and of the beam starting from the origin at the angle $\beta_< = \arctan\beta$. It is clear from Fig. 2 that for $\kappa = 1, 2$, $\beta = 0.4$ one observes radiation of the harmonics with $n = -3$ normal to the focal plane and with $n = -2$ at the angle $\psi_{-2} = 33.5$ (in the direction of flow's movement); the harmonic with $n = -4$ radiates in the opposite direction at the angle 123.5.

One can establish by means of Brilluin diagram that the space harmonics only with one index radiate for $\kappa < \frac{1}{2}$. These values of κ are of considerable interest since here the number of space harmonics of diffractional radiation is minimal what enables one to form the fields optimal in practical applications; besides, for $\kappa < \frac{1}{2}$ the value $\beta = \frac{v}{c} << 1$, i. e., one can use non-relativistic flows. And finally, this case corresponds to long-wave asymptotics in the scattering theory what makes possible to explicitly analyze energy characteristics of radiation.

3. Let us determine the field of diffractional radiation for $\kappa < \frac{1}{2}$. Here the radiation condition holds when the number n of radiating harmonic is equal to n_0 and $\nu \leq \kappa$ ($\kappa\alpha = n_0 + \nu$, n_0 is the integer closest to $\kappa\alpha$; $|\nu| < \frac{1}{2}$).

If, in addition to this, $e^{-k\sqrt{\delta^2\kappa^2-1}} << 1$ (here $\delta = 1 - |\nu|$), then

$$\vec{H} = \vec{x}_0 \mathrm{sign}(z+a)\frac{2\pi\rho_0\beta Re^{-\frac{a}{\beta\lambda}\sqrt{1-\beta^2}}}{1+i\kappa \mathrm{P}\sin\psi_{-\mathrm{n}0}}e^{ik(|z+a|\sin\psi_{-n_0}+y\cos\psi_{-n_0})};$$

$$\vec{E} = [\vec{y}_0\sin\psi_{-n_0} + \vec{z}_0\mathrm{sign}(z+a)\cos\psi_{-n_0}]\times$$
$$\frac{2\pi\rho_0\beta Re^{-\frac{a}{\beta\lambda}\sqrt{1-\beta^2}}}{1+i\kappa \mathrm{P}\sin\psi_{-\mathrm{n}0}}e^{ik(|z+a|\sin\psi_{-n_0}+y\cos\psi_{-n_0})}. \tag{1.10}$$

where $\cos\psi_{-n_0} = \frac{\nu}{\kappa}$; $\sin\psi_{-n_0} = \sqrt{1-\nu^2\kappa^{-2}}$; ψ_{-n_0} is the angle between wave vector of radiating harmonic and the axis Oy.

Average energy radiated by the unit square of grating's surface, over which the flow flies, is determined by the Pointing vector

$$\vec{S} = [\vec{y}_0\cos\psi_{-n_0} + \vec{z}_0\mathrm{sign}(z+a)\sin\psi_{-n_0}]\times$$
$$\frac{\pi\rho_0\beta^2cR^2e^{-2\pi\frac{a}{\beta\lambda}\sqrt{1-\beta^2}}}{1+\kappa^2\mathrm{P}^2\sin^2\psi_{-\mathrm{n}0}}e^{ik(|z+a|\sin\psi_{-n_0}+y\cos\psi_{-n_0})}. \tag{1.11}$$

where

$$R = \frac{P_{n_0}(u)P_\nu(-u) + P_{n_0-1}(u)P_{-\nu}(-u)}{P_\nu(-u)+P_{-\nu}(-u)};$$
$$\mathrm{P} = \frac{P_\nu(-u)-P_{-\nu}(-u)}{\nu[P_\nu(-u)+P_{-\nu}(-u)]}, \tag{1.12}$$

$P_{n_0}(u)$ and $P_\nu(u)$ are Legendre polynomials and functions, respectively; $u = \cos\frac{\pi d}{l}$.

As it follows from (1.10)–(1.12) in the limiting cases $u \to 1$ (the slotwidth tends to zero) and $u \to -1$ (the stripwidth tends to zero) radiation vanishes; optimal value d of the stripwidth corresponding to maximal radiation lies in the interval $0 < d < l$. When the wave with $n = -1$ radiates in the direction perpendicular to the grating approximate optimal value d_{opt} is determined by the formula

$$d_{opt} \approx l\frac{2}{\pi}\arcsin\sqrt{\zeta[1+\zeta^2(\zeta^2-\kappa^2)^{-2}]} \tag{1.13}$$

under the condition $|\zeta^2(\zeta^2-\kappa^2)^{-2}| << 1$, where ζ is a root of transcendental equation

$$exp\left(-\frac{\zeta}{(\kappa\sin\psi_{-1})^2}\right)=\zeta.$$

The normalized values of d_{opt} related to the grating's period for different values of κ are given in the Table 1. The magnitude d_{opt} decreases when the beam deviates from the normal.

For d_{opt} the power of the flow of diffractional radiation

$$\vec{S}_{opt}\approx[\vec{y}_0\cos\psi_{-1}+\vec{z}_0\mathrm{sign}(z+a)\sin\psi_{-1}]I^2\frac{94.5(1-\zeta)e^{-2\pi\frac{a}{\beta\lambda}\sqrt{1-\beta^2}}}{\sqrt{1+(\zeta\kappa^{-1}\sin\psi_{-1})^2}} \tag{1.14}$$

where $|\vec{S}|$, Wt cm^{-2}, I, A cm^{-1} is the linear current's density of a source; scalar coefficient by I^2 has the dimension in Ohm and it may be considered as radiation's resistance. It is clear that the last formula implies the growth of radiation's energy when the angle deviates from the normal and reaches its maximal value for $\psi_{-1}=0$ (the slotwidth decreases due to (1.13)).

It has been already indicated above the specific behaviour of SPR with respect to λ. This effect may be clearly followed if to analyze the representation (1.14). In the real situation $(\zeta\kappa^{-1})^2 << 1$ hence the factor by the exponent in (1.14) weakly depends on λ and radiation power practically does not change for small values of a as a function of λ. In this way one may conclude that for certain geometrical parameters of a grating and the value of aimed distance the flow of radiation's power does not depend on the wavelength what justifies the use of the effect of diffractional radiation for generating *millimeter and submillimeter* waves.

4. As an example let us consider the results of calculation by (1.10) the tensity of magnetic field in the case when radiation condition (1.7) does not hold (Fig. 3) and when (-1) harmonic radiates perpendicular to the grating (Fig. 4; the constant phase is depicted there by bold lines and absolute values of amplitude – by dotted lines). The fields have vortices on both sides of the grating. When there is no radiation they are situated exactly over the middle of the grating; amplitude's distribution is symmetrical with respect to strip's center and phase's distribution is asymmetrical with respect to it. Field's vortices shift in the direction of the movement of electrons' flow when diffractional radiation appears: field's symmetry related to strip's

center is violated (but the total field's symmetry with respect to grating's plane $z = -a$ holds in both cases; it follows from the condition $a_n = -b_n$).

It is of interest to calculate the average value of Pointing vector $\vec{S}_n$, Wt cm^{-2}, of the n-th space harmonic on the unit square what determines radiation's power from the area of 1 cm^2 of grating's surface

$$\vec{S}_n \approx [\vec{y}_0 \cos\psi_{-n_0} + \vec{z}_0 \mathrm{sign}\,(\mathrm{z}+\mathrm{a}) \sin\psi_{-\mathrm{n_0}}]\mathrm{I}^2 \frac{2.39\mathrm{S_p}}{\beta^2} \qquad (1.15)$$

The value

$$S_n = 4\pi\sqrt{1-\beta^2}\left|\frac{A_{n+n_0}}{\sqrt{1-\alpha_n^2}}\right|^2 e^{-4\pi\frac{a}{\beta\lambda}\sqrt{1-\beta^2}}, \qquad (1.16)$$

Ohm, may be called radiation's resistance; the coefficient $A_n = a_{n-n_0}\frac{\gamma_{n-n_0}}{\alpha_{n-n_0}}$ is found from the solution to corresponding system of linear algebraic equations, to which the boundary value problem describing diffractional radiation is reduced (here n is the number of space harmonic determined from the radiation condition (1.7), n_0 is the digit closest to $\kappa\alpha$).

The dependence of S_n, $n = -1, -2, -3, \ldots$ on dl^{-1} in the case when every harmonic propagates perpendicular to the grating is shown on Fig. 5, a,b (the obtained optimal slotwidth for the minus first harmonic well corresponds to the one predicted by formula (1.13)). The higher order space harmonics have $|n|$ maximums in the interval $0 \leq dl^{-1} \leq 1$ and S_n reaches its maximal value as a rule for small dl^{-1}.

Analogous dependence of S_n for different fixed radiation angels is presented on Fig. 5, c,d. As it is predicted by formula (1.13) the optimal slotwidths shift towards small dl^{-1} when radiation angle deviates from $\frac{\pi}{2}$. Radiation energy falls when radiation angle grows.

Fig. 6, a illustrates exponential decay of radiation energy with growing distance between electron flow and the grating. The domain of optimal values of the parameter

$$a < \frac{\beta\lambda \ln 10}{4\pi\sqrt{1-\beta^2}}. \qquad (1.17)$$

S_n as a function of flow's velocity $\beta = vc^{-1}$ is presented on Fig 6, b–e. The curves are denoted by corresponding index n; bold line determines the domains of variation of (1.7) where radiation condition holds for the given n, κ; dotted lines – where it does not hold. In all cases diffractional radiation vanishes when flow's velocity approaches the velocity of light. It is a formal consequence of (1.16): the coefficient A_{n+n_0} of S_n remains bounded for $\beta \to 1$ and the factor $1 - \beta^2$ tends to zero. It is clear from the physical point of view that the longitudinal component of source's eigenfield, which induces currents (on strips) generating the radiation, undergoes Lorentz reduction and tends to zero when $\beta \to 1$, hence, radiation naturally vanishes.

When several (not less than two) space harmonics simultaneously radiate, radiation energy of one of them increases in a narrow interval of β. It happens at the moment when other space harmonic slips along the grating and may be discovered only if to use the diffraction conception for description of the radiation of charges moving over periodical structures.

Fig. 7 illustrates the dependence of the average value of Pointing vector of the n-th space harmonic (equal to radiation power from the area of 1 cm^2 of the surface of a strip grating) on the parameter κ (Fig. 7, a,b shows long-wave range, the curves are related to different values of dl^{-1}; Fig. 7, c corresponds to the case similar to the one occurring in optics when many space harmonics radiate simultaneously). One can see sharp changes of radiation power of one of the harmonics close to integer values of κ, when simultaneous radiation of several harmonics in the narrow interval of β takes place. This effect indicated above has the same nature as the known Wood anomalies [68] in optics.

It is clear from Fig. 7, a that Wood anomalies S_{-1} are different for odd and even generated space harmonics. For the new propagating space harmonic with odd index S_{-1} anomalously decreases and with even index it grows. Hence, the absolute value of SPR in the peaks of anomalous parts of the curve does not grow but slightly fall with increasing κ, i. e., energy of diffractional radiation does not fall but grow with increasing wavelength when transferring from optical (big κ) to long-wave (small κ) range. It is the result of the rigorous solution to the boundary value problem of diffractional radiation disproving the known conclusion, which follows from the geometrical-optical approximation and "blinking" dipole theory [1,56,69] stating that energy of diffractional radiation may be considerably strong only in the optical wave range and falls with growing λ as λ^{-2}. It confirms the prospective use of the effect of diffractional radiation for generating millimeter and submillimeter waves.

5. In order to obtain analytical expression for the density of the current induced by a source on grating's strips one must apply boundary value con-

dition for the normal components of the tensity of electrical field. Let us limit ourselves to the case of narrow slots assuming that

$$\frac{\pi}{\sqrt{2}}\frac{d}{l} << 1, \quad \nu\Big(\frac{\pi}{\sqrt{2}}\frac{d}{l}\Big)^2 \ln\Big(\frac{\pi}{2\sqrt{2}}\frac{d}{l}\Big) << 1.$$

This assumption yields the required expression (verified under the condition that single space harmonic with the index $n = -1$ radiates)

$$\sigma = -\rho_0\sqrt{2(1-\beta^2)}e^{-2\pi\frac{a}{\beta\lambda}} \times$$

$$\frac{\Big[1 + 2\nu A_0 \ln\big(\frac{\pi}{2\sqrt{2}}\frac{d}{l}\big)\Big] \sin\frac{\pi y}{2l} + iA_0)\cos\frac{\pi y}{2l}}{\sqrt{\cos\frac{\pi d}{l} - \cos\frac{\pi y}{l}}} e^{ik\cos\psi_{-1}} + O(1), \tag{1.18}$$

where

$$A_0 \approx i\frac{\beta\sin\psi_{-1}}{\sqrt{1-\beta^2}}\frac{\cos^2\frac{\pi d}{2l} - 2\nu\sin^2\frac{\pi d}{2l}\ln\sin\frac{\pi d}{2l}}{1 + i2\kappa\sin\psi_{-1}\ln\sin\frac{\pi d}{2l}}.$$

In the given formulas the density of surface charge has the required singularity $r^{-0.5}$ on strip's edges and the density of the induced surface current decreases as $r^{0.5}$ towards the edges (r is the distance from strip's edge). Therefore, these magnitudes are not symmetrical with respect to the middle of the strip and they are shifted in the direction of movement of an electron flow. The results of calculations on Fig. 4 for the tensity of magnetic field determining the density of surface current confirm the conclusions following from (1.18).

6. The solution to the problem about radiation of plane monochromatic electron flow may be generalized for the case of radiation of a charged thread parallel to the axis Ox and moving towards the axis Oy.

Radiation diagrams of a charged thread moving close to a strip grating with relative velocity $\beta = 0.4$ are presented on Fig. 8, a–d. The thread which moves uniformly and straightforwardly excites diffractional radiation corresponding to the minus first harmonic; its wavelength λ_{-1} lies in the interval $1.5l < \lambda_{-1} < 3.5l$ and its radiation angle ψ_{-1} related to Oy belongs to the interval $0 < \psi_{-1} < \pi$; $\cos\psi_{-1} = (\kappa\alpha - 1)\kappa^{-1}$. The radiated minus second harmonic has the spectrum $0.75l < \lambda_{-2} < 1.75l$. When this harmonic with $\lambda_{-2} = 1.75l$ radiates at the angle $\psi_{-2} = \pi$, the minus first radiates at the angle $\psi_{-1} \approx 41$. Close to this angle the radiation pattern undergoes a strong change caused by the Wood anomaly.

One can see on Fig. 8, a–d that the character of Wood anomaly considerably depends on geometrical parameters of the grating: radiation energy grows for wide slots; when a slot narrows this anomaly leads to the fall of energy of diffractional radiation; for the very small slot radiation at the angle $\psi_{-1} \approx 41$ practically vanishes. The efficiency of the use of diffractional radiation for generating millimeter and submillimeter waves depends on the domains of the presence of Wood anomalies.

§2. Smith-Purcell radiation for the gratings made of perfectly conducting rectangular bars.

As it is shown in the previous paragraph, energy characteristics of SPR are mainly determined by the geometrical properties of grating's cross-section (by the filling coefficient in the case of strip grating). Gratings made of perfectly conducting rectangular bars as well as reflection grating of the same cross-section present the more general family of gratings [70]; they are widely used in vacuum electronics.

1. If an electron flow having the charge density (1.3) moves with constant velocity at the distance a over semi-transparent grating made of perfectly conducting rectangular bars (with the period $2l$, slotwidth d, bars' height h), the electromagnetic field of diffractional radiation may be represented in the form

$$\vec{H}^{+} = \vec{x}_0 A \sum_{n=-\infty}^{\infty} a_n e^{ik[\gamma_n(z-h)+\alpha_n y]}, \ z > h; \tag{1.19}$$

$$\vec{H}^{in} = -\vec{H}^0 + \frac{1}{2}\vec{x}_0 A \sum_{n=-\infty}^{\infty} (c_n \cos q_n z + i d_n \sin q_n z) \cos \frac{\pi n}{2d}(y + d - 2N),$$

$$|z| < h, \ |y - 2Nl| < d;$$

$$\vec{H}^{-} = -\vec{H}^0 + \vec{x}_0 A \sum_{n=-\infty}^{\infty} b_n e^{ik[-\gamma_n(z+h)+\alpha_n y]}, \ z < -h;$$

$$\vec{E}^{-} = \frac{1}{ik}\left(\vec{y}_0 \frac{\partial H_x^-}{\partial z} + \vec{z}_0 \frac{\partial H_x^-}{\partial y}\right), \tag{1.20}$$

where the source's eigenfield

$$\vec{H}^0 = \vec{x}_0 \mathrm{sign}\,(\mathrm{z} - \mathrm{z}_1) 2\pi \rho_0 \beta \mathrm{e}^{-\mathrm{k}\alpha\sqrt{1-\beta^2}|\mathrm{z}-\mathrm{z}_1|} \mathrm{e}^{\mathrm{i}(\mathrm{k}\alpha \mathrm{y}-\omega \mathrm{t})};$$

$$\vec{E}^0 = [\vec{y}_0 i\sqrt{1-\beta^2} + \vec{z}_0 \mathrm{sign}\,(\mathrm{z}-\mathrm{z}_1)] 2\pi \rho_0 \mathrm{e}^{-\mathrm{k}\alpha\sqrt{1-\beta^2}|\mathrm{z}-\mathrm{z}_1|} \mathrm{e}^{\mathrm{i}(\mathrm{k}\alpha \mathrm{y}-\omega \mathrm{t})}; \tag{1.21}$$

For the sake of convenience we assume that the origin is shifted along the axis Oz to the point $z_1 = h + a$; the notation coincides with the one introduced in the previous section and

$$q_n = k\sqrt{1 + \left(\frac{n}{2\kappa\theta}\right)^2}; \ \theta = \frac{d}{l}; \ \delta = \frac{h}{l}. \tag{1.22}$$

The sign in the expressions for γ_n, q_n is chosen in such a way that $Im\, q_n \geq 0$, $Im\, \gamma_n \geq 0$; a_n, b_n, c_n, d_n are the sought for amplitudes of the space harmonics of diffractional radiation.

It follows from (1.19) that the scattered field contains diffractional radiation consisting of superposition of non-attenuating waves, if there exist the digits $-n = 1, 2, \ldots$ providing the fulfillment of inequality $\alpha_n^2 < 1$. From the other side, it means that either κ for given β or β for fixed κ must lie in the intervals (1.7). Radiation wavelength λ is connected with grating's period $2l$, relative flow's velocity and radiation angle ψ_n of the n-th space harmonic by the same relation (1.8) as in the case of the strip grating (one should replace l by $2l$ in the formula for λ). This assertion follows from the field's representation as a Fourier series by the system of functions which is complete on the interval coinciding with structure's period and it is not connected with the solution to boundary value problem.

The symmetry of the considered grating with respect to the plane $z = 0$ enables one to obtain expressions for the total field when the combs have the height h and the basis made of perfect magnetic lying on the plane $z = 0$.

2. The magnitudes a_n, b_n, c_n, d_n, which are the sought for amplitudes of the space harmonics of diffractional radiation, are determined after applying the boundary value conditions on the grating's surface for the total field. It yields the system of functional equations with respect to the linear combinations of amplitudes of field's harmonics in external domain

$$X_n = a_n + b_n, \quad Y_n = a_n - b_n \tag{1.23}$$

and amplitudes of field's harmonics inside the slots.

Let us note that X_n, Y_n coincide with the amplitudes of field's space harmonics of reflection grating with the basis made of perfect conductor and of the grating made of perfect magnetic, respectively. One can obtain the field inside the slots for the first grating if to put $c_n^{con} = 2c_n$, $d_n \equiv 0$ and for the second grating if to put $d_n^M = 2d_n$, $c_n \equiv 0$.

The system of functional equations for X_n and Y_n may be reduced to the infinite systems of linear algebraic equations of the second kind using the method of re-distribution of a system of functions complete on one interval by a system of functions complete on the other interval [71]:

$$X_n - \sum_{s=-\infty}^{\infty} X_s P_{ns} = \Phi_n,$$

$$Y_n - \sum_{s=-\infty}^{\infty} Y_s Q_{ns} = \Psi_n, \quad (n = 0, \pm 1, \pm 2, \ldots), \tag{1.24}$$

where

$$P_{ns} = P^0_{ns} + P^1_{ns}; \Phi_n = \Phi^0_n + \Phi^1_n; \Phi^0_n = -\delta^n_0 - P^0_{n0}; \Phi^1_n = -P^1_{n0};$$

$$\mu_n = \kappa\alpha + n; \; \nu_{ns} = n + 2\theta\mu_s; \; \eta_{ns} = n - 2\theta\mu_s; \; \chi_n = n^2 - (2\kappa\theta)^2;$$

$$P^1_{ns} = i\theta^2 \frac{4}{\kappa\gamma_n}\mu_n\mu_s \sum_{m=1}^{\infty} \frac{\sqrt{\chi_m}\tanh\left[\pi\delta(2\theta)^{-1}\chi_m^{-\frac{1}{2}}\right]}{\nu_{mm}\nu_{ms}} S_{mn}S_{ms};$$

$$P^0_{ns} = i\theta\frac{1}{\gamma_n}\tan(\kappa\pi\delta) S_{0n}S_{0s}; \; S_{mn} = \frac{\sin\frac{\pi}{2}\eta_{mn}}{\frac{\pi}{2}\eta_{mn}};$$

$$Q_{ns} = Q^0_{ns} + Q^1_{ns}; \Psi_n = \Psi^0_n + \Psi^1_n; \Psi^0_n = -\delta^n_0 - Q^0_{n0}; \Phi^1_n = -Q^1_{n0}; \tag{1.25}$$

$$Q^1_{ns} = i\theta^2 \frac{4}{\kappa\gamma_n}\mu_n\mu_s \sum_{m=1}^{\infty} \frac{\sqrt{\chi_m}\coth\left[\pi\delta(2\theta)^{-1}\chi_m^{-\frac{1}{2}}\right]}{\nu_{mm}\nu_{ms}} S_{mn}S_{ms};$$

$$Q^0_{ns} = i\theta\frac{1}{\gamma_n}\cot(\kappa\pi\delta) S_{0n}S_{0s}; \; \delta^n_0 = \begin{cases} 1, & n = 0; \\ 0, & n \neq 0. \end{cases}$$

The systems of equations (1.24) are suitable for determination of X_n, Y_n by numerical methods for arbitrary values of $\kappa, \beta, \theta, \delta$. In the case of narrow slots one can establish from (1.24) approximate formulas to within the accuracy of the order θ^2:

$$X_n \approx -\delta^n_0 - \frac{2i\kappa\theta\tan(\kappa\pi\delta)}{1 - iS\theta\tan(\kappa\pi\delta)}[\kappa^2 - \mu_n^2]^{-\frac{1}{2}} \frac{\sin\pi\theta\mu_n}{\pi\theta\mu_n}\frac{\sin\pi\theta\kappa\alpha}{\pi\theta\kappa\alpha}, \tag{1.26}$$

where

$$S = S_1 - iS_2 = \sum_{n=-\infty}^{\infty} \frac{\sin^2[\pi\theta\mu_n]}{\pi^2\theta^2\mu_n^2\sqrt{1 - \left(\frac{\mu_n}{\kappa}\right)^2}}. \tag{1.27}$$

The coefficient Y_n may be calculated in a similar way:

$$Y_n \approx -\delta^n_0 + \frac{2i\kappa\theta\cot(\kappa\pi\delta)}{1 + iS\theta\cot(\kappa\pi\delta)}[\kappa^2 - \mu_n^2]^{-\frac{1}{2}} \frac{\sin\pi\theta\mu_n}{\pi\theta\mu_n}\frac{\sin\pi\theta\kappa\alpha}{\pi\theta\kappa\alpha}. \tag{1.28}$$

Formulas (1.26), (1.27) hold only when one principal wave propagates in the slots between bars, i. e., if $2\kappa\theta < 1$. If $\kappa < 0.5$, one can approximately calculate the sum of the series for S. In order to verify this approximate representation let us use the notation $\kappa\alpha = n_0 + \nu$, $|\nu| < 0.5$ and replace the index n by $m = n + n_0$; finally we get the required result

$$S = \frac{1}{\sqrt{1 - \frac{\nu^2}{\kappa^2}}} + 2i\kappa \ln \sin \frac{\pi d}{2l}. \tag{1.29}$$

In this case explicit expressions for the amplitudes of space harmonics for SPR field of the comb with the perfectly conducting basis have the form

$$a_n = X_n - \delta_0^n + 2P \frac{\sqrt{1 - \frac{\nu^2}{\kappa^2}}}{\sqrt{1 - \frac{\mu_n^2}{\kappa^2}}} \frac{\sin \pi\theta\mu_n}{\pi\theta\mu_n} \frac{\sin \pi\theta\kappa\alpha}{\pi\theta\kappa\alpha};$$

$$c_0 \approx -\frac{2\sqrt{1 - \frac{\nu^2}{\kappa^2}}}{\sqrt{(1 - \frac{\nu^2}{\kappa^2}} \cos \pi\kappa\delta - i\theta(1 + i\sqrt{\kappa^2 - \nu^2} \ln \sin \frac{\pi d}{2l}) \sin \pi\kappa\delta} \frac{\sin \pi\theta\kappa\alpha}{\pi\theta\kappa\alpha}; \tag{1.30}$$

$$b_n \equiv 0; \;\; d_0 \equiv 0,$$

where

$$P = \left[1 + i\sqrt{1 - \frac{\nu^2}{\kappa^2}} \left(2\kappa \ln \sin \frac{\pi d}{2l} + \frac{1}{\theta \tan \pi\kappa\delta}\right)\right]^{-1}. \tag{1.31}$$

3. Let us analyze physical properties of the diffractional radiation using the results of numerical solution to the initial equations. Let us determine the average value of Pointing vector over the unit square of arbitrary plane $z = const$ of the n-th space harmonic satisfying radiation condition. In the case of reflection grating made of bars with perfectly conducting basis the radiation is possible only to the upper halfspace; the flow of power from the unit square has the form

$$\vec{S}_n = (\vec{y}_0 \cos \psi_n + \vec{z}_0 \sin \psi_n) \frac{c}{4\pi} \rho_0^2 S_n, \tag{1.32}$$

where

$$S_n = 4\pi^2 \beta^2 |a_n|^2 e^{-4\pi \frac{a}{\beta\lambda} \sqrt{1-\beta^2}}. \tag{1.33}$$

The expression for the flow of power from the unit square may be rewritten in the form suitable for engineering calculations, Wt cm^{-2}:

$$\frac{c}{4\pi} \rho_0^2 S_n = I^2 R_n, \tag{1.34}$$

where I^2, A cm^{-1}, R_n, Ohm, I is the linear current's density;

$$R_n \approx 94.5|a_n|^2 e^{-4\pi \frac{a}{\beta\lambda}\sqrt{1-\beta^2}} = \frac{2.38}{\beta^2} S_n. \tag{1.35}$$

Calculations show that the radiated power is a periodical function of grating's depth. For $h = 0$ the grating made of bars is transformed to the strip grating and calculated values of S_n graphically coincide with the numerical data obtained by means of the Riemann-Hilbert method (§1). Characteristic resonances of $S_n^{\pm}$ occur when the grating's depth is approximately divisible by 0.5λ and the optimal depth is somewhat less than $(2N+1)0.5\lambda$ ($N = 0, 1, 2, \ldots$). Radiation power both over and under semi-transparent grating made of bars is minimal when the slotdepth is close to $(2N+1)0.25\lambda$ ($N = 0, 1, 2, \ldots$). Variation of grating's depth has much stronger influence on the power of diffractional radiation going down from the grating than the one going up.

For fixed β, κ and δ there are definite values of the grating's filling coefficient when S_n reaches its maximum. There exist optimal values $\theta_{opt}, \delta_{opt}$ when $S_n = S_n(\theta, \delta)$ has maximal possible value. For $\delta = \delta_{opt}$ (δ_{opt} is a little less than $(2N+1)0.5\lambda$ $N = 0, 1, 2, \ldots$) the optimal slotwidth is small and lies in the interval $0.15 \leq \theta_{opt} \leq 0.25$. Hence, the formulas (1.30)–(1.33) obtained for $\theta^2 << 1$ are important for constructing explicit expressions for the grating with optimal geometrical parameters and for analytical investigation of the properties of diffractional radiation in the optimal regime. If the slotwidth is not optimal, the filling coefficient is shifted in the direction of greater values of θ and lies in the interval $0.3 < \theta < 0.5$.

It has been calculated also S_n as function of κ for the fixed depth of a grating and different values of the filling coefficient. Resonance growth of radiation power for the specially chosen grating's depth is noticeable only when the slotwidth is less than half of the period; it vanishes for $\theta > 0.5$. The sharp change of radiation intensity is observed in the vicinity of the values

$$\kappa_a^+ = \frac{|n|\beta}{1+\beta}; \ \kappa_a^- = \frac{|n|\beta}{1-\beta}, \ n = -1, -2, \ldots, \tag{1.36}$$

which is connected with resonance phenomena similar to Wood anomalies in optics. The formulas (1.36) exactly predict the conditions when these anomalies occur in diffractional radiation. Its character is determined by a number of other circumstances and become rather complicated when radiation conditions hold simultaneously for several space harmonics.

The numerical results for the periodical structure made of semi-infinite bars with arbitrary thickness $2(l-d)$ are given on Fig. 9 (these data enable, in particular, to judge to what extent the structure consisting of half-planes models a real structure). When $2(l-d) < 0.1$, this structure may be simulated by the grating of half-planes and accuracy of calculation of the energy of diffractional radiation will not exceed 10numerical data obtained by the Wiener-Chopf technique graphically coincide on Fig. 9, b, e. Their difference grows while approaching the limiting values of κ (1.36) lying on the boundary of radiation zone. The energy going off along the waveguides formed by slots between the bars is denoted by the index $m = 0$.

Fig. 9 illustrates also the variety of Wood anomalies for the given type of periodical structures when the only one space harmonic radiates as well as in the case of instantaneous radiation.

The numerical results for the reflection grating with perfectly conducting basis are given on Fig. 10. If the depth of the slots is a little less than $0.25\lambda(2N+1)$, $N = 0, 1, 2, \ldots$, the radiation power reaches its maximum and vanishes for $h = 0.5\lambda N$. One period of variation of S as the function of δ is shown on Fig. 10, a–c. If to choose optimal depth and then to consider radiation power with respect to the filling coefficient, it turns out that radiation is maximal for rather narrow slots (see Fig. 9, d, e). It means that (1.30), which determines the amplitudes of space harmonics to within the corrections proportional to θ^2, describes the effect of diffractional radiation on reflection gratings in the domain which is the most interesting for practical applications. Dependence of the power of diffractional radiation on the frequency (on the parameter κ) is shown on Fig. 10, f–k. If the combs' width $2(l-d)$ tends to zero, radiation is absent in the direction perpendicular to the grating (in the plane $z = h$), the condition $\psi_n = 0.5\pi$ yields $\kappa = \beta$. It holds for semi-infinite structures (Fig. 9, a) and also for the gratings made of bars having zero thickness. This physical feature may be easily explained. If the bars' thickness vanishes, the currents generated by a source on structure's edges in the plane $z = h$ (and $z = -h$ in the case of transparent grating) disappear and hence disappear the fields produced by them. It is clear that the currents generated on elements with zero thickness and flow parallel to the axis Oz may excite radiation in any direction except the one along Oz.

4. Analytical solution to the problem of SPR essentially complements the

data describing physical properties of this effect.

a. The Pointing theorem for the electromagnetic field (1.19), (1.20) has the form

$$\sum_{(n)} (S_n^+ + S_n^-) = W. \qquad (1,37)$$

Here S_n^+, S_n^- are the intensities of diffractional beams going through the square 1 cm^2 on the planes $z = h + \tilde{h}$, $z = -(h + \tilde{h})$, respectively (the value of $\tilde{h}$ is arbitrary); ψ_n is the radiation angle of n-th diffractional beam equal to the angle between beam's direction of propagation and its projection on the plane $z = const$ ($\psi_n = \arccos \frac{\kappa\alpha+n}{\kappa}$). Summation in (1.37) is carried out by the same values of n satisfying radiation conditions

$$\kappa\alpha(1-\beta) \leq n \leq \kappa\alpha(1+\beta). \qquad (1.38)$$

The magnitude W is the time average of energy taken from the source and spent for radiation through the unit square. S_n^+, S_n^-, W are proportional to the dimensional factor $2\pi\rho_0^2\beta c$ which is determined by the density of surface current I, A cm^{-1} as follows: $2\pi\rho_0^2\beta c = I^2 R_0$, Wt cm^{-2}, where $R_0 \approx 189$ Ohm. Then

$$W = wI^2 R_0 e^{-4\pi\frac{a}{\beta\lambda}\sqrt{1-\beta^2}}\mathrm{Wtcm}^{-2}; \qquad (1.39)$$

$$S_n^{\pm} = \frac{1}{2} s_n^{\pm} I^2 R_0 e^{-4\pi\frac{a}{\beta\lambda}\sqrt{1-\beta^2}}\mathrm{Wtcm}^{-2};\ \mathrm{s}_\mathrm{n}^+ = |\mathrm{a}_\mathrm{n}|^2, \mathrm{s}_\mathrm{n}^- = |\mathrm{b}_\mathrm{n}|^2, \qquad (1.40)$$

where

$$w = -\sqrt{\alpha^2 - 1}\,\Im a_0. \qquad (1.41)$$

$\Im a_0$ is the imaginary part of the complex amplitude of zero space harmonic in that domain of the space where the flow flights. Hence, it follows from (1.39) that the total SPR energy W is determined by the linear part of non-propagating zero space harmonic running along the grating with phase velocity equal to the one of the source.

The conservation law (1.37) is obtained for "transparent" gratings; for reflection diffractional gratings it has the form

$$\sum_{(n)} S_n \sin\psi_n = W, \qquad (1.42)$$

where

$$S_n = |a_n|^2 I^2 R_0 e^{-4\pi\frac{a}{\beta\lambda}\sqrt{1-\beta^2}}, \qquad (1.43)$$

and expression for W formally coincides with (1.39), (1.40).

Let us determine with the help of the formulas (1.30), (1.31) maximal value of the total energy W of diffractional radiation only for reflection diffractional grating having the perfectly conducting basis:

$$w = \left(\frac{\sin \pi\theta\kappa\alpha}{\pi\theta\kappa\alpha}\right)^2 \frac{2S_1}{S_1^2 + \left(S_2 - \frac{\cot \pi\kappa\delta}{\theta}\right)^2}, \tag{1.44}$$

where $S = S_1 - iS_2$ is determined in (1.27) for $2\kappa\theta < 1$ and in (1.29) for $2\kappa < 1$.

If $\kappa < \frac{1}{2}$, then for the considered grating

$$w_{\max} = 2\sin\psi_{-n_0}\left(\frac{\sin \pi\theta\kappa\alpha}{\pi\theta\kappa\alpha}\right)^2 < \frac{2}{n_0^2}\ (n_0 = 1, 2, \ldots), \tag{1.45}$$

what holds when

$$\cot \pi\kappa\delta = \theta S_2. \tag{1.46}$$

Now one can conclude that the maximal energy which a flow may spend for SPR is given, in Wt cm^{-2}, by the expression

$$W = W_{\max} = 2I^2 R_0 e^{-4\pi\frac{a}{\beta\lambda}\sqrt{1-\beta^2}}. \tag{1.47}$$

This estimate holds for arbitrary values of κ. Taking into account the global condition $W \lesssim W_{\max}$, it is easy to see that due to (1.38), (1.42) the radiation energy of some of diffractional beams S_n^+, S_n^-, S_n decreases with growing κ (i. e., with decreasing λ) since here the number of beams grows and the total radiation energy is distributed among them in the definite way.

b. It is rather important to find the dependence of SPR energy versus λ in the long-wave and short-wave ranges, $\lambda > 4l$ and $\lambda << 2l$, respectively ($2l$ is the grating's period). For radiation energy in the short-wave range the known estimate [56, 57] has been obtained: $S_n \sim \lambda^{-2}$, which was usually extrapolated on the whole wavelength range. This estimate implies the considerable level of SPR energy only in the optical wave range. In reality this extrapolation is illegal.

Let us generalize the obtained solution for modulated charged plane and consider the case of uniformly charged thread with charge's density $\rho =$

$e\delta(z-z_1)\delta(y-vt)$. Let us formulate the general expression for the energy given by a part of the thread with the length 1 cm to the n-th diffractional beam $dS_{n\lambda}$ moving along the grating on the unit distance in the interval of wavelengths $d\lambda$ in the vicinity of λ:

$$dS_{n\lambda} = \frac{\pi e^2}{2\lambda^2}|a_n|^2 \sin\phi e^{-4\pi\frac{a}{\beta\lambda}\sqrt{1-\beta^2}}d\lambda. \tag{1.48}$$

The factor $|a_n|^2$ is a function of λ, β and geometrical parameters of a grating; it is found from the solution to the boundary value problem. For the perfectly conducting grating in the long-wave range

$$|a_n|^2 = 4\mid P\mid^2\left(\frac{\sin\pi\theta\kappa\alpha}{\pi\theta\kappa\alpha}\right)^2, \tag{1.49}$$

where $n=-n_0$; n_0 is the digit closest to $\kappa\alpha$, P is determined in (1.31) and

$$|a_n|^2 = 4\sin^2\pi\kappa\delta\left(\frac{1}{\beta^2}-1\right)\left(\frac{\sin\pi\theta n}{\pi n}\right)^2. \tag{1.50}$$

Formulas (1.49), (1.50) must satisfy inequalities (1.38). Since κ is big in (1.50), here $|n|$ must also be big. The expression (1.50) is obtained from the system of functional equations for the coefficients a_n, b_n, c_n, d_n without limitations for the parameters β, δ, θ.

Let us compare the dependence of $dS_{n\lambda}$ versus λ in these ranges. The relations (1.48)–(1.50) give radiation energy in the long-wave and optical ranges, respectively:

$$dS_{n_0\lambda} = \frac{\pi e^2}{2\lambda^2}\left(\frac{\sin\pi\theta\kappa\alpha}{\pi\theta\kappa\alpha}\right)^2\frac{4\sin\psi_{-n_0}}{1+\left(\kappa Q-\frac{1}{\beta^2}\right)^2\sin\psi_{-n_0}}e^{-4\pi\frac{a}{\beta\lambda}\sqrt{1-\beta^2}} \tag{1.51}$$

$$dS_{n\lambda} = \frac{\pi e^2}{2\lambda^2}4\sin^2\pi\kappa\delta\left(\frac{1}{\beta^2}-1\right)\left(\frac{\sin\pi\theta n}{\pi n}\right)^2\sin\psi_n e^{-4\pi\frac{a}{\beta\lambda}\sqrt{1-\beta^2}} \tag{1.52}$$

If $a\to 0$, then $dS_{n_0\lambda}$ is determined only by non-exponential factor and hence it is bounded in the optical range by the value

$$\frac{\pi e^2}{2}\frac{1+\beta}{1-\beta}\frac{1}{\pi^2 l^2}, \tag{1.53}$$

since indices n must satisfy radiation conditions (1.38). In the long-wave range this factor has the order

$$\frac{\pi e^2}{2}\frac{1}{\pi^2 l^2}, \left(\frac{\sin \pi\theta n_0}{\pi\theta n_0}\right)^2 \quad (n_0 = 1, 2, \ldots), \tag{1.54}$$

One may conclude that the intensities of diffractional beams are comparable in the short-wave and optical ranges.

c. Let us find optimal profiles of diffractional gratings, which provide maximum of diffractional radiation with the wavelength λ in the given direction. For $\lambda < 0.5\theta$, $\theta^2 << 1$ the optimal profile is determined by (1.46) and by the condition

$$\frac{d}{d\theta}\left(\frac{\sin \pi\theta\kappa\alpha}{\pi\theta\kappa\alpha}\right) = 0.$$

Let us find optimal profiles in the long-wave range. The profile of the considered gratings is characterized by the two parameters: relative width of the slots θ and relative thickness δ. The strip grating is a particular case and it is determined by one parameter θ (here $\delta = 0$). For the comparison let us give the amplitudes of the waves of diffractional radiation produced by this grating

$$|a'_{-n_0}|^2 = |b'_{-n_0}|^2 = \frac{\sin \pi\theta\kappa\alpha}{\pi\theta\kappa\alpha}\left(1 + 4\kappa^2 \ln^2 \frac{\pi\theta}{2} \sin^2 \psi_{-n_0}\right)^{-\frac{1}{2}} \tag{1.55}$$

The values $S_n^{\pm}$ are determined here by the formulas (1.40). One must take into account that $|a'_n| = |b'_n|$. The optimal values of θ lie in the interval $[0.1, 0.2]$ for the strip grating.

The formula (1.55) is not only rather exact quantitative characteristic of waves' amplitudes for $\theta \lesssim 0.33$ but is also a good reflection of their qualitative behaviour in the entire interval $0 \leq \theta \leq 1$. It follows from (1.55) that the harmonic with index $n = -n_0$ must have n_0 zeros and the same number of maximums in the interval $0 \leq \theta \leq 1$. This fact is completely confirmed by numerical calculations. The values of θ where the indicated harmonics reaches these zeros and maximums are turned out to be slightly shifted in the direction of the parameters greater than the ones of the formula (1.5). The waves' amplitudes versus radiation angle are maximal in the case of sliding propagation ($\sin \psi_{-n_0} = 0$) and are minimal for the "normal" radiation when $\sin \psi_{-n_0} = 1$.

For the semi-transparent grating made of rectangular bars the optimal values of θ lie approximately in the same interval $0.1 \lesssim \theta \lesssim 0.3$ as for strip grating. However the given structure has some specific features: one can choose optimal bars' thickness for the fixed $\theta, \kappa\alpha$ when $S_n^{\pm}, W$ are maximal. The maximums of $S_{-n_0}^{+}, S_{-n_0}^{-}$ do not depend in the long-wave range on radiation's direction to within the accuracy θ^2 and they are reached for the thickness

$$2h = N\frac{\lambda}{2} + d\frac{4}{\pi}\ln\sin\frac{\pi\theta}{2} \quad (N = 1, 2, \ldots). \tag{1.56}$$

Here and in what follows we use the approximate formula (1.29) for S.

The numerical data enable us to assert that the number of half-waves per bars' thickness $2h$ must be a little less than a whole number in the grating with optimal profile. The correction to the half-wave thickness $\Delta h = d\frac{4}{\pi}\ln\sin\frac{\pi d}{2}$ are equal for all harmonics; it is determined only by geometrical parameters of the grating and it does not depend on the wavelength and radiation angle to within the accuracy given by terms proportional to θ^2. This correction is maximal for the slotwidth $\theta \approx 0.3$ and reaches the value $\Delta h_{\max} \approx -0.77d = -0.23l$. The correction falls with growing or decreasing slotwidth. When the bars' thickness is equal to the whole number of half-waves ($2h = 0.5\lambda N$), complex amplitudes of the waves of diffractional radiation are equal to within the accuracy θ^2 to the complex amplitudes of the waves of radiation from the strip grating. If the odd number of quarter-waves go in bars' thickness, i. e., if

$$2h = (2N+1)\frac{\lambda}{4} + d\frac{2}{\pi}\ln\sin\frac{\pi\theta}{2} \quad (N = 0, 1, 2, \ldots), \tag{1.57}$$

then the waves' amplitudes are minimal.

The energy of diffractional radiation of the semi-transparent grating made of bars having optimal and non-optimal profiles considerably differs (except of the special case when the wave slides along the grating, $\psi_{n_0} = 0, \pi$).

For the perfectly conducting comb one can choose resonance depth of the slots

$$h = (2N+1)\frac{\lambda}{4} + d\frac{2}{\pi}\ln\sin\frac{\pi\theta}{2} \quad (N = 0, 1, 2, \ldots) \tag{1.58}$$

for every value of θ, when the amplitude of radiated wave is maximal. Radiation minimum is observed for the depth of the slots

$$h = N\frac{\lambda}{2} \quad (N = 1, 2, \ldots). \tag{1.59}$$

Here there is no radiation at all: $W = S_n^+ = 0$. Hence , the difference between the radiation energy of the gratings with optimal (1.58) and non-optimal (1.59) depth of the slots is even greater than for the semi-transparent grating made of bars.

The maximal acceptable deviation $\pm\Delta h$ of the depth of the slots from the optimal value, when $S_{-n_0}^{\pm}$ is not less than $0.5(S_{-n_0}^{+})_{\max}$ and for radiation angles $\mid \psi_n - \frac{\pi}{2} \mid \lesssim \frac{\pi}{4}$ is determined by the relation

$$\mid \Delta h \mid \lesssim \lambda \frac{\theta}{2\pi \sin \psi_n}. \tag{1.60}$$

This limitation is used for determination of the accuracy in the process of manufacturing the real gratings.

d. The study of Wood anomalies is if great importance for creating the theory of SPE. It is known that diffraction of plane wave by a grating yields resonance scattering leading to sharp variation of intensity of the diifractional beam in the narrow range of angles or wavelengths. This effect is called the Wood anomalies. They occur also in the process of SPR. The field's resonances with respect to the slotdepth considered above are the Wood anomalies.

Let us consider the anomalies connected with appearance of fats diffractional harmonics sliding along a grating with the velocity close to c (such anomalies in optics have been predicted for the first time by Lord Rayleigh). Let us consider them in the range $\kappa < 0.5$ where the expressions for the waves' amplitudes of the scattered field have the simplest form and the use of the phenomenon of diffractional radiation is the most purposeful.

The condition $\sin \psi_{-n_0} \to 0$ corresponds to this type of Wood anomalies. The formulas (1.30), (1.31) yield existence of the resonance ($\sin \psi_{-n_0} = 0$) scattered field with the wavelength λ only in the half-space where the source moves ($z > h$). It is proportional to θ^2 in the slots between the grating's elements (it is negligible in the half-space $z < -h$ for the semi-transparent grating made of bars). In the domain $z > h$ all energy of the source is concentrated in the single harmonic with the number $n = -n_0$ and other harmonics have small amplitudes.

The values $|a_n|^2$, $|b_n|^2$, which determine energetic characteristics of radiation , may be calculated in the vicinity of resonances by the formulas (1.50). They are evidence of the very narrow band of such resonances. The proper choice of the parameters δ, θ yield the arbitrary big q-factor of the resonance curve $\frac{\lambda}{\Delta\lambda}$; the value of q-factor is bounded only by the ohmical losses in a metal (here $\Delta\lambda$ is the wavelength's band close to λ where the value $S_n^{\pm}$ is reduced in two times). The formula for $\frac{\lambda}{\Delta\lambda}$ of the perfectly conducting reflection grating has the form

$$\frac{\lambda}{\Delta\lambda} = 2\frac{\left[\left(-2\ln\sin\frac{\pi\theta}{2}\right)\kappa\theta\tan\pi\kappa\delta - 1\right]^2}{\theta^2\tan^2\pi\kappa\delta} \tag{1.61}$$

The properties of the Wood anomalies in the vicinities of sliding points are treated in the similar way.

5. Our investigations have shown that the characteristics of diffractional radiation were determined by the properties of an electron flow and geometrical parameters of periodical structures which participate in transformation of the surface eigenfield of a flow to the plane waves of radiation. The gratings face a number of requirements when these waves are used for creating generators of coherent electromagnetic oscillations of the millimeter and submillimeter ranges: high degree of transformation of the flow's energy to the radiation's energy, manufacturing simplicity et al. Different configurations of gratings do not equally satisfy these demands. It is important to carry out comparative analysis of periodical structures in order to find optimal profile of the grating which should be chosen in a given generator where SPE is used for generating oscillations. Such an analysis is necessary for clearing up all possibilities for the study of various types of the gratings. It enables one to choose the optimal type of diffractional grating for the design of devices where SPE is used. The criterion for the optimal choice of a grating is the value of maximum of the amplitude of homogeneous harmonics, i. e., maximum of the energy of plane waves going off the grating when an electron flow flights over it.

The choice of the most acceptable grating's profile is determined also by the possibility to use SPE at higher space harmonics since here the grating's period grows proportional to the number of working harmonic. The latter is advantageous for millimeter and submillimeter technology.

Three types of reflection gratings are considered: the grating made of rectangular bars; echelette with rectangular teeth; the gratings made of circular half-cylinders lying on metallic plane (Fig. 11). Energetic characteristics of the diffractional radiation of every grating is found by solving the corresponding problems of diffraction of inhomogeneous plane waves on the gratings with further computer simulations [72].

Determining the conditions when the energy of waves going off the grating is maximal in the given directions is based on reciprocity interrelations for the amplitudes of space harmonics produced as a result of plane waves diffraction on periodical structures. It enables one to use the data obtained for the problems of plane homogeneous waves diffraction on the same gratings in order to find optimal profiles of the gratings.

Reciprocity interrelation for arbitrary periodical structures have the form [73]

$$\frac{a_{-qp}}{\Gamma_p} = \frac{\hat{a}_{-pq}}{\hat{\Gamma}_p}, \tag{1.62}$$

where a_{mn} is the amplitude of the m-th space harmonic in the case of diffraction of the n-th harmonic (m and n may be the numbers of homogeneous and inhomogeneous harmonics), i. e., when the incident field and the field scattered by the grating are presented in the formula

$$u = e^{-i\Gamma_m z + i\Phi_m y} \sum_{n=-\infty}^{\infty} a_{nm} e^{i\Gamma_n z + i\Phi_n y}, \quad z > 0. \tag{1.63}$$

Here $\Gamma_n = \sqrt{k^2 - \Phi_n^2}$; $\Phi_n = n + \frac{k}{\beta}$, the grating's period is chosen to be equal to 2π. The hat over a_{nm} in (1.62) means that the sign of Φ_n in (1.63) is changed to the opposite one.

Hence, reciprocity interrelations unite the amplitudes of space harmonics of the two cases of waves diffraction on a grating. Here the wavelength of the scattered electromagnetic fields must be the same and every p-th harmonic in the first case of diffraction must correspond to a certain q-th harmonic of the second case such that $| \Phi_p^1 | = | \Phi_q^2 |$, i. e., the moduli of propagation constants of the harmonics running along the grating coincide. Let us consider two cases of diffraction.

a. Diffraction of inhomogeneous plane wave (of the type (1.21)) of the unit amplitude corresponding to the eigenfield of a monochromatic electron flow with the charge density (1.3). We assume that this wave is the zero harmonic. Let $k = n\beta$ and $2\pi\rho_0\beta = 1\,(n = 1, 2, \ldots)$. Then it follows from the Brilluin diagram (see Fig. 9) that the $(-n)$ harmonic is homogeneous radiating in the vertical direction off the grating. Here $\Phi_0^{(1)} = k\beta^{-1} = n$; $\Phi_p^{(1)} = n + p\,(p = 0, \pm 1, \pm 2, \ldots)$.

b. Diffraction of homogeneous plane wave in the case of normal incidence. We assume that this wave is the n-th harmonic, hence $\Phi_n^{(2)} = 0$; $\Phi_0^{(2)} = -n$; $\Phi_p^{(2)} = -n + p\,(p = 0, \pm 1, \pm 2, \ldots)$. Of course, the wavelength of the scattered field in both cases is the same. Since $\Phi_p^{(1)} = \Phi_{p+2n}^{(2)}$; $\left| \Phi_p^{(1)} \right| = \left| \Phi_{-p}^{(2)} \right|$, the considered two cases of diffraction are connected by the reciprocity interrelation (1.62) which yields the condition

$$\frac{a_{-n0}^{(1)}}{\Gamma_0^{(1)}} = \frac{a_{0n}^{(2)}}{\Gamma_n^{(2)}}.$$

Hence, the amplitude $a_{-n0}^{(1)}$ of the n-th harmonic radiating in the vertical direction, when the grating is excited by a monochromatic electron flow, may be determined by the zero field's harmonic obtained as a result of the diffraction of the n-th homogeneous harmonic in the case of normal incidence:

$$a_{-n0}^{(1)} = a_{0n}^{(2)} \frac{\Gamma_0}{\Gamma_n} = a_{0n}^{(2)} \sqrt{1 - \frac{1}{\beta^2}}.$$

Similar expressions may be obtained also for the non-vertical radiation:

$$a_{-n0}^{(1)} = a_{0n}^{(2)} \frac{\sqrt{1 - \frac{1}{\beta^2}}}{\cos\phi},$$

where ϕ is the radiation angle of the harmonic with number $(-n)$ in the first case and the angle of incidence in the second case.

It is important for further analysis to calculate the energy of the harmonic having the number $(-n)$ for the case of radiation of an electron flow flying over the grating:

$$S_{-n}(k, \phi, d) = \left| a_{0n}^{(2)} \right|^2 \left(1 - \frac{1}{\beta^2}\right) \cos^{-1}\phi \mathrm{B}^2.$$

B^2 is introduced for taking into account the normalizing factor of an incident field. The magnitude S_{-n} depends on three parameters: $k = 2\pi\lambda^{-1}$; ϕ – radiation angle and d which characterizes the grating's geometry (for the echelette it coincides with ψ, the inclination angle of one of the sides of the tooth with respect to the direction perpendicular to the grating; for the grating made of half-cylinders it is equal to al^{-1}, the relative radius of half-cylinders; for the reflection grating made of rectangular bars – for the comb – it is equal to d, the relative slotwidth and relative height of the lamellae).

Since it is considered diffractional radiation at one harmonic and radiation energy of an electron flow is determined by the coefficient a_{00} which does not depend on the sign of Φ_0 (on the sign of the velocity) due to the reciprocity interrelations [73], one can limit oneself to the interval $\psi \in [45°, 90°]$ when varying geometrical parameters of an echelette.

It has been studied above the properties of diffractional radiation generated as a result of the movement of an electron flow close to the grating made of rectangular bars.

The analysis of the angle radiation patterns of the flows flying over echelette, comb and reflection grating made of half-cylinders, respectively, shows that radiation angle ϕ, flow's velocity β and wavenumber k are connected by the relation

$$k = \frac{-n}{\sin\phi + \beta^{-1}} \tag{1.64}$$

(n is the number of radiated harmonic). It is assumed in the case of a comb that $h = 0.22\lambda = 0.44\pi k^{-1}$.

It turns out that the comb has considerable vertical radiation. For the echelette and for the grating made of half-cylinders this radiation is negligible; the lateral radiation is rather intensive but it is directed at a very small angle to the grating.

6. Let us analyze the results of numerical simulations in the resonance domain of variation of the relative electrons' velocity $\beta = \kappa(n - \kappa\sin\phi)^{-1}$, where ϕ is the angle counted in the positive direction of the tearing off the grating a part of energy lost by electrons at the principal space harmonic ($p = 0, -n$) when β varies close to the velocity of light in vacuum. The interest to the consideration of this domain is caused by the study of the possibility to create relativistic devices of diffractional electronics.

For such values of β excitation wavelength is comparable with characteristic grating's size and various resonance effects, connected with excitation of electromagnetic oscillations which are close to eigenoscillations of the structures, begin to play the basic role in forming the secondary field [62].

The dependence of the energy of radiation to the principal harmonic of the space spectrum W_0^A on the relative grating's height δ for $n = 1$ and for different values of θ, ϕ, β is presented on Fig. 12. The functions $W_0^A(\delta)$ are almost periodical. With growing β the period decreases while increasing ϕ leads to the growth of oscillations' period. The maximal value of W_0^A grows with increasing β and decreasing relative slotwidth θ. One must choose the gratings with less height of the lamellae for big β in order to guarantee the maximal intensity of radiation to zero harmonic of diffractional spectrum. This height is less for the vertical radiation in comparison with the inclined radiation.

Let us note the tendency of decreasing the depth of grating's profiles for obtaining optimal characteristics of diffractional radiation with growing β (growing flow's velocity) for relativistic devices, which is especially favorable concerning manufacturing technology. When the radiation is almost vertical ($\phi \approx 1°$), the corresponding ratio of ditches' depth h to the wavelength of flow's modulation $\lambda\,(h\lambda^{-1} = \delta\kappa)$ is approximately 0.15 for $\beta = 0.90$ and 0.121 for $\beta = 0.95$. The closer β to unit, the stronger this effect, what is well illustrated on Fig. 13. It also shows that the growth of β leads to stronger requirements concerning stability of the velocity of electron flow, whose small variations in the domain $\beta > 0.99$ for fixed δ, ϕ may cause sharp fall of the intensity of diffractional radiation. For $\beta < 0.99$ these requirements are much lower (because the lines of equal level of $W_0^A(\delta, \beta)$ are practically parallel to the axis of β).

Analysis of the lines $W_0^A(\beta, \theta) = \text{const}$ on Fig. 14 enables us to study in details the characteristics of radiation for the combs with varying width of the slots. For example, if $\beta < 0.93$, the decrease of the width of the slots is accompanied at first by gradual and monotonous growth of radiation intensity and then it falls after going through the maximum of $W_0^A(\theta)$ reached in the interval $0.2 \leq \theta \leq 0.4$. For $0.93 < \beta < 0.98$ the functions $W_0^A(\theta)$ have several local extremes. The work in optimal regimes for wide slots are possible only when electron flows with highly stabilized velocities are used. The stability requirements may be weakened when one obtains optimal characteristics of radiation by using the combs with narrow slots. For $\beta > 0.98$ the tendency of the growth of maximums of W_0^A with increasing β for small values of θ changes to the opposite one. Here it is preferable to choose the gratings with the filling coefficient close to 0.5 in order to achieve high levels of radiation intensity in the almost vertical directions.

It is pointed out above the decrease of the depth of gratings' profiles which are necessary for obtaining the optimal values of W_0^A with growing β. This appropriateness, which is especially distinct for $\beta > 0.99$, is characteristic only for the pencils whose velocity is close to the velocity of light. Let us recall that for $\beta < 0.5$ it is not found the noticeable deviations of the parameter $h\lambda^{-1}$, providing the maximal level of W_0^A, from the value $h\lambda^{-1} = 0.22$ [22]. The use of relativistic pencils leads to arising of one more feature, which considerably influences radiation far field pattern. The number of sliding points in the interval $|\phi| < \frac{\pi}{2}$ (the points where the Wood anomalies take place) grows with increasing β. The degree of the "indented edge" of the radiation far field pattern correspondingly grows. The lines of the equal level of W_0^A in the case of comb's excitation by the plane wave are presented on Fig. 15. The dot lines indicate the lower boundaries of radiation zones for the minus first, minus second and minus third harmonics. In the interval $-2° < \phi < 90°$, where the radiation occurs only at the principal harmonic, the maximal intensity corresponds to the directions close to vertical. The curves on Fig. 15 confirm the appropriateness mentioned above: the closer radiation direction to the vertical one, the smaller must be the depth of grating's ditches to provide the optimal regime. The peak in the radiation pattern shifts to greater values of ϕ with growing β. The fall is observed close to the boundary of the appearance of radiation at the minus first harmonic ($\phi = -2.14°$) and the depth of this fall increases when δ approaches its optimal values. The similar falls indicate the appearance of radiation at the minus second ($\phi = -22.5°$) and at the minus third ($\phi = -33.8°$) harmonics. The value of W_p^A sharply grows for these harmonics with almost sliding direction of propagation.

One of the main features of the scattering of plane inhomogeneous H-polarized waves on the echelette-type grating (Fig. 16) is its ability to form intensive radiation in the direction close to vertical for big values of β. For $\beta < 0.4$ one can obtain satisfactory levels only in the range of angles $\phi > 80°$ [62]. The lines of the equal level of $W_0^A(\psi, \beta)$ on Fig. 16 pick out two basic zones where the radiation level of W_0^A may be greater than $W_0^A = 6$ (here $\phi = 1°$, $n = 1$). It means that an echelette can generate the radiation comparable with the one produced by a comb. This result is important since with shortening the wavelength of excitation the manufacturing of the combs with optimized profiles becomes rather problematic while the echelette manufacturing technology is well developed both in optical and microwave ranges.

One of the zones of radiation with high intensity on Fig. 16 corresponds to the echelette having deep profile ($\psi \approx 45° - 55°$) and another one – to the practically scaly surface ($\psi \approx 75° - 82°$). Echelettes with intermediate value $\psi \approx 65°$ have the stable characteristics of SPR for varying β. When $\beta < 0.95$ the maximal intensity of vertical radiation is achieved for the gratings with $\psi \approx 75°$.

Echelette enables one to satisfy in the most natural way the usual requirements of the relativistic diffractional electronics connected with generating intensive almost vertical radiation as a result of interaction of electron flows with low-profile periodical structures. It is confirmed on Fig. 17 by the lines of equal level of $W_0^A(\phi, \psi)$; they distinctly pick out radiation zones with maximal values $W_0^A > 4$ in the region $\phi = 0$, $\psi \approx 78°$.

The given results enable us to choose optimal parameters of a comb and echelette which are the most accessible types of the gratings for application in the devices of relativistic diffractional electronics. The indicated appropriatenesses simplify the synthesis of the structures by the output data and may form the basis for creating new powerful radiation sources and units of the microwave range.

In the conclusion let us briefly analyze the reasons of anomalous growth of radiating capability of reflection gratings in the resonance domain of variation of the parameters. This anomalous level of radiation is caused by excitation of field's quasi-eigenoscillations in the structures. Namely, it is E_{001}-oscillations with the diffractional q-factor $\Re\kappa(-2\Im\kappa)^{-1}$ of about ten for the first set of splashes of W_0^A on Fig. 12. The distribution of field's tensity of one of such oscillations is presented on Fig. 18 (the lines $|H_x(y,z)| = \text{const}$, the direction of the decrease of $|H_x(y,z)|$ is indicated by short lines). The choice of the values θ, Φ, δ is determined by the new splash of the curve 6 on Fig. 12 ($\Phi = \kappa \sin\phi$; $\kappa = (\sin\phi + \beta^{-1})^{-1}$; $\phi = 35°, \beta = 0.95$) with taking into account the reciprocity principle [62,73], which enables one to reduce the diffractional problems with negative phase raid $\hat{\Phi}$ to the standard case $\Phi = -\hat{\Phi}$. The scattering of the field of homogeneous plane wave with $\Phi = -\hat{\Phi}$ at the frequency close the eigenfrequency of the structure is accompanied by the growth of tensity of the grating's near field which due to the conservation law may hold only because of the increasing amplitudes of non-propagating harmonics, in particular, of a_{-1} (in the considered case only one wave propagates in the reflection zone). The reciprocity principle yields also the growth of the level of a_0 determining radiation characteristics of the grating excited by inhomogeneous wave with $\Phi_1 = 1 - \Phi$. Let us note that the partial components with $n = 0, -1$ prevail in the field of free oscillation

(Fig. 18). The ratio of the moduli of their amplitudes is 2:4.7. The lines of equal phases of $H_x(y, z)$ of a free oscillation in the direction off the grating stand perpendicular to the direction of radiation occurring at the principal harmonic of the space spectrum. Thus, the field of a free oscillation (with taking into account reciprocity interrelations) represent all singularities of the secondary field generated as a result of the scattering of inhomogeneous plane wave.

§3. Quasi-optical Smith-Purcell wave pencils

The study of SPE carried out in §1,2 is limited by the use of two-dimensional model where the source is taken either as a plane monochromatic electron flow or a thread with a current (charged thread), which are homogeneous along the coordinate perpendicular to the vector of velocity.

Here it is considered the radiation of the four types of sources: a plane monochromatic electron flow bounded by its width and length; the point charge; unit charged clot; finite sequence of charged clots. We study the diffraction radiation generating in the process of movement of these sources along various infinite diffractional gratings. The fields of diffractional radiation are obtained in analytical form as functions of space coordinates and parameters of the problem, what enables us to determine in the complete volume the physical properties of the radiation of a source occupying bounded region in the space.

The formal scheme of the solution to the problem of diffractional radiation within the given current approximation for such types of sources bounded in the space is in the following. The functions of distributions of a charge or of a current which take into account the finiteness of their size in the space is represented in the form of Fourier integral over the plane waves. The problem of diffractional radiation is reduced thus to the correctly stated problem of the plane wave diffraction on the given structure at a certain complex incident angle. Hence, if for the given periodical structure it is known the solution to the problem of diffractional radiation in the case of plane monochromatic flow, whose charge's density is additionally modulated in the harmonic way along the coordinate perpendicular to the vector of its velocity, one can obtain the formal solution to the problem of radiation for arbitrary source bounded in the space. Integral representations for radiation fields are rather bulky and they cannot be simplified even for those particular cases when the amplitudes are determined explicitly for a given structure [74]. One meets of principle difficulties when transferring from the fields' spectral densities to the fields themselves. The approximate method has been proposed for the analytical calculation of integral representations for the field of diffractional radiation of a bounded electron flow. This method is based on the use of parabolic equation. It has been established that a bounded source produced diffractional radiation in the form of superposition of quasi-optical wave pencils having orthogonal polarizations.

The study of a single electron clot and a sequence of clots, which form the grouped electron flow, enables one to consider the problem of impulse SPR excitation.

Diffractional radiation of an unbounded monochromatic electron flow determined the structure of radiation field, its properties for the given radiation frequency equal to the modulation frequency of the charge's density of a flow. The frequency spectrum of the the diffractional radiation is excited for the point charge, charged clot or finite sequence of charged clots. Hence, for such sources one must also establish frequencies and distribution of energy at these frequencies. The radiation spectrum of such sources is discrete consisting of spectral lines of divisible frequencies. The wave field on each spectral line is presented by a superposition of three-dimensional wave pencils.

1. The plane monochromatic electron flow of a finite width moves in the direction Oy with the constant velocity $\vec{v} = \vec{y}_0\beta c$ along the unbounded strip grating perpendicular to its strips (see Fig. 19 where the origin is placed in the middle of one of the slots). Instantaneous value of the varying component of the charge's density of the electron flow

$$\rho(x, y, z, t) = \rho_0 f(x)\delta(z - a)e^{i(k\alpha y - \omega t)} \tag{1.65}$$

where $f(x)$ is the given function describing the distribution of charge's density in the transversal direction.

The completeness of the system of Hermit functions $f(x)$ may be represented in the form of Gramm-Charlier series [75]. The electron flow (1.65) consists of partial flows whose charge's densities varies along the axis Ox as the Hermit function. In accord with superposition principle it is sufficient to find diffractional radiation of a separate flow in order to solve the problem for the given function of distribution $f(x)$.

The eigenfield of a partial flow is represented by the magnetic $\vec{\Pi}^0_M$ and electrical $\vec{\Pi}^0_E$ Hertz vectors (with the only one non-zero x-component).

The formulas representing $\vec{E}$, $\vec{H}$ via $\vec{\Pi}^0_M$, $\vec{\Pi}^0_E$ show that $\vec{\Pi}^0_M$ gives H-polarized and $\vec{\Pi}^0_E$ – E-polarized parts of the source's eigenfield.

Now the total field may be represented with the help of Hertz vectors $\vec{\Pi}^0_M = \vec{x}_0(\Pi^0_M + \Pi^1_M)$; $\vec{\Pi}^0_E = \vec{x}_0(\Pi^0_E + \Pi^1_E)$, where the Hertz potentials of the scattered field

$$\vec{\Pi}^0_M = \pm\frac{k}{2\pi}\int_{-\infty}^{\infty}\sum_{-\infty}^{\infty}\Pi^{\pm}_M(m,\xi)e^{-k\gamma a}e^{ik(\xi y + \alpha_m y \pm \gamma_m z)}d\xi \tag{1.66}$$

$$\vec{\Pi}_E^0 = \frac{k}{2\pi} \int_{-\infty}^{\infty} \sum_{-\infty}^{\infty} \Pi_E^{\pm}(m,\xi) e^{-k\gamma a} e^{ik(\xi y + \alpha_m y \pm \gamma_m z)} d\xi \tag{1.67}$$

where $\alpha_m = \alpha + m\kappa^{-1}$; $\gamma_m = \sqrt{1 - \alpha_m^2 - \xi^2}$ and it is chosen the branch of the root with $\Im \gamma_m \geq 0$. The sign + corresponds to the region $z > 0$ and the sign − − to the region $z < 0$.

The potentials (1.66) satisfy Helmholtz equation, the principle of limiting absorption and they are periodical functions of y; the field's energy is finite in every finite volume of the space. The unknown Fourier coefficients $\Pi_M^{\pm}(m,\xi)$, $\Pi_E^{\pm}(m,\xi)$ are the sought for quantities. Explicit expressions for Π_M^1, Π_E^1 are obtained after calculating the integrals over ξ.

The radiation conditions have the form

$$\beta^2(1 - \xi^2) - (1 + m\beta\kappa^{-1})^2 > 0 \tag{1.67}$$

The formula (1.67) yields the possibility of radiation only at the space harmonics with negative indices. The main contribution to the field at a certain distance from the grating is made by the domain of integration $|\xi| < 1$.

The boundary value conditions applied for $\Pi_M^{\pm}(m,\xi)$, $\Pi_E^{\pm}(m,\xi)$ yield two systems of functional equations which may be replaced by the method of Riemann-Hilbert problem [61] by the equivalent infinite system of linear algebraic equations. These system is solved explicitly by the method of successive approximations for $|\kappa| < 0.5$ as well as in the case of sufficiently narrow slots.

If $|\kappa| < 0.5$ and $m + n_0 \neq 0$, all the terms in the sums in (1.66) except of $\Pi_M^{\pm}(-n_0,\xi)$, $\Pi_E^{\pm}(-n_0,\xi)$ exponentially decrease with growing $|z|$ and the single-wave radiation regime takes place, which is observed at the only one space harmonic with $m = -n_0$ (the harmonic with $m = 0$). This regime determines the surface wave whose field propagates along the grating with the source's velocity and has an exponential decay in the direction off the grating's surface.

We reduce the fields of diffractional radiation in the single-wave regime with the help of the method of stationary phase and Laplace method to the form convenient for further analysis

$$\Pi_M^1 \approx \frac{c}{k^2} W_M^{\pm}(x,z) e^{-2\pi \frac{a}{\beta\lambda}\sqrt{1-\beta^2}} e^{ik(\nu y \pm \sqrt{1-\nu^2} z)};$$

$$\Pi_E^1 \approx \frac{c}{k^2} W_E^{\pm}(x,z) e^{-2\pi \frac{a}{\beta\lambda}\sqrt{1-\beta^2}} e^{ik(\nu y \pm \sqrt{1-\nu^2} z)}, \tag{1.68}$$

where

$$W_M^{\pm}(x,z) = \pm T_1(-n_0)\frac{k}{2\pi}\int_{|\xi|<1} F_p(\sqrt{kA_0}\xi)F_0(\sqrt{kB_1(z)}\xi)e^{ik\xi x}\,d\xi;$$

$$W_E^{\pm}(x,z) = \frac{T_2(-n_0)}{2\sqrt{kA_0(1-\beta^2)}}\frac{k}{2\pi}\times$$

$$\times\int_{|\xi|<1}\Big[F_{p+1}(\sqrt{kA_0}\xi) - 2pF_{p-1}(\sqrt{kA_0}\xi)\Big]F_0(\sqrt{kB_2(z)}\xi)e^{ik\xi x}\,d\xi;$$

$$T_1(-n_0) = \frac{P_{n_0-1}(u)P_{n_0}(u) + \nu Q_1[P_{n_0-1}(u) - P_{n_0}(u)]}{2(1 - i\kappa Q_1\sqrt{1-\mu^2})};$$

$$T_2(-n_0) = \frac{P_{n_0-1}(u) - P_{n_0}(u) + \nu Q_2[P_{n_0-1}(u) + P_{n_0}(u)]}{2(1 - i\kappa Q_2\sqrt{1-\mu^2})};$$

$$D_j = D_j' + iD_j'' = -\frac{k\tilde{Q}_j^2}{1+\kappa^2 Q_j^2} + i\frac{\tilde{Q}_j}{1+\kappa^2 Q_j^2};\ j = 1,2; \qquad (1.69)$$

$$B_j'(z) = B_j' + iB_j'' = D_j' + \frac{\beta a}{\sqrt{1-\beta^2}} + i\Big(D_j'' \pm \frac{z}{\sqrt{1-\nu^2}}+\Big);$$

$$\tilde{Q}_j = \frac{1}{2\pi}Q_j;\ \mu = \frac{\nu}{\kappa};\ Q_1 = -2\ln\sin\frac{\pi d}{2l};\ Q_2 = -2\ln\cos\frac{\pi d}{2l};\ A_0 = \frac{kb^2}{2}.$$

If to spread the integration in (1.69) on the entire infinite axis, it will be rather easy to evaluate them explicitly. The justification of this technique is given in [74].

The obtained asymptotics, which is uniform with respect to $k\sqrt{A_0^2+z^2}$, [76], may be considered as the modification of the method of stationary phase when the non-exponential factor of the integrand contains Hermit function $\psi_p(\sqrt{kA_0}\xi)$. It combines the method of stationary phase and the Laplace method. As it is pointed out in [77], the possibilities of these methods are diverse. The considered determination of asymptotics for $kz \to \infty$ and small kA_0 give the results coinciding with the ones obtained by the method of stationary phase; for $kA_0 \to \infty$ and small kz the results in the essential domain $x^2 << A_0^2+z^2$ coincide with the data obtained by the Laplace method. The main advantage of this modofication of the method of stationary phase is in the fact that it gives asymptotics with respect to the parameter $k\sqrt{A_0^2+z^2}$. It keeps then the dependence on ξ in the interval of integration close to

$\xi = 0$ of the non-exponential factor, what enables one to determine with high accuracy the phase raids in the field of radiation; these phase raids are equal to $\exp(-i(p+\frac{1}{2})\frac{\pi}{2})$ for $kz \to \infty$ or $kA_0 \to \infty$. The presented method enables us to find them as a function of z asymptotically converging to the given number (determination of phase raids of the diffractional radiation is of independent interest for practical applications of the theory, especially in the studies of OR excitation).

The Fourier integrals in the functions $W_M^{\pm}(x,z)$, $W_E^{\pm}(x,z)$, which describe slow amplitude's variation in the direction perpendicular to the propagation of diffractional radiation are the convolutions of the initial Hermit functions. The convolutions of two Hermit functions of the order p is also the Hermit function of the same order. These two properties yield analytical expressions

$$W_M^{\pm}(x,z) = \pm T_1(-n_0)[\cos u_1(z)]^{-0.5} e^{ik\frac{x^2}{2R_1(z)}} \psi_p\left(\sqrt{\frac{k}{A_1(z)}}x\right) e^{-i(p+0.5)u_1(z)}$$

$$W_E^{\pm}(x,z) = \frac{1}{2} T_2(-n_0)[kA_0(1-\beta^2)\cos u_2(z)]^{-0.5} e^{ik\frac{x^2}{2R_2(z)}} \times$$

$$\left[\psi_{p+1}\left(\sqrt{\frac{k}{A_1(z)}}x\right) e^{-i(p+0.5)u_2(z)} - 2p\psi_{p-1}\left(\sqrt{\frac{k}{A_1(z)}}x\right) e^{-i(p-0.5)u_2(z)}\right] \quad (1.70)$$

Here it is used the notion of variance [78] known in quasioptics:

$$V_j(z) = A_0 + i\left(D_j'' \pm \frac{z}{\sqrt{1-\nu^2}}\right); \; (j=1,2) \quad (1.71)$$

where $u_j(z)$ is the phase of complex variance $V_j(z)$ which as well as the variance itself is different for the E- and H-polarized field's components:

$$u_j(z) = \frac{1}{A_0}\left[\arctan\left(D_j'' \pm \frac{z}{\sqrt{1-\nu^2}}\right)\right]; \; (j=1,2) \quad (1.72)$$

Hence, the partial electron flow of the order p produces diffractional radiation in the form of wave pencils (1.70). The field of SPR is localized in the bounded volume of space and its distribution along the coordinate x depends on the charge's density of the electron flow in this direction. It may be represented as a superposition of partial Gauss pencils of both polarizations:

$$W_M^{\text{tot}}(x,z) = \sum_{p=0}^{\infty} W_{Mp}(x,z); \; W_E^{\text{tot}}(x,z) = \sum_{p=0}^{\infty} W_{Ep}(x,z).$$

The presence of the phase factors

$$e^{-i(p+0.5)u_1(z)},\ e^{-i(p+1.5)u_2(z)},\ e^{-i(p-0.5)u_3(z)}$$

in the representations for $W_M^{\pm}(x,z)$, $W_E^{\pm}(x,z)$ for the total radiation field of a bounded electron flow yields qualitative changes in the function describing amplitude's distribution along the axis Ox from one plane z = const to another.

The parameter $A_j(z)$, $(j = 1, 2)$ of the partial wave pencil of the order p determines the size of "field's spot" of the wave pencil

$$w_j(z) = \sqrt{2k^{-1}(2p+1)A_j(z)}, R_j(z)$$

– curvature radius of the phase front in the middle of wave pencil (for each polarization $R_j(z)$ is a constant value which does not depend on the order of wave pencil; the size of the spot of the p-th order is in $\sqrt{2p+1}$ times greater than the spot of the principal wave pencil). (1.68) in combination with (1.70) is the main term of asymptotic expansion with respect to inverse orders of the parameter $k\sqrt{A_0^2+z^2} >> 1$. The next term contains the derivative of $\psi_p(\sqrt{kA_0}\xi)$ and has the order $k\sqrt{A_0^2+z^2}^{-1.5}$.

The functions $W_M^{\pm}(x,z)$, $W_E^{\pm}(x,z)$ satisfy parabolic equation. Hence the described way of calculating the radiation fields may be called the method of parabolic equation in the theory of SPR of the sources bounded in the space. This method enables one to explicitly determine radiation field not only in the wave zone but also in intermediate domains where $k\sqrt{A_0^2+z^2} >> 1$, $|z| > \lambda$. The field is determined in rather simple way up to the terms of the order $k\sqrt{A_0^2+z^2}^{-1.5}$ in the domains with strong radiation ($x^2 << A_0^2 + z^2$) (such formulas lose the accuracy in the domains with weak radiation). Applicability of this method is determined by the condition $kA_0 \geq 20$ what is characteristic for the quasi-optical range where the use of the effect of diffractional radiation is the most expedient (similar situation occurs in the theory of OR where the wave field is described not by the Hermit functions but by the Eiri functions in the domain of low intensity: in the acoustic area or out of it).

2. It has been obtained the solution to the problem of the radiation for the electron flow of the bounded width (1.65) moving along reflection gratings in the form of combs and semi-transparent gratings made of rectangular bars. Hertz potentials Π_M^1, Π_E^1 of the electromagnetic field of the source's eigenfield

scattered by the grating are determined in the domains over and under the grating by the formulas (1.66) (one should replace z by $z_{(\pm)}$; upper signs correspond to the half-space over the grating, $z > 0$, and lower signs – to the half-space under the grating, $z < 0$; for the grating made of bars $z_{(+)} = z$ for $z > 0$, $z_{(-)} = z + h$ for $z < -2h$; $z_{(+)} = z$ for $z > 0$, for reflection gratings $z_{(+)} = z$ for $z > 0$, $\Pi_M^- = \Pi_E^- = 0$ for $z < 0$).

It is rather important to consider the gratings with narrow slots ($\theta^2 = d^2 l^{-2} << 1$) when the H-polarized radiation reaches its maximum [70, 79, 80]. For $\theta^2 << 1$ $\Pi_E^{\pm}(m, \xi)$ (when $H_x \equiv 0$, $E_x \neq 0$) has the order θ^2, hence the entire E-polarized field has the same order; $\Pi_M^{\pm}(m, \xi)$, which determines the H-polarized part of radiation, has the order $c(\xi) + O(\theta^2)$ where the minimum of $c(\xi)$ is rather big, as well as when the propagation of diffractional radiation over a strip grating is considered (see §1). The estimates of $\Pi_M^{\pm}(m, \xi)$ $\Pi_E^{\pm}(m, \xi)$ show that in the case of narrow slots of volume gratings the integral in (1.66) makes a negligible contribution to the total field of diffractional radiation.

For $\kappa < 0.5$ the only one harmonic with the index $m = -n_0\,(n_0 \neq 0)$ propagates off the gratings; $\Pi_M^{\pm}(-n_0, \xi)$ for it has the form $\Pi_M^{\pm}(-n_0, \xi) = \Pi_M^0(\xi) T_1(-n_0, \xi)$, where $T_1(-n_0, \xi)$ is again treated as a spectral coefficient characterizing the transformation of source's field $\Pi_M^0(\xi)$ to the field of (-n)-th propagating harmonic. Expressions for $T_1(-n_0, \xi)$ for each of the gratings are presented in [74].

The obtained formulas (1.68), (1.70) for the radiation field of a partial electron flow are applicable for every periodical structure. One must only specify the necessary coefficient $T_j(-n_0, \xi) \approx T_j(-n_0) F_0(\sqrt{kD_j}\xi)$ and the parameter $B_j(z)$.

Optimal values of θ for the grating made of bars lie as well as for the strip grating in the interval $0.1 \leq \theta \leq 0.3$. In this case one can choose the optimal bars' thickness for the fixed θ, κ, α when $T_j(-n_0)$ are maximal for every direction of radiation. If (1.58) holds, $T_j(-n_0)$ reaches its maximal value

$$T_j^{\max}(-n_0) = 2 \frac{\sin \pi\theta\kappa\alpha}{\pi\theta\kappa\alpha} \frac{\sin \nu\pi\theta}{\nu\pi\theta}$$

for the comb with perfectly conducting basis and

$$D_1 = iD_1'' = i\sqrt{1 - \nu^2}\left(Q_1 + \frac{h}{\theta}\right). \tag{1.73}$$

Parameter $B_j(z)$ entering the integrand $F_0(\sqrt{kB_1(z)}\xi)$ is determined by the approximate expression

$$B_1(z) \approx iB_1''(z) = i\left(D_1'' \pm \frac{z_\pm}{\sqrt{1-\nu^2}}\right). \tag{1.74}$$

The final expressions for Π_M^1 are determined by the formulas (1.68), (1.70). The H-polarized field of diffractional radiation of the partial electron flow of order p flying over the volume gratings is similar to the plane wave with the only difference concerning inhomogeneous distribution of intensity in the flow: the power of radiation field is concentrated close to the axis of propagation, the phase front slightly deviates from the plane one; the field is a two-dimensional wave pencil whose order coincides with the initial order of the flow.

The diffractional radiation of a partial flow for the optimal gratings made of bars is a superposition of two wave pencils with different parameters propagating along the same direction; these pencils (having the amplitude reduced it two times) are generated over corresponding combs having perfectly conducting and magnetic basis, respectively.

3. It follows from (1.70) that there is one H-polarized pencil having the amplitude varying along the axis Ox as an Hermit function with the index p coinciding with the order of the charge's density of the source; the E-polarized part of radiation field consists (for $p \neq 0$) of the sum of two pencils. The amplitude of one of them is proportional to the Hermit function (of x) with the index $p+1$ and the amplitude of other pencil is proportional to the Hermit function with the index $p-1$. If the source is characterized by the Gauss function ($p = 0$), the E-polarized radiation consists of the only one pencils to which corresponds the Hermit function with the index 1. the Hermit function with the index p has $p+1$ variation along corresponding coordinate [81]; wave pencils have the same number of half-wave oscillations along the axis Ox. All wave pencils have different phase raids in the direction of propagation.

In the general case all three wave pencils of radiation for a partial flow propagate in the direction of propagation of diffractional radiation which is excited as a result of the movement of infinite homogeneous along the axis Ox electron flow. The direction of propagation of the pencils for partial flows with different p does not depend on p (i. e., on the order of Hermit function describing its distribution), lies in the plane yOz and has the angle $\phi_{-n_0} = \arccos\nu$ with the vector of velocity. For $\kappa\beta^{-1} = n_0$ the parameter $\nu = 0$

and wave pencils propagate in the direction perpendicular to the grating. In the waveguide regime radiation is observed only in partial intervals in accord with radiation conditions (1.7) and the wavelength of diffractional radiation is connected with its direction by the formula (1.9).

The limitation of an electron flow leads only to the change in the function of the amplitude of radiation field (along the axis Ox). The fields in all regions of the space where the radiation noticeably differs from zero are calculated by (1.70) with the accuracy of the order $(k\sqrt{A_0^2 + z^2})^{-1.5}$. Unlike the H-polarized wave pencil the pencils E-polarized with respect to the amplitude decrease with broadening the flow proportional to $[2\sqrt{kA_0(1-\beta^2)}]^{-1}$. For $A_0 \to \infty$ the field of E-polarization vanishes and entire radiation field becomes the plane H-polarized wave.

The coefficients $T_1(-n_0), T_2(-n_0)$ determine the dependence of the pencils' amplitudes on the grating's parameters. $T_j(-n_0)$ $(j = 1, 2)$ as a function of dl^{-1} for the strip grating and three first space harmonics of diffractional radiation ($\kappa = 0, 1$; $\nu = 0$) are presented on Fig. 20. The similar graphs for the comb with perfectly conducting basis ($\kappa = 0, 1$; $\nu = 0$; $\delta = 2.4$) are given on Fig. 21, a and on Fig. 21, b it is shown $T_1(-n_0)$ versus $\delta = hl^{-1}$ for the same grating (the amplitude of H-polarized wave pencil is maximal for narrow slots and the amplitudes of E-polarized pencils are small: $T_2(-n_0) = O(\theta^2[2\sqrt{kA_0(1-\beta^2)}]^{-1})$).

The behaviour of wave pencils of radiation when they propagate (in the process of pencil's broadening or of variation of the curvature of its phase front) is accumulated in the parameter $V_j(z)$ $(j = 1, 2)$ entering the function $W_j^{\pm}(x, z)$. This parameter will be called variance by analogy with quasi-optics and it will be found with the help of (1.71).

If to introduce two systems of beam coordinates

$$x' = x;\; y'_{(+)} = y\sqrt{1-\nu^2} - \nu z_{(+)};\; z'_{(+)} = z_{(+)}\sqrt{1-\nu^2} + \nu y;$$

$$x' = x;\; y'_{(-)} = -y\sqrt{1-\nu^2} - \nu z_{(-)};\; z'_{(-)} = -z_{(-)}\sqrt{1-\nu^2} + \nu y; \qquad (1.75)$$

where the axis $z'_{(+)}$ coincides with the direction of radiation of the wave pencil over the grating and the axis $z'_{(-)}$ – under the grating, their variances are written in the canonical form:

$$V_j(z') = A_0 + i(z'_{(\pm)} + D''_j).(1.76)$$

This formula envelopes all the considered gratings; corresponding parameters D_j'' are determined by (1.69). The width $A_j(z)$ of the wave pencils is connected [82] with the size of the spot $w_j(z)$ by the expression

$$A_j(z) = \frac{kw_j^2(z)}{2(2p+1)}, \; j = 1,2 \tag{1.77}$$

and the curvature radius $R_j(z)$ of the phase front is found by the formula

$$V_j^{-1}(z) = A_j^{-1}(z) = iR_j^{-1}(z) \tag{1.78}$$

If to extrapolate the representation (1.76) of radiation field by wave pencils from the domain where it holds ($|z_{(\pm)}| > \lambda$ for $kA_0 >> 1$) to the pencil's "throat", then the latter must be in the plane

$$z_{(\pm)} = -D_j''. \tag{1.79}$$

Hence, D_j'' is the measure of diffractional shift of the plane of pencil's throat from the grating's surface. (1.69), (1.70) yield the different shifts of the plane of pencil's throat for E- and H-polarizations depending on the type of the grating: they are minimal for the strip grating and grow in many times for the space gratings with optimal profile (see Fig. 22, a b); this effect is important for the study of excitation of diffractional radiation in OR by the wave pencils.

One obtains the formulas for $A_j(z)$ and the curvature radius $R_j(z)$ of the phase front from (1.76) and (1.78):

$$A_j(z) = A_0\left[1 + \frac{\left(D_j'' + z'_{(\pm)}\right)^2}{A_0^2}\right];$$

$$R_j(z) = \left(D_j'' + z'_{(\pm)}\right)\left[1 + \frac{A_0^2}{\left(D_j'' + z'_{(\pm)}\right)^2}\right]. \tag{1.80}$$

For $R > 0$ the wave pencil of radiation is a radiating one and due to (1.80) this condition always holds. This formula enables one to follow by means of geometrical constructions the variation of the spot of the pencil of radiation and of the curvature of its front in the process of propagation (Fig. 23). The segment with the length equal to the width of the pencil A_0 in the throat is marked in the plane $z'_{(+)} = -D_j''$ along the axis Ox what indicates the point

F, which is the lateral focus of the wave pencil. In order to construct the phase front in the vicinity of arbitrary point $z' = O_1$ one must join this point with F by a segment and draw the perpendicular to this segment through the point F up to its intersection with the axis Oz'. The obtained point C_1 is the curvature center of the pencil's phase front in O_1. The graph showing the variation of w versus z' is constructed by the known size of pencil's spot w_0 and with the help of the condition stating that beams' propagation occurs perpendicular to the wave front. The curves $w'(z)$ which describe the pencil's broadening are hyperbolas and their asymptotes are straight lines joining the points on $z'_{(+)} = -D''_j$ and $z'_{(+)} = A_0 - D''_j$, $x = \pm w_0$.

The knowledge of $V_j(z')$ allows to follow the transformation of radiation pencil in every complicated optical system, which is completely described by the variation of complex variance $V_j(z')$ as a function of z' similar to the transformation of impedance in multi-link chain [78].

The components of the vectors E and H of wave radiation pencils are found with the use of known potentials Π^1_M, Π^1_E under the condition $|H_{z'}| <$ $< |H_{x'}|, |E_{y'}|, |E_{z'}| << |E_{x'}|, |H_{y'}|$.

The Pointing theorem yields the expression for the time-average value of energy

$$W = \sum_{(m)} (W_m^+ + W_m^-) \sin \phi_m. \tag{1.81}$$

Here W_m^+, W_m^- are the intensities of the wave pencils of diffractional radiation propagating over and under the grating, respectively, calculated in Wt cm^{-2}.

In the case of single-wave radiation regime

$$W_m^\pm = \frac{1}{2} R_0 I^2 \Big[|W_M^\pm(x,z)|^2 + |W_E^\pm(x,z)|^2 \Big] e^{-\frac{2ka}{\beta}\sqrt{1-\beta^2}}. \tag{1.82}$$

The "diffusion" of wave pencils along the axis Ox yields decrease of intensity of two-dimensional beams proportional to $\cos u_j(z)$. The value of energy W also depends on the coordinate of the point where unit square is chosen. The energy $W_m^\pm$ of the m-th wave radiation pencil falls with growing κ (i. e., with decreasing λ) since the number of these pencils grows for the given partial flow.

The radiation field excited by a partial electron flow has the form of two TEM-waves having different polarizations propagating along the same direction of diffractional radiation. Superposition of two such waves is the TEM-wave with elliptic polarization (for the example of a strip grating). The ellipticity coefficient ϵ and azimuth of the main axis of polarization ellipse Φ with respect to the axis Oy are determined for $\theta^2 << 1$ on the boundary of field's spot by the formulas

$$\epsilon = \frac{x}{\sqrt{A_0^2 + z'^2}} \frac{|T_2(-n_0)|}{|T_1(-n_0)|} \sin\phi;$$

$$\Phi = \frac{x}{\sqrt{A_0^2 + z'^2}} \frac{|T_2(-n_0)|}{|T_1(-n_0)|} \cos\phi; \tag{1.83}$$

where $\phi = \phi_{x'} - \phi_{y'}$ is the phase shift between the components $E_{x'}, E_{y'}$.

The expressions (1.83) yield the definite type of H-polarization of diffractional radiation in observation points situated on the plane yOz. We will call yOz the plane of linear polarization. Out of this plane radiation has elliptic polarization (the direction of rotation of $\vec{E}$ and the big axis of polarization ellipse turn in different ways on both sides of the plane of linear polarization). Radiation always has the right polarization in the half-space $z > 0$ for $x > 0$ and the left polarization for $x < 0$; it has opposite polarizations in corresponding domains under the grating (for $z < 0$). The direction of the turn of big axis of the polarization ellipse depends on the sign of x and on the sign of $\cos\phi$. The latter changes sign in the planes $z'_{(\pm)} = \tilde{z}'_{(\pm)}$ which are determined by the condition

$$\arctan\frac{\kappa\sqrt{1-\mu^2}(Q_1 - Q_2)}{1 + \kappa^2(1-\mu^2)Q_1Q_2} + u(z'_{(\pm)} = \frac{\pi}{2})$$

and hence on both sides of these planes the direction of the turn of ellipse has different signs.

ϵ and Φ proportional to x within the limits of the field's spots; however, the broader electron flow and the father observation point from the grating, the weaker its growth with increasing x. It is caused by the fact that the phase front of wave pencil becomes more plane and the amplitude of E_x decreases faster than E_y. If $A_0 \to \infty$, then $\epsilon \to 0, \Phi \to 0$, i. e., radiation becomes elliptically polarized; for small A_0 this coefficient may reach more noticeable values within the limits of the field's spots.

For wide slots of the grating $T_2(-n_0)T_1^{-1}(-n_0) >> 1$ and $\delta = E_2E_1^{-1} \sim$ 1on the boundary of the field's spot. The coefficient of ellipticity may reach here considerable values within the limits of the field's spots for big A_0.

The radiation with circle polarization ($|\epsilon| = 1$) may appear on the spot (for certain values of κ, θ, A_0). The coefficient of ellipticity is maximal for $|x| = \sqrt{A_0^2 + z'^2_{(\pm)}} T_1(-n_0) T_2^{-1}(-n_0)$. Here $\delta = 1$, ϵ is determined only by the value $\sin\phi$ and the azimuth of the big axis of ellipse is $\Phi = \pm\frac{\pi}{4}$. If $T_1(-n_0)T_2^{-1}(-n_0) < \sqrt{2(kA_0)^{-1}}$, the point of maximal ellipticity lie within the limits of the spot of the field of radiation wave pencil.

The total radiation field is bounded by the width of electron flow; as a superposition of elliptically polarized partial TEM-waves, it also has elliptic polarization and corresponding parameters ϵ, Φ are determined by the ones of partial waves with the help of known interrelations [83].

It is also of great interest to study SPE within the frame of the model of given current for 1) plane monochromatic electron flow bounded in its width and length, 2) point charge, 3) single charged clot, 4) finite sequence of charged clots flying along an infinite grating. The influence of the finite size of a grating on SPR has been also studied.

§4. Smith-Purcell radiation of the point charge, single clot and finite sequence of charged clots.

1. The study of diffractional radiation of the point charge (charged electron) enables one to establish the properties of radiation on elementary level, since for the clots or grouped electron flows the properties of a set of electrons are developed. The solution to this problem forms the basis for investigating more complicated sources.

We will use in what follows the notation of §1.

Let us define the density of electron's charge

$$\rho(x, y, z, t) = e\delta(x)\delta(y - vt)\delta(z - a). \tag{1.84}$$

Scattering of the eigenfield (1.84) on a grating yields SPR which may be represented in far zone for $\beta^2 << 1$ in the form

$$\Pi_M^1 = \mathrm{P}_M + \mathrm{P}_M^*; \ \Pi_E^1 = \mathrm{P}_E + \mathrm{P}_E^*, \tag{1.85}$$

where

$$\mathrm{P}_M = \pm e\beta \frac{\cos\theta}{1 - \sin^2\theta\cos^2\phi} \sum_{n_0=1} T_1(-n_0) e^{-\frac{ka}{\beta}\sqrt{1-\beta^2} - ikct} \frac{e^{i(kR^{(\pm)} + \frac{\pi}{2})}}{kR^{(\pm)}};$$

$$\mathrm{P}_E = \frac{e\beta}{\sqrt{1-\beta^2}}\frac{\sin 2\theta\cos\phi}{1-\sin^2\theta\cos^2\phi}\sum_{n_0=1} T_2(-n_0)e^{-\frac{ka}{\beta}\sqrt{1-\beta^2}-ikct}\frac{e^{ikR^{(\pm)}}}{kR^{(\pm)}}; \quad (1.86)$$

Here it is introduced the (accompanying) spherical coordinate system moving with the velocity of charge

$$x = R^{(\pm)}\sin\theta\cos\phi;\ y - vt = R^{(\pm)}\sin\theta\sin\phi;\ z^{(\pm)} = R^{(\pm)}\cos\theta. \quad (1.87)$$

Coefficients $T_1(-n_0), T_2(-n_0)$ have the following representation:

a) for the strip grating

$$T_1(-n_0) = \frac{1}{2}\Big\{P_{n_0-1}(u) + P_{n_0}(u) + \beta n_0 Q_1 \sin\theta\sin\phi\Big[P_{n_0-1}(u) - P_{n_0}(u)\Big]\Big\}\times$$
$$(1 - i\beta n_0 Q_1\cos\theta)^{-1};$$

$$T_2(-n_0) = \frac{1}{2}\Big\{P_{n_0-1}(u) - P_{n_0}(u) + \beta n_0 Q_2 \sin\theta\sin\phi\Big[P_{n_0-1}(u) + P_{n_0}(u)\Big]\Big\}\times$$
$$(1 - i\beta n_0 Q_2\cos\theta)^{-1};$$

b) for the reflection grating with perfectly conducting basis

$$T_1(-n_0) = 2G(n_0,\theta,\phi)\frac{\sin[n_0\pi\theta(1+\beta\sin\theta\sin\phi)]}{n_0\pi\theta(1+\beta\sin\theta\sin\phi)}\frac{\sin[\beta n_0\pi\theta\sin\theta\sin\phi]}{\beta n_0\pi\theta\sin\theta\sin\phi} \quad (1.89)$$

where

$$G(n_0,\theta,\phi) = \Big[1 - i\beta n_0\cos\theta(Q_1 - \frac{1}{q(\phi,\theta)})\Big]^{-1};$$

$$q(\phi,\theta) = \beta n_0\sqrt{1-\sin^2\theta\cos^2\phi}\tan[2\pi\delta\beta n_0\sqrt{1-\sin^2\theta\cos^2\phi}].$$

It follows from the representation in the form (1.85) that radiation fields are connected by the condition stating that they are purely real for purely real source for negative and positive frequencies.

(1.86) yields spherical form of every space harmonic in the accompanying coordinates and such spherical waves propagate off the grating with the intensity having an exponential decay when the source go off the grating (the factor giving the degree of the fall of intensity is proportional to the number of radiation harmonic).

Radiation spectrum does not depend on the shape of the grating and it is determined only by its period $\lambda_n = l(\beta n)^{-1}$ $(n = 1, 2, \ldots)$. Only those frequencies radiate for which the ratio $\frac{\lambda_n}{\beta\lambda}$ is a whole number and it determines the condition for the maximums of interference of the scattered waves of given frequency. It is equivalent to the conditions $\Delta s = n\lambda$ and $\Delta s = l\cos\alpha$. All scattered waves propagate along the axis Oy with the source's velocity v and with the velocity of light c along the direction of propagation, where $l = v\Delta t$, $s = c\Delta t$ what yields $\cos\alpha = \Delta s l^{-1} = \beta^{-1}$ and hence $\lambda_n = l(\beta n)^{-1}$.

In the laboratory coordinate system radiation spectrum consists of radiation zones because radiation occurs at different angles for different frequencies:

$$\lambda_n = \frac{l}{n}\left(\frac{1}{\beta} - \sin\theta\sin\phi\right). \tag{1.90}$$

For the observation points in the plane yOz the radiation wavelength

$$\lambda_n = \frac{l}{n}\left(\frac{1}{\beta} \mp \sin\theta\right). \tag{1.91}$$

where the sign minus is taken for the first quadrant and the sign plus – for the second quadrant of the plane yOz. In this very plane Smith and Purcell registered the radiation of non-modulated electron flow and obtained the formula for radiation wavelength [1]

$$\lambda_n = \frac{l}{n}\left(\frac{1}{\beta} - \cos\gamma\right), \tag{1.92}$$

where γ is the angle between the vector of velocity and the wave vector of radiation. It is clear now that (1.90) and (1.91) coincide. The formula (1.92) may be rewritten in the form

$$\omega_n = \frac{2\pi n\frac{v}{l}}{1 - \beta\cos\gamma}; \tag{1.93}$$

it expresses the dependence of radiation frequency on the source's velocity, observation angle and grating's period. This formula has rather simple physical sense: the time which is necessary for the source to cover one space period of the structure $T = lv^{-1}$; the covering frequency $\Omega = 2\pi T^{-1} = 2\pi v l^{-1}$. The expression $2\pi n\frac{v}{l}$ is the whole number divisible by this frequency and $1 - \beta\cos\gamma$ gives the Doppler shift of the frequency of moving source.

The electron's energy losses spent for the radiation in the free half-space at the lowest frequencies ($\kappa < 0.5$) and on the structure's period in the angles' range $\phi, \phi + d\phi$; $\theta, \theta + d\theta$ are determined in the case of narrow slots for $\beta^2 << 1$ by the formula

$$W(\theta,\phi) = \frac{cl}{8\pi} e^2\beta^2 \sum_{n_0=1}^{N} k^2 Q^e(\theta,\phi) \mid T_1(-n_0,\theta,\phi) \mid^2_{\theta=0} e^{-\frac{2ka}{\beta}\sqrt{1-\beta^2}} \quad (1.94)$$

with the far field pattern

$$Q^e(\theta,\phi) = \frac{\cos^2\theta}{1-\sin^2\theta\cos^2\theta} \frac{\mid T_1(-n_0,\theta,\phi) \mid^2}{\mid T_1(-n_0,0,\phi) \mid^2}. \quad (1.95)$$

Far field patterns (1.95) of the radiation over the strip grating at the principal frequency ($n_0 = 1$) for the particle's velocity $\beta = 0.1$ are shown on Fig. 24, a in the planes corresponding to different fixed values of the angle ϕ. One can see on the figure rather intensive radiation to a small sector close to the axis Oz of the plane yOz.

The far field pattern in observation planes $\phi = 0$, $\phi = \frac{\pi}{2}$ coincides with the analogous one in the two-dimensional case (see Fig. 8, the charged thread). The radiation in the direction $\theta = 0$ perpendicular to the grating is calculated by the formula

$$W(\theta,\phi) = \frac{cl}{8\pi} e^2\beta^2 \sum_{n_0=1}^{N} k^2 \mid T_1(-n_0,\theta,\phi) \mid^2 e^{-\frac{2ka}{\beta}\sqrt{1-\beta^2}}. \quad (1.96)$$

The radiation intensity here is maximal for $\left(\frac{d}{l}\right)^2 << 1$ and $\beta^2 << 1$ while it always vanishes in the direction $\theta = 0$ for the grating made of half-planes.

In the experiment Smith and Purcell observed [1] that the light radiated in the direction $\phi = 0$, θ close to $\frac{\pi}{2}$ was strongly polarized in z-direction. One can see it from the angle diagram for the components of radiation field. It turns out that only the following components are non-zero: $\mid H_x \mid \sim \cos\theta$; $\mid E_z \mid \sim \sin 2\theta$; $\mid E_y \mid \sim \cos^2\theta$; and $\mid E_y \mid << \mid H_x \mid, \mid E_z \mid$ for $\theta \to \frac{\pi}{2}$. The radiation field is polarized presumably in z-direction (θ may be anyhow close to $\frac{\pi}{2}$ but the condition $|z| > \lambda$ must hold for arbitrary value of R).

When the electron moves at a sharp angle to the grating's axis, the wave vector of radiation is determined by the formula $k = 2\pi l^{-1}\beta n_0 \cos\alpha$ in the accompanying coordinate system

$$x-\beta ct\sin\alpha = R^{(\pm)}\sin\theta\cos\phi;\ y-\beta ct\cos\alpha = R^{(\pm)}\sin\theta\sin\phi;\ \pm z_{(\pm)} = R^{(\pm)}\cos\theta.$$

In this coordinate system the radiation spectrum

$$\lambda_n = \frac{l}{\beta n}\frac{1}{\cos\alpha}, \quad (n = 1, 2\ldots) \tag{1.97}$$

In the laboratory coordinate system the spectrum of electron's radiation frequencies moving obliquely to the grating is given by the formula

$$\lambda_n = \frac{l}{n\cos\alpha}\Big[\frac{1}{\beta} - \sin\theta\sin(\alpha+\phi)\Big]. \tag{1.98}$$

For observation points lying on the plain which goes through the electrons' trajectories perpendicular to the grating ($\phi = \frac{\pi}{2} - \alpha$)one can rewrite (1.98) in the form

$$\lambda_n = \frac{l}{n\cos\alpha}\Big(\frac{1}{\beta} - \cos\gamma\Big)$$

what completely coincides with the formula of Smith and Purcell [1]. It is easy to see comparing (1.90) with (1.97), (1.98) that the latter are obtained from (1.90) for $l \to l(\cos\alpha)^{-1}$, $\phi \to \alpha + \phi$. Hence one can effectively change l rotating the grating around the axis Oz and keeping the fixed direction of electron flow; $l(\cos\alpha)^{-1}$ will be the new effective grating's period.

2. Let us determine the influence of the form of a charged clot on field's structure and radiation energy losses, i. e., let us establish how essential is interference fall of radiation caused by different clot's elements. Let us consider the problem on diffractional radiation of the electron clot having charge's density

$$\rho(x,y,z,t) = \rho_0\psi\Big(\frac{x}{\sqrt{A_0}}\Big)\psi\Big(\frac{y-vt}{\sqrt{B_0}}\Big)\delta(z-a). \tag{1.99}$$

The Gauss functions

$$\psi\Big(\frac{x}{\sqrt{A_0}}\Big) = \frac{1}{\pi A_0}e^{-\frac{x^2}{A_0}};\ \psi\Big(\frac{y-vt}{\sqrt{B_0}}\Big) = \frac{1}{\pi B_0}e^{-\frac{(y-vt)^2}{B_0}} \tag{1.100}$$

determine effective width of a clot $2\bar{x} = 2\sqrt{A_0}$ and its effective length $2\bar{y} = 2\sqrt{B_0}$.

The physical nature of the phenomenon yields the impossibility for the distributions of charge's density to be spasmodic functions due to the Coulomb repulse forces. It is also clear that interaction of a flow with the slow wave, which forms the clot, yields the wave function which is a Gauss function with respect to the longitudinal coordinate $y - vt$.

One can take into account finite thickness of a clot by analogy with §3. For the sake of simplicity charge's density is again approximated by a Gauss function with respect to x. Frequency characteristics of the clot's radiation and its dependence on the coordinate coincide with corresponding parameters of a single electron (1.90) and the "spread" charge leads only to the change of space structure of the radiation field. In the case when $\left(\frac{d}{l}\right)^1 << 1$ and $\beta^2 << 1$ the radiation far field pattern the clot has the form

$$Q^{\rm cl} = Q^e(\theta,\phi)e^{-\frac{\bar{A}_0}{2}\beta^2 n_0^2 \sin^2\theta\cos^2\phi}e^{-\frac{\bar{B}_0}{2}n_0^2(1+\beta\sin\theta\sin\phi)^2+\frac{\bar{B}_0}{2}n_0^2} \tag{1.101}$$

where $Q^e(\theta,\phi)$ is the radiation far field pattern of a single electron (1.95); $\bar{A}_0 = (\frac{2\pi}{l})^2 A_0$; $\bar{B}_0 = (\frac{2\pi}{l})^2 B_0$.

Radiation of the point charge in the plane $\phi = 0$ is weakly directed; the second factor in (1.101) causes appearance of a petal in the clot's pattern in this plane whose width is determined by $2\bar{x}\lambda^{-1}$: greater this parameter, narrower the petal.

The odd character of the third factor in (1.101) as a function of θ (for $y > 0 \phi = \frac{\pi}{2}$, $y < 0 \phi = \frac{3\pi}{2}$) yields asymmetry of the radiation far field pattern with respect to the axis $\theta = 0$. For $\left(\frac{d}{l}\right)^1 << 1$ the angle θ giving the maximum of the radiation far field pattern is found as a root of the equation

$$\sin\theta = \frac{\bar{B}_0 n_0^2 \beta^2}{2}\left(\frac{1}{\beta} - \sin\theta\right)\cos^2\theta.$$

Unlike the radiation of the point charge, the far field pattern of a clot reaches its maximums at different frequencies when θ takes different values. With growing n_0, i. e., with growing radiation frequency, the maximum of a pattern approaches the grating's plane. The analogous effect takes place for the fixed frequency when the length of a clot increases, i. e., when B_0 grows.

The character of the direction of the clot's radiation over the strip grating is illustrated by Fig. 24, b where far field patterns at the principal frequency $n_0 = 1$ for $\beta = 0.1$ and fixed values of ϕ, $2\bar{x} = 5l, 2\bar{y} = l$ (as well as for the point charge) are presented. Radiation far field pattern of a clot looks like a three-dimensional petal whose maximum lies in the back halfspace and it is reached for $\theta = 25°$.

In order to explicitly calculate the clot's radiation field one can use the method of parabolic equation giving the possibility to obtain the field in the wave zone and for arbitrary $z > \lambda$ in the vicinity of the normal direction, where $\sin^2\theta << 1$. Limiting ourselves to the case $kz >> 1$ we obtain in spherical coordinate system the formulas analogous to (1.86) for small angles θ. The method of parabolic equation in the problem describing diffractional radiation of a clot yields the results coinciding in the wave zone with the data obtained by the method of stationary phase only in the vicinity of the direction perpendicular to the grating and going through the center of a moving clot (i. e., for $\sin^2\theta << 1$); in addition to this, the method of parabolic equation enables one to take into account more precisely phase interrelations in the radiation field.

The spectrum of clot's radiation is completely determined in the accompanying coordinate system by the structure's period, velocity and the number n_0 of the space harmonic where the radiation occurs with $\lambda_n = l(\beta n)^{-1}$. The beam with the strongest intensity ($n_0 = 1$) has the longest wavelength; its intensity falls with growing n_0 together with the wavelength which remains divisible by the latter index. For $(\frac{d}{l}) << 1$ radiation is concentrated practically in one wave beam ($n_0 = 1$).

Radiation frequency of a clot is found in the laboratory coordinate system with the help of time-depending factors

$$\omega_n = \frac{2\pi n \frac{v}{l}}{1 - \cos^2 u_y(z) - \beta\frac{y}{R_y}}. \qquad (1.102)$$

If to introduce observation angle ϕ such that $\cos\phi = yR_y^{-1}$, one obtains the formula for the spectrum of diffractional radiation of a source bounded in the space, which takes into account the Doppler's effect:

$$\lambda_n = \frac{l}{n}\left[\frac{1}{\beta}(1 - \cos^2 u_y) - \cos\phi\right].$$

Expression (1.102) differs from (1.92), which is introduced for the point charge, by the factor $1 - \cos^2 u_y$; in the wave zone $u_y \to \frac{\pi}{2}$, $\cos^2 u_y \to 0$ and they coincide. For the wave zone radiation waves depend only on observation angle ϕ, while for intermediate regions (formation zone) they also depend on the coordinate z of observation point.

3. Charge's density for the sequence of N identical clots has the form

$$\rho(x,y,z,t) = \rho_0\delta(z-a)\psi\Big(\frac{x}{\sqrt{A_0}}\Big)\sum_{n=0}^{N}\int_{-\frac{L}{2}}^{\frac{L}{2}}\psi\Big(\frac{\eta}{\sqrt{B_0}}\Big)\delta(y-vt-nL-\eta)\,d\eta, \tag{1.103}$$

where L is the period of clots' movement (for $t=0$ the center of the first clot $n=0$ is in the point $y=0$ and other clots follow it on the positive half-axis Oy). Representing the right-hand side of (1.103) in the form of Fourier integral with respect to time and coordinate x we obtain

$$\rho = \iint_{-\infty}^{\infty}\rho(\omega)F(\xi)\Phi(\omega)f_N(\omega)e^{i\xi x-i\omega t}\,d\xi d\omega$$

where

$$\rho(\omega) = \frac{\rho_0}{v(2\pi)^2}e^{i\frac{y}{v}\omega}\delta(z-a)$$

is the Fourier-component of the charge's density of a point particle having the charge ρ_0; $F(\xi)\Phi(\omega)$ is the form-factor determining by the same formulas as the single clot. The value

$$f_N(\omega) = \frac{1}{2}\Big[\Phi\Big(\frac{L}{2\sqrt{B_0}}+\frac{i}{2}\sqrt{B_0}\frac{\omega}{2}\Big)$$

$$+\Phi\Big(\frac{L}{2\sqrt{B_0}}-\frac{i}{2}\sqrt{B_0}\frac{\omega}{v}\Big)\frac{\sin\frac{\omega L}{2v}N}{\sin\frac{\omega L}{2v}}\Big]e^{-i\frac{\omega L}{2v}(N-1)} \tag{1.105}$$

is the form-factor carrying out the transfer from the integral representation of the charge's density and radiation field of the single charged clot to the formulas for the sequence of N identical clots. $\Phi(x)$ in (1.105) is the probability integral. In the moving coordinate system $y_1 = y - \frac{L(N-1)}{2-\beta ct}$ every separate clot radiates discrete spectrum of frequencies $\omega_n = \frac{2\pi v}{l}n$ $(n=1,2\ldots)$.

It follows from (1.105) that in the laboratory coordinate system the interference of radiation of separate clots leads to vanishing the radiation at certain frequencies and to amplifying it at other frequencies. For example, if $\omega_m = \frac{2\pi m v}{Nl}$ $(m=1,2\ldots)$ and $\frac{m}{N}$ is not a whole number, the spectral density of radiation energy is zero and these frequencies are absent in the radiation; the fall in the far field pattern corresponds to this frequency. If $\omega_m = \frac{2\pi m v}{L}$ $(m=1,2\ldots)$, the radiation energy is proportional to the squared value of the number of clots N.

Far field pattern of the radiation field of a sequence of identical clots in the moving coordinate system has the form

$$Q^{\rm seq}(\theta,\phi) = Q^{\rm cl}(\theta,\phi)\frac{1}{4}\left|\left[\Phi\left(\frac{L}{2\sqrt{B_0}}+\frac{i}{2}\sqrt{B_0 n_0 q_0}\right)+\Phi\left(\frac{L}{2\sqrt{B_0}}-\frac{i}{2}\sqrt{B_0 n_0 q_0}\right)\right]\right|^2 \times \frac{\sin^2\frac{1}{2}L_1 n_0 N q_0}{\sin^2\frac{1}{2}L_1 n_0 q_0} \qquad (1.106)$$

where

$$q_0 = q_0(\theta,\phi) = 1 + \beta\sin\theta\sin\phi.$$

Unlike $Q^{\rm cl}$, the shape of the far field pattern (1.106) becomes very broken for $L >> l$ in all planes except of $\phi = 0$ as a result of interference of separate clots. Interference of separate clots is described by the factor

$$\frac{\sin^2\frac{1}{2}L_1 n_0 N q_0}{\sin^2\frac{1}{2}L_1 n_0 q_0}, \quad L_1 = \frac{2\pi L}{l}. \qquad (1.107)$$

Observation angles θ_m, ϕ_m (on the n-th spectral line), which correspond to the radiation maximums on the far field pattern, are found from the interrelation

$$Ln(1 + q_0(\theta_m,\phi_m)) = lm \quad (m = 1, 2, \ldots). \qquad (1.108)$$

If the condition $Ln = ml$ holds for $m(m = N)$, the M-th maximum lies in the plane $\phi = 0$. Minimums on the far field pattern of the n-th radiation spectral curve are determined by the interrelation

$$Ln(1 \mid q_0(\theta_m,\phi_m)) = l\frac{m}{N} \quad (m - 1, 2, \ldots). \qquad (1.109)$$

where $\frac{m}{N}$ may be a whole number. If the condition $Ln = \frac{M}{N}l$ holds for $m = n$, the M-th minimum lies in the plane $\phi = 0$ (it is a purely interference effect which is impossible for the radiation of a single clot).

For $L \sim l$ and big velocity of the clots (when their number is small) the angle pattern has no blips as it follows from (1.107); but if $Ln = \frac{M}{N}l$ holds for the given n, additional minimum appears in the plane $\phi = 0$. It is clear that for $L = l$ (or if L is divisible by l) the condition $Ln = \frac{M}{N}l$ cannot be fulfilled since $\frac{M}{N}$ is not a whole number (1.109) (here additional minimum appears in the plane $\phi = 0$).

4. The interesting problem of the study of SPR for a sequence of clots is the one describing radiation over unbounded grating of the sequence of clots which "is born" in the plane $y = -0.5D$, moves with the velocity v and "annihilates" in the plane $y = 0.5D$. The charge's clots appear on the background of the constant component of charge's density of the grouped electron flow. We assume that variation of constant component yields such variation of the amplitude of charge's density of the clot ρ_0 in the whole interval $(-0.5D, 0.5D)$ which guarantees the fulfillment of conservation law. The variable component of the charge's density of such flow has the form

$$\rho_\sim = \rho_0 \delta(z-a) f(x)\phi(y) \sum_{n=-\infty}^{\infty} \int_{-\frac{L}{2}}^{\frac{L}{2}} \psi\Big(\frac{\eta}{\sqrt{B_0}}\Big) \delta(y - vt - nL - \eta)\, d\eta, \quad (1.110)$$

where $f(x)$ is given in the interval $(-b, b)$ and it describes transversal variation of the charge's density of clots. The function $\phi(y)$ given in the interval $(-0.5D, 0.5D)$ is a filtering function; $\phi(y) = 0$ out of the interval $(-0.5D, 0.5D)$, hence, it filters only those clots from the infinite periodical sequence which lie in this interval in every moment of time. In the same interval the function $\rho_\sim$ gives the amplitude's variation of the variable component ρ_0.

For $\sqrt{B_0} \geq \frac{L}{\pi\sqrt{3}}$ one can use the following formula in place of (1.110)

$$\rho_\sim = A_1 \rho_0 \delta(z-a) f(x)\phi(y) e^{i(k\alpha y - \omega t)} \quad (1.111)$$

(one must add the complex conjugation to the real-valued expressions). In (1.111)

$$A_1 = \frac{1}{2L} e^{-\frac{\pi^2 B_0}{L^2}} \left[\Phi\Big(\frac{L}{2\sqrt{B_0}} + \frac{i}{L}\pi\sqrt{B_0}\Big) + \Phi\Big(\frac{L}{2\sqrt{B_0}} - \frac{i}{L}\pi\sqrt{B_0}\Big)\right]$$

where $\Phi(x)$ is the integral of probability.

The completed calculations enable us to conclude that the wave field of diffractional radiation of a partial ps-th electron flow having the charge's density given by (1.111) is the superposition of three -dimensional wave pencils with variances

$$V_{jx}(z') = A_{0x} + i(z'_{(\pm)} + D_{jx});$$

$$V_{jy}(z') = A_{0y} \sin^2 \phi_{-n_0} + i(z'_{(\pm)} + D_{jy}) \ \ (j = 1, 2) \quad (1.112)$$

The phases u_{jx}, u_{jy} of these complex variances

$$u_{jx} = \arctan \frac{z'_{(\pm)} + D_{jx}}{A_{0x}}; \; u_{jy} = \arctan \frac{z'_{(\pm)} + D_{jy}}{A_{0y} \sin^2 \phi_{-n0}} \qquad (1.113)$$

The given expressions describe phase raids in the wave pencils when their propagation is caused by the curvature of the phase fronts. In the general case the spot of the wave pencil of radiation has ellipsoidal form and the phase front is spheroidal surface. The width $A_j(z')$ of the pencil of radiation and the curvature radius of the phase front $R_j(z')$ in the direction of axes Ox', Oy' are determined by formulas

$$A_{jx}(z') = \frac{A_{0x}}{\cos^2 u_{jx}}; \; R_{jx}(z') = \frac{2A_{0x}}{\sin 2u_{jx}};$$

$$A_{jy}(z') = \frac{A_{0y} \sin^2 \phi_{-n0}}{\cos^2 u_{jy}}; \; R_{jy}(z') = \frac{2A_{0y} \sin^2 \phi_{-n0}}{\sin 2u_{jy}}. \qquad (1.114)$$

Reconstruction of the geometrical parameters of wave pencils by the formula (1.114) is carried out similarly to the procedure described in §3 with the only difference: the lateral focus F_y of the pencil in the plane $y'Oz'$ lies on the axis Oy' at the distance $A_{0y} \sin^2 \phi_{-n_0}$ from the origin (see Fig. 25). The direction of radiation of wave pencils occurs at the angle ϕ_{-n_0} to the vector of pencil's velocity and it is called radiation angle ($\sin^2 \phi_{-n_0} = \sqrt{1 - \nu^2}$). The change of flow's velocity yields the change of radiation angle and hence it is carried out electrical scanning of the radiation wave pencils in the plane yOz.

In the beam coordinates (1.75) electromagnetic radiation field is determined by the Hertz potentials with the help of formulas obtained from the Maxwell's equations, if not to take into account in these equations the terms which are negligible in the method of parabolic equation. They hold, if the following interrelation are fulfilled for A_{0x}, A_{0y} and the radiation frequencies

$$4kA_{0x} >> 1; \; 4kA_{0y} \sin^2 \phi_{-n_0} >> 1, \qquad (1.115)$$

which are always true in the quasioptical range. If (1.115) holds, the field of quasioptical radiation pencil is approximately the TEM-wave. For $A_{0x} \to \infty$, $A_{0y} \to \infty$ all components of radiation field vanish except of $H_{x'}$, $E_{y'}$. The field transforms to the plane H-polarized wave.

§5. Influence of the finite size of gratings on the Smith-Purcell radiation.

Infinite size of periodical diffractional gratings is assumed in the statements of the problems of diffractional radiation considered above. Such structures have been chosen which admitted the use of rigorous mathematical methods of the diffraction theory for solving such problems and exposing justified physical properties of radiation. In the real situations one must deal with gratings of finite size. Therefore, it is important (especially in various applications) to know the influence of this factor on physical properties of SPR. Of principle mathematical difficulties arise when one solves the problems in such statement. Approximate and rigorous methods have been sufficiently well developed for infinite periodical structures, while rather modest results have been obtained for finite gratings [85-88]. The reason of this discrepancy is extraordinary complexity of the given diffraction problem caused by the necessity of finding the field which is a superposition of a continuum of plane waves together with superposition of waves corresponding to the discrete spectrum.

When solving the problems on diffractional radiation with finite gratings, one should separate the gratings with infinite elements along their axes and the gratings whose elements have finite size along the axes and their infinite sequence forms bounded periodical structure. The grating with finite area may be obtained from the infinite one by simultaneous use of both limitations. Essential contribution to the mathematically justified studies of finite periodical gratings has been made by the works [89-91].

The problems describing the influence of finite size of the gratings on diffractional radiation will be considered below in rigorous statement and with the help of approximate methods of quasioptics. The results of the first part of the book will be widely used. Approximate methods of the study of bounded gratings are developed in [92].

1. Let the electron flow with the charge's density given in (1.3) moves over the periodical structure consisting of $2N$ infinitely thin perfectly conducting strips lying in the plane $z = 0$ (the coordinate system is chosen in such a way that strips are symmetrical with respect to the origin, Fig. 25). The flow excites electromagnetic field in the free space, which is determined by the only one non-zero component $H_x^0(y, z)$ of magnetic field defined in (1.4). The problem is to determine the field scattered by the bounded strip grating excited by inhomogeneous plane wave $H_x^0(y, z)$. Its solution is analogous to the one for problem of the scattering of plane wave by this structure [92].

As well as in [92] we will look for the only one non-zero component of the magnetic field scattered by the finite number of strips in the form

$$H_x^1 = 2\pi \rho_0 e^{i\kappa p\gamma} \frac{|z|}{z} \int_{-\infty}^{\infty} Y(\lambda) e^{i\sqrt{\kappa_L^2-\lambda^2}|\xi|} e^{i\lambda\eta} d\lambda, \qquad (1.116)$$

where $\kappa_L = kL$ ($2L$ is the longest distance between grating's edges); $\eta = yL^{-1}$; $\xi = zL^{-1}$ are dimensionless coordinates of observation point; $p = aL^{-1}$ is dimensionless aimed parameter. It is assumed in (1.116) that $\sqrt{\kappa_L^2 - \lambda^2} > 0$ for $\lambda \to \infty$ along the real axis.

Boundary value conditions yield the following system of coupled integral equations with respect to the sought for function $Y(\lambda)$

$$\int_{-\infty}^{\infty} Y(\lambda) e^{i\lambda\eta} d\lambda = 0, \;\; \beta_{n-1} < |\eta| < \alpha_n \;\; (n = 1, 2, \ldots, N+1)$$

$$\int_{-\infty}^{\infty} Y(\lambda) K(\lambda) e^{i\lambda\eta} d\lambda = f(\eta), \;\; \alpha_m < |\eta| < \beta_m \;\; (m = 1, 2, \ldots, N), \qquad (1.117)$$

where $K(\lambda) = -i\sqrt{\kappa_L^2 - \lambda^2}$; $f(\eta) = i\kappa_L \gamma e^{i\kappa_L \alpha\eta}$; $\eta = \alpha_m, \eta = \beta_m$ are related to L coordinates of the edges of m-th strip such that $\beta_0 = 0, \beta_N = 1, \alpha_{N+1} = \infty$. The function $Y(\lambda)$ must satisfy in addition to (1.117) the condition

$$\int_{-\infty}^{\infty} |Y(\lambda)|(|\lambda| + 1) d\lambda < \infty, \qquad (1.118)$$

which follows from the demand concerning finiteness of energy of the scattered wave in every bounded domain of the space. The function $Y(\lambda)$ with the accuracy up to a constant factor is the Fourier transform of the density of surface current in the plane $z = 0$. Hence, its values in the interval $-\kappa_L < \lambda < \kappa_L$ determine without additional integration the radiation far field pattern

$$S(\phi) = 8\pi^3 \rho_0^2 \beta^2 \kappa_L L e^{-\frac{2\kappa_L p}{\beta}\sqrt{1-\beta^2}} \cos^2\phi \mid Y(\kappa_L \sin\phi) \mid^2 .$$

The total radiated energy (energy losses) is found from the equality

$$S^{\text{tre}} = 2\int_{-\frac{\pi}{2}}^{\frac{\pi}{2}} S(\phi) d\phi = -8\pi^2 \rho_0^2 \beta^2 \sqrt{1-\beta^2} L e^{-\frac{2\kappa_L p}{\beta}\sqrt{1-\beta^2}} \Im Y\left(\frac{\kappa_L}{\beta}\right).$$

Further investigations are mainly connected with the analysis of such characteristics of diffractional radiation.

Let us continue the study of equations (1.117) by the change of variables $y(\lambda) = \lambda Y(\lambda)$ and introducing the notation

$$g(\eta) = i\kappa_L \gamma e^{i\kappa_L \alpha \eta} + \int_{-\infty}^{\infty} y(\lambda)\epsilon(\lambda)e^{i\lambda\eta}\frac{d\lambda}{\lambda}, \tag{1.119}$$

where $\epsilon(\lambda) = |\lambda| + i\sqrt{\kappa_L^2 - \lambda^2}$.

Calculation of the first derivative of the first equation in (1.117) with respect to η yields equations

$$\int_{-\infty}^{\infty} y(\lambda)e^{i\lambda\eta}d\lambda = 0, \ \beta_{n-1} < |\eta| < \alpha_n;$$

$$\int_{-\infty}^{\infty} y(\lambda)\frac{|\lambda|}{\lambda}e^{i\lambda\eta}d\lambda = g(\eta), \ \alpha_m < |\eta| < \beta_m, \tag{1.120}$$

which are equivalent to (1.117) under the conditions

$$y(0) = 0; \ \int_{-\infty}^{\infty} y(\lambda)e^{i\lambda\eta_n}\frac{d\lambda}{\lambda}, \ (n = \pm 1, \pm 2, \dots, \pm N) \tag{1.121}$$

Here

$$\eta_1 = 0; \ \eta_n = \frac{1}{2}(\alpha_n + \beta_{n-1}); \ \eta_{-n} = -\eta_n \ (n = 2, 3, \dots, N).$$

Let us introduce two functions $\tilde{y}_{\pm}(\eta)$ analytical in the upper and lower half-planes of the complex variable η, respectively:

$$\tilde{y}_{\pm}(\eta) = \int_0^{\infty} y(\lambda)e^{i\lambda\eta_n}d\lambda.$$

The first equation of the system (1.120) determines

$$\tilde{y}(\eta) = \begin{cases} \tilde{y}_+ & \text{if } \Im\eta > 0; \\ \tilde{y}_- & \text{if } \Im\eta < 0 \end{cases}$$

as analytical function of η in the whole complex plane except of $2N$ segments of the real axis $\alpha_m < |\eta| < \beta_m \ (m = 1, 2 \dots, N)$ where due to the second equation (1.120) the sum of limiting values of $\tilde{y}(\eta)$ from the upper and lower half-planes is equal to $g(\eta)$: $\tilde{y}_+(\eta) + \tilde{y}_-(\eta)g(\eta)$; $\alpha_m < |\eta| < \beta_m\,(m = 1, 2\ldots, N)$. The system of coupled integral equation (1.120) is thus equivalent to the Riemann-Hilbert problem of reconstructing the function $\tilde{y}(\eta)$ analytical everywhere except of $2N$ segments $\alpha_m < |\eta| < \beta_m\,(m = 1, 2\ldots, N)$ and decreasing in infinity by the given sum of its limiting values from the upper and lower half-planes on these segments. It is known [93] that the solution to this problem has the form

$$y(\lambda) = \frac{2}{2\pi^2 l}\int_L \frac{e^{-i\lambda\eta}}{\sqrt{Q(\eta)}}d\eta \text{v.p.} \int_L \frac{\sqrt{Q(\tau)}}{\tau - \eta} g(\tau)d\tau + \frac{1}{\pi}\int_L \frac{R_{2N-1}}{\sqrt{Q(\eta)}} e^{-i\lambda\eta}d\eta \tag{1.122}$$

where L is unclosed contour consisting of the segments $\alpha_m < |\eta| < \beta_m\,(m = 1, 2\ldots, N)$; $Q(\eta) = \Pi_{n=1}^N(\eta^2 - \alpha^2)(\eta^2 - \beta_n^2)$; $R_{2N-1} = \Sigma_{n=0}^{2N-1} i^{n-1}A_n\eta^n$ and $\sqrt{Q(\eta)}$ denotes the branch of the function $\sqrt{Q(\eta)}$ determined on the complex plane η cut along the contour L for which $\sqrt{Q(\eta)} = \eta^{2N} + O(\eta^{2N-2})$ for $\eta \to \infty$. The unknown constants $A_n\,(n = 0, 1, 2, \ldots 2N - 1)$ entering the expression for $R_{2N-1}(\eta)$ are found with the help of $2N$ conditions (1.121).

The following integral Fredholm equation of the first kind with respect to the sought for function is obtained after rewriting (1.222) on the real axis and taking into account the definition of $g(\eta)$:

$$Y(\lambda) = i\kappa_L \frac{\gamma}{\lambda} W^{\kappa_L\alpha} + \frac{1}{\lambda}\int_{-\infty}^{\infty} Y(\mu)\epsilon(\mu)W_\lambda^\mu d\mu +$$

$$\frac{i}{\lambda}\Big[\sum_{n=0}^{N-1}(-1)^n A_{2n}S^{(2n)}(\lambda) + \sum_{n-0}^{N-1}(-1)^n A_{2n-1}S^{(2n-1)}(\lambda)\Big] \tag{1.123}$$

where

$$W_\lambda^\mu = \frac{1}{2\pi^2 i}\int_L \frac{e^{-i\lambda\eta}}{\sqrt{Q(\eta)}}d\eta \text{v.p.} \int_L \frac{\sqrt{Q(\tau)}}{\tau - \eta} e^{i\mu\tau}d\tau;$$

$$S(\lambda) = -\frac{i}{\pi}\int_L \frac{e^{-i\lambda\eta}}{\sqrt{Q(\eta)}}d\eta;\ S(\lambda) = \frac{d^n}{d\lambda^n}S(\lambda). \tag{1.124}$$

The entire function of exponential type $S(\lambda)$ vanishes in the point $\lambda = 0$ together with all its derivatives up to the order $2N - 1$ and satisfies linear differential equation of the order $4N$

$$\sum_{m=0}^{2N} q_m \left[S^{(2m)}(\lambda) + \frac{m}{\lambda} S^{(2m-1)}(\lambda) \right] = 0; \tag{1.125}$$

q_m are the coefficients of the polynomial $Q(i\eta)$ related to η^2. The expansion of the function $S(\lambda)$ in the power series has the form

$$S(\lambda) = \sum_{n=0}^{\infty} (-1)^{n+N-1} S_n \frac{\lambda^{2(n+N)-1}}{[2(n+N)-1]!}$$

where S_n are the expansion coefficients of the function $\eta^{2N}(Q(\eta))^{-0.5}$ in the vicinity of infinity.

The definition of $S(\lambda)$ yields simple representation of the function W_λ^μ versus $S(\lambda)$, $S(\mu)$ and their derivatives:

$$W_\lambda^\mu = \frac{1}{2(\lambda-\mu)} \sum_{m=0}^{2N} q_m \left[\sum_{n=0}^{2m-2} (-1)^n (n+m-1) S^{(n)}(\lambda) S^{(2(m-n)}(\mu) - \right.$$

$$\left. \lambda \sum_{n=0}^{2m-1} (-1)^n S^{(n)}(\lambda) S^{(2m-n-1)}(\mu) \right].$$

Substitution to (1.121) the right-hand side of (1.123) in place of $Y(\lambda)$ yields two systems of linear algebraic equations with respect to unknown coefficients A_{2n}, A_{2n-1}:

$$\sum_{n=0}^{N-1} A_{2n} Q_s^{(2n)} = \sum_{m=0}^{2N-1} c_s^m \left[\frac{\gamma}{\alpha} S^{2m}(\kappa_L \alpha) - i \int_{-\infty}^{\infty} Y(\mu) \epsilon(\mu) S^{(2m)}(\mu) \frac{d\mu}{\mu} \right]; \tag{1.126}$$

$$\sum_{n=1}^{N-1} A_{2n-1} Q_s^{(2n-1)} = \sum_{m=1}^{2N-1} d_s^m \left[\frac{\gamma}{\alpha} S^{2m-1}(\kappa_L \alpha) - i \int_{-\infty}^{\infty} Y(\mu) \epsilon(\mu) S^{(2m-1)}(\mu) \frac{d\mu}{\mu} \right], \tag{1.127}$$

$(s = 1, 2, \ldots N)$

$$Q_s^{(n)} = \frac{1}{2\pi} \int_{\alpha_s}^{\beta_s} \eta^n \mid Q(\eta) \mid^{-0.5} d\eta;$$

$$c_s^m = \frac{1}{2} \sum_{n=m}^{2N-1} (-1)^{n-m} (n-2m) q_{n+1} Q_s^{2(n-m)};$$

$$d_s^m = \frac{1}{2}\sum_{n=m}^{2N-1}(-1)^{n-m}(n+1-2m)q_{n+1}Q_s^{(2n-2m+1)}.$$

The determinants of these systems are not equal to zero [93]. Therefore, one can express the coefficients A_n ($n = 0, 1, 2, \ldots, 2N-2$) versus the right-hand sides of (1.126), (1.127) and derive the integral equation with respect to the sought for function $Y(\lambda)$ by substituting these coefficients to (1.123):

$$Y(\lambda) = \frac{1}{\lambda}\int_{-\infty}^{\infty} Y(\mu)\epsilon(\mu)U_\lambda^\mu d\mu + i\kappa_L\alpha\frac{\gamma}{\lambda}U_\lambda^{\kappa_L\alpha}, \; (-\infty < \lambda < \infty) \qquad (1.128)$$

where

$$U_\lambda^\mu = W_\lambda^\mu + \frac{1}{\mu}\sum_{n=0}^{N-1}(-1)^n S^{(2n)}(\lambda)\sum_{m=0}^{2N-1} a_{2n}^m S^{(2m)}(\lambda) +$$

$$\frac{1}{\mu}\sum_{n=1}^{N-1}(-1)^n S^{(2n-1)}(\lambda)\sum_{m=1}^{2N-1} a_{2n-1}^m S^{(2m-1)}(\lambda);$$

$$a_{2n}^m = \det{}^{-1} \| Q_s^{(2n)} \| \sum_{s=1}^{N} c_s^m A_s^{2n}; \; a_{2n-1}^m = \det{}^{-1} \| Q_s^{(2n-1)} \| \sum_{s=1}^{N} d_s^m A_s^{2n-1}$$

and A_s^{2n}, A_s^{2n-1} are algebraic supplements of the elements of the s-th string and n-th column of the determinants $\det \| Q_s^{(2n)} \|$, $\det \| Q_s^{(2n-1)} \|$. In the case of odd number of strips one must put $\alpha_1 = 0$ in the equation (1.128). If to assume then that $N = 1$, (1.128) will coincide with the equation of the problem of scattering of inhomogeneous plane wave (1.116) by one strip. If to take into account the definition of functions $W_\lambda^\mu, S(\lambda)$ then, as well as in the case of single strip, one can show the solvability of (1.128) for arbitrary parameters of the problem in the class of functions satisfying the condition (1.128) and the possibility to find the solution by the method of successive approximations for $\kappa_L\delta << \kappa_L \max(\beta_s - \alpha_s) << 1\,(1 \leq s \leq N)$. If to take the first approximation, then

$$Y(\lambda) = i\kappa_L\gamma U^{\kappa_L\alpha} + O(\kappa_L^3\delta^2 \ln\kappa\delta)).$$

Hence, the radiation far field pattern $S(\phi)$ of a system of finite number of narrow strips is a superposition of independent radiation far field patterns of every separate strip determined within the accuracy $O(\kappa_L\delta^2)$ for $\delta << 1, \kappa_L \sim 1$. Within the same accuracy this superposition gives the total energy scattered by the strips, which is close to the sum of energies scattered by every strip independently of the presence of other strips if $\kappa_L << 1$ ($\kappa_L \delta << 1$).

If the width of at least one strip becomes commensurable with the wavelength, the method of successive approximations cannot be applied for solving the equation (1.128). In this case the problem may be analyzed on the basis of numerical solution to (1.128). The calculations are carried out by the following scheme. Infinite set of integration in (1.128) is replaced by the finite interval $(-M, M)$ with taking into account the decay of integrand which has the order $O(\mu^{-4})$ for $|\mu| \to \infty$. The obtained "cut" equation is solved numerically with the help of Sympson quadrature formula which yields finite system of linear algebraic equations with respect to the values of unknown functions in the knots of quadrature formula. The approximate values of the considered characteristics of diffractional radiation are determined then by the obtained approximate solution. The rate of convergence of the solution of the "cut" equation to the one of exact equation may be evaluated similarly to the case of one strip [90].

Within the frame of the concrete realization of given scheme it is intended to solve two equivalent independent equation on the half-axis with respect to odd and even parts of $Y(\lambda)$ in place of (1.128) on the whole axis. It enables one to reduce the order of the systems of linear algebraic equations in two times for the same calculation accuracy. The guaranteed accuracy of determining diffractional radiation by the total scattered energy and far field pattern has the order 0.1

As it is shown in §1, the radiation field for the infinite periodical strip grating consists of the superposition of plane waves (1.4) going off the grating at the angles given in (1.8). If the radiation condition (1.7) holds for none of the harmonics (it may take place only when $\kappa = l\lambda^{-1} < 0.5$), the field of SPR is identically zero. In this case the field scattered by the grating consists of infinite number of inhomogeneous plane waves propagating along it with phase velocities $v_\Phi = cl[\lambda(n + l(\lambda\beta)^{-1})]^{-1}$, whose moduli are less than the velocity of light in vacuum, and exponentially decreasing in the direction perpendicular to the grating. With growing frequency $(\omega l(2\pi c)^{-1} \to \infty)$ the number of non-damping plane waves increases but as it follows from (1.15) the total energy radiated by them falls proportional to $2\pi c(\omega l)^{-1}\exp(-2ka\beta^{-1}\sqrt{1-\beta^2})$.

Before to consider the diffractional radiation excited in the process of movement of an electron flow close to a bounded strip grating, let us first of all indicate its features in the case of single strip having the width $2L$.

For $\kappa_L << 1$ (narrow slot) and $\epsilon = 1$ the radiation far field pattern has the same character as the one of the field of homogeneous plane wave scattered by the strip [90]. The total radiated energy here is given by the expression

$$S^{\text{tre}} = 2\pi^4 \rho_0^2 \kappa_L \beta^2 L(1-\beta^2) J_1^2\Big(\frac{\kappa_L}{\beta}\Big)\Big[1 + O(\kappa_L^2)\Big] e^{-\frac{2\kappa_L p}{\beta}\sqrt{1-\beta^2}}. \qquad (1.129)$$

In the case of wide slot ($\kappa_L >> 1$) the angle distribution of the energy radiated by the strips described by the same formula as in the problem of scattering of the obliquely falling plane wave [89]. However, due to the condition $\alpha > 1$ the sharp maximums of the order $O(\kappa)$ are absent here in the far field pattern, which are observed in the direction of propagation of incident and reflected waves. The total radiation energy losses in the considered case ($\kappa(1-\beta^2) >> 1$) are given by the formula

$$S^{\text{tre}} = \frac{8\pi^2 \rho_0^2 \beta^2 L}{\kappa_L \sqrt{1-\beta^2}} e^{-\frac{2\kappa_L p}{\beta}\sqrt{1-\beta^2}} \Big\{1 + \frac{1}{\sqrt{\pi \kappa_L}}\Big[\sqrt{\frac{1-\beta}{1+\beta}} \cos\big(2\kappa_L(\frac{1}{\beta}+1) - \frac{\pi}{4}\big)$$

$$-\sqrt{\frac{1+\beta}{1-\beta}} \cos\big(2\kappa_L(\frac{1}{\beta}-1) - \frac{\pi}{4}\big)\Big]\Big\}\Big[1 + O\Big(\frac{1}{\kappa_L}\Big)\Big] \qquad (1.130)$$

Hence, with growing frequency the energy radiated by the strip oscillates and falls as $\kappa_L^{-1}\exp(-\frac{2\kappa_L p}{\beta}\sqrt{1-\beta^2})$; the depth of oscillations tends to zero as $\kappa_L^{-1.5}\exp(-\frac{2\kappa_L p}{\beta}\sqrt{1-\beta^2})$.

The dependence of

$$\tilde{S}^{\text{tre}} = \frac{S^{\text{tre}}}{8\pi \rho_0^2 \beta^2 \sum_{n=1}^{N}(\beta_n - \alpha_n)}$$

versus frequency is shown on Fig. 27, b (for $\beta = 0.2, p = 0$) obtained as a result of numerical solution of the integral equation (1.128). Behaviour of the curve for $\kappa_L << 1$, $\kappa_L >> 1$ completely agrees with asymptotic expression presented above. Maximal losses spent for diffractional radiation occurs at the wavelength which is greater the the double stripwidth ($\kappa_L \approx 0.45$; $\kappa_L \approx 1.1$). It is interesting to point out that in the case of the highest total radiation losses the direction of radiation with maximal intensity coincides with normal direction to the plane containing the strip (Fig. 27, a).

The dependence of energy radiated by the wide strip ($\kappa_L = 25$) versus relative velocity of electron flow is shown on Fig. 27; it well agrees with the formula (1.130) up to $\beta = 0.8$. The radiated energy reaches its maximal values for $\kappa_L \beta^{-1}(1-\beta) \approx 1$.

When finite number of identical strips are equally spaced in the plane $z = 0$ (forming the bounded periodical strip grating), the general character of energy losses spent on diffractional radiation as functions of frequency and flow's velocity is similar to their behaviour for one strip (Fig. 28). As well as in the previous case, the radiated energy versus frequency has the order $O(\kappa_L^3)$ for low frequencies and the order $\kappa_L^{-1} e^{-\frac{2\kappa_L p}{\beta}\sqrt{1-\beta^2}}$ for high frequencies. But now intensive radiation is closely connected with the domains of existence of plane waves going off an infinite grating. The first such domains of maximal radiation for the curves of S^{tre} versus frequency for five and ten strips ($\beta = 0.2$, Fig. 28, a) are close to the domains of existence of the first and the second harmonics, respectively, in the expression (1.4). The next frequency domain of intensive radiation weakly differs from the sum of the domains of existence of the third and the fourth non-dumping harmonics in (1.4) which in this case intersect. Greater the number of strips forming the grating, more accurately the domains of maximal radiation are connected with radiating harmonics in (1.4). Unlike the case when representation of the field scattered by an infinite grating does not contain radiating harmonics, the radiation far field patterns corresponding to these domains have sharp maximums of intensity close to the directions of propagation of plane waves going off the grating (Fig. 29, a).

The dependence of radiated energy on the ratio of the width of the strips and the distance between them for fixed length of the grating is shown on Fig. 30. There exist optimal values of this ratio when radiated energy is maximal. The number of such values is determined by the index of the harmonic in (1.4) which propagates for the given parameters of the problem.

Growing the number of strips forming the grating leads to the proportional increasing of energy radiated by them in the domain of existence of non-dumping harmonics in (1.4) with simultaneous narrowing of the petals of radiation pattern close to the directions of propagation of these harmonics (Fig. 31). The ratios of maximums of intensity approaches here the interrelations between the squared values of the moduli of amplitudes of plane waves going off the grating (see §1).

2. Let us present the solution to the problem of diffractional radiation produce by the unbounded plane monochromatic flow moving over a bounded

diffractional grating with elements which are finite along its axis. Simplifications which form the basis of the approximate methods for the solution to this problem mainly affect the part of the field connected with lateral cylindrical waves going off the grating's boundaries. The sought for quantity is the distribution of intensity of electromagnetic field in the diffractional beams (space harmonics of the field of diffractional radiation). The present method brings in considerably less inaccuracy in the obtained function of field's intensity. Since in the practical applications it is used the part of the field of diffractional radiation, which is concentrated in space harmonics, this way of solving the given problem may be considered acceptable, all the more it enables one to verify simple analytical formulas for the function of intensity distribution in diffractional beams going off the grating [92].

The method of parabolic equation is used for finding the distribution of field's intensity in the diffractional beams. By means of this method one can study both types of bounded gratings. Further results are completely applicable for the gratings made of elements for which it is known the solution to diffractional problem in the case of infinite periodical structure because the functions describing the distribution of field's intensity in the diffractional beams are determined by the finite size of grating's aperture and do not depend on its transversal profile, i. e., on the period's configuration of corresponding infinite grating.

As well as in all previous cases, it is convenient to treat the problem of radiation within the frame of the model of given current as the problem of diffraction on this structure of the flow's eigenfield having the form of inhomogeneous plane H-polarized wave. As the result of diffraction of this wave on the grating, the $\vec{H}, \vec{E}$ field is completely determined by the H_x-component which by definition may be written as the superposition of the field of incident wave and the scattered field, $H_x = H_x^0 + H_x^1$ where H_x^1 is determined by (1.4). H_x^1 is the sought for field of diffractional radiation. We will look for the latter in the form

$$H_x^1 = HW(x, y, z)e^{i(k\alpha y - \omega t)}, \tag{1.131}$$

where

$$H_= - 2\pi\rho_0\beta e^{-k\gamma p} \tag{1.132}$$

is the tensity of incident field for $z = -p$; $W(x, y, z)$ is found under the assumption that the scattered field is the superposition of diffractional beams caused by the periodical configuration of structure's elements and the set of lateral cylindrical waves caused by the finite size of the grating along the axis Ox. The structure is infinite and periodical along the axis Oy with the period l, hence the function

$$W(x,y,z) = \sum_{-\infty}^{\infty} \left\{ \begin{matrix} A_n \\ B_n \end{matrix} W_n(x,z) e^{ik\left[\left(\alpha + \frac{2\pi n}{kl}\right)y + \gamma_n |z+p|\right]} \right. . \tag{1.133}$$

Here A_n, B_n are the amplitudes of diffractional harmonics (beams) for the elements unbounded along Ox and having the same profile as for the finite grating; A_n corresponds to the half-space $z > -p$ and B_n – to the half-space $z < -(p+h)$. For "non-transparent" gratings $B_n \equiv 0$. A_n, B_n are assumed to be known from the solution to the boundary value problem for infinite gratings [61,62]. The values α_n, γ_n are the cosines giving the direction of the wave vector of the n-th space harmonic with respect to the axes Oy, Oz, respectively, and they are also known in the theory of diffractional radiation (1.6).

In what follows we are interested only in propagating diffractional harmonics for which the radiation condition $|\alpha_n| < 1$ holds. Since $\alpha = \beta^{-1}$, $\beta = vc^{-1} < 1$, this condition holds only for harmonics with negative index n.

The problem is to find the functions $W_n(x,z)$ describing the distribution of field's tensity over the transversal beam's cross-section. In the representation of the sought for field (1.139) there is no the part which corresponds to the lateral cylindrical waves. The amplitudes of cylindrical waves decrease proportional to $(kR)^{-0.5}$ with growing distance R from corresponding edge of the grating to the observation point. There exist domains in the space where the field of diffractional beams is prevalent; hence, there it is possible to look for the field in the form (1.133) neglecting the lateral cylindrical waves.

Diffractional field in every point of the space may be easily found if it is known the tangential field's components on grating's surface. When looking for the field in those domains where the influence of lateral cylindrical waves is slightingly small we assume that the field on grating's surface coincides with the one on the surface of infinite structure and that it vanishes in complementary part of the plane $z = -p$ where the finite grating is situated. Otherwise, the functions $W_n(x,z)$ must satisfy the following boundary value conditions on the grating having the width L:

$$W_n(x,z)\Big|_{z=-p} = \begin{cases} 1, & \text{if } |x| < L; \\ 0, & \text{if } |x| > L. \end{cases} \tag{1.134}$$

Differential equation for the function $W_n(x, z)$ may be obtained by subordinating the field $H_x = H_x^0 + H_x^1$, which is known on the grating's surface due to the condition (1.134), to the Helmholtz equation

$$\frac{\partial^2 W_n}{\partial x^2} + 2ik\gamma_n \text{sign}(z+p)\frac{\partial W_n}{\partial x} + \frac{\partial^2 W_n}{\partial z^2} = 0. \tag{1.135}$$

(for arbitrary fixed index n) everywhere outside the grating.

Let us introduce the orthogonal "beam" coordinate system (χ, η_n, ζ_n) connected with the initial coordinates by the formulas

$$\chi = x; \quad \eta_n = y\gamma_n \text{sign}(z+p) - (z+p)\alpha_n;$$

$$\zeta_n = y\alpha_n + \gamma_n(z+p)\text{sign}(z+p). \tag{1.136}$$

for every space harmonic with the index n. The axis $O\zeta_n$ coincides with the direction of beam's propagation and the axis $O\eta_n$ is orthogonal to the axes $O\zeta_n$, Ox.

Let us make one more simplification which is commonly used in the method of parabolic equation [94]. We will look for the solution in those domains of the space where the following condition holds

$$\frac{1}{2k}\left|\frac{\partial^2 W_n(\chi, \zeta_n)}{\partial \zeta_n^2}\right| << \left|\frac{\partial W_n(\chi, \zeta_n)}{\partial \zeta_n}\right|. \tag{1.137}$$

Finally we obtain the parabolic equation

$$\frac{\partial W_n}{\partial \zeta_n} = D\frac{\partial^2 W_n}{\partial \chi^2}; \quad D = \frac{i}{2k}. \tag{1.138}$$

The condition (1.137) determined the boundaries of the possible use of the solution to parabolic equation. The Green's function of the equation (1.138) has the form

$$G(\chi - \chi', \zeta_n) = \frac{1}{\sqrt{4\pi D\zeta_n}} e^{-\frac{(\chi-\chi')^2}{4D\zeta_n}}, \tag{1.139}$$

where by definition $\zeta_n \geq 0$. Hence the sought for solution, satisfying the boundary value conditions (1.134), has the form

$$W_n(\chi, \zeta_n) = \int_L^L G_n(\chi - \chi', \zeta_n) d\chi'. \tag{1.140}$$

After transforming the integral in (1.140) the solution may be rewritten with the help of special functions. In particular, using the Fresnel integral

$$F(\tau)=\frac{e^{-\frac{\pi}{4}}}{\sqrt{2\pi}}\int_{infty}^{\pi}e^{i\frac{t}{2}}dt, \tag{1.141}$$

we obtain

$$W_n(\chi,\zeta_n)=F\left(\sqrt{\frac{k}{\zeta_n}}\left(\chi+L\right)\right)-F\left(\sqrt{\frac{k}{\zeta_n}}\left(\chi-L\right)\right), \tag{1.142}$$

or we may write the equivalent expression for $W_n(\chi,\zeta_n)$ versus the Fresnel integral

$$c(\tau)=\sqrt{\frac{2}{\pi}}\int_0^{\tau}\cos^2 tdt;\ \ s(\tau)=\sqrt{\frac{2}{\pi}}\int_0^{\tau}\sin^2 tdt \tag{1.143}$$

in the following form

$$W_n(\chi,\zeta_n)=\frac{1-i}{\sqrt{2}}\left\{c\left(\sqrt{\frac{k}{\zeta_n}}\left(\chi+L\right)\right)-c\left(\sqrt{\frac{k}{\zeta_n}}\left(\chi-L\right)\right)+\right.$$

$$\left.i\left[s\left(\sqrt{\frac{k}{\zeta_n}}\left(\chi+L\right)\right)-s\left(\sqrt{\frac{k}{\zeta_n}}\left(\chi-L\right)\right)\right]\right\}. \tag{1.144}$$

It it easy to establish the equivalence of (1.142) and (1.144) if to take into account functional connection between complex Fresnel integrals and functions (1.143)

$$F(\tau)=\frac{1-i}{\sqrt{2}}\left\{\frac{1}{2}+c\left(\frac{\tau}{\sqrt{2}}\right)+i\left[\frac{1}{2}+s\left(\frac{\tau}{\sqrt{2}}\right)\right]\right\}. \tag{1.145}$$

Let us write one more equivalent representation for the function $W_n(\chi,\zeta_n)$ using the probability integral

$$\Phi(\tau)=\frac{2}{\sqrt{2\pi}}\int_0^{\tau}e^{-t^2}dt, \tag{1.146}$$

$$W_n(\chi,\zeta_n)=\frac{1}{2}\left[\Phi\left(\sqrt{\frac{k}{2i\zeta_n}}\left(\chi+L\right)\right)-\Phi\left(\sqrt{\frac{k}{2i\zeta_n}}\left(\chi-L\right)\right)\right]. \tag{1.147}$$

Let us consider in the similar way the diffractional radiation of a plane monochromatic flow generated as a result of its flight along a grating consisting of a finite number of elements which are infinite along grating's axis. We assume that the number of elements is big enough providing thus the forming of diffractional beams of a grating. As it follows from the experimental data [95], which are confirmed theoretically in [86], the qualitative features of the diffractional field scattered by a bounded grating are developed already for comparatively small number of elements ($N = 3 - 7$).

The component of total field

$$H_x = H_x^0 + H \sum_{-\infty}^{\infty} \left\{ \begin{matrix} A_n \\ B_n \end{matrix} V_n(y, z) e^{ik(\alpha_n y + \gamma_n |z+p|) - i\omega t}. \right. \tag{1.148}$$

where H_x^0, H are determined by (1.4); $V_n(y, z)$ are the sought for functions; A_n, B_n are the Fourier coefficients of the field of inhomogeneous plane wave scattered by the unbounded grating: A_n are the amplitudes of space harmonics radiated to the half-space $z > -p$, B_n – to the half-space $z < -(p+h)$ and the latter are non-zero only for the "transparent" gratings; h is the size of elements along the axis Oz. It is evident that $V_n(y, z) \equiv 1$ for the unbounded grating.

Differential equation for the function $V_n(y, z)$ for the fixed n is obtained by subordinating (1.148) to the Helmholtz equation:

$$\frac{\partial^2 V_n}{\partial y^2} + \frac{\partial^2 V_n}{\partial z^2} + 2ik \left[\alpha_n \frac{\partial V_n}{\partial y} + \gamma_n \mathrm{sign}(z+p) \frac{\partial V_n}{\partial z} \right] = 0. \tag{1.149}$$

Differential equation for the function describing the distribution of field's tensity in the transversal cross-section of diffractional beam is more complicated in the case of a grating made of a finite number of elements than the equation (1.135) for the grating bounded along the axis of its elements. However, it may be simplified if to rewrite it in the beam coordinates (1.136):

$$\frac{\partial^2 V_n(\eta_n, \zeta_n)}{\partial \eta_n^2} + 2ik \frac{\partial V_n(\eta_n, \zeta_n)}{\partial \zeta_n} + \frac{\partial^2 V_n(\eta_n, \zeta_n)}{\partial \zeta_n^2} = 0 \tag{1.150}$$

and under the condition

$$\frac{1}{2k} \left| \frac{\partial^2 V_n(\eta_n, \zeta_n)}{\partial \zeta_n^2} \right| << \left| \frac{\partial V_n(\eta_n, \zeta_n)}{\partial \zeta_n} \right| \tag{1.151}$$

it may be justified the transference to parabolic equation

$$\frac{\partial V_n(\eta_n, \zeta_n)}{\partial \zeta_n} = D \frac{\partial^2 V_n(\eta_n, \zeta_n)}{\partial \eta_n^2}; \quad D = \frac{i}{2k}, \tag{1.152}$$

The boundary value conditions which are necessary for its solution may be obtained if to assume that on the surface of bounded grating, $|y| \leq L$, the field coincides with the one on the infinite grating and vanishes on the complementary part of this strip on the planes $z = -p$, $z = -(p+h)$ tangent to the surface of the elements of diffractional grating

$$V_n(y,z) = \begin{cases} 1, & \text{if } |y| < L; \\ 0, & \text{if } |y| \geq L. \end{cases} \tag{1.153}$$

One can rewrite the solution of parabolic equation by means of the formula analogous to (1.140) transferring to the beam coordinates in the boundary value conditions and using the Green's function of parabolic equation

$$V_n(\eta_n, \zeta_n) = \text{sign}(p+z) \int_{-L\gamma_n}^{L\gamma_n} G(\eta_n - \eta_n', \zeta_n) d\eta_n' \tag{1.154}$$

which is reduced to the same special functions.

Now we can evaluate explicit expression for the function of field's tensity distribution in the transversal cross-section of the n-th diffractional beam

$$\text{sign}(p+z)V_n(\eta_n, \zeta_n) = F\Big(\sqrt{\frac{k}{\zeta_n}}\Big(\eta_n + L\gamma_n\Big)\Big) - F\Big(\sqrt{\frac{k}{\zeta_n}}\Big(\eta_n - L\gamma_n\Big)\Big) =$$
$$\frac{1-i}{\sqrt{2}}\Big\{c\Big(\sqrt{\frac{k}{2\zeta_n}}\Big(\eta_n + L\gamma_n\Big)\Big) - c\Big(\sqrt{\frac{k}{2\zeta_n}}\Big(\eta_n - L\gamma_n\Big)\Big) +$$
$$i\Big[s\Big(\sqrt{\frac{k}{2\zeta_n}}\Big(\eta_n + L\gamma_n\Big)\Big) - s\Big(\sqrt{\frac{k}{2\zeta_n}}\Big(\eta_n - L\gamma_n\Big)\Big)\Big]\Big\} =$$
$$\Phi\Big(\sqrt{\frac{k}{2i\zeta_n}}\Big(\eta_n + L\gamma_n\Big)\Big) - \Phi\Big(\sqrt{\frac{k}{2i\zeta_n}}\Big(\eta_n - L\gamma_n\Big)\Big). \tag{1.155}$$

Analytical solutions (1.142) (1.147) (1.155) have been obtained as a result of taking into account only the first terms in the equations (1.135), (1.149) what gave us the possibility to treat parabolic equations which are simpler than the initial elliptic equations. By now there is now rigorous justification of this procedure. Hence, in order to determine the domain of applicability of the obtained solution let us substitute (1.142) (1.147) (1.155), respectively, to the inequalities (1.137), (1.151):

$$\frac{3}{4k\zeta_n}\Big|1 + i\frac{k(\chi \pm L)^2}{3\zeta_n}\Big| << 1; \quad \frac{3}{4k\zeta_n}\Big|1 + i\frac{k(\eta_n \pm \gamma_n L)^2}{3\zeta_n}\Big| << 1. \tag{1.156}$$

The condition $k\zeta_n >> 1$ holds for both cases; therefore, one can use the solutions to (1.142) (1.147) (1.155) at the distances along the beam coordinate $O\zeta$ which are much greater than the wavelength. The conditions for the transversal beam coordinates for the grating made of the elements having finite size differ from such conditions for the grating made of finite number of elements:

$$\left|\frac{\chi \pm L}{\zeta_n}\right|^2 << 1; \quad \left|\frac{\eta_n \pm \gamma_n L}{\zeta_n}\right|^2 << 1. \tag{1.157}$$

In the first case the ratio of the interval of admissible values of the transversal beam coordinate χ to the grating's width $2L$ does not depend on the directions of radiation with respect to the axis Oy. In the second case the interval of the coordinate η_n depends on the "effective" grating's width $2\gamma_n L$ where $\gamma_n = \sin\psi_n$ (ψ_n is the radiation angle of the n-th space harmonic relative to Oy). The interval of coordinate's admissible values decreases as longitudinal coordinate deviates from the direction perpendicular to the grating and it is caused by the decrease of the effective grating's width counted along the transversal coordinate.

4. Let us consider several numerical examples. Let the relative flow's velocity $\beta = 0.1$, the wavelength of its modulation (and hence its radiation wavelength) $\lambda = 9.5l$, $10l$, $10.5l$. Then it is possible the radiation of the single space harmonic with the index $n = -1$ (the radiation conditions for space harmonics formulated in §1 remain valid in this case). The direction of radiation may be characterized by the angle ψ_n between wave vector of this harmonic and the axis Oy; this vector is collinear with the unit vector of the longitudinal beam coordinate $O\zeta$ and it is equal to $\pm 45°, \pm 90°, \pm 135°$, respectively, for the indicated wavelengths. The beam coordinates of the radiating harmonic with $n = -1$ are determined by the following formulas for every chosen wavelength

$$\chi = x; \quad \eta_{-1}^{(1)} = \frac{\pm y - z}{\sqrt{2}}; \quad \zeta_{-1}^{(1)} = \frac{y \pm z}{\sqrt{2}};$$

$$\chi = x; \quad \eta_{-1}^{(2)} = \pm y; \quad \zeta_{-1}^{(2)} = \pm z; \tag{1.158}$$

$$\chi = x; \quad \eta_{-1}^{(3)} = \frac{\pm y + z}{\sqrt{2}}; \quad \zeta_{-1}^{(1)} = \frac{-y \pm z}{\sqrt{2}}.$$

Here the upper index i ($i = 1, 2, 3$) corresponds to the values of the angle $\psi_{-1} = \pm 45°, \pm 90°, \pm 135°$. The sign plus should be taken for the upper half-space and the sign minus – for the lower half-space. The beam coordinates in the plane $\chi = \text{const}$ are shown on Fig. 32.

Let us take the width of gratings $2L = 20l$ (Fig. 33). The calculated amplitudes of the diffractional harmonic in the beam's transversal cross-section at the distances $\zeta_{-1}^{(i)} = 200$; 400; 625 and functions $W_{-1}^{(i)}(\chi, \zeta_{-1})$ ($i = 1, 2, 3$) are shown on Fig. 34. For the fixed "electrical" transversal size of the grating along the axis Ox the distribution of field's tensity in the beam's cross-section does not depend on the direction of radiation of space harmonic. The function $V_{-1}(\eta_{-1}^2, , \zeta_{-1}^2)$ corresponding to the radiation angle $\psi_{-1} = \pm 90°$ coincides with the functions $W_{-1}(\chi, \zeta_{-1})$, hence, it is also described by Fig. 34. The variation of field's tensity in the transversal beam's cross-section is the same at the angles $\pm 45°$, $\pm 135°$. One must take into account that different wavelengths correspond to these angles and here the size of gratings are different, since $2L = 20l$ by the initial condition.

Let us indicate some characteristic features of the effect of diffractional radiation caused by the limitation of transversal size of a grating [92].

First, the function of the distribution of field's tensity in a beam are symmetrical with respect to the middle of a grating bounded in the direction of its axis (i. e., to the plane $\chi = 0$) or symmetrical with respect to the transversal beam coordinate for a grating consisting of a finite number of elements. These functions have qualitative behaviour similar to the Gauss-Hermit functions and may be well approximated by them in the domains with sufficiently high radiation intensity. Far from the grating radiation intensity along the transversal coordinate varies in accord with the Gauss law.

The Gaussian character of field's variation in diffractional beams determines the area of possible practical application. It is natural to expect that the radiation with such field's distribution may be effectively used for excitation of open resonators having eigenfields with similar distribution. This assumption has been confirmed by the experimental studies of excitation of open resonators with the help of Smith-Purcell effect. These results are analyzed below.

Secondly, the field's distribution in the transversal cross-section of diffractional beam depends only on the relative velocity of the source's movement, radiation angle and interrelation between the bounded transversal size of a grating and the wavelength. The profile of grating's transversal cross-section practically does not influence the field's distribution in a beam and determines only the amplitude of field's tensity.

These conclusions enable us to present the following way for solving the problem of obtaining maximal amplitude of diffractional radiation for bounded structures. The optimal profile of transversal grating's cross-section corresponding to the maximal amplitude of the chosen space harmonic is found as a result of the mathematically rigorous solution to the problem of diffractional radiation for the infinite grating. In order to take into account the influence of the finite size of a grating on the distribution of field's tensity in a beam one should use the data of the present paragraph which gives the possibility to determine the required function of intensity distribution in a beam by the method of parabolic equation.

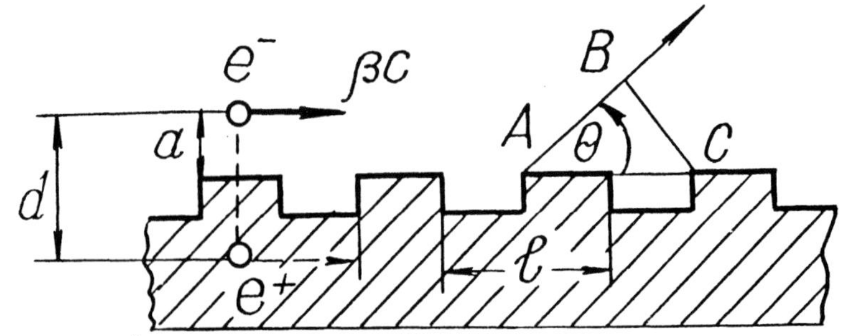

Fig. 1. The model of blinking dipole in the Smith-Purcell effect.

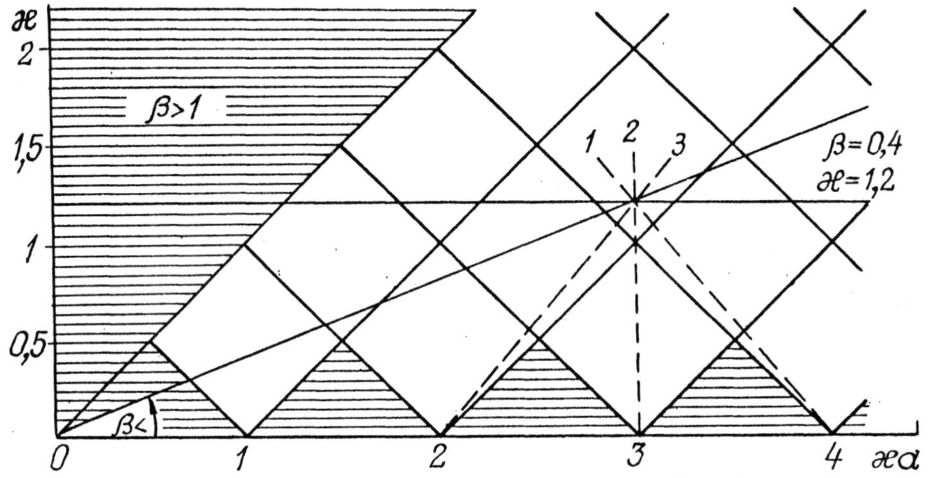

Fig. 2. The Brillouin diagram for Smith-Purcell radiation (1 – the beam $\psi_{-4} = 123.5°$; 2 – $\psi_{-3} = 90°$; 3 – $\psi_{-2} = 33.5°$.

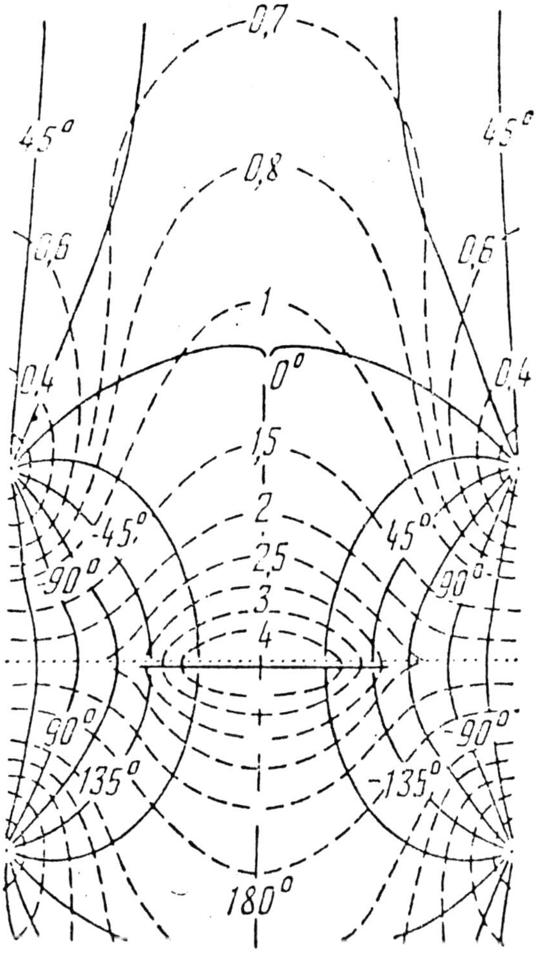

Fig. 3. The lines of equal amplitudes and phases of magnetic field (1.10) (the condition (1.7) does not hold, $\beta = 0.4$; $\kappa = 0.285$).

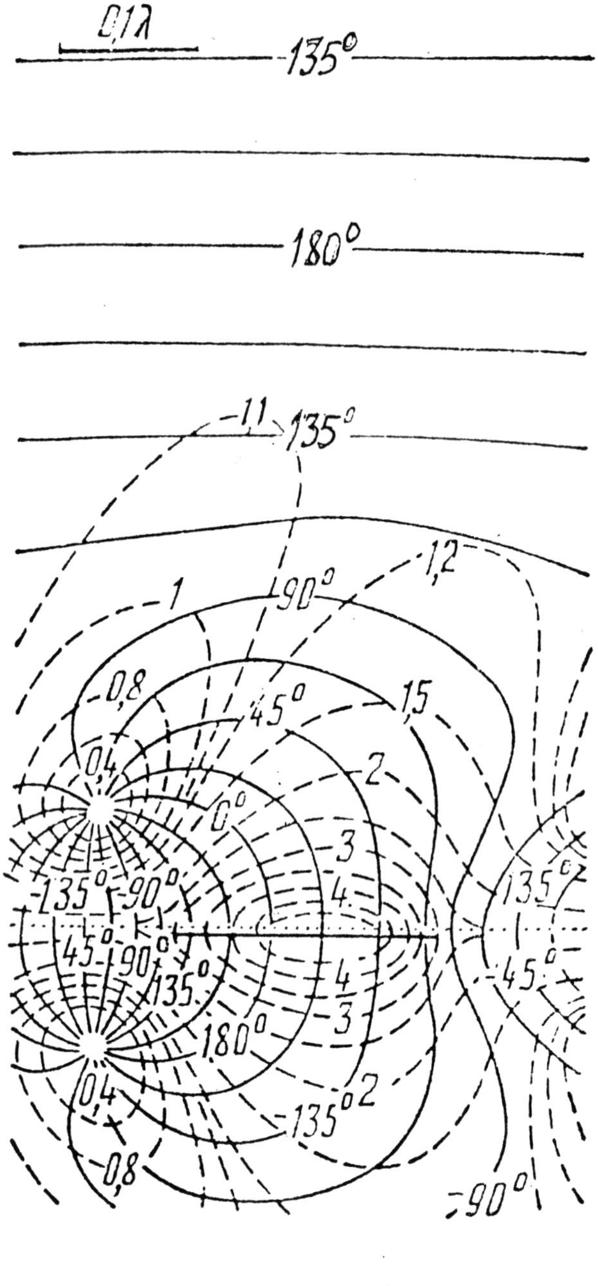
0,1λ
135°
180°
1,1
135°
1,2
1
90°
0,8
45°
1,5
0,4
2
0°
3
135°
90°
4
135°
45°
90°
135°
4
3
45°
180°
0,4
2
135°
0,8
90°

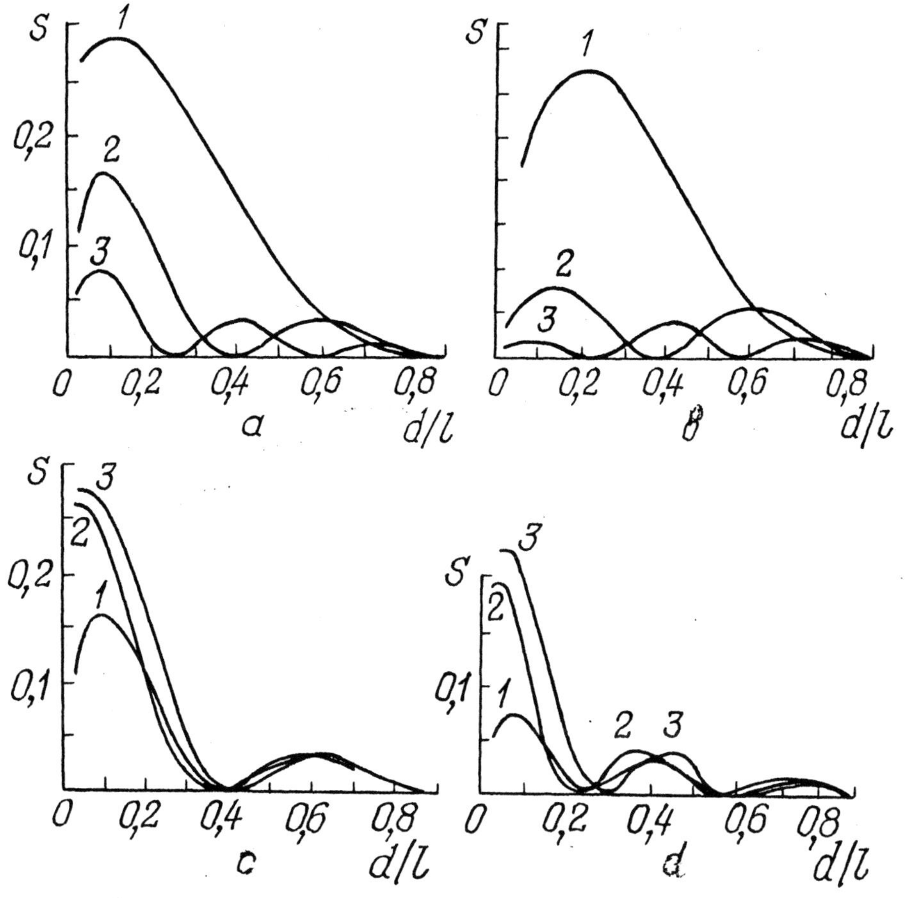

Fig. 5. Influence of the filling coefficient of strip grating on the power flow of Smith-Purcell radiation:

a) 1 – $\kappa = 0.1$; $n = -1$; 2 – $\kappa = 0.2$; $n = -2$; 3 – $\kappa = 0.3$; $n = -3$;

b) 1 – $\kappa = 0.25$; $n = -1$; 2 – $\kappa = 0.5$; $n = -2$; 3 – $\kappa = 0.75$; $n = -3$;

c) $n = -2$; $\beta = 0.1$; 1 – 90°; 2 – 160°; 3 – 10°;

d) $n = -3$; $\beta = 0.1$; 1 – 90°; 2 – 160°; 3 – 10°;

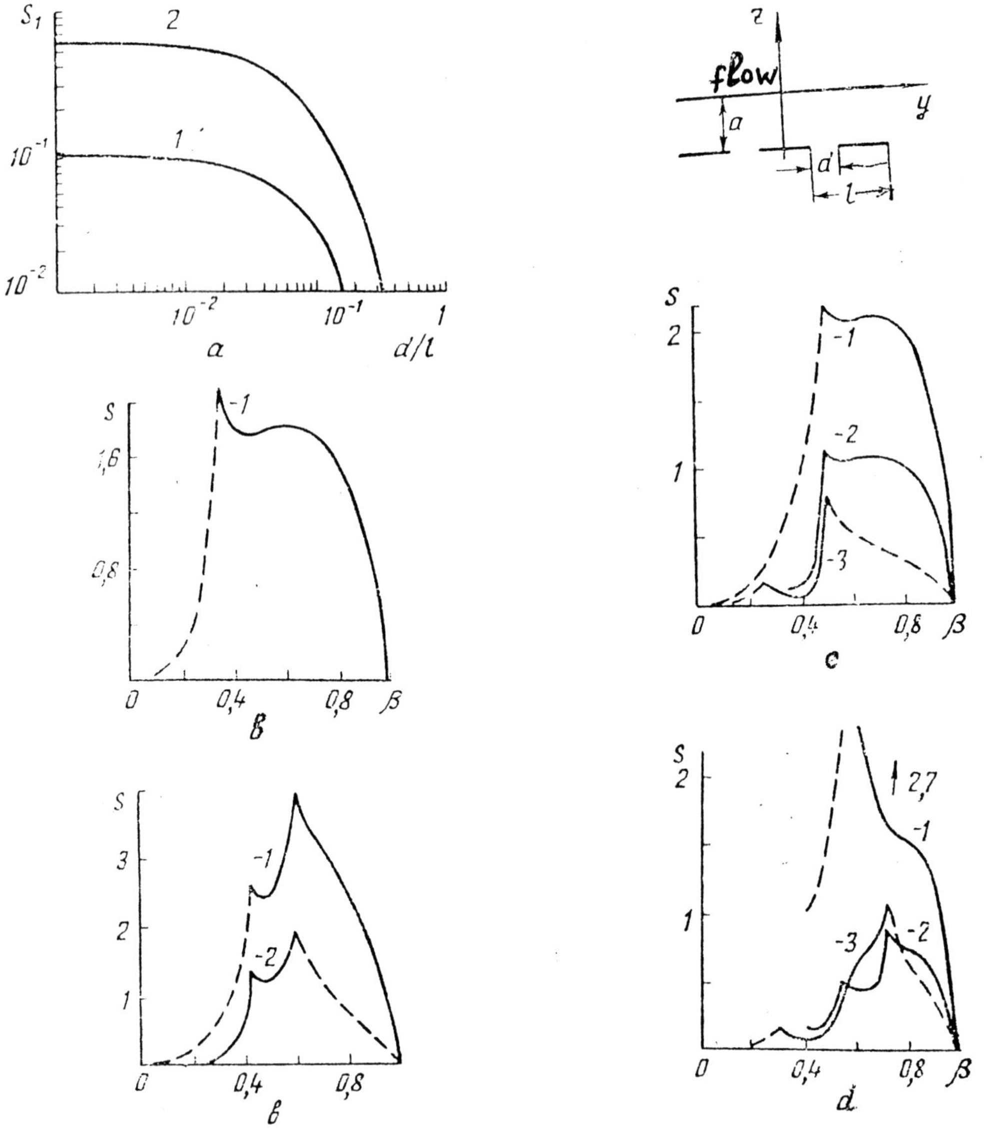

Fig. 6. Influence of the velocity of electron flow on the power flow of Smith-Purcell radiation:

a) $\frac{d}{l} = 0.5$; $1 - \beta = \kappa = 0.1$; $2 - \beta = \kappa = 0.25$;
b) $\kappa = 0.25$; c) $\kappa = 1.0$; d) $\kappa = 0.15\ (\frac{d}{l} = 0.5)$;

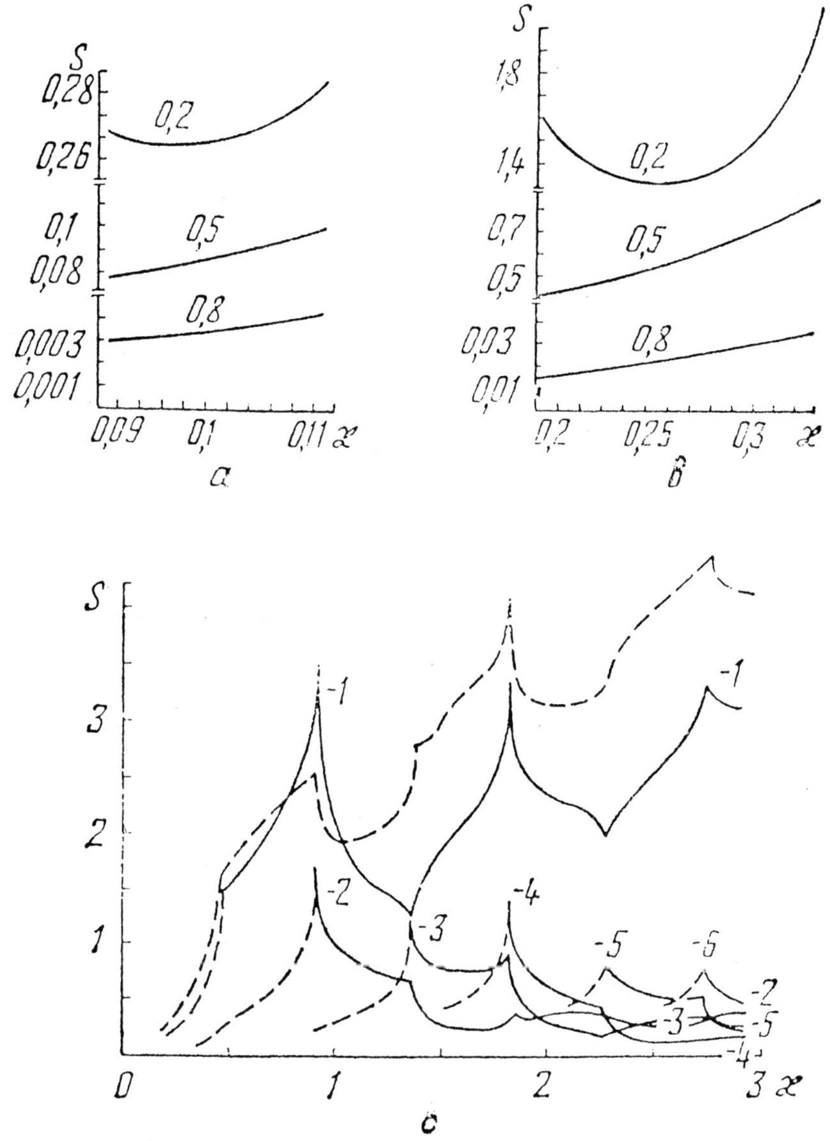

Fig. 7. Influence of the frequency on the power flow of Smith-Purcell radiation. The domains of emergence of Wood anomalies are observed on the last figure.

a) $\beta = 0.1$; $n = -1$; b) $\beta = 0.25$; $n = -1$; c) $\frac{d}{l} = 0.5$; $\beta = 0.83$;

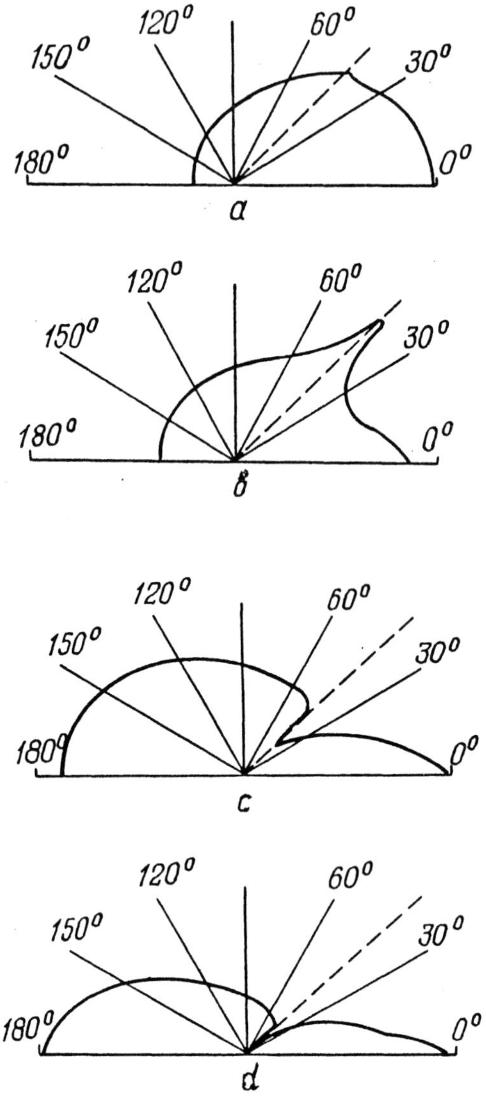

Fig. 8. Radiation patterns of charged thread moving over strip grating: a) $n = -1$; $\beta = 0.4$; $\frac{d}{l} = 0.8$; b) – $\frac{d}{l} = 0.4$; c) – $\frac{d}{l} = 0.2$; d) – $\frac{d}{l} = 0.1$;

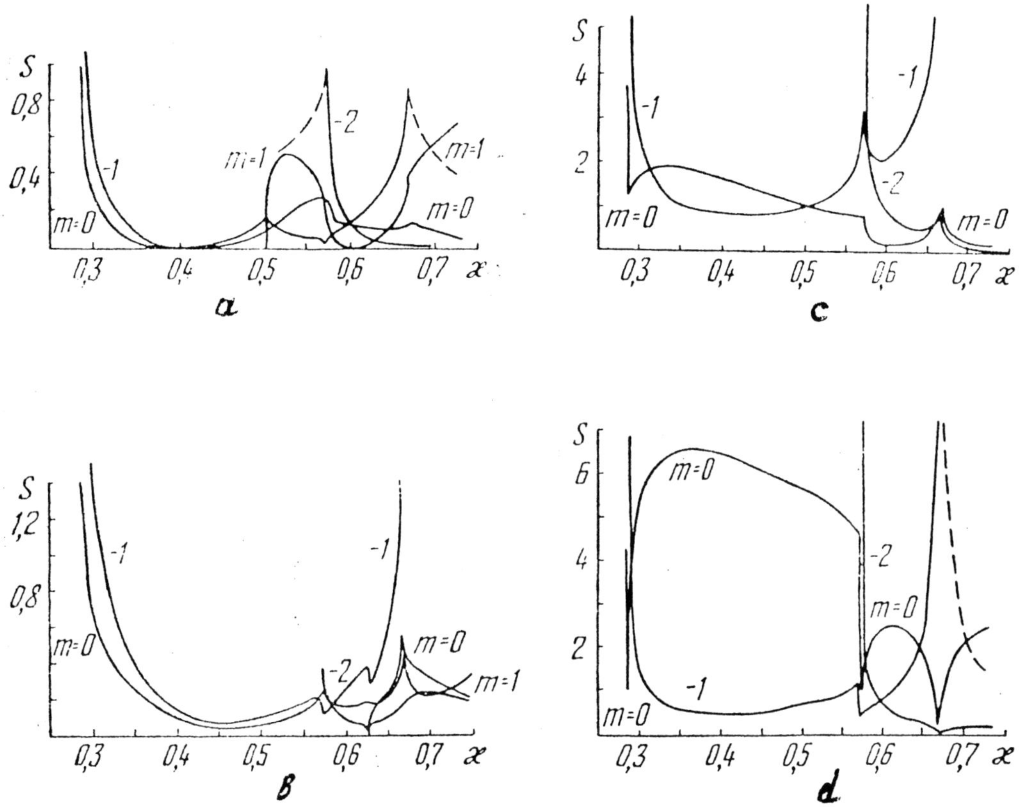

Fig. 9. Wood anomalies in the Smith-Parcell radiation for semi-transparent gratings made of rectangular bars:

a) $\frac{d}{l} = 1.0$; b) – $\frac{d}{l} = 0.8$; c) – $\frac{d}{l} = 0.5$; d) – $\frac{d}{l} = 0.2$ ($\beta = 0.4$, $\frac{h}{l} = \infty$);

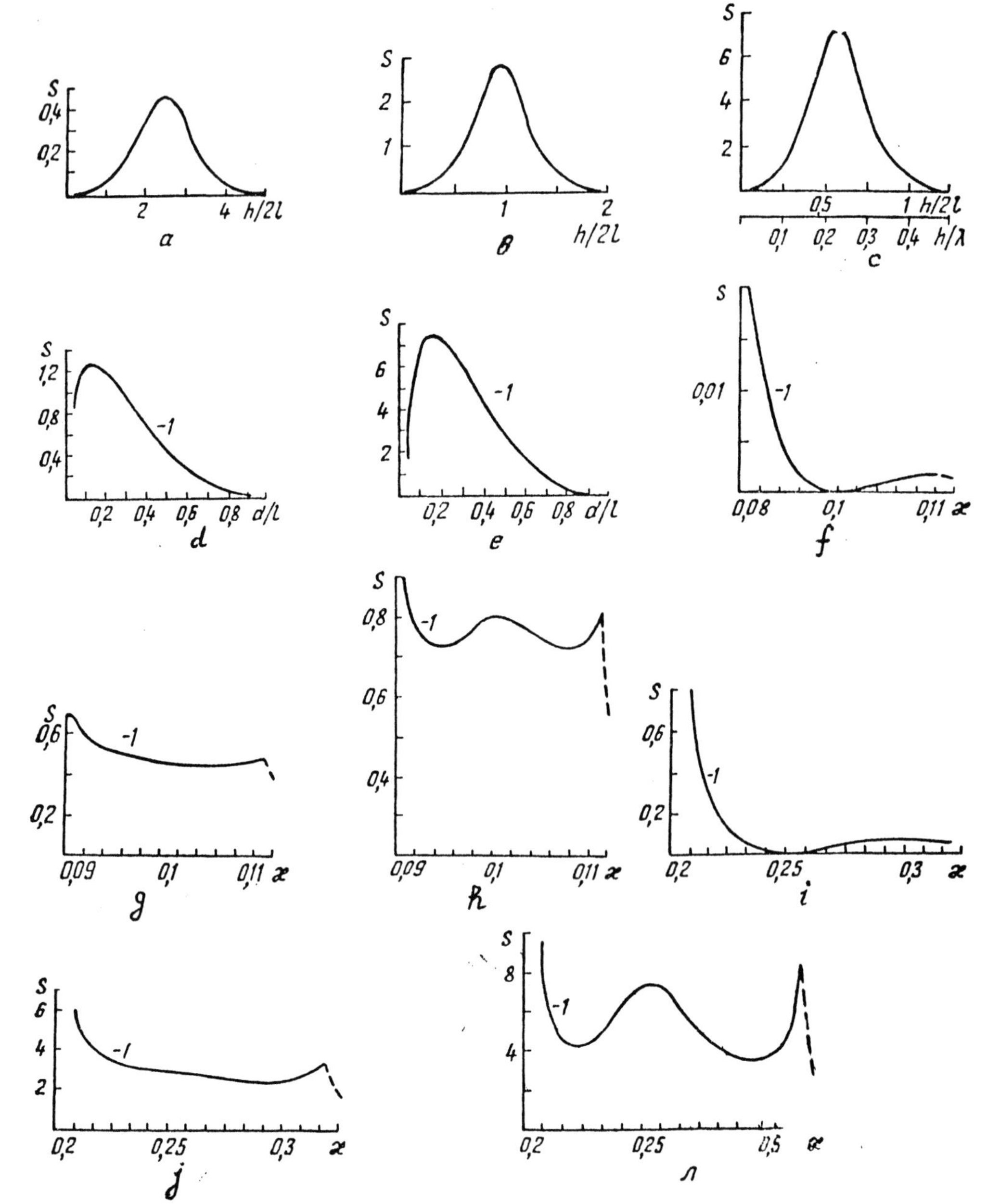

Fig. 10. Power flow of Smith-Purcell radiation versus frequency, width and depth of reflection grating:

a) $\beta = \kappa = 0.1$; b) $\beta = \kappa = 0.25$; c) $\beta = \kappa = 0.1\,(\frac{h}{l} = 0.24)$; d) $\beta = \kappa = 0.25$; $\frac{h}{l} = 0.9$; e) $\frac{d}{l} = 1.0$; f) $\frac{d}{l} = 0.5$; g) $\frac{d}{l} = 0.2$ ($\beta = 0.1$; $\frac{h}{l} = 2.4$) h) $\frac{d}{l} = 1.0$; i) $\frac{d}{l} = 0.5$; j) $\frac{d}{l} = 0.2$ ($\beta = 0.25$; $\frac{h}{l} = 0.9$).

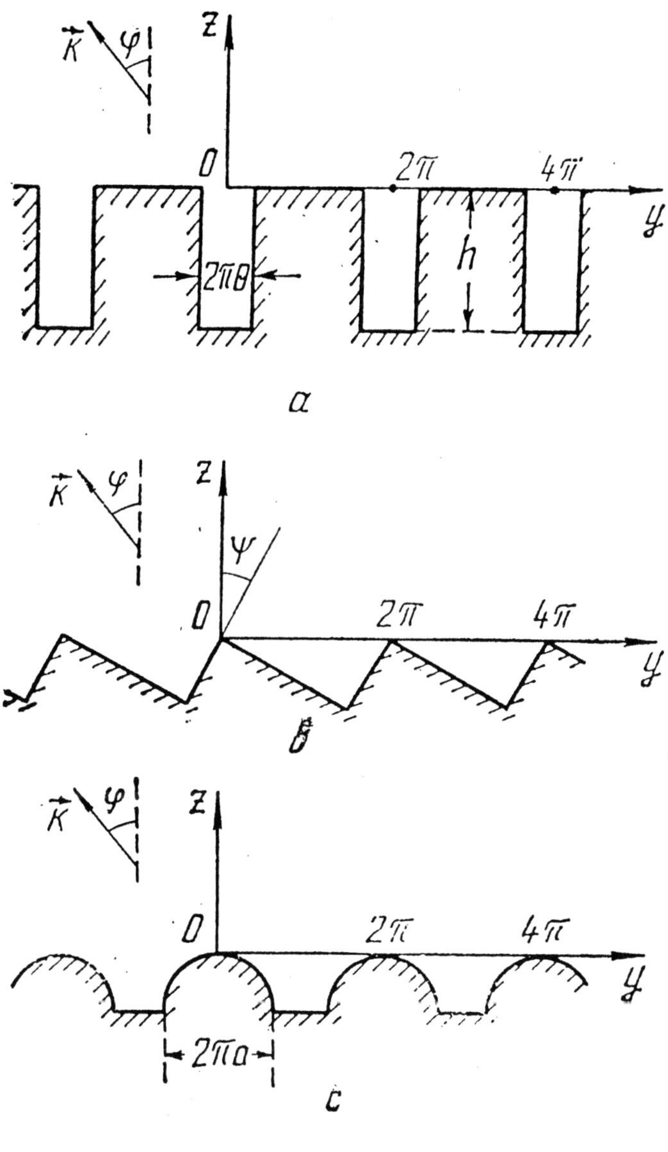

Fig. 11. Different types of reflection gratings.

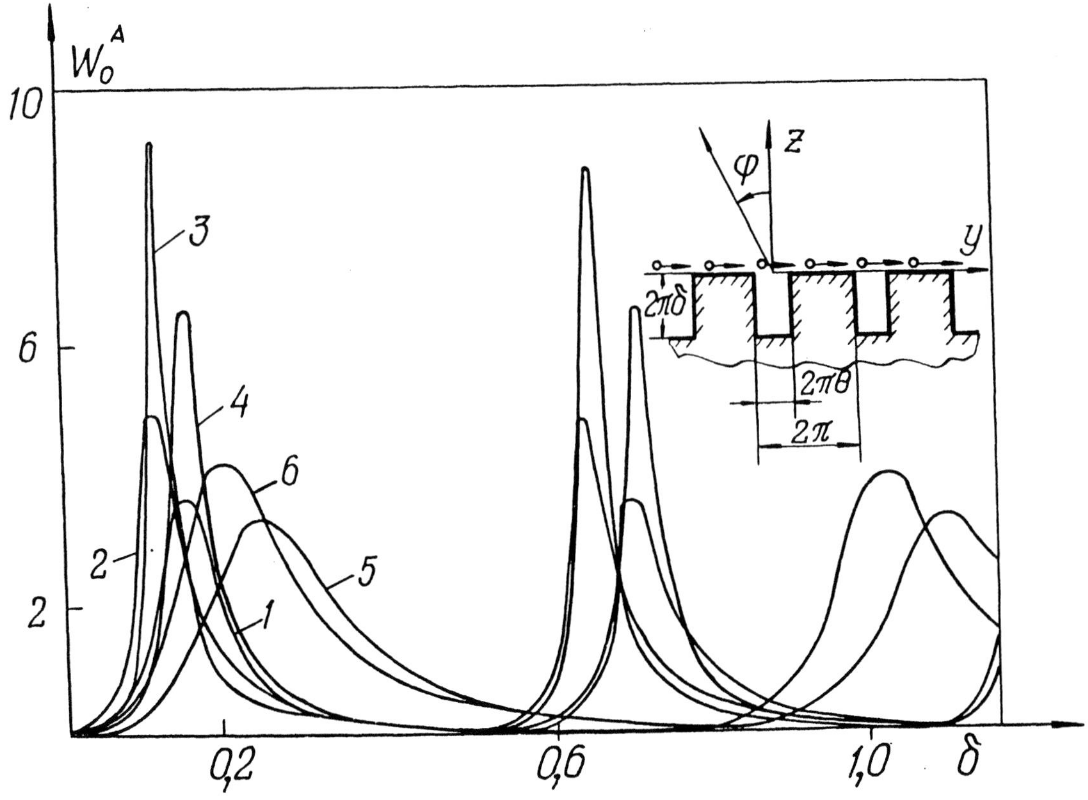

Fig. 12. Energy radiated to the free space at the principal harmonic of diffractional spectrum versus relative grating's height δ:

1 – $\phi = 1°$; $\theta = 0.4$; $\beta = 0.90$;

2 – $\phi = 1°$; $\theta = 0.4$; $\beta = 0.95$;

3 – $\phi = 1°$; $\theta = 0.2$; $\beta = 0.95$;

4 – $\phi = 1°$; $\theta = 0.2$; $\beta = 0.90$;

5 – $\phi = 35°$; $\theta = 0.4$; $\beta = 0.90$;

6 – $\phi = 35°$; $\theta = 0.4$; $\beta = 0.95$.

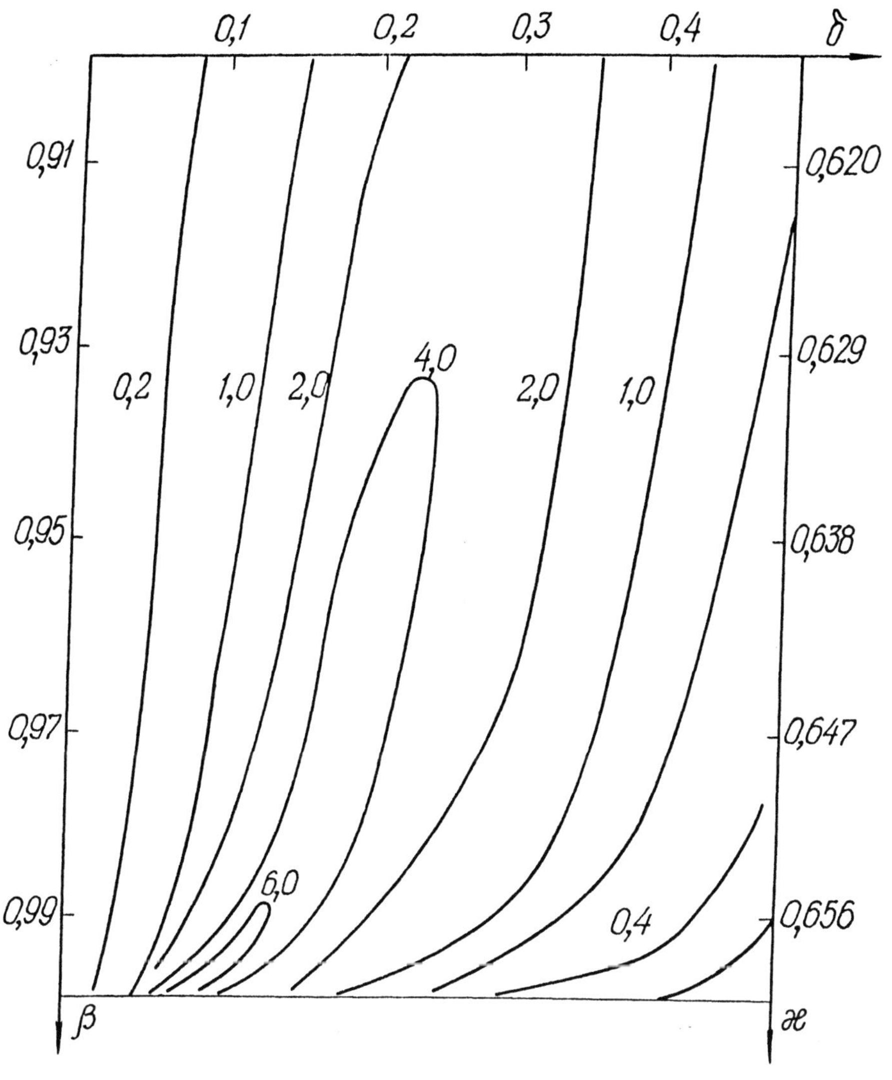

Fig. 13. Lines of equal energy radiated to the free space at the principal harmonic of diffractional spectrum, $\phi = 31°$; $\theta = 0.4$; $n = 1$.

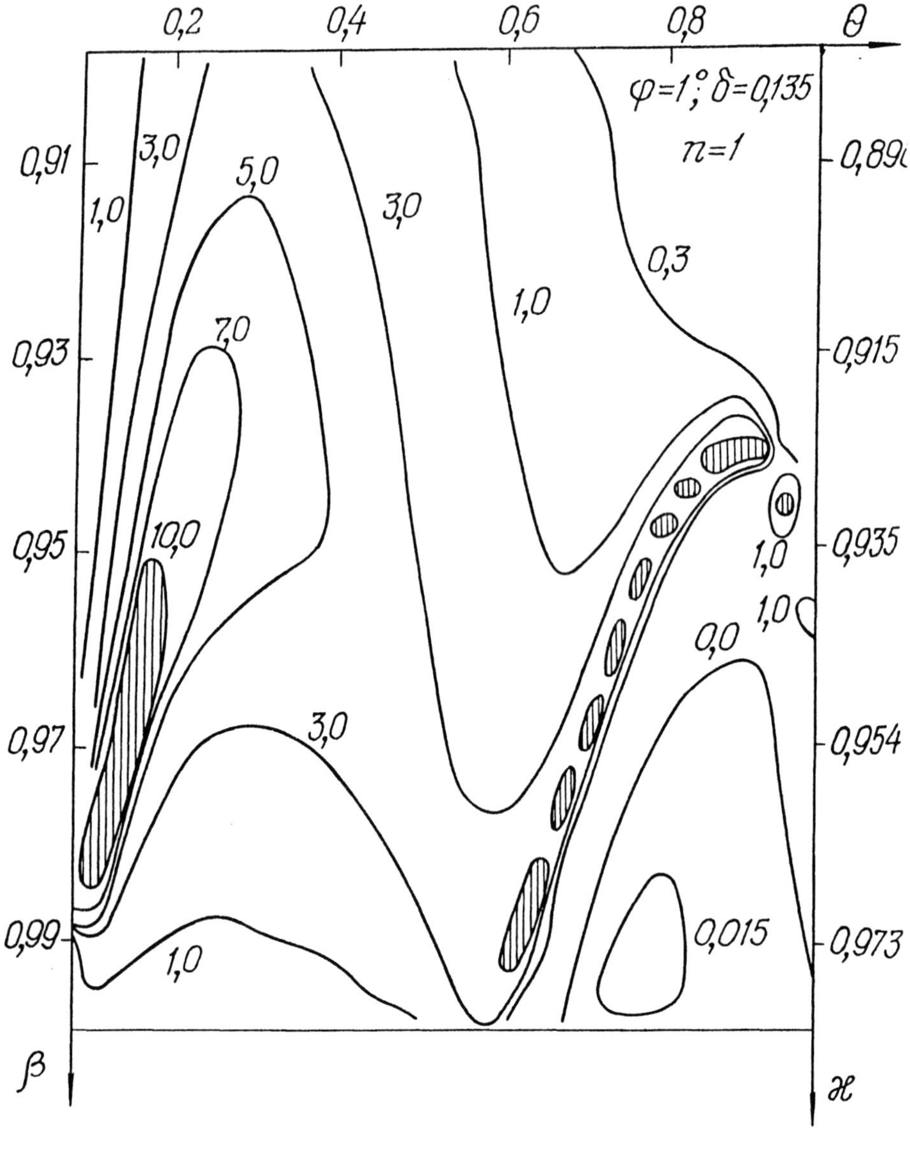

Fig. 14. Radiation intensity in the direction close to normal. Lines of equal energy level are given in the coordinates β, δ; in the shaded domains energy level is greater than 10; $\phi = 1°$; $\delta = 0.135$; $n = 1$.

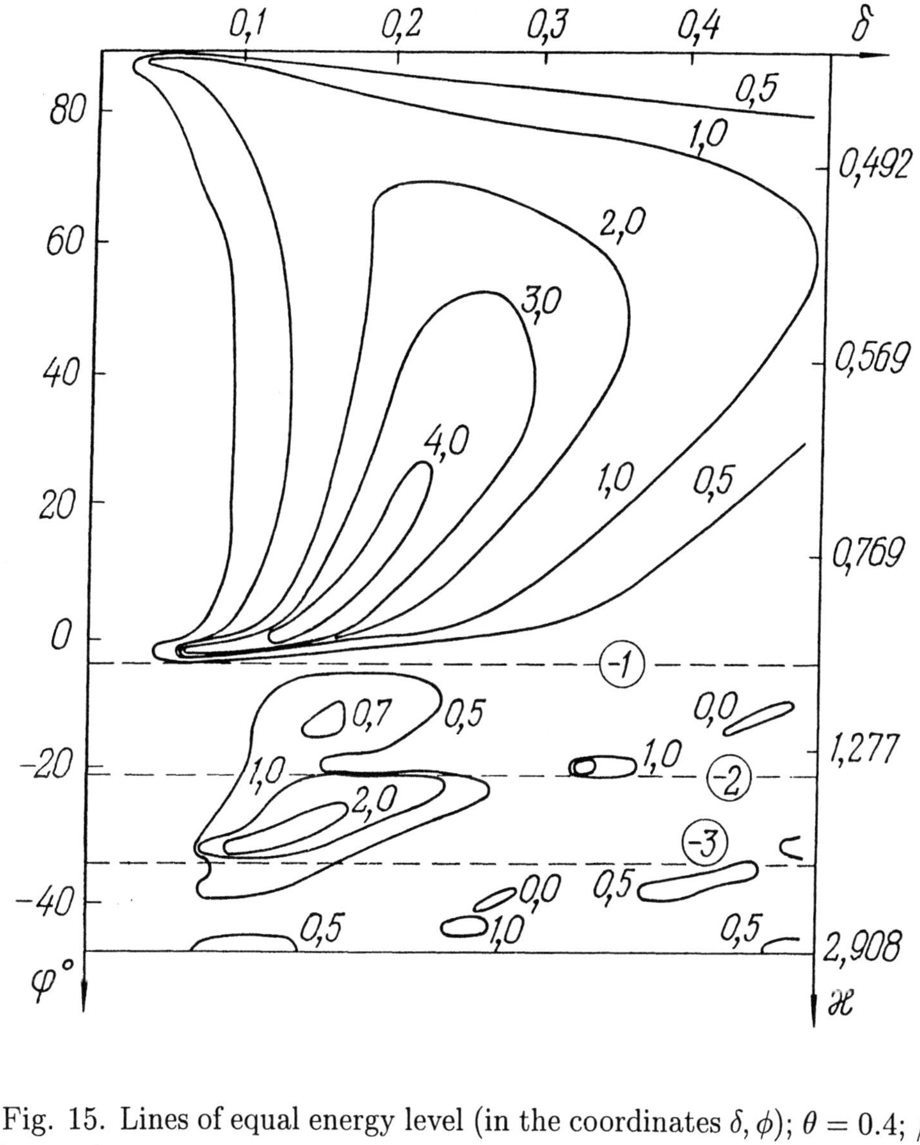

Fig. 15. Lines of equal energy level (in the coordinates δ, ϕ); $\theta = 0.4$; $\beta = 0.93$; $n = 1$.

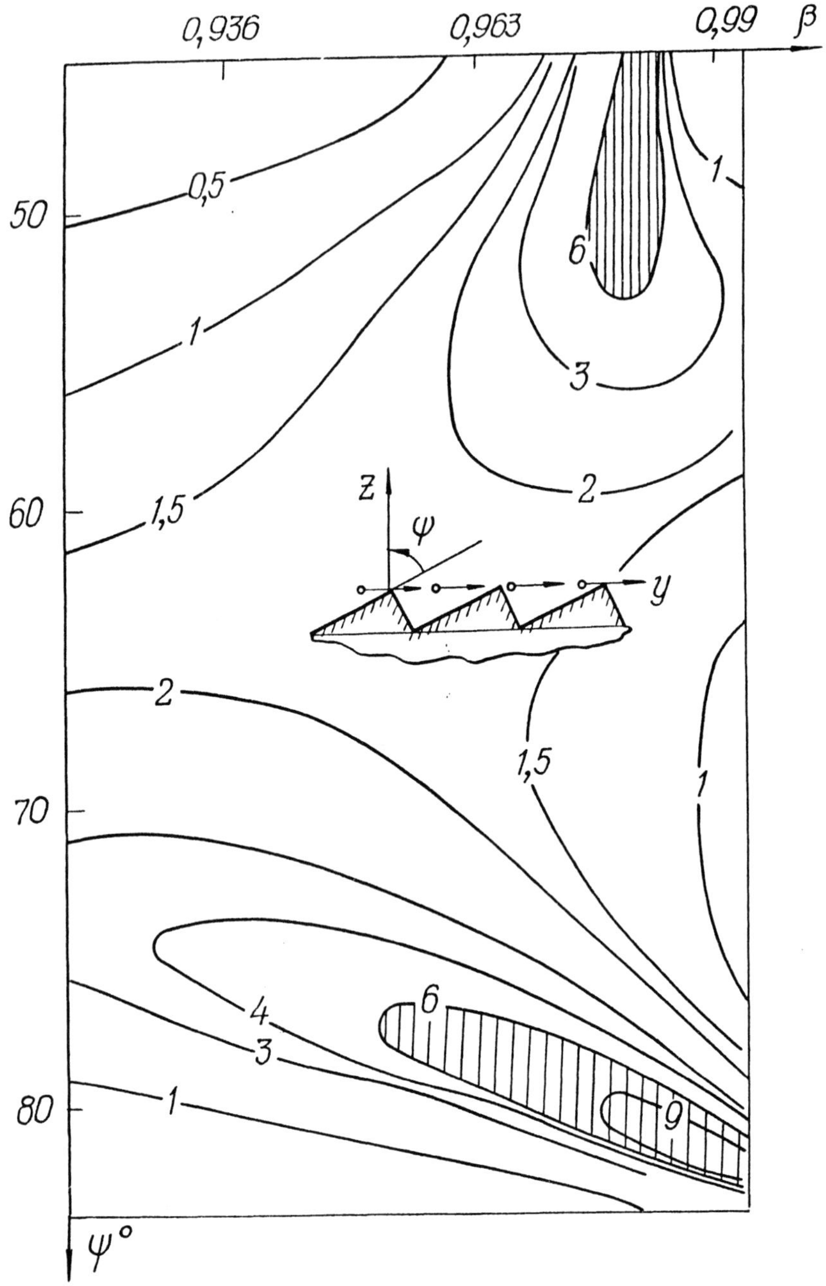

Fig. 16. Zones of high intensity of the Smith-Purcell radiation for echelettes.

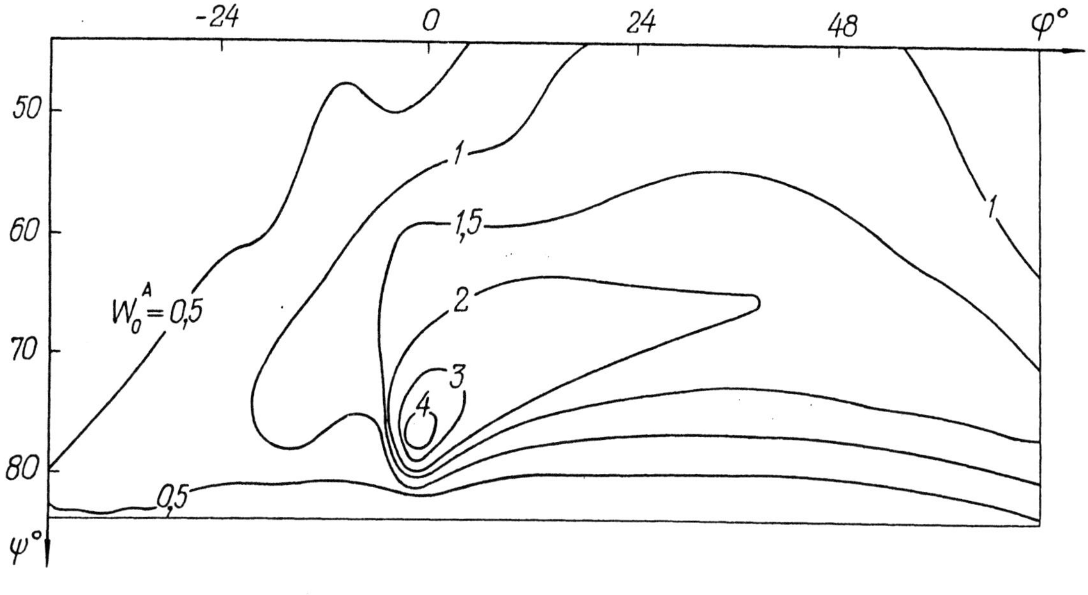

Fig. 17. Intensive almost vertical radiation emerging when electrons' flow interacts with low-profiled echelette: $\beta = 0.93$; $n = 1$.

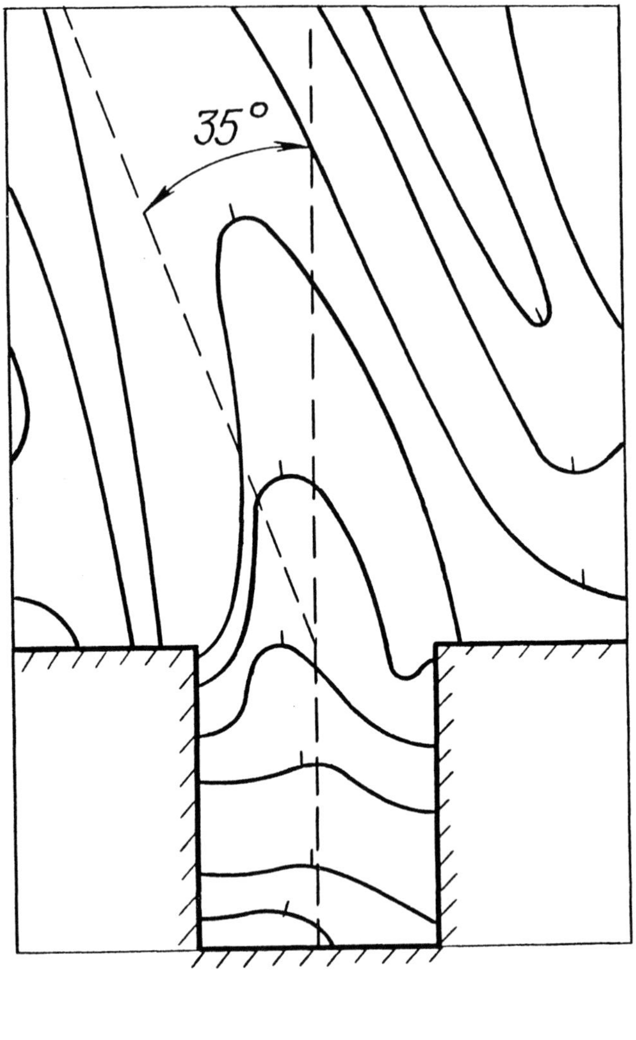

Fig. 18. On determination of communication of the regimes of resonanse scattering of inhomogeneous plane waves with excitation of quasi-eigenoscillations: $\Re \bar{\kappa} = 0.5939$; $\Im \bar{\kappa} = -0.0462$; $\delta = 0.005$; $\theta = 0.4; \Phi = 0.353$.

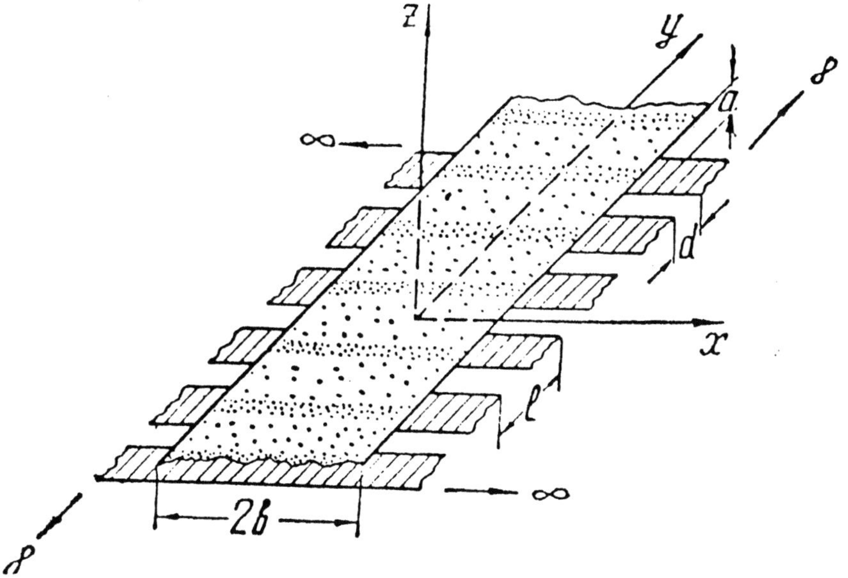

Fig. 19. Bounded electron flow moving over infinite strip grating.

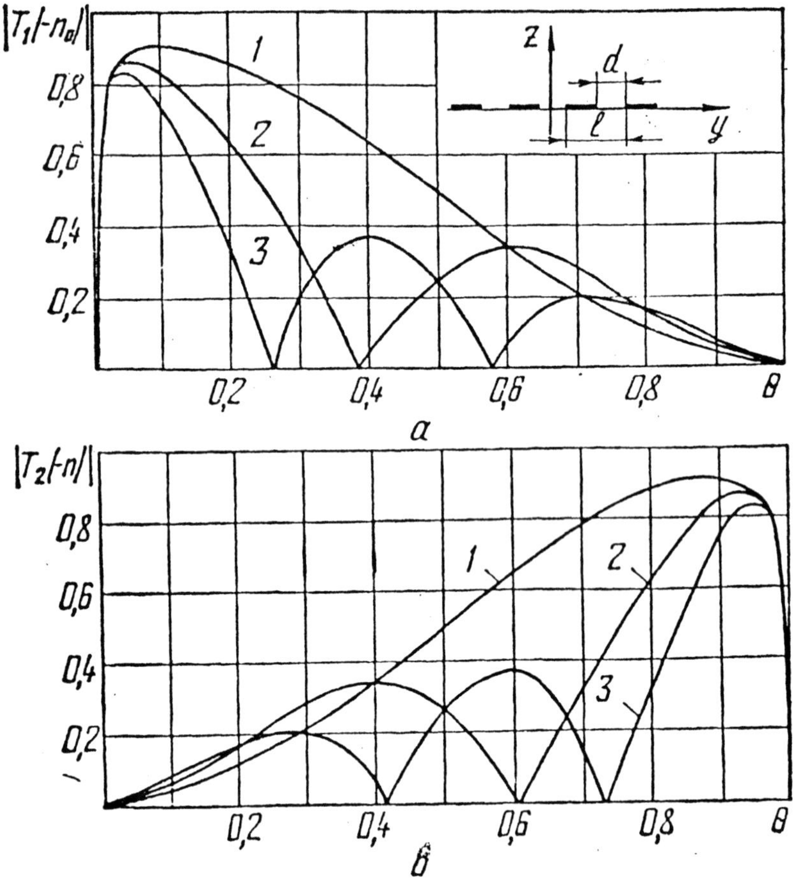

Fig. 20. Diffraction transformation coefficients versus relative width of strip grating: 1 – $(-n_0) = 1$; 2 – $(-n_0) = 2$; 3 – $(-n_0) = 3$;

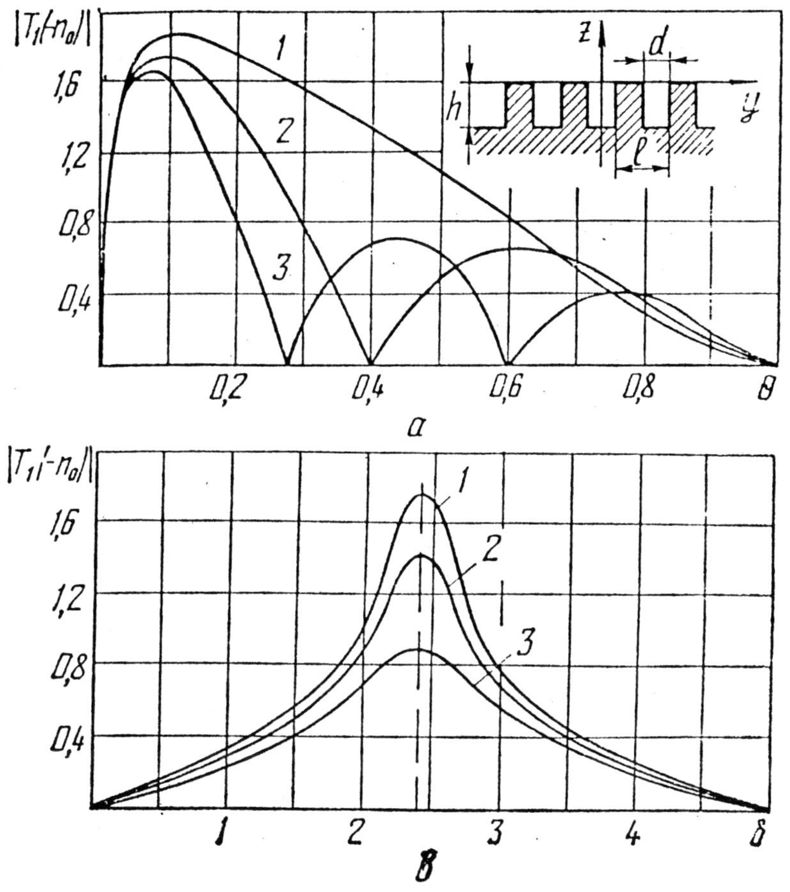

Fig. 21. Diffraction transformation coefficients versus relative width and height of reflection grating made of rectangular bars: 1 – $(-n_0) = 1$; 2 – $(-n_0) = 2$; 3 – $(-n_0) = 3$;

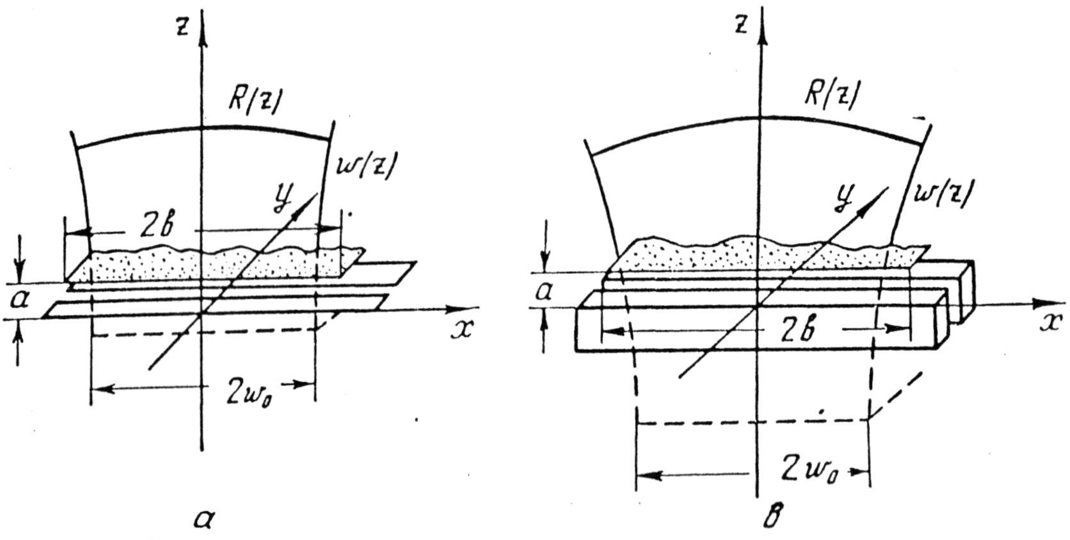

Fig. 22. Broadening of wave pencils of Smith-Purcell radiation, variation of the curvature of phase front for strip gratings and gratings made of rectangular bars.

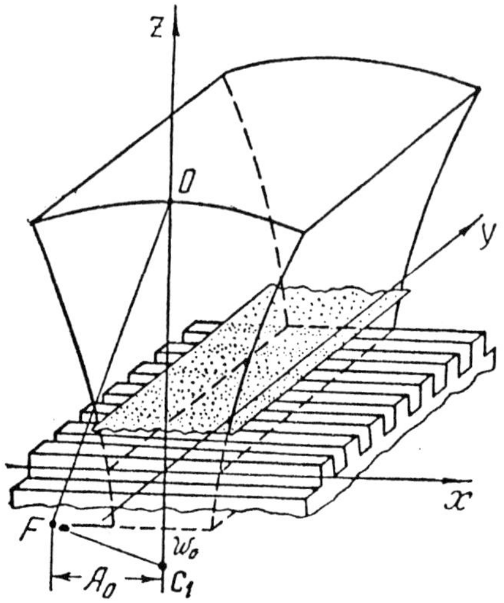

Fig. 23. Variation of the size of field's spot of Smith-Purcell radiation and the curvature of phase front for the electron flow having bounded width.

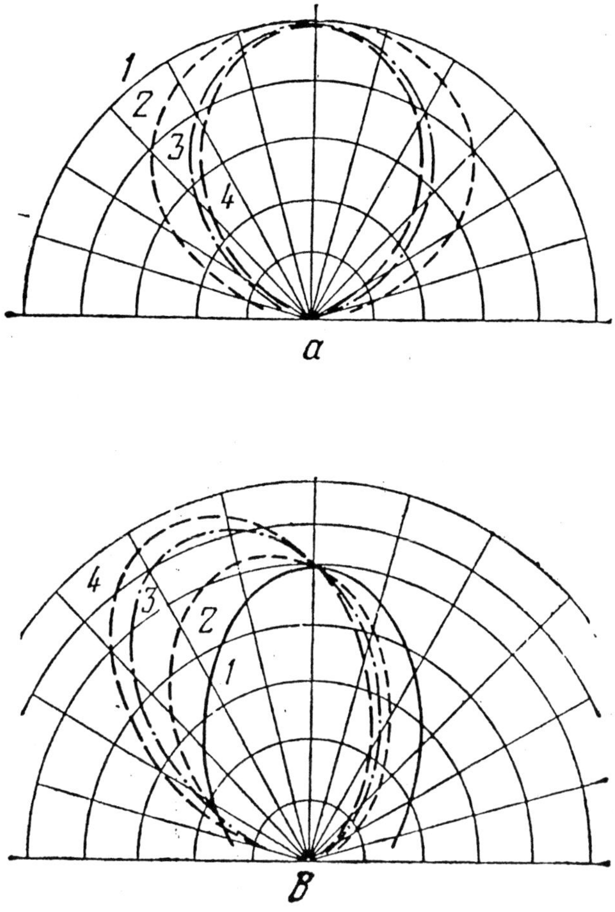

Fig. 24. Radiation far field patterns of a separate electron of Smith-Purcell radiation moving over strip grating: $1-\phi=0$; $2-\phi=\frac{\pi}{6}$; $3-\phi=\frac{\pi}{3}$; $4-\phi=\frac{\pi}{2}$;

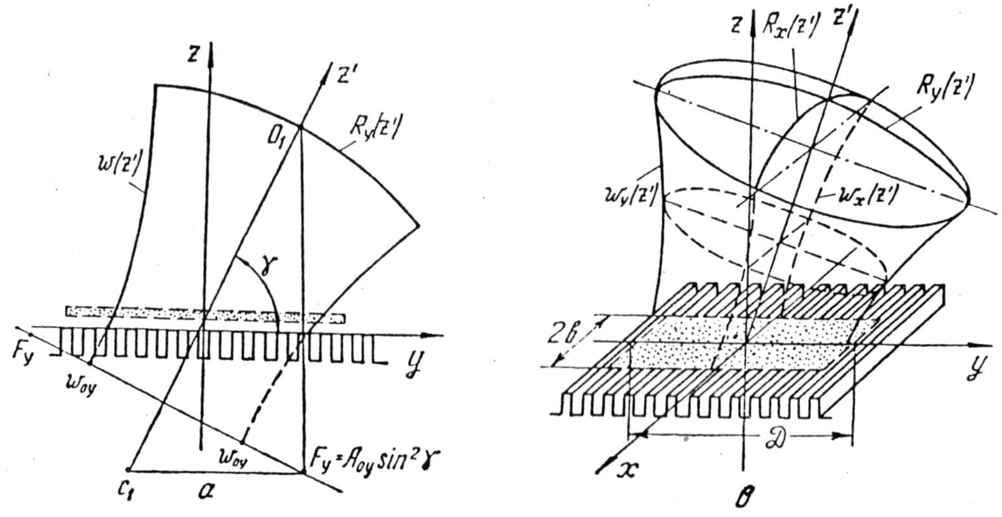

Fig. 25. Variation of the size of field's spot of diffractional radiation and the curvature of phase front for the electron flow having bounded width and length.

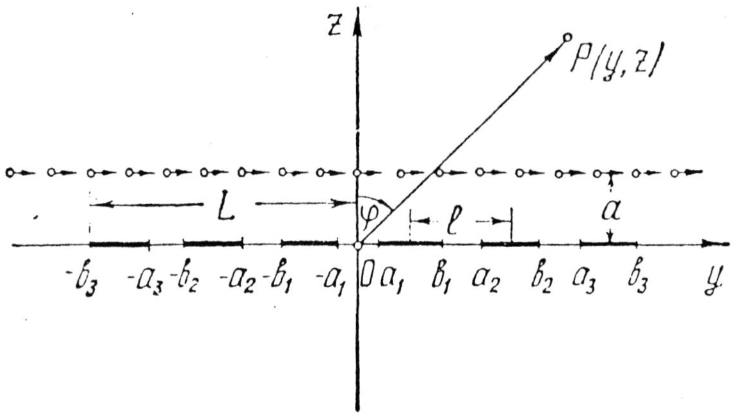

Fig. 26. Bounded strip grating and infinite electron flow moving close to it.

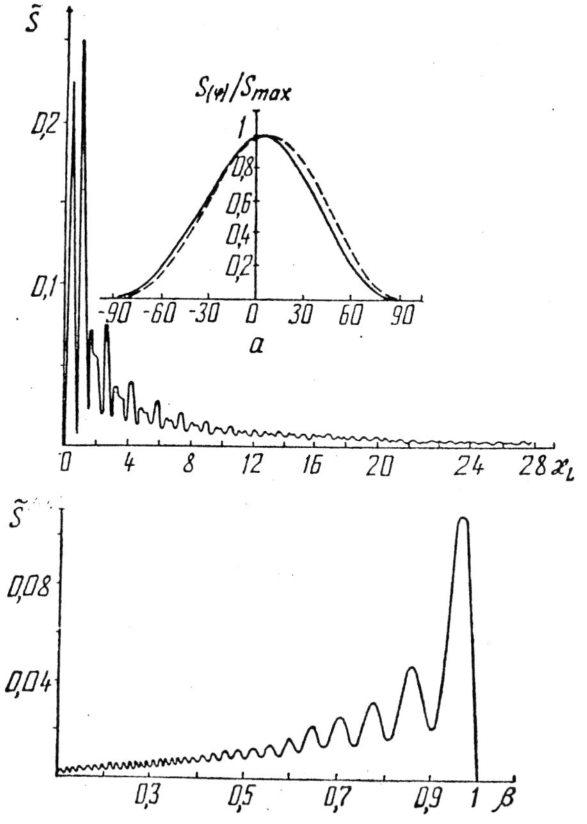

Fig. 27. Radiation losses of Smith-Purcell radiation versus frequency and velocity of electron flow moving over the metal strip. Radiation far field patterns (a) corresponds to the values $\kappa_L = 0.45$ (bold line) and $\kappa_L - 1.10$.

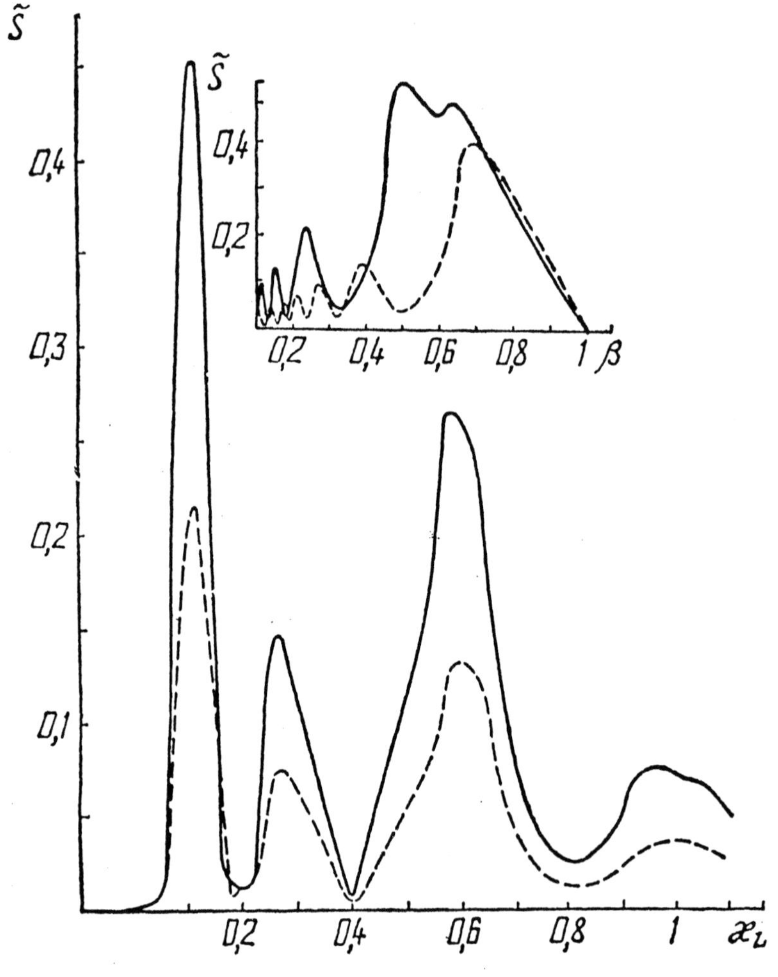

Fig. 28. Radiation losses of Smith-Purcell radiation versus frequency and velocity of electron flow moving over bounded grating consisting of five strips (dot line) and ten strips. Slots and strips have equal width.

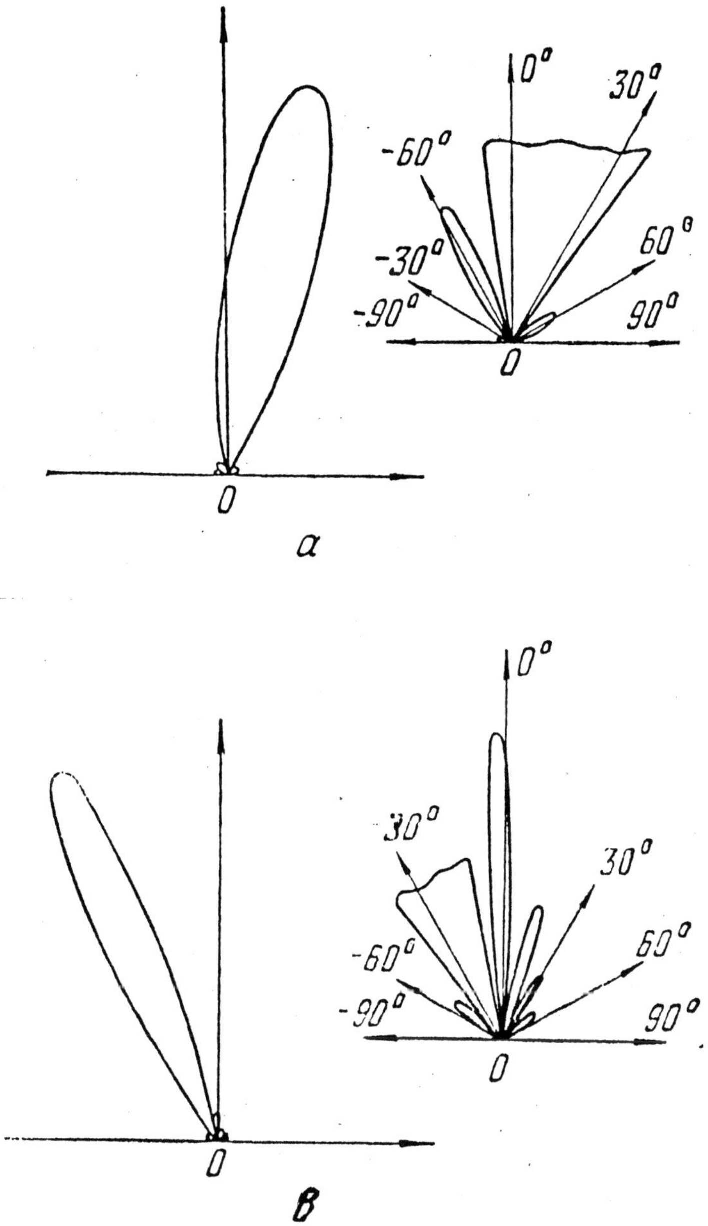

Fig. 29. Normalized far field patterns of diffractional radiation for the bounded grating consisting of ten strips (slots and strips have equal width), $\beta = 0.2$: a) – $\kappa = 0.210$; b) – $\kappa = 0.369$.

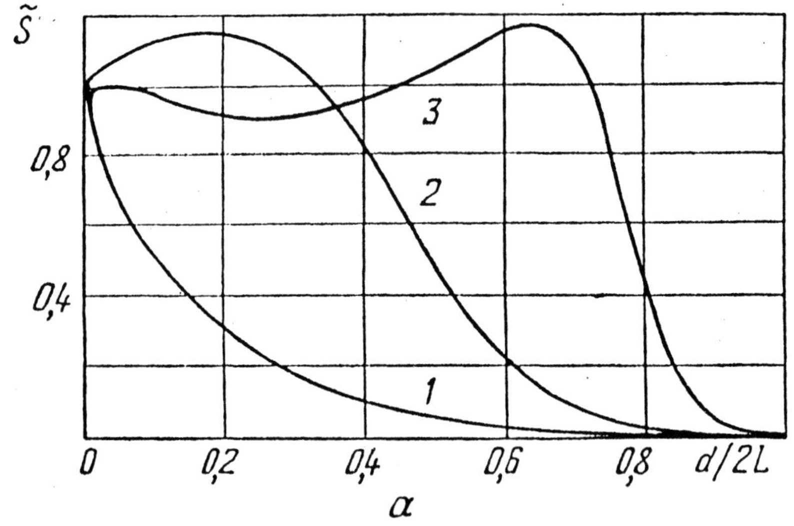

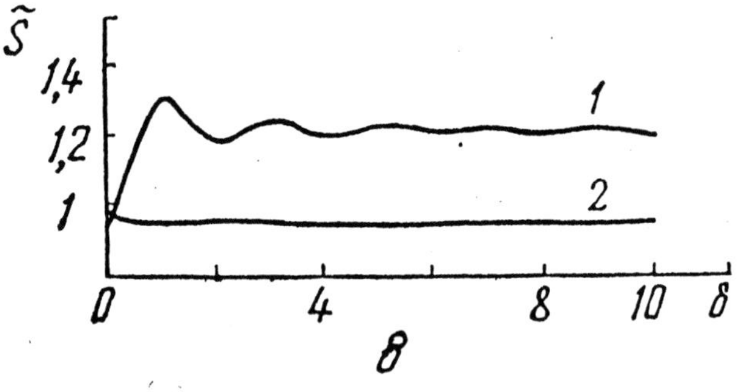

Fig. 30. Radiation losses of diffractional radiation versus ratio of the strips' width and the distance between them for the grating with fixed length: a) $1 - \kappa = 2$; $2 - \kappa = 4$; $3 - \kappa = 10$; b) $1 - \kappa = \pi$; $2 - \kappa = 2\pi$.

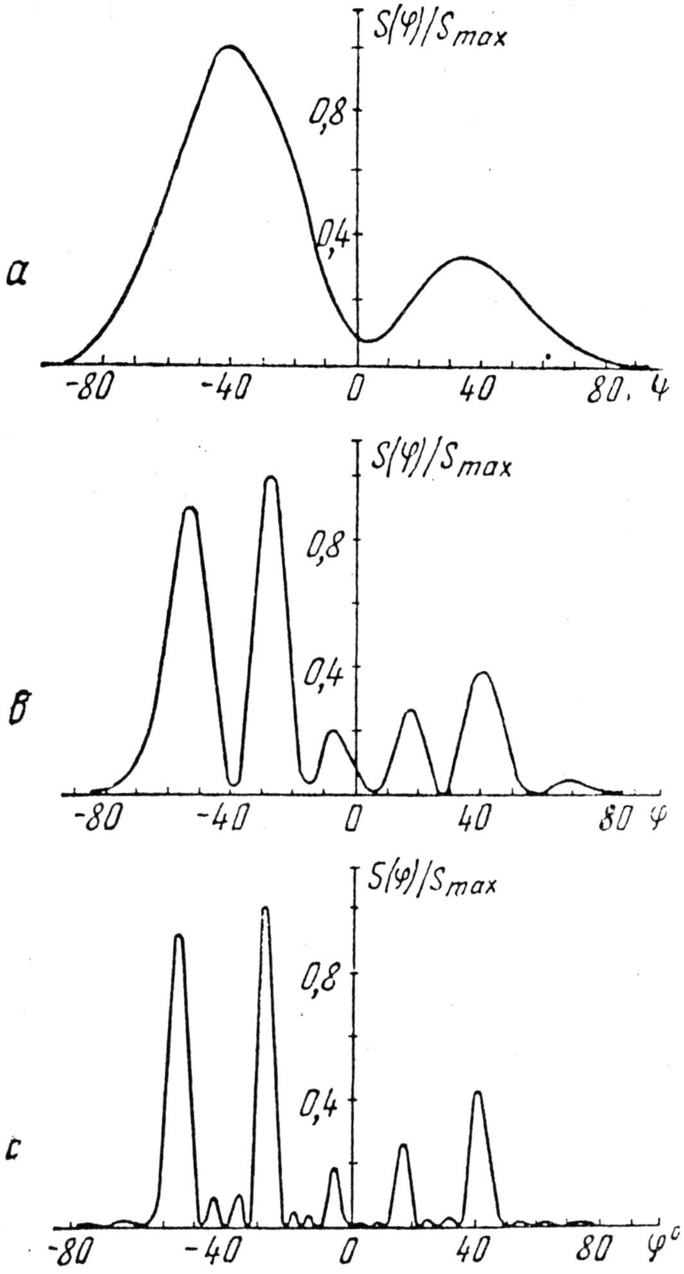

Fig. 31. Far field patterns of Smith-Purcell radiation in the cases of different number of strips, $\beta = 0.84$ and $\kappa = 2$: a) – one strip; b) – two strips; c) – four strips (slots and strips have equal width).

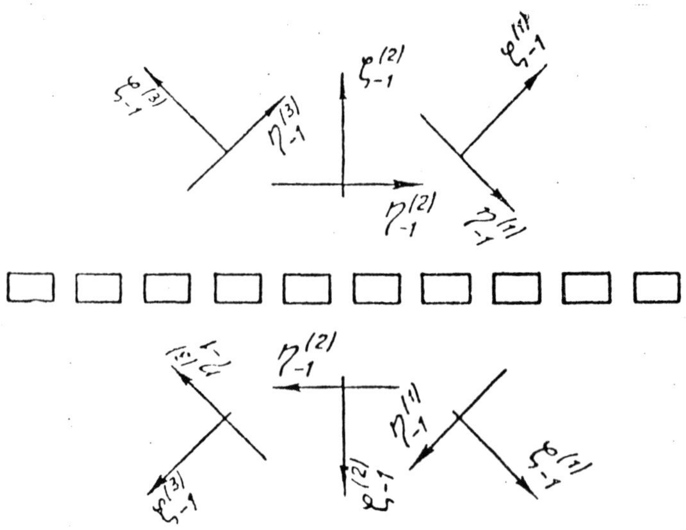

Fig. 32. Beam coordinates of the minus first harmonic of Smith-Purcell radiation.

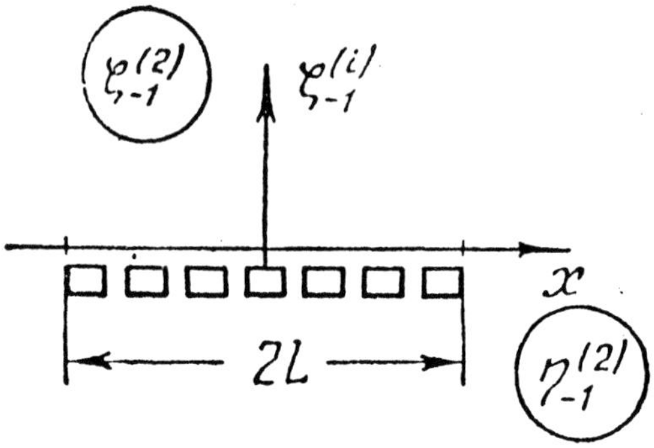

Fig. 33. Bounded grating and its beam coordinates.

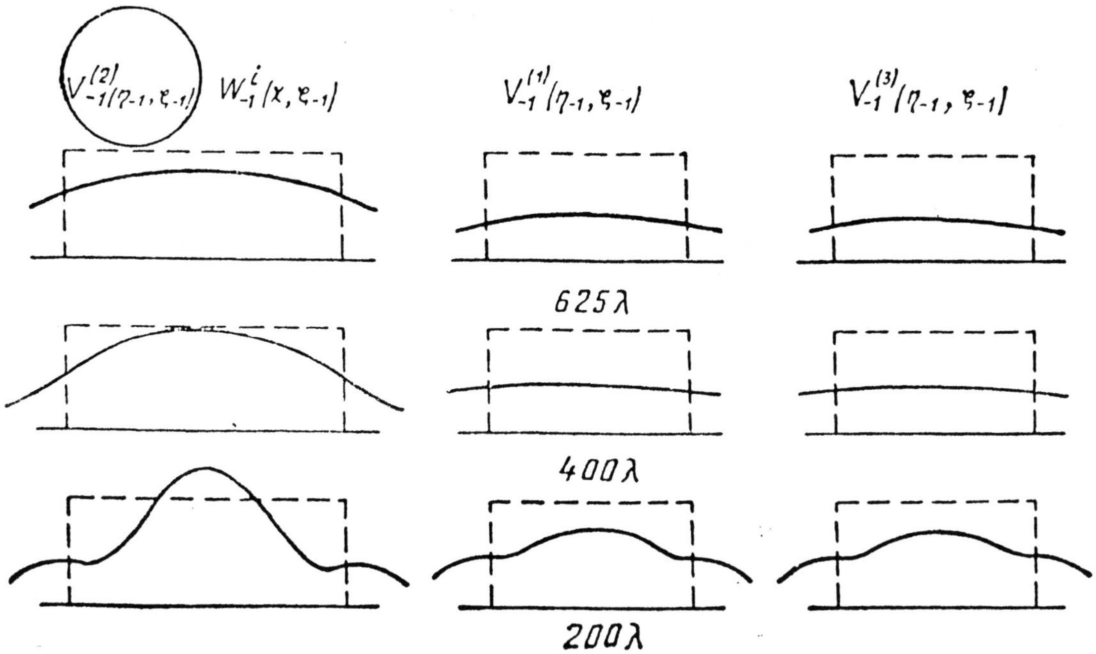

Fig. 34. Distribution of the amplitude of diffractional radiation on the bounded grating.

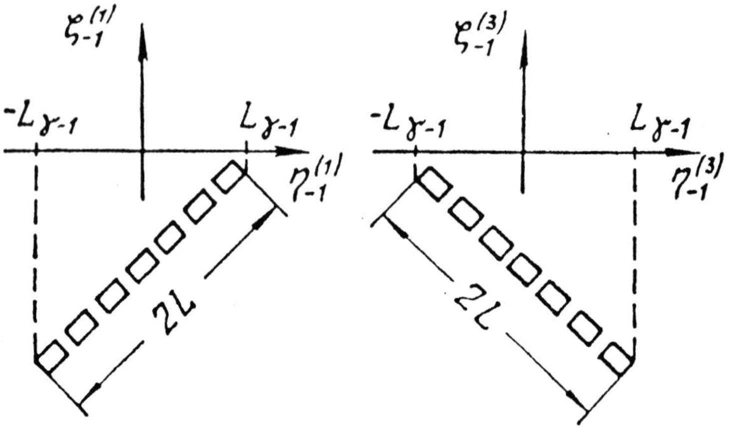

Fig. 35. Grating consisting of finite number of elements in the diffractional radiation at the angles ±45° and ±135°.

Table 1.

x	0.1	0.141	0.2	0.316
d_{opt}	0.12	0.13	0.17	0.15

CHAPTER 2

Smith-Purcell and Cherenkov radiation. Electron-wave model of the Smith-Purcell effect.

In the first chapter it has been expounded the theory of Smith-Purcell effect within the frame of the approximate model of given current for the case of excitation of "pure" diffractional radiation on the gratings having different geometrical shape. It has been established that the energy density of SPR varied negligibly and was practically constant if the current's density increased proportional to λ^{-1}. Under the certain conditions a flow excites intensive SPR at the angle 90° and one can control the intensity and direction of radiation by varying the grating's geometry.

The given chapter is devoted to the comparative analysis of the Cherenkov radiation and Smith-Purcell radiation excited by the monochromatic electron flow moving close to a dielectric medium shielded by diffractional grating. These problems are of practical interest for creating the devices of diffractional electronics with quasi-optical output of energy, for experimental modeling of excitation and radiation processes in the volume integrated schemes et al. [5, 96-99]. Electrodynamical structures together with electron flow moving close to them, which are used for solving the indicated problems, are shown on Fig. 36, a-d.

The thin structure of SPR has been also investigated. Unlike the one-wave model of SPR (where the source of radiation is one longitudinal electron wave of space charge propagating close to the grating with constant phase velocity equal to the velocity of a flow), in real situation electron flow is a complicated active medium which within the frame of linear approximation may be described by a superposition of longitudinal and transversal current's electron waves propagating with different phase velocities. Such a flow causes new characteristics of SPR: splitting of the radiation far field pattern, change of the polarization of excited field, additional SPR and some other phenomena which should be taken into account when constructing the theory. Knowledge of the thin structure of SPR enables one to create the methods for remote measurements of the dynamical parameters of the source. These questions are studied in [5, 7-18].

Here we will consider two classes of problems: excitation of SPR by longitudinal and transversal waves of an electron flow together with absorption and amplifying of electromagnetic waves by a flow interacting with the field scattered by a grating.

§6. Theory of Smith-Purcell and Cherenkov radiations.

1. Let us consider the electron flow with charge's density (1.3) moving with constant velocity $\vec{v} = \vec{j}_0 c\beta_e c$ close to the strip grating situated in the plane $z = -a$ on the surface of isotropic dielectric medium (see Fig. 36, a-d). One must determine electromagnetic field, excitation conditions and energy density of SPR and Cherenkov radiation. Such problems are solved in rigorous mathematical statement by the methods presented in the works [3,61]: at first the eigenfield of a flow of electrons is found by solving the D'Alembert equation for the electron flow in free space and then it is solved the boundary value problem of determining the amplitudes of scattered fields in the considered systems; the latter is reduced by the method of Riemann-Hilbert problem to the system of linear algebraic equations. Numerical solution to these systems gives unknown coefficients of electromagnetic field and other characteristics of radiation obtained with required accuracy. High rate of convergence enables us to construct explicitly the solution in the long-wave approximation. Let us consider the results of the solution of the stated problems.

The key problem for the comparative analysis of the properties of Cherenkov and diffractional radiations is the problem of excitation of electromagnetic waves by the monochromatic current of space charge (1.3) moving close to the dielectric half-space shielded by a diffractional grating (see Fig. 36, a). If to take into account the dielectric medium and periodicity of the structure along the axis Oy, one can write down electromagnetic field in vacuum (the domain I, $z > -a$) and in dielectric (the domain II, $z < -a$):

$$\vec{H}^{\mathrm{I}} = \vec{H}_c + \vec{i}\sum_n A_n e^{iq_n(z+a)} e^{ih_n y}; \quad z > -a;$$

$$\vec{E}^{\mathrm{I}} = \vec{E}_c + \sum_n (-\vec{j}\sqrt{1-\tau_n^2} + \vec{k}\tau_n) A_n e^{iq_n(z+a)} e^{ih_n y}; \quad z > -a;$$

$$\vec{H}^{\mathrm{II}} = \vec{i}\sum_n B_n e^{-iq_{n\epsilon}(z+a)} e^{ih_n y}; \quad z < -a; \tag{2.1}$$

$$\vec{E}^{\mathrm{II}} = \sum_n \left(\vec{j}\frac{1}{\epsilon}\sqrt{\epsilon-\tau_n^2} + \vec{k}\frac{\tau_n}{\epsilon}\right) B_n e^{-iq_{n\epsilon}(z+a)} e^{ih_n y}; \quad z < -a;$$

$$\vec{H}_c = i2\pi\beta_0\beta_e \mathrm{sign} z e^{-q|z|} e^{iky};$$

$$\vec{E}_c = 2\pi\rho_0(-\vec{j}i\theta + \vec{k}\mathrm{sign} z) e^{-q|z|} e^{iky},$$

where H_c, E_c is the eigenfield of electron flow; A_n, B_n are unknown coefficients of the Fourier-amplitudes of scattered field; $e^{-i\omega t}$ is the time factor in the field in (2.1); $q_n = k\theta_n = k\beta_e\sqrt{1-\tau_n^2}$; $q_{n\epsilon} = k\theta_{n\epsilon}k\beta_e\sqrt{\epsilon-\tau_n^2}$; $q = k\theta = k\sqrt{1-\beta_e^2}$; $h_n = k+2\pi n l^{-1}$; $\tau_n = \kappa^{-1}(\eta+n)$; $\eta = \kappa\beta_e^{-1}$; $\kappa = l\lambda^{-1}$; $\beta_e = v_e c^{-1}$.

In accord with the principle of limiting decay of electromagnetic waves only those branches of the roots q_n, $q_{n\epsilon}$ have physical sense whose imaginary part is positive and $\Re q_n > 0$, $\Re q_{n\epsilon} > 0$ for $\Im q_n = 0$, $\Im q_{n\epsilon} = 0$.

It follows from the equalities (2.1) that the scattered field is the superposition of propagating (volume) along the axis Oz and leaky (surface) electromagnetic waves of diffractional harmonics. The wavenumbers $q_n, q_{n\epsilon}, h_n$ in the free space (in vacuum) and in dielectric form the vectors $\vec{r} = \vec{i}h_n + \vec{k}q_n$, $\vec{r}_\epsilon = \vec{j}h_n + \vec{k}q_{n\epsilon}$ whose moduli are equal to ωc^{-1}, $\omega(c\epsilon)^{-1}$. The directions of these vectors depend on the parameters $\kappa, \beta_e, n, \epsilon$ and determine the radiation angle of volume waves in vacuum

$$\gamma_n = \arccos\tau_n, \quad z > -a \tag{2.2}$$

and in dielectric

$$\gamma_{n\epsilon} = \arccos\frac{\tau_n}{\sqrt{\epsilon}}, \quad z < -a \tag{2.3}$$

where the angles with the sign "+" are counted counter-clockwise and with the sign "-" – clockwise (in dielectric) related to the direction of movement of electron flow.

Electromagnetic waves radiate to the free space under the condition $|\tau_n| < 1$ and to the dielectric medium – for $|\tau_n| < \sqrt{\epsilon}$, i. e., when the velocities of electron flow satisfy the inequalities

$$\frac{\kappa}{|n|+\kappa} \le \beta_e \le \frac{\kappa}{|n|-\kappa}, \quad z > -a; \tag{2.4}$$

$$\frac{\kappa}{|n|+\kappa\epsilon} \le \beta_e \le \frac{\kappa}{|n|-\kappa\epsilon}, \quad z < -a. \tag{2.5}$$

It is clear from the analysis of (2.2)–(2.4) that for the given parameters $\kappa, \beta_e, \epsilon$ only "negative" space harmonics ($n = -1, -2, -3, \ldots$) are excited in the free space and the harmonics with indices $n = 0, \pm 1, \pm 2, \ldots$ are excited in the dielectric medium. Diffractional radiation occurs for zero ($n = 0$) and positive ($n = 1$) space harmonics only at a sharp angle in the presence of dielectric medium (if $\epsilon = 1$, the pure diffractional radiation is excited). Here zero space harmonic is excited under the same conditions as the Cherenkov radiation of an electron flow moving close to dielectric, i. e., for electrons' velocities $\epsilon\beta_e^2 > 1$ and with radiation angle $\cos\gamma_{0\epsilon} = (\beta_e\sqrt{\epsilon})^{-1}$. Hence in what follows we will assume that zero harmonic of SPR coincides with the Cherenkov radiation, where the role of a grating is reduced to the shielding factor of the communication coefficient of electron flow interacting with dielectric medium in the process of excitation of Cherenkov radiation.

Coefficients A_n, B_n are determined from the solution to the boundary value electrodynamical problem with exact boundary value conditions: vanishing of tangential components of the field $\vec{E}$ on the strip and continuity of the tangential components of $\vec{E}, \vec{H}$ on the grating's slots and on the surface of dielectric layer. The solution is given in [69]. We will consider here only the long-wave approximation ($\kappa < 0.5$), which may be written for the amplitudes of the field of diffractional radiation of the principal space harmonics ($n = 0; -1$) in the form

$$A_{-1} = -B_{-1}\frac{\theta_{-1\epsilon}}{\epsilon\theta_{-1}}; \quad B_0 = \frac{C_\epsilon}{1 + iG_0\Phi_G};$$

$$B_{-1} = -\frac{i}{2}C_\epsilon\theta_{0\epsilon}(1+\sqrt{\epsilon})\theta_{-1}\frac{\Phi_{-1}}{\epsilon\theta_{-1} + \theta_{-1\epsilon}}, \tag{2.6}$$

where $u = \cos\frac{\pi d}{l}$ is the grating's filling coefficient, C_ϵ is the field's amplitude of Cherenkov radiation without a grating [100]:

$$C_\epsilon = -4\pi\rho_0\beta_e\frac{i\epsilon\theta}{i\epsilon\theta + \theta_{0\epsilon}}e^{-q|a|};$$

$$\Phi_{-1} = \frac{u - 1 + \zeta(u+1)\ln\frac{u+1}{2}}{1 + iG_{-1}\zeta\ln\frac{u+1}{2}}; \quad G_n = \frac{\theta_n(1+\sqrt{\epsilon})\theta_{n\epsilon}}{\beta_e\tau_n(\theta_{n\epsilon} + \epsilon\theta_n)}; \tag{2.7}$$

$$\Phi_G = \frac{(u-1)(3 - 2\eta - \zeta\ln\frac{u+1}{2})}{1 - u + 2\eta u - (1+u)\zeta\ln\frac{u+1}{2}}, \quad \zeta = \eta - 1.$$

It directly follows from (2.6), (2.7) that for $u = 1$ (when there is no grating on the surface of dielectric) $\Phi_{-1} = \Phi_G = 0$, hence, $A_{-1} = B_{-1} = 0$; $B_0 = C_\epsilon$, i. e., electron flow excites only the Cherenkov radiation with field's amplitude C_ϵ. If $u = -1$ (the surface of dielectric is completely covered by a metal layer), $\Phi_{-1} = 0$, $\Phi_G = \infty$, $A_{-1} = B_{-1} = B_0 = 0$ what yields absence of the energy radiation losses of electron flow.

Energy density determined by the average value of the flow of Pointing vector $\vec{S}_n = \frac{c}{8\pi\sqrt{\epsilon}}[\vec{E}, \vec{H}^*]$ gives energetic characteristic of the excited radiation. For a separate space radiation harmonic this value is found by the formula $\vec{S}_n = \vec{l}_n \frac{c}{8\pi\sqrt{\epsilon}} \mid x_n \mid^2$, where x_n is one of the coefficients of the field (2.1) which is determined by (2.6) for $n = 0, -1$; $\vec{l}_n$ is the unit vector characterizing the direction of radiation determined by the angles (2.2), (2.3) and by the expression

$$\vec{l}_n = \vec{j}\frac{\tau_n}{\sqrt{\epsilon}} \pm \vec{k}\frac{\sqrt{\epsilon - \tau_n^2}}{\sqrt{\epsilon}}.$$

The energy density of diffractional radiation of the principal space harmonics excited in vacuum and in dielectric medium may be easily evaluated by means of (2.6):

$$S_0 = S_\epsilon \frac{\theta_{0\epsilon}^2 + \epsilon^2\theta^2}{\epsilon^2\theta^2 + \theta_{0\epsilon}^2[1 - (1+\sqrt{\epsilon})\theta\Phi_G]^2}; \quad z < -a;$$

$$S_{-1}^{\mathrm{I}} = S_{-1}^{\mathrm{II}} \frac{\theta_{-1\epsilon}^2}{\epsilon^2\theta_{-1}^2}; \quad z > -a; \tag{2.8}$$

$$S_{-1}^{\mathrm{II}} = \frac{1}{4} S_\epsilon \left[\frac{(1+\sqrt{\epsilon})\theta_{0\epsilon}\theta_{-1}}{\epsilon\theta_{-1} + \theta_{-1\epsilon}}, \middle| \Phi_{-1} \middle| \right]^2; \quad z < -a. \tag{2.9}$$

Here S_ϵ is the density of energy of the Cherenkov radiation of monochromatic electron flow moving close to the surface of isotropic dielectric without diffractional grating [100]:

$$S_\epsilon = 2\pi c \rho_0^2 \beta_e^2 \frac{\epsilon^{\frac{3}{2}}\theta^2}{\epsilon^2\theta^2 + \theta_{0\epsilon}^2}. \tag{2.10}$$

Let us point out that if there is no dielectric medium ($\epsilon = 1$), the formulas (2.9) transfer to interrelations for the pure diffractional radiation. Since the factor $2\pi c\rho_0^2\beta_e^2 = J_0^2 R$, [Wt cm^{-2}], the energy density has the form $S = RJ_0^2$, where J_0, [A cm^{-1}], is the surface density of pencil's current, R, [Ohm], is radiation resistance [3], which for the Cherenkov radiation has the form

$$R_\epsilon = 189 \frac{\epsilon^{\frac{3}{2}}(1-\beta_e^2)}{\epsilon\beta_e^2 - 1 + \epsilon^2(1-\beta_e^2)} e^{-2q|a|}, \tag{2.11}$$

and radiation resistance for zero space harmonic of diffractional radiation

$$R_0 = R_\epsilon \frac{\epsilon\beta_e^2 - 1 + \epsilon^2(1-\beta_e^2)}{\epsilon^2\theta^2 + \theta_{0\epsilon}^2[1-(1+\sqrt{\epsilon})\theta\Phi_G]^2}. \tag{2.12}$$

It is easy to determine R for other space harmonics of diffractional radiation by means of formulas (2.11), (2.8) and (2.9).

It follows from (2.8), (2.9) that, unlike the formula (1.5) obtained within the frame of blinking dipole theory, in the given case the energy density of diffractional radiation grows proportional to J_0^2 and it is directed at the angle $\gamma_n = 90°$ to the grating's surface when $\kappa = |n|\beta_e$. It also follows from these relations that the dependence of the energy density of diffractional radiation on the radiation wavelength is presumably determined by the non-exponential factor in (2.11) which practically does not depend on λ for small values of a. Since $\zeta^2\kappa^{-2} << 1$ in the representation for Φ, the energy density of diffractional radiation does not depend on radiation wavelength for certain geometrical parameters of the grating what justifies the use of diffractional radiation for generating millimeter and submillimeter oscillations.

2. Let us consider the space distribution of the energy density of Cherenkov and diffractional radiation. As it was pointed out above, the electron flow (1.3) moving close to dielectric shielded by diffractional grating for definite values of the parameters $\kappa, \beta_e, \epsilon$ excited either volume (going off the source) or surface electromagnetic waves. The domains of variation of these parameters corresponding to the excitation of volume and surface waves are easily found from the Brilluin diagrams (Fig. 37) for different values of ϵ. These diagrams are obtained by the graphical solution of inequalities $|\tau_n| < 1$, $|\tau_n| < \sqrt{\epsilon}$ determining the conditions of excitation of diffractional radiation. Diagrams consist of discrete domains bounded by straight lines $\kappa = \pm(\eta + n)$, $\kappa\sqrt{\epsilon} = \pm(\eta + n)$ which are marked by figures $I_0 - -5_s^m$; they determine the most characteristic cases of excitation of electromagnetic waves by an electron flow: I – Cherenkov radiation; 2 – surface waves; 3 – diffractional radiation only to the dielectric medium; 4 – simultaneous diffractional and Cherenkov radiations; 5 – diffractional radiation to dielectric ($z < -a$) and to the free space ($z > -a$). Lower indices $s = 0, \pm1, \pm2, \ldots$ indicate the numbers of space harmonics radiating to dielectric and upper indices $m = 0, \pm1, \pm2, \ldots$ – to vacuum. For example, the domain $4_{0,-1,-2}^{-1}$ corresponds to the minus first harmonic radiating to the free space and to three

harmonics with indices $n = 0, -1, -2$ excited in the dielectric medium for all parameters $\kappa, \beta_e, \epsilon$ hitting this domain.

One can see form the given diagrams that if the parameters of electrodynamical system satisfy the condition $\kappa < (2\sqrt{\epsilon})^{-1}$, an electron flow with fixed velocity excites one space harmonic to the dielectric medium or two diffractional harmonics such that one of them radiates to vacuum and the other – to dielectric. Only zero diffractional harmonic is excited in the domain 1_0 (or the Cherenkov radiation); surface waves – in the domain 2; diffractional radiation to dielectric at a sharp angle (the domain 3^{δ}_{-1}) or at the angle $\gamma_{-1\epsilon} < -\frac{\pi}{2}$ (the domain 3^{a}_{-1}) – in the domain 3_{-1}; negative diffractional harmonics simultaneously in vacuum and dielectric – in the domain 5^{-1}_{-1}. If $\kappa > (1 + \sqrt{\epsilon})^{-1}$, several space harmonics of diffractional radiation are excited simultaneously in dielectric and vacuum. For the given κ only diffractional radiation is observed when electrons' velocities $\beta_e < (\epsilon)^{-0.5}$ and for $\beta_e > (\epsilon)^{-0.5}$ in addition to it the Cherenkov radiation is excited ($n = 0$). The radiation of zero and positive space diffractional harmonics are always directed at a sharp angle (as well as the Cherenkov radiation [101]), while one can obtain the diffractional radiation at an angle from 0° to 180° if to vary the parameters $\kappa, \beta_e, \epsilon$.

If metallic grating with the period $l < \lambda(2\sqrt{\epsilon})^{-1}$ is drifted on the surface of dielectric, then only Cherenkov radiation is excited in the dielectric medium (the domain 1_0) for the flow's velocities $\beta_e > (\epsilon)^{-0.5}$. In order to clear up the influence of the shielding effect of a grating on the Cherenkov radiation, resistances R_ϵ, R_0 as functions of dielectric permittivity ϵ and grating's parameters κ, u were calculated in the case of excitation of Cherenkov radiation on the pure dielectric (Fig. 38, a,b) and on the dielectric shielded by a grating (Fig. 38, c,d).With growing ϵ the energy density of Cherenkov radiation falls; the degree of shielding the radiation by diffractional grating essentially depends on κ, β_e and the filling coefficient $u = \cos\frac{\pi d}{l}$. It is especially clearly reflected on Fig. 39,a where the the density of radiation energy of zero space harmonic versus u for different $\epsilon, \kappa, \beta_e$ is shown (normalized by the S_ϵ of Cherenkov radiation excited on the pure dielectric). For small velocities of the electron flow ($\beta_e < 0.25$) and dielectric permittivities $\epsilon > 25$ diffractional gratings with filling coefficient $u \geq 0.8$ negligibly influence the energy density of Cherenkov radiation for certain values of κ. With growing β_e and decreasing ϵ shielding effect of a grating on the Cherenkov radiation increases.

The dependence of $R_0 R_\epsilon^{-1}$ on the velocity of electron flow and parameter κ is shown on Fig. 39, b, c. It is clear that Cherenkov radiation is completely shielded by a grating for definite values of κ. The corresponding grating's parameters are easily determined by the formula (1.12) for $S_0 = 0$:

$$\kappa_\epsilon = \beta_e \frac{(u-1) + (1+u)\ln\frac{u+1}{2}}{2u + (1+u)\ln\frac{u+1}{2}}. \tag{2.13}$$

Hence, the growth of β_e yields development of the shielding effect of Cherenkov radiation for greater values of κ.

Let us carry out comparative analysis of the energetic characteristics of SPR and Cherenkov radiation. It follows from the Brilluin diagram that if $\epsilon\beta_e < 1$, the electron flow moving close to a periodical structure placed on the surface of dielectric excites only the diffractional radiation. For this case it is presented on Fig, 40 the variation of energy density of diffractional radiation in vacuum ($z < -a$) and in dielectric in the process of gradual increasing of κ for constant flow's velocity and different values of ϵ (here and on further graphs the energy density of space harmonics of diffractional radiation is normalized by $|S_\epsilon|$). It follows from the graphs that the growth of ϵ yields the fall of energy density of diffractional radiation and more intensive radiation is excited to the dielectric medium. In the considered variant the third and the fifth cases of the transformation of energy of electron flow to the diffractional radiation take place (the domains $3_{-1}, 5_{-1}^{-1}$ on the Brilluin diagram). When $\epsilon\beta_e < 1$, $|\eta+n|\epsilon^{-0.5} < \kappa < |\eta+n|$, the diffractional radiation is excited only to the dielectric medium; when $\epsilon\beta_e < 1$, $|n\beta_e|(1+\beta_e)^{-1} \leq \kappa \leq |n\beta_e|(1-\beta_e)^{-1}$, the diffractional radiation is excited simultaneously to dielectric and to vacuum. While κ grows, in the beginning (for small κ) the diffractional radiation is excited to dielectric (at the angle $\gamma_{0\epsilon} < -\frac{\pi}{2}$), then it is excited to vacuum and to dielectric (with variation of radiation angle in vacuum in the interval $0 \leq \gamma_{-1} \leq \pi$) and finally – to the dielectric medium but at a sharp angle with respect to the direction of electrons' movement (the domain 3_{-1}^{δ}). Here with growing κ the range of variation of the parameters, for which the diffractional radiation is excited in the domain 3_{-1}, increases.

The dependences of $S_{-1}S_\epsilon^{-1}$, $S_{-1\epsilon}S_\epsilon^{-1}$ on the parameter κ in the case of simultaneous excitation of diffractional and Cherenkov radiations are presented on Fig. 41. Unlike the situation considered above, the two-petal field pattern corresponding either to Cherenkov ($n = 0$) or to diffractional radiation ($n = -1$) in the domain $4_{0,-1}$ is excited in dielectric. For definite values of κ the diffractional radiation of the minus first space harmonic is excited to vacuum what causes forming the three-petal pattern of the diffractional-Cherenkov radiation (the domain $4_{0,-1}^{-1}$). Fig. 41 is the evidence of the fact that intensity of Cherenkov radiation prevails the one of diffractional radiation. With growing κ the energy density S_0 at first falls up to zero level, where the effect of total shielding of Cherenkov radiation by the grating is observed under the condition (2.13), and then it grows for $\kappa > \kappa_\epsilon$.

Let us point out that for $\kappa > (1+\sqrt{\epsilon})^{-1}$ the higher order space harmonics are also excited and their number increases with growing κ. Hence, in practical applications realization of one of the channels of excitation of volume waves should be carried out by variation of the velocity of electron flow for the given parameters of periodical structure (in particular, κ and γ).

The density of radiation energy versus velocity β_e for the principal space diffractional harmonics are presented on Fig. 42. One can see that by variation of flow's velocity and the given parameter κ it can be easily excited only the diffractional (the domain $3_{-1}^{a}, 3_{-1}^{\delta}$, $\beta_e = 0.14$), diffractional-Cherenkov (the domain $4_{0,-1}$, $0.14 \leq \beta_e \leq 0.25$) and Cherenkov (the domain 1_0, $\beta_e > 0.25$) radiations. It is also presented there the values of accelerating voltage for which these cases of radiation's excitation are realized in practice.

One may conclude that, unlike the blinking dipole theory, the solution to the problem within the frame of the model of given current gives the possibility to deeper analyze the excitation of radiation by an electron flow moving close to a diffractional grating and to establish the main properties of the process of excitation of diffractional and Cherenkov radiations.

3. Let us determine the influence of the thickness of dielectric layer on diffractional and Cherenkov radiations. The established characteristics of transformation of energy of electron flow to the volume waves of radiation on the boundary of dielectric half-space shielded by a diffractional grating remain valid for the dielectric layer having finite thickness (see Fig. 36, b). But, unlike the considered case, here the bounded size of dielectric yields propagation of radiation's energy along the layer in the direction of flow's movement due to the total internal reflection of Cherenkov radiation from the layer's boundaries in the planes $z = -a, z = -\delta$. The law of total internal reflection does not hold for the space harmonics of diffractional radiation ($n \geq 1$); therefore, diffractional radiation partially goes through the layer and propagates in the free space ($z < -\delta$) at the angle (2.2). Hence, finite thickness of dielectric presumably changes the amplitude and phase of excited

electromagnetic field what influences energetic characteristics of radiation. Variation of these characteristics may be found from the solution to the problem within the given current approximation [99].

In the general case electromagnetic field excited by the source (1.3) may be written down in the form similar to (2.1) in the domains over periodical structure (I, $z > -a$), inside the dielectric plate (II, $-\delta < z < -a$) and under the dielectric layer (III, $z < -\delta$).

Unknown field's amplitudes $A_n^\delta, B_n^\delta, D_n^\delta, F_n^\delta$ are found by the rigorous method from the solution to boundary value electrodynamical problem. In the long-wave approximation for the Cherenkov ($n = 0$) and diffractional ($n = -1$) radiations we have

$$A_{-1}^\delta = \frac{1+\sqrt{\epsilon}}{\epsilon} C_\epsilon^\delta \theta_{0\epsilon} \sin(q_{0\epsilon}\Delta - \psi)\Phi_{-1}^\delta e^{(q_{0\epsilon}\Delta-\psi)},$$

$$B_0^\delta = \frac{1}{2} C_\epsilon e^{-i(q_{0\epsilon}\Delta-\psi-\frac{\pi}{2})}\Big[(1+M^2-2M\cos\psi)^{\frac{1}{2}}\sin(q_{0\epsilon}\Delta-\phi_0)\Big]^{-1}, \quad (2.14)$$

$$B_{-1}^\delta = \frac{1}{2} V_{-1} F_{-1}^\delta; \quad D_{-1}^\delta = \frac{1}{2} F_{-1}^\delta \Big(1 - \frac{\epsilon\theta_{-1}}{\theta_{-1\epsilon}}\Big);$$

$$F_{-1}^\delta = 2A_{-1}^\delta \frac{\epsilon\theta_{-1}}{\theta_{-1\epsilon} V_{-1}} \Big(1 - f_{-1}\Big)^{-1},$$

where $C_\epsilon, C_\epsilon^\delta$ are the field's amplitudes of Cherenkov radiation for the dielectric half-space (2.7) and for the dielectric having finite thickness without diffractional grating:

$$\psi = \arctan\frac{\epsilon\theta}{\theta_{0\epsilon}}; \quad M = \frac{(1+\sqrt{\epsilon})\theta\Phi_G}{\sqrt{1+\epsilon^2\theta^2\theta_{0\epsilon}^{-2}}};$$

$$\phi_0 = \arctan\frac{\frac{\epsilon\theta}{\theta_{0\epsilon}}[2-(1+\sqrt{\epsilon})\theta\Phi_G]}{1-\left(\frac{\epsilon\theta}{\theta_{0\epsilon}}\right)^2-(1+\sqrt{\epsilon})\theta\Phi_G};$$

$$\Phi_{-1}^\delta = \frac{u-1+\zeta(u+1)\ln\frac{u+1}{2}}{1+\frac{\epsilon\theta_{-1}}{\theta_{-1\epsilon}}\frac{1+f_{-1}}{1-f_{-1}}+i\eta\theta_{-1}(1+\sqrt{\epsilon})\ln\frac{u+1}{2}};$$

$$V_{-1} = \Big(1+\frac{\epsilon\theta_{-1}}{\theta_{-1\epsilon}}\Big)e^{iq_{-1\epsilon}(a-\delta)}; \quad f_{-1} = \frac{1-\frac{\epsilon\theta_{-1}}{\theta_{-1\epsilon}}}{1+\frac{\epsilon\theta_{-1}}{\theta_{-1\epsilon}}}e^{-2iq_{-1\epsilon}(a-\delta)}.$$

Now it is easy to find the energy densities of Cherenkov and diffractional radiations with the help of (2.14):

$$S_0^\delta = \frac{1}{4} S_\epsilon \left[(1 + M^2 - 2M\cos\psi)\sin^2(q_{0\epsilon}\Delta - \phi_0) \right]^{-1}; \tag{2.15}$$

$$S_{-1}^{\delta,\mathrm{I}} = S_\epsilon^\delta \left[\frac{1+\sqrt{\epsilon}}{\epsilon} \theta_{0\epsilon} \mid \Phi_{-1}^\delta \mid \sin^2(q_{0\epsilon}\Delta - \psi) \right]^2; \quad z > -a;$$

$$S_{-1\epsilon}^{\delta,\mathrm{II}} = S_{-1}^\delta \left| \frac{\frac{\epsilon\theta_{-1}}{\theta_{-1\epsilon}}}{1 - f_{-1}} \right|^2; \quad -\delta < z < -a; \tag{2.16}$$

$$S_{-1}^{\delta,\mathrm{III}} = 4 S_{-1}^{\delta,\mathrm{I}} \left| \frac{\frac{\epsilon\theta_{-1}}{\theta_{-1\epsilon}}}{V_{-1}(1 - f_{-1})} \right|^2; \quad z < -\delta.$$

If there is no diffractional grating on the surface of dielectric plate ($u = 1$), then $S_{-1}^\delta \equiv 0$, $S_0^\delta \equiv S_\epsilon^\delta$ and electron flow excites in the dielectric medium only Cherenkov radiation with energy density

$$S_\epsilon^\delta = \frac{1}{4} \frac{S_\epsilon}{\sin^2(q_{0\epsilon}\Delta - 2\psi)}, \tag{2.17}$$

forming the flow of power in the dielectric plate (of unit width)

$$P_\epsilon = R_\epsilon J_0^2 \frac{1 + \frac{\sin q_{0\epsilon}\Delta}{q_{0\epsilon}\Delta} \cos(q_{0\epsilon}\Delta - 2\psi)}{2\sqrt{\epsilon}\sin^2(q_{0\epsilon}\Delta - 2\psi)}. \tag{2.18}$$

It follows from the formulas (2.17), (2.18) that if the grating is absent, the maximal energy density of Cherenkov radiation is excited when dielectric plate has the resonance width

$$\Delta_\epsilon = \frac{\lambda}{2} \frac{\beta_e}{\sqrt{\epsilon\beta_e^2 - 1}} \left(k + \frac{2}{\pi} \arctan \frac{\epsilon\theta}{\theta_{0\epsilon}} \right), \quad k = 0, 1, 2, \ldots. \tag{2.19}$$

If the diffractional grating with the filling coefficient $u = \cos\frac{\pi d}{l}$ is drifted on the dielectric plate, the resonance thickness of dielectric

$$\Delta_u = \frac{\lambda}{2} \frac{\beta_0}{\sqrt{\epsilon\beta_e^2 - 1}} \left(k + \frac{\phi_0}{\pi} \right). \tag{2.20}$$

Comparison of the formulas (2.19), (2.20) is evidence of the fact that the strip grating on the surface of dielectric plate changes its resonance thickness. The growth of u (or decreasing of the width of metallic strips) causes increasing of Δ_u and $\Delta_u = \Delta_\epsilon$ for $u = 1$.

One can also increase the efficiency of diffractional radiation varying the thickness of dielectric plate. It follows from (2.14)–(2.16) that resonance diffractional radiation of the space harmonic with $n = -1$ in the domain 3_{-1} is discovered when the thickness is given by the expression

$$\Delta_{-1\epsilon} = \frac{\lambda}{2} \frac{\beta_0}{\sqrt{\epsilon\beta_e^2 - (1 - \frac{1}{\eta})^2}} \left[k + \frac{1}{\pi} \arctan \frac{m(2 - M_u\sqrt{1+m^2})}{1 - m^2 - M_u\sqrt{1+m^2}} \right],$$

where

$$m = \frac{\epsilon\hat{\theta}_{-1}}{\theta_{-1\epsilon}}; \quad \hat{\theta}_{-1} = \sqrt{1 - \frac{1}{\tau_{-1}^2}};$$

$$M_u = \frac{\eta\hat{\theta}_{-1}(1 + \sqrt{\epsilon}) \ln \frac{u+1}{2}}{\sqrt{1 + m^2}}.$$

It clearly follows from Fig. 43 that resonance thickness of the plate $\Delta_u\lambda^{-1}$ essentially depends on system's parameters.

Hence, the appropriate choice of the thickness of dielectric layer may provide optimal conditions for excitation of intensive Cherenkov and diffractional radiation. These theoretical results are effectively used in experimental modeling of the processes of excitation of diffractional radiation of monochromatic flow [142] and also for investigations of the characteristics of Cherenkov radiation in the millimeter range [101].

4. In the analyzed electrodynamical systems the energy density S_0 of Cherenkov radiation mainly exceeds the energy density S_{-1} of diffractional radiation. The latter, however, may be essentially increased if to use the reflection diffractional grating [3] or other special electrodynamical systems (reflection mirrors [99], Fig. 36, c, double periodical structures, Fig. 36, d, et al).

Let us consider the influence of reflection mirror on the characteristics of diffractional radiation excited by the monochromatic electron flow moving in an open electrodynamical system (see Fig. 36, c). If dielectric medium is absent, then only pure diffractional radiation is excited in such a system. Varying the distance h between the grating and the mirror one can change intensity of diffractional radiation in the given direction. Maximal density of energy radiating out of the system (through the grating to the free space) is excited by a flow when $h \approx \frac{\lambda}{4}(2k+1)$; if $h \approx \frac{\lambda}{2}k$, the pencil's radiation energy concentrates mainly inside the system with minimal radiation to the free space [102, 103] (i. e., open electrodynamical system formed by the screen and strip grating possesses resonance properties).

Analogous changes of the characteristics of SPR with respect to varying h occur when the strip grating lies on a dielectric layer (see Fig. 36, c). However, here the control is possible in the channels of the type $4_{0,-1}^{-1}, 5_{-1}^{-1}$ when the Cherenkov radiation is excited simultaneously to the dielectric medium and resonance system (over diffractional grating, $z > -a$). When radiation is excited to the dielectric medium (the channels $1_0, 3_{-1}, 4_{0,-1}$, the reflection screen considerably influences its characteristics.

Excitation of diffractional radiation of the resonance system at the angle $\gamma_{-1} = \frac{\pi}{2}$ in the case when dielectric permittivity and the distance h between the grating and reflection mirror vary is shown on Fig. 44 (here and in what follows the values of S_{-1}^{n} within the accuracy up to norming factor $w = c\rho_0^2(8\pi)^{-1}$, i. e., the magnitude $S_{-1}(w)$, are marked along the axis of ordinates). One can see that with varying h the energy density of diffractional radiation essentially increases and the value of ϵ influences not only the radiated power but also the space distribution of S_{-1}.

One should separate on Fig. 45 the cases when the reflection screen is close to diffractional grating (Fig. 45, b, $h < 0.5\lambda$) and when it is far from the grating (Fig. 45, a, $h > 0.5\lambda$). In the first case the energy density of diffractional radiation gradually varies with growing κ in the whole range determined by the radiation condition (2.4). If $\epsilon = 1$, the power of radiation out of the system falls at a certain wavelength λ_1 with increasing h (Fig. 45, b, the curve 1, $\kappa \approx 0.1$), what is connected with the growth of resonator's q-factor at this wavelength. For $h \approx 0.5\lambda$ the q-factor of this system is maximal and radiation out of the system at this frequency is minimal, although two new maximums are developed at the edges of the range of κ corresponding to the optimal radiation out of the system at other frequencies for which the condition $h \approx 0.25\lambda(2k+1)$ holds, when resonator's q-factor is minimal. The presence of dielectric medium violates these characteristics. In the case given on Fig. 45, b, the curve 2, it is observed maximal diffractional radiation out of the system at the angle $\frac{\pi}{2}$ ($\kappa = \beta_e = 0$). In the second case ($h > 0.5\lambda$) the number of resonance frequencies in the functions determining radiation energy density versus wavelength (the parameter κ, Fig. 45, a) grows due to the condition $h = 0.5\lambda k$. Here one can change the filling coefficient of diffractional grating and, hence, resonance characteristics of electrodynamical system by varying the value of ϵ.

The degree of influence of the ϵ of dielectric substrata on the grating's filling coefficient u is shown on Fig. 46. The growth of ϵ shifts the optimal value of u in the domain of narrow metallic strips and wide slots of the grating (for $u = 1$ the stripwidth $d = 0$, i. e., the grating is absent).

It is of practical interest to consider the density of energy of diffractional radiation out of the system $(S\,S_{\max}^{-1})$ versus observation angle in the polar coordinates. Such far field patterns of SPR for the minus first space diffractional harmonic are shown on Fig. 47, where it is assumed that electron flow moves in the direction of axis Oy. The given far field patterns are evidence of the fact that, unlike the grating situated in free space [3] (where considerable part of energy is radiated close to sliding regimes), for the considered system it is characteristic resonance sharply directed radiation with growing number and spectral-space saturation of resonance petals. In fact, if for small distances between the mirrors of resonance system (Fig. 47, a) the radiation is formed as a one-petal pattern, with growing h this pattern gradually collapses and transfers to multi-petal pattern for $h > 0.5\lambda$, in particular, it transfers to the two-petal pattern for $h = 0.5\lambda$ (Fig. 47,b), to the six-petal pattern for $h = 15l$ (Fig. 47, c) and to the fan-type multi-petal pattern with opening angle of about 180° for $h = 252l$ (Fig. 47, d).

Diffractional radiation out of a resonance system for the case when strip grating is placed on the surface of dielectric is presented on Fig. 48. Far field patterns show that radiation's intensity distribution in the space depends on dielectric permittivity.

Let us point out in the conclusion that, unlike the classical far field patterns of single-frequency antenna, independent radiation wavelength corresponds to every resonance petal in the presented far field patterns of SPR. In practice one can obtain fan-type pattern of diffractional radiation if to let the electron flow modulated by wide-band signal fly through the considered resonance system [102]. The same principle may be applied for creating the antenna with broad view angle if to use in place of an electron flow the planar dielectric waveguide fed by wide-band generator [3,4].

§7. Electron-wave excitation of Smith-Purcell radiation.

1. In the general case interaction of electron flow with electromagnetic field yields forming the space-time spectrum of electron waves propagating with different phase velocities [5-8, 104]. Space harmonics of convectional current of such electron waves may be represented in the form

$$\vec{I} = \sum_m \vec{I}_m(\vec{r}) e^{i(\vec{k}_m \vec{r} - \omega t)}, \tag{2.21}$$

where $\vec{I}_m$, $\vec{k}_m = \vec{e}_m \omega v_m^{-1}$ are the amplitude and wave vector of a separate harmonic of current's density; v_m is the phase velocity of propagation of electron wave; ω is modulation frequency; t is the time parameter; $vecr$ is the coordinate and $\vec{e}_m$ is the unit vector.

Longitudinal and transversal electron waves are excited in the flow (2.21) under the influence of super-high frequency (modulating) field. SWSC and FWSC are related to longitudinal waves propagating with different phase velocities

$$\vec{v} = \frac{\vec{j}\beta_0 c}{1 \pm \omega_R \omega^{-1}}, \tag{2.22}$$

where β_0 is the relative constant component of the velocity of non-perturbed electrons equal to $v_0 c^{-1}$; ω_p, [Hz], is the plasma frequency equal to $8.137 \times 10^8 \sqrt{i_0 \beta_0^{-1}}$; i_0, [A cm^{-2}], is the current's density; ω_R is the effective (reduced) plasma frequency equal to $R\omega_p$; R is the reduction coefficient.

SCW and FCW are related to transversal waves which are excited by an electron flow focused by magnetic field. Their phase velocities

$$v_c = \frac{v_0}{1 \pm s\omega_c \omega^{-1}}, \tag{2.23}$$

where $s = 0, 1, 2, \ldots$ is the number of cyclotron wave; ω_c is the cyclotron frequency equal to $\eta_e B$; B is the induction of focusing magnetic field; e, m are the charge and mass of electron; the signs $+$, $-$ correspond to the velocities of SCW and FCW, respectively; $\eta_e = em^{-1}$.

Electron beam excited at one of transversal waves takes the form of a spiral with the step $h_s = (-1)^s 2\pi k_s^{-1}$ moving onward with the velocity v_0. When a beam is excited at cyclotron wave this spiral also rotates with the frequency ω_c. For $s = 0$ spiral's rotation is absent and the phase velocity $v_c = v_0$. Unlike longitudinal, the transversal waves have polarization properties [104]: FCW and FSEW are right-polarized, SCW and SSEW are left-polarized. In the classical devices with long interaction the efficiency of energy exchange is determined by the condition of quasi-synchronization of one or several electron waves with the field of surface wave of decelerating system.

In the devices of diffractional electronics the condition of coherent radiation must hold, in addition to the condition of quasi-synchronization of electrons. Such conditions together with energetic characteristics of diffractional radiation of electron waves may be found as a result of the solution to the problem of diffractional electronics within the frame of the model of given current. In this case an electron flow represented in the form of superposition of electron waves considerably simplifies the solution to diffractional problem, if to separately consider excitation of diffractional radiation by longitudinal and transversal waves.

2. Assume that modulation by high-frequency signal in an electron flow moving close to diffractional grating yields the formation of mainly longitudinal waves. Then the density of convectional current of the whole pencil may be represented as the sum of partial monochromatic flows

$$\vec{I} = \sum_p \vec{I}_p \delta(z) e^{-i(k_p y - \omega t)}, \tag{2.24}$$

where the parameters $\vec{I}_p, k_p, v_p$ essentially depend on the type of modulator. For the principal SWSC and FWSC the amplitudes I_p are determined by the formulas

$$I^{\pm} = \frac{1}{2}\left[\hat{\imath}(0) \mp \rho_0 \omega \frac{\hat{v}(0)}{\omega_R}\right],$$

where $\hat{\imath}(0), \hat{v}(0)$ are varying output components of current's density and electrons' velocity. If the clystron generator (resonator) is used as electrons' modulator, $\hat{\imath}(0) = 0$ at modulator's output, and if it is used the modulator with long interaction (GDR or LRW), then one must take into account in I_p the electrons' modulation with respect to velocity and charge's density ($\hat{\imath}(0) \neq 0$, $\hat{v}(0) \neq 0$).

The main characteristics of SPR of electron WSC may be found after the solution to the problem of diffractional electronics within the frame of the model of given current (2.24). They are easily determined for a separate electron wave with the help of the results of the solution to the problem on monochromatic flow [3, 105-107] if to replace the electrons' velocity v_0 and current's amplitude by corresponding values for electron WSC. In this case the analysis of radiation conditions yields excitation of diffractional radiation by SWSC and FWSC for the electrons' velocities satisfying inequalities [9]

$$\left(1 \pm \frac{\omega_R}{\omega}\right)\frac{\kappa c}{|n| + \kappa} \leq v_0 \leq \left(1 \pm \frac{\omega_R}{\omega}\right)\frac{\kappa c}{|n| - \kappa}, \tag{2.25}$$

and energy flow of this radiation is directed at the angles

$$\gamma_n^{\pm} = \arccos\left(\frac{1}{\beta_0} \pm \frac{\omega_R}{\omega} - \frac{|n|}{\kappa}\right), \tag{2.26}$$

where angle is counted counter-clockwise related to the positive direction of electrons' movement; $n = 0, -1, -2, \ldots$ is the number of space harmonic of diffractional radiation.

It follows from the condition (2.25), (2.26) that electron waves formed in the flow excite diffractional radiation at different angles. One can control the fulfillment of radiation conditions for one or simultaneously for several electron waves by varying the grating's period, electrons' velocity and modulation frequency. The difference between radiation angles essentially depends on the density of pencil's current. The general feature of theoretical behaviour of the angles of diffractional radiation of SWSC and FWSC versus current's density in the case of unit reduction coefficient $R = 1$ (Fig. 49) is the growth of radiation angle with increasing i_0. For $i \geq 10$ A cm^{-2} the difference between radiation angles of SWSC and FWSC in the four-millimeter range may reach essential values what indicates the possibility to discover these effects experimentally. Let us note that this data differs from the results presented in [6] where (according to theoretical calculations) the difference between radiation angles of these waves is equal to fractions of degree. It impedes in practice revealing of the thin structure of radiation.

Varying the system's parameters one can control radiation direction of electron WSC or provide radiation of a certain wave in the given direction. For example, diffractional radiation at the angle 90° is excited by SWSC on the grating with the period $l = |n|\beta_0\lambda(1 + \omega_0\omega^{-1})^{-1}$. Taking into account that usually in the short-wave part of millimeter range the plasma wavelength $\lambda_p >> \lambda$ we obtain

$$l = l_0 - |n|\beta_0\frac{\lambda^2}{\lambda_p} \approx l_0(1 + \frac{\omega_p}{\omega}). \tag{2.27}$$

Hence, the grating's period may differ from the value $l_0 = |n|\beta_0\lambda$ of monochromatic flow for high current's density. These estimates are used in working out the GDR of raised power.

3. Transversal electron waves appear in a flow under the influence of different factors connected with dynamical and statical shifts of the trajectory of pencil's electrons. One can also determine the influence of these waves on diffractional radiation by solving the problem within the frame of given current approximation.

Let us consider at first the problem of excitation of radiation by oscillating electron flow and then let us indicate the main features of excitation of diffractional radiation by cyclotron electron waves. Consider monochromatic electron flow with the current's density [8]

$$\rho = \rho_0\delta[(z + z_0) - R_P \cos(k_P y + \phi_0)]e^{i(ky-\omega t)} \tag{2.28}$$

moving with constant velocity $\vec{v} = \vec{j}\beta_0 c$ along the wave-type trajectory over the strip diffractional grating (Fig. 50, a). Here R_P, λ_P are the amplitude and wavelength of the pulsations of electrons' trajectory; ϕ_0 is the initial phase of flow's trajectory; $k = \omega v_0^{-1}$, $k_P = 2\pi\lambda_P^{-1}$ are the wavenumbers; z_0 is the flow's equilibrium trajectory. The rest of the notations are the same as in $ 6.

The components of current's density $\vec{I} = \vec{j}I_y + \vec{k}I_z$ are connected with charge's density (2.28) by interrelations

$$I_y = \rho\beta_0 c; \quad I_z = -\rho\beta_0 c R_P k_P \sin(k_P y + \phi_0). \tag{2.29}$$

It follows from (2.29) that synchronous ($m = 0$) slow and fast ($m \neq 0$) transversal electron waves are formed in the electron flow with wave-type trajectory, which propagate along the axis Oy with phase velocities

$$v_m = \frac{v_0}{1 \pm s\beta_0\lambda\lambda_P^{-1}}, \quad m = \pm s. \tag{2.30}$$

If there is no diffractional grating, the "eigenfield" of the flow (2.28) has the form

$$\vec{H}_c = -\vec{i}2\pi\rho_0\beta_0 \sum_m (\mp i)^m F_m e^{i\phi_m^\pm};$$

$$\vec{E}_c = -\frac{2\pi\rho_0\beta_0}{\omega} \sum_m (\mp i)^m F_m[\vec{j}(\pm q_m) + \vec{k}k_m]e^{i\phi_m^\pm}; \tag{2.31}$$

$$F_m = J_m(r)\Big(1 + mk_P\frac{k_m}{q_m^2}\Big);\ k_m = k + mk_P;\ r = q_m R_P;$$

$$q_m = k\theta_m = k\Big[\beta_0^2 - \Big(1 + m\frac{k_P}{k}\Big)^2\Big]^{\frac{1}{2}};\ \phi_m^{\pm} = \pm q_m(z + z_0) + m\phi_0 + k_m y;$$

$J_m(r)$ is the Bessel function; the signs + and − correspond to the pencil's field in the domains $z > z_0$, $z < z_0$, respectively.

In accord with formulas (2.31) the field of electron flow with curvilinear trajectory is the superposition of leaky and propagating space harmonics along the axis Oz . If

$$\beta_0^2 > \Big(1 + m\beta_0\frac{\lambda}{\lambda_P}\Big)^2, \tag{2.32}$$

the wave-type electron flow radiates "eigenoscillations" at the wavelength

$$\frac{(1-\beta_0)\lambda_P}{-m\beta_0} \le \lambda \le \frac{(1+\beta_0)\lambda_P}{-m\beta_0},\quad m = -1, -2, \ldots. \tag{2.33}$$

In this case the density of radiation energy excited by a separate transversal current's harmonic has the form

$$S_m = \frac{1}{2}\pi\beta_0^2 c\rho_0^2 J_m^2(r)\Big[1 + \frac{mk_P}{q_m^2}(k + mk_P)\Big]^2 \tag{2.34}$$

and the direction of radiation is determined by the formula

$$\phi_m = \arccos\Big(\frac{1}{\beta_0} + m\frac{\lambda}{\lambda_P}\Big). \tag{2.35}$$

If the are no pulsations in electron flow ($R_P = 0$), then $S_m = 0$, i. e., the pencil does not radiate the energy. Under the condition $\lambda_P = m\beta_0\lambda$, $m > 0$ the pencil's radiation is perpendicular to the plane $z = z_0$ and its energy density

$$S_m^- = \frac{1}{2}\pi\beta_0^2 c\rho_0^2 J_m^2(r). \tag{2.36}$$

In accord with (2.36), the flow radiates maximal energy density for the amplitudes of pulsations of electrons' trajectories $R_P = \lambda(4\pi)^{-1}(\mu_{m,k} + \mu_{m,k-1})$, where $\mu_{m,k}$, $\mu_{m,k-1}$ are the roots of the Bessel function of order m.

Energy density (2.36) versus amplitudes of pulsations of flow's trajectory for the harmonics with indices $m = -1, -2$ are given on Fig. 51. It is clear that when the amplitudes $R_P \approx 0.3\lambda$, electron flow excites the maximal energy.

The presence of periodical structure leads to scattering the flow's eigenfield (2.31) and excitation of the spectrum of space harmonics of diffractional radiation with indices m, n. In the general case the radiation electromagnetic field in the domains I ($z > -a$) and II ($z < -a$) may be represented as follows

$$\vec{H}^{\mathrm{I}} = \vec{H}_c + \vec{i}\sum_m \sum_n A_{mn} e^{i\Gamma^+_{mn}}; \; \vec{H}^{\mathrm{II}} = \vec{H}_c + \vec{i}\sum_m \sum_n B_{mn} e^{i\Gamma^-_{mn}};$$

$$\vec{E} = i\frac{c}{\omega}\mathrm{rot}\vec{H}. \tag{2.37}$$

Here

$$k_{mn} = k + mk_P + \frac{2\pi n}{l} = k\beta_0\tau_{mn}; \; \kappa_P = \frac{l}{\lambda_P}; \; \tau_{mn} = \frac{1}{\kappa}(\eta + m\kappa_P + n);$$

$$q_{mn} = k\theta_{mn} = k\beta_0\sqrt{1 - \tau^2_{mn}}; \; \Gamma^{\pm}_m = \pm q_{mn}(z + a) + k_{mn}y;$$

A_{mn}, B_{mn} are the unknown coefficients which are found from the solution to the boundary value electrodynamical problem by the method [61].

It follows from (2.37) that excitation of mn-th space harmonic of diffractional radiation is observed under the condition

$$\kappa^2 > (\eta + m\kappa_P + n)^2, \tag{2.38}$$

and the flow of radiation energy density is directed at the angle

$$\gamma_{mn} = \pm \arccos \tau_{mn}, \tag{2.39}$$

where the "current" and diffractional indices $m = 0, \pm 1, \ldots; n = 0, \pm 1, \ldots$ and $m^2 + n^2 \neq 0$.

In the long-wave approximation the energy density of diffractional radiation of transversal electron waves with index $n = -\mu$is calculated by the formulas [8]

$$S^{\mathrm{I,II}}_{m,-\mu} = \frac{\pi c}{8}\left(\frac{\beta_0\rho_0|\hat{F}_m|}{\nu_m}\right)^2 (\nu^2_m - \kappa^2) \mid \Phi_{m,-\mu} \mid^2 e^{-2\hat{q}mp}, \; \mu \neq 0; \tag{2.40}$$

$$S^{\mathrm{I}}_{m,0} = -S^{\mathrm{II}}_{m,0}(\nu^2_m - \kappa^2)\ln^2\frac{1+u}{2}; \tag{2.41}$$

$$S^{\mathrm{II}}_{m,0} = \frac{\frac{\pi c}{2}\beta_0^2 \rho_0^2 F_m^2}{1 - (\nu_m^2 - \kappa^2)\ln^2 \frac{1+u}{2}}, \tag{2.42}$$

where $p = |a - z_0|$ is the aimed parameter;

$$\Phi_{m,-\mu} = \frac{P_{\zeta_m,-\mu} P_\mu - P_{-\zeta_m,-\mu} P_{-\mu}}{1 + i\sqrt{\kappa^2 - \zeta^2_{m-\mu}} \ln \frac{u+1}{2}},$$

$$\nu_m = \eta + m\kappa_P = \mu + \zeta_{m,-\mu};$$

μ is the whole number closest to ν_m; $|\zeta_{m,-\mu}| < 0.5$; $P_\mu(u)$, $P_{\pm\zeta_m,-\mu}(u)$ are the Legendre polynomials and functions, respectively; $u = \cos \pi d l^{-1}$. If $R_P = 0$, $S^{\mathrm{I,II}}_{m,0} = 0$ and (2.42) transfers to interrelation for the monochromatic flow [3].

With the help of (2.31), (2.42) one can study the influence of the parameters of oscillating flow and electrodynamical structure on the energy density of diffractional radiation of the transversal electron waves. Let us note that if the amplitude and pulsation's wavelength in a flow are independent magnitudes, the level of radiation energy density versus R_P are determined by the values of Bessel functions $J_m(R_P), I_m(R_P)$ and by the exponent with aimed parameter p [8]. However, in practice the parameters R_P, λ_P are not independent (if an electron flow is focused by magnetic field, the amplitude R_P is proportional to λ_P). Therefore, one should know how these parameters vary when applying the formulas (2.40)-(2.42).

When an electron flow is focused by magnetic field, the energy density of diffractional radiation of transversal harmonics decrease with growing number (m, n) of space harmonics. The principal space harmonics of the current of transversal waves radiate maximal power $(m, n = 0, -1;\ 1, -1;\ -1, 1;\ -1, 0)$. Here the value $S_{0,1}$ of the current's electron waves is in several times higher than the radiation of other (slow and fast) space harmonics. At the same time diffractional radiation of zero harmonic $(m, 0)$ grows with increasing pulsations' amplitude and may exceed $S_{0,1}$ for certain values of R_P.

4. The results obtained for the wave-type electron flow may be used for the analysis of diffractional radiation of the "thick" pulsating flow (see Fig. 50, b). Such electron flows are formed in the interaction space of "O"-type devices (in particular, in GDR), where the calculation of statical electron cannon and electrodynamical structure is reduced to transformation and integration of the equations describing the movement of electrons in statical magnetic or electrical fields. In the general case, irrespective of the

type of focusing the flow, one can write down the statical electrons' trajectory in the form

$$z(y) = z_0 + R_P \cos(k_P y + \phi_0); \quad R_P = g z_0; \quad g = \text{const}, \tag{2.43}$$

where the parameter g equal to $\frac{R_g}{z_g}$ is determined by the pulsations' amplitude R_g and equilibrium half-thickness z_g of the boundary electrons. In accord with interrelation (2.38), the boundary electrons have maximal pulsations' amplitude, while the electrons moving along the flow's axis have minimal (zero, $R_P = 0$) amplitude. It is important that in such model the conditions of excitation of diffractional radiation by the transversal waves of pulsating flow are described by the same formulas (2.38), (2.39) as for oscillating electrons, and the energy density of diffractional radiation of pulsating flow may be found as a result of integrating the obtained interrelations for the energy of oscillating pencil along equilibrium electrons' trajectories (under the assumption that the flow's thickness is in many times less than its width and pencil's pulsations negligibly violate statical density of space charge). In this case determination of energy of diffractional radiation is reduced to the calculation of integrals

$$\int_{-z_1}^{z_g} J_m^2(r)\,dz_0, \quad \int_{-z_1}^{z_g} I_m^2(r) e^{-2q_m z_0}\,dz_0$$

where the upper limit of integration corresponds to z_g of boundary electrons and lower limit z_1 depends on the correlation between the aimed parameter p (distance from the grating to the lower pencil's boundary, $a = p + z$) and the amplitude of pulsations of the trajectory of boundary electrons: if $p \geq R_g$, then $z_1 = z_g$ and electron flow moves over the grating without subsidence; for $z_1 = z_g - R_g - p$ the subsidence of electrons on periodical structure is taken into account.

It follows from the analysis of the obtained formulas that, as well as for oscillating flow, the level of energy of diffractional radiation of the space harmonics with indices m, n falls with growing m, n. The most interesting from the point of view of energy characteristics is the diffractional radiation of electron waves exciting the radiation at the same levels as the monochromatic electron flow and the flow's pulsations mainly influence the energy of diffractional radiation.

It is shown on Fig. 52 the variation of the value $S_{0,-1}$ of the pulsating flow normalized with respect to the density $\bar{S}_{0,-1}$ equal to $S_{0,-1}$ for $R_P = 0$

versus the amplitude of pulsations of the trajectory of boundary electrons for different values of the aimed parameter. For $p = 0$, when the pulsations' amplitude grows (in this case the subsidence of electron flow on diffractional grating increases), the power of diffractional radiation decreases in comparison with non-pulsating flow ($R_g = 0$). The radiation power versus pulsations' amplitude has different character for the lifted flow. The power of diffractional radiation increases in comparison with non-pulsating flow when the pulsations' amplitude grows up to $R_g = p$ (electron flow touches the grating by its pulsations); however, for $R_g > p$ electron flow subsides on the grating and hence the radiation power falls with growing R_g, as well as for $p = 0$. For the lifted pencil ($qp = 0.4\pi$) the pulsations' amplitude of boundary electrons for $g = 0.8$ is 80 per cent of the flow's half-thickness and the radiation power of the space diffractional harmonic $0, -1$ has the 20 per cent growth in comparison with radiation power of non-pulsating flow. It is explained by the fact that the decrease of pencil's aimed parameter is equivalent to the growth of the pulsations' amplitude of boundary electrons.

5. Longitudinal magnetic field is used for creating strip electron flow in the space over diffractional grating, while pencil's electrons moving along spiral trajectories form the pulsating flow with the amplitude R_P and pulsations' wavelength λ_P (see Fig. 50, b). In practice the spiral movement of electrons in a thin strip flow in longitudinal plane (in the direction of the axis Oy) is modeled by wave-type trajectories with the amplitude R_g and wavelength λ_P of the pulsations of boundary electrons determined by the formulas [108]

$$R_g \approx 1.12\dot{1}0^3 \frac{\beta_0}{B}\frac{z_g}{d_1}, \quad \lambda_P \approx 10.6\frac{\beta_0}{B} = 2\pi\beta_0\frac{c}{\omega_c}, \tag{2.44}$$

where R_g, λ_P, [mm]; z_g is the equilibrium trajectory of boundary electrons; d_1 is the distance from cathode to anode of the diode cannon; B, [Tl], is the magnetic induction of focusing field. In this case the formula (2.30) transfers to the expression (2.23) for phase velocities of cyclotron electron waves and radiation condition (2.38) yields excitation of diffractional radiation by cyclotron waves for electrons' velocities satisfying inequalities

$$\left(1 \pm s\frac{\omega_c}{c}\right)\frac{\kappa}{|n| + \kappa} \leq \beta_0 \leq \left(1 \pm s\frac{\omega_c}{c}\right)\frac{\kappa}{|n| - \kappa}; \tag{2.45}$$

the direction of radiation of these waves is determined by the angle

$$\gamma_{m,n} = \arccos \frac{(1 \pm s\frac{\omega_c}{c})\eta + n}{\kappa}, \tag{2.46}$$

where the signs + and − are related to the diffractional radiation of SCW and FCW, respectively; if $m = 0$ (or $s = 0$, since $m = \pm s$), the diffractional radiation is excited by SCW.

In accord with (2.46), FCW and SCW excite diffractional radiation only for the specific value of focusing magnetic field

$$\frac{2\pi c}{\eta_e}\Big[\frac{(1-\beta_0)\kappa}{m\beta_0} + \frac{n}{m}\Big] \le B \le \frac{2\pi c}{\eta_e}\Big[\frac{(1+\beta_0)\kappa}{m\beta_0} + \frac{n}{m}\Big] \tag{2.47}$$

Angles of diffractional radiations of cyclotron electron waves as functions of magnetic field calculated by the formulas (2.46) for different values of accelerating voltage U_0 ($\lambda = 4$ mm; $\kappa = 0.1$) are given on Fig. 53, a. One can see that the angle $\gamma_{0,-1}$ of the radiation of SEW (marked by dot horizontal lines) is mainly determined by the parameters κ, β and does not depend on magnetic field, while the angle of diffractional radiation of FCW grows (point lines) and the angle $\gamma_{1,-1}$ of SCW decreases with growing B. Here one must reduce U_0 for excitation of the diffractional radiation of FCW and increase the velocity of electron flow in the case of excitation by SCW.

Fig. 53 illustrates the character of variation of radiation resistance $R_{0,-1}(m = 0, n = -1)$ calculated by (2.40) for SEW generated by focusing magnetic field. With growing B the energy density of diffractional radiation increases up to the maximal value and then negligibly changes due to harder focusing and moving the lower boundary of electron flow off the diffractional grating.

It follows from the formulas (2.39), (2.45) that the number of cyclotron electron waves exciting the diffractional radiation considerably depends on the interrelation between pulsations' wavelength λ_P and the wavelength λ of the HF modulation of electron flow. It is shown on Fig. 53, a the variation of the space splitting of radiation far field pattern caused by the excitation of higher order transversal (cyclotron) waves under the condition $\kappa = \beta_0$ and with growing the wavelength of pulsations λ_P. One can see that for $\lambda_P < \lambda$the diffractional radiation is mainly excited by SEW; for $\lambda_P > \lambda$this radiation is also excited by SCW and FCW. The radiation of cyclotron electron waves occurs at constant frequency but at different angles with respect to the direction of flow's movement.

We point out in the conclusion that if the diffractional grating is absent, the cyclotron waves of the electron flow modulated by HF signal, unlike WSC, also excite eigenradiation at the frequencies divisible by ω_c in the range

$$\frac{|m|\omega_c}{1+\beta_0} \leq \omega \leq \frac{|m|\omega_c}{1-\beta_0}. \tag{2.48}$$

Such radiation is observed for magnetic fields

$$B \approx \frac{107}{|m|\lambda}(1 - \beta_0 \cos\phi_m), \tag{2.49}$$

where B, [Tl]; $\phi_m = \gamma_{m0}$ is the angle of cyclotron radiation (2.35).

If $\phi_m = 90°$, cyclotron radiation is observed at the frequency $\omega = m\omega_c$ and (2.49) transfers to the known classical formula for MCR-type devices [109] where modulation of electron flow is provided in the process of electrons' interaction with the field of eigenoscillations of OR. For $\phi_m = 0$ cyclotron radiation is directed along the pencil and it is excited when the magnetic fields are in $1 - \beta_0$ times less than in the case of $\phi_m = 90°$.

6. Let us separate the parameters of electron flow and periodical structure for which diffractional radiation is excited by longitudinal and transversal electron waves. The domains of variation of these parameters, where different cases of excitation of electromagnetic waves are realized, are determined from the Brilluin diagrams (Fig. 54) obtained with the help of radiation conditions (2.25), (2.38), (2.45).

The system's parameters from the non-shaded domains bounded by the lines $\beta = 1$ and by the beams of sectors' angles with apices $\eta = \eta_p$, $1 \pm \eta_p$ (marked by dot lines) on Fig. 54, a correspond to the fulfillment of radiation condition for SWSC and FWSC; SCW and FCW are observed in the sectors with apices $\eta = s\kappa_P$, $\eta = 1 \pm s\kappa_P$. One can easily find the system's parameters for which he diffractional radiation is excited by one ($\kappa < \kappa_1$) or simultaneously by several longitudinal and transversal electron waves. The point A on Fig. 54, a corresponds to the case of excitation of SWSC, FWSC and SEW; here SWSC excite the radiation directed at a sharp angle, FWSC – at an angle greater than 90° and SEW – in the direction perpendicular to grating's surface.

One can determine the interval of variation of the velocities of electron waves for which the diffractional radiation is excited by means of the modified

Brilluin diagram on Fig. 54, b. The parameters on dot lines correspond to pencil's radiation perpendicular to the grating; the value of radiation angle γ_{mn} are related to the following beams: $\gamma_{mn} = \pi$ for $\beta = \beta_1^{\pm}$, $\beta = a_1$, $\gamma_{mn} = 0$ for $\beta = \beta_2^{\pm}$, $\beta = a_2$ ($\beta_1^+ = a_1 b^+, \beta_2^+ = a_2 b^+$; $\beta_1^- = a_1 b^-, \beta_1^- = a_1 b^-$; $b^{\pm} = 1 \pm s\nu$; $a_1 = \kappa(|n| - \kappa)^{-1}, a_2 = \kappa(|n| + \kappa)^{-1}$; $\nu = \omega_{p,c}\omega^{-1}$). Six characteristic domains (I – VI) are indicated on this diagram which are bounded by the lines $\beta = \beta_{1,2}^{\pm}$; $\beta = a_{1,2}$ intersecting in the points $A(\beta = \kappa|n|^{-1}), \nu = \kappa|n|^{-1})$, $B(\beta = a_1, \nu = 2a_1)$, $C(\beta = a_2, \nu = 2a_2)$ and correspond to excitation of diffractional radiation by one or simultaneously by several electron waves. The intervals of variation of the parameters β_0, ν and of the focusing field B where this radiation is possible are given in Table 2. This data are evidence of excitation of diffractional radiation in the domain I simultaneously by all low order electron waves; similarly, in the domain II radiation is excited only by SWSC or SCW, in the domain III – by FWSC or FCW. In the millimeter range one has $\omega >> \omega_p$, $\omega >> \omega_c$, hence, WSC excite diffractional radiation for $\frac{\omega_p}{\omega} << 1$, i. e., in the beginning of the domains I, II, III, V and VI; cyclotron and synchronous waves excite radiation both for strong ($B > B_A$) and weak ($B < B_A$) magnetic fields. It is of practical interest to consider the excitation of radiation by FCW (the domain III) which is realized for comparatively low accelerating voltages (or velocities of electron flow).

The given Brilluin diagrams enable one to determine parameters of the grating and electron flow which provide excitation of diffractional radiation by a given current's electron wave. We have not considered here polarization characteristics of radiation in the case of varying structure of electron flow. These effects will be studied experimentally in what follows although they may be also considered within the model of given current by representing a source as a superposition of spiral electron waves propagating close to periodical structure.

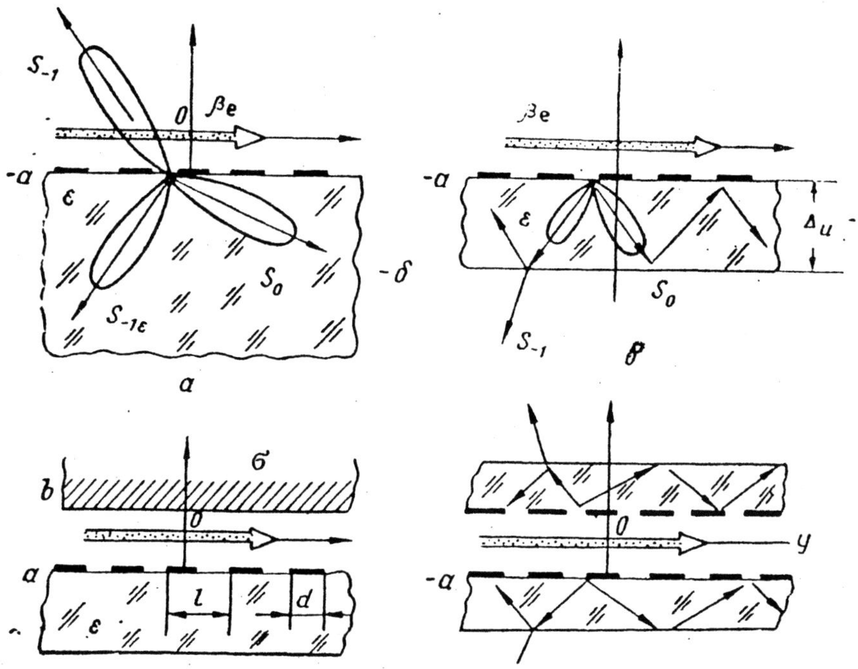

Fig. 36. Theoretical models of the excitation of Cherenkov radiation and Smith-Purcell radiation.

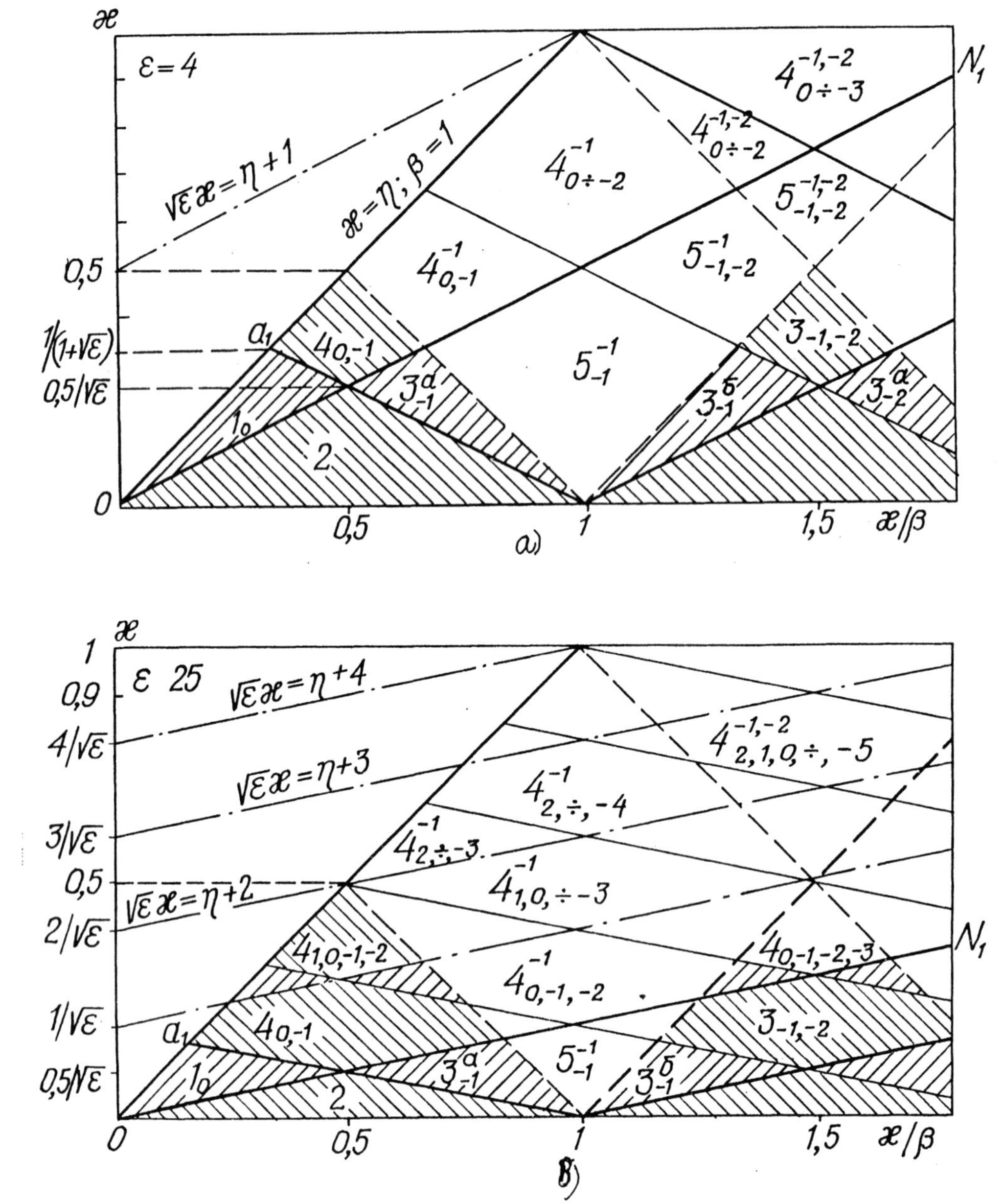

Fig. 37. Brillouine diagrams for the excitation of Cherenkov radiation and diffractional radiation: a) – $\epsilon = 4$; b) – $\epsilon = 25$.

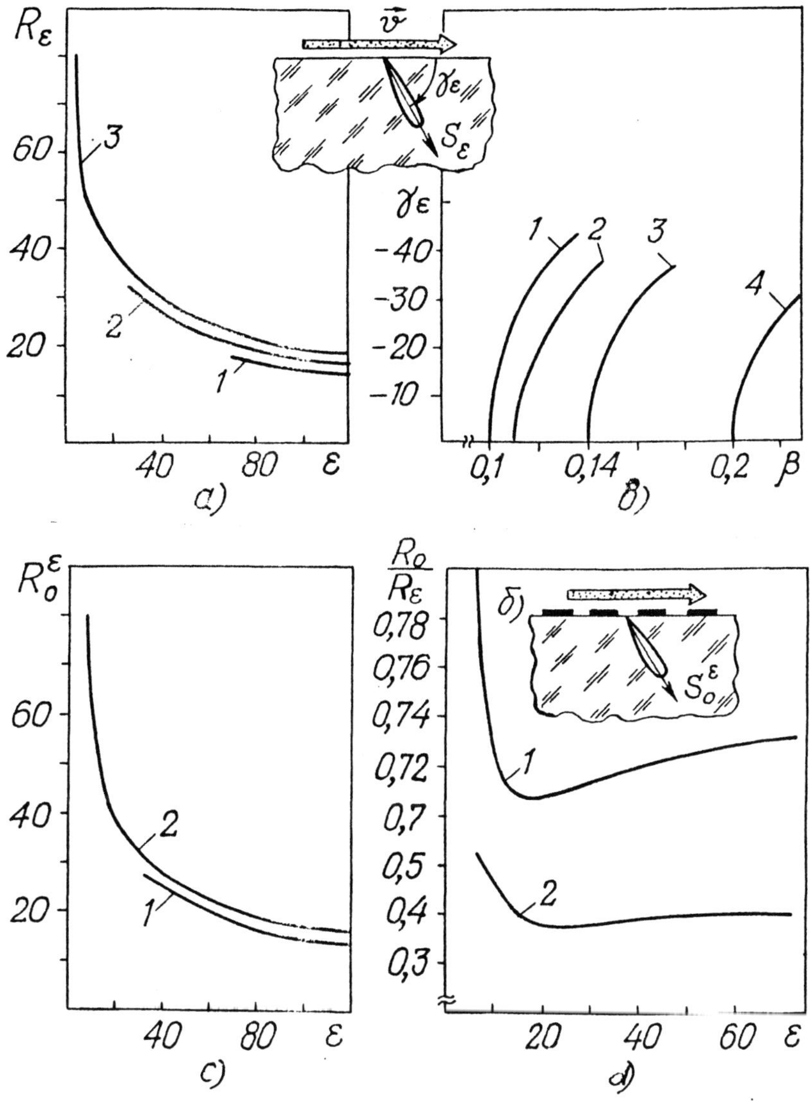

Fig. 38. Influence of dielectric permittivity on the energy density of Cherenkov radiation for the dielectric which is non-shielded ($u = 1$, a, b) or shielded ($u = 0$, c, d) by the diffractional grating: a) 1 – $\beta_e = 0.12$; 2 – $\beta_e = 0.20$; 3 – $\beta_e = 0.50$;

b) 1 – $\epsilon = 100$; 2 – $\epsilon = 75$; 3 – $\epsilon = 50$; 4 – $\epsilon = 25$;

c) 1 – $\kappa = 0.2$, $\beta_e = 0.20$; 2 – $\kappa = 0.2$, $\beta_e = 0.50$;

d) 1 – $\kappa = \beta_e = 0.50$; 2 – $\kappa = \beta_e = 0.50$ for $u = -0.8$.

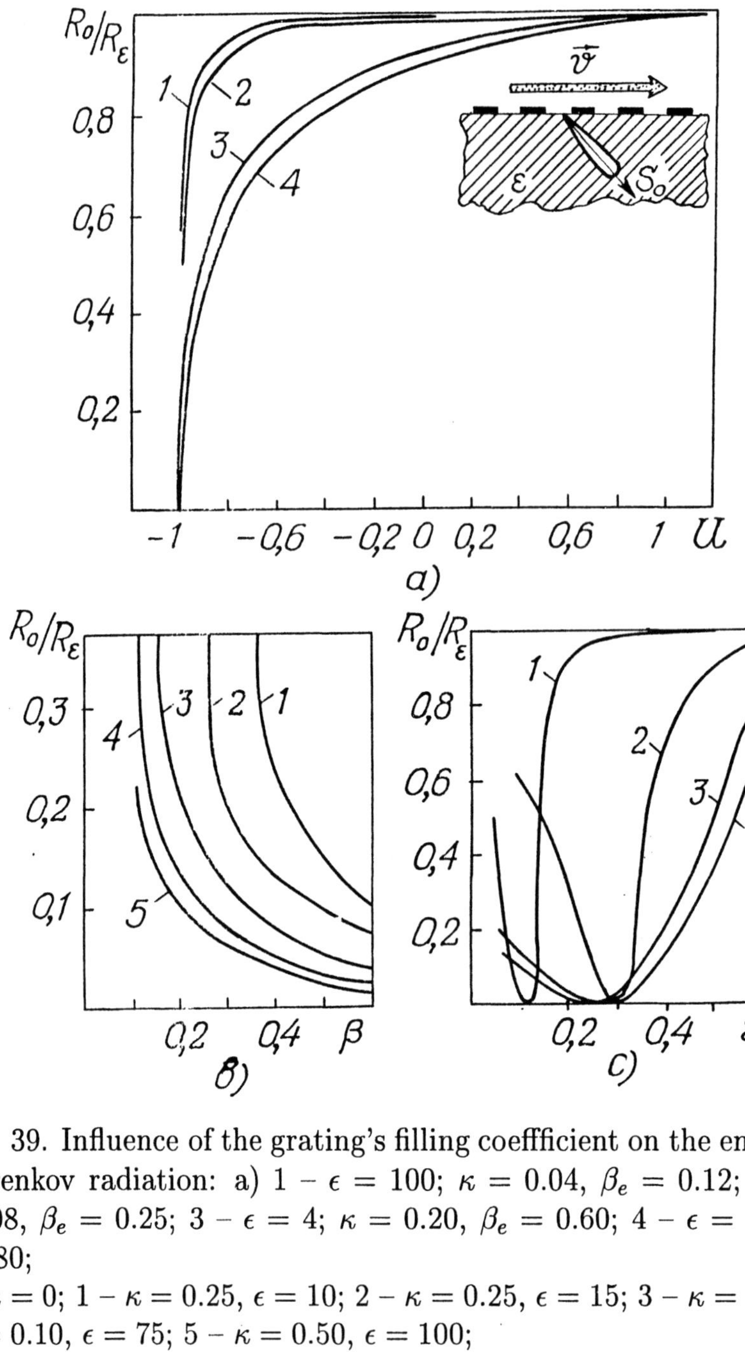

Fig. 39. Influence of the grating's filling coeffficient on the energy density of Cherenkov radiation: a) 1 – $\epsilon = 100$; $\kappa = 0.04$, $\beta_e = 0.12$; 2 – $\epsilon = 25$; $\kappa = 0.08$, $\beta_e = 0.25$; 3 – $\epsilon = 4$; $\kappa = 0.20$, $\beta_e = 0.60$; 4 – $\epsilon = 2$; $\kappa = 0.20$, $\beta_e = 0.80$;

b) $u = 0$; 1 – $\kappa = 0.25$, $\epsilon = 10$; 2 – $\kappa = 0.25$, $\epsilon = 15$; 3 – $\kappa = 0.10$, $\epsilon = 50$; 4 – $\kappa = 0.10$, $\epsilon = 75$; 5 – $\kappa = 0.50$, $\epsilon = 100$;

c) $u = -0.8$; 1 – $\beta_e = 0.10$, $\epsilon = 125$; 2 – $\beta_e = 0.25$, $\epsilon = 20$; 3 – $\beta_e = 0.25$, $\epsilon = 75$; 4 – $\beta_e = 0.25$, $\epsilon = 50$.

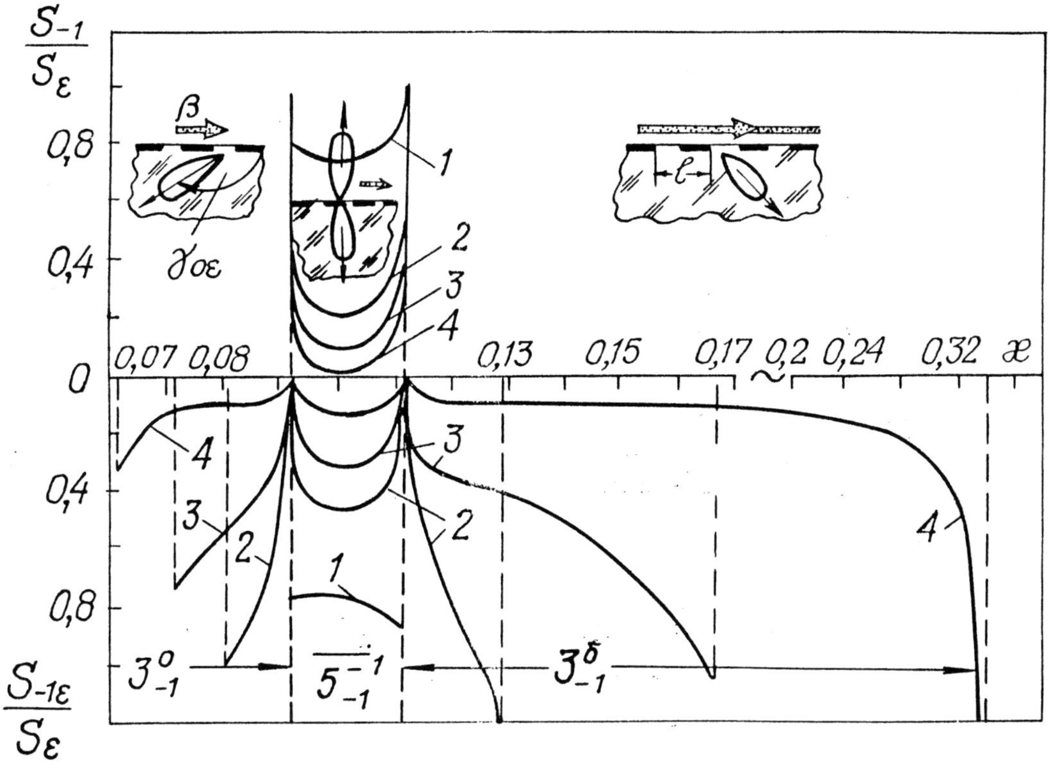

Fig. 40. Channels of the transformation of energy of electron flow to the "pure" Smith-Purcell radiation when κ varies for $\beta_e = 0.1$, $d = 0.8l$ and different values of $\epsilon = 1\,(1)$, $5\,(2)$, $15\,(3)$, $50\,(4)$.

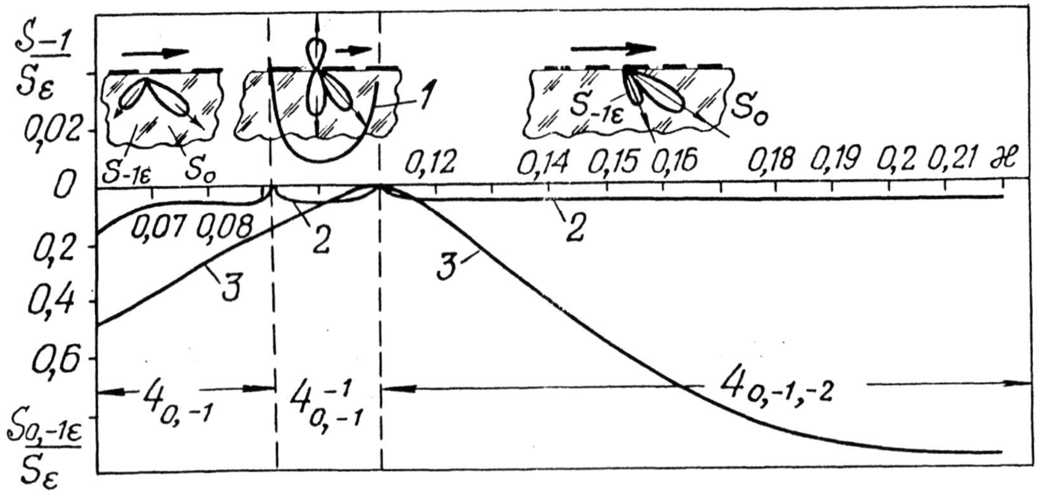

Fig. 41. Energy density of diffractional radiation and Cherenkov radiation versus κ ($\epsilon = 125$; $d = 0.8l$; $\beta_e = 0.1$; $\epsilon\beta_e^2 > 1$): 1 – diffractional radiation in vacuum (S_{-1}); 2 – diffractional radiation in dielectric ($S_{0,-1\epsilon}$); 3 – Cherenkov radiation (zero harmonic of the diffractional radiation S_0).

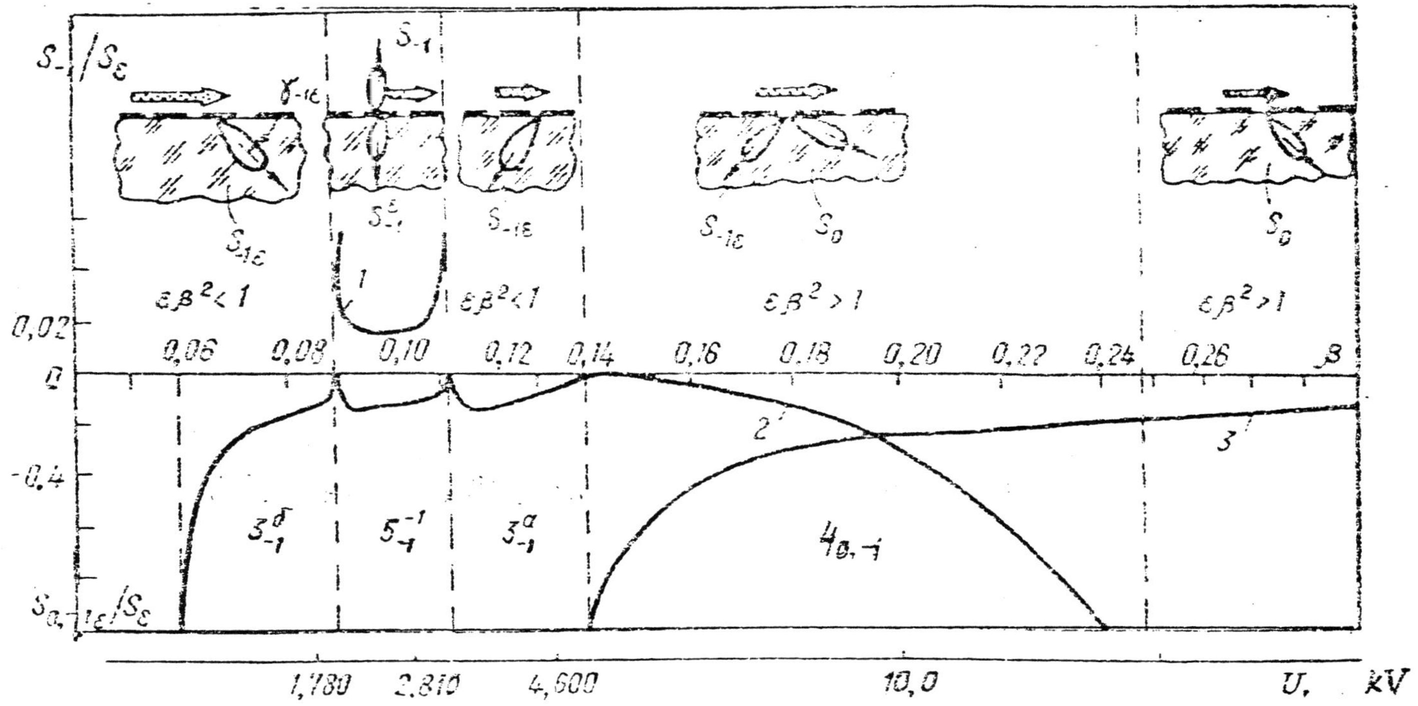

Fig. 42. Transformation of energy of electron flow to Smith-Purcell radiation, diffractional-Cherenkov radiation and Cherenkov radiation in the case of varying pencil's velocity ($\kappa = 0.1$; $d = 0.8l$; $\epsilon = 50$): 1 – energy of diffractional radiation to vacuum (S_{-1}); 2 – energy of diffractional radiation to dielectric ($S_{-1\epsilon}$); 3 – energy of Cherenkov radiation (zero diffractional harmonic, S_0).

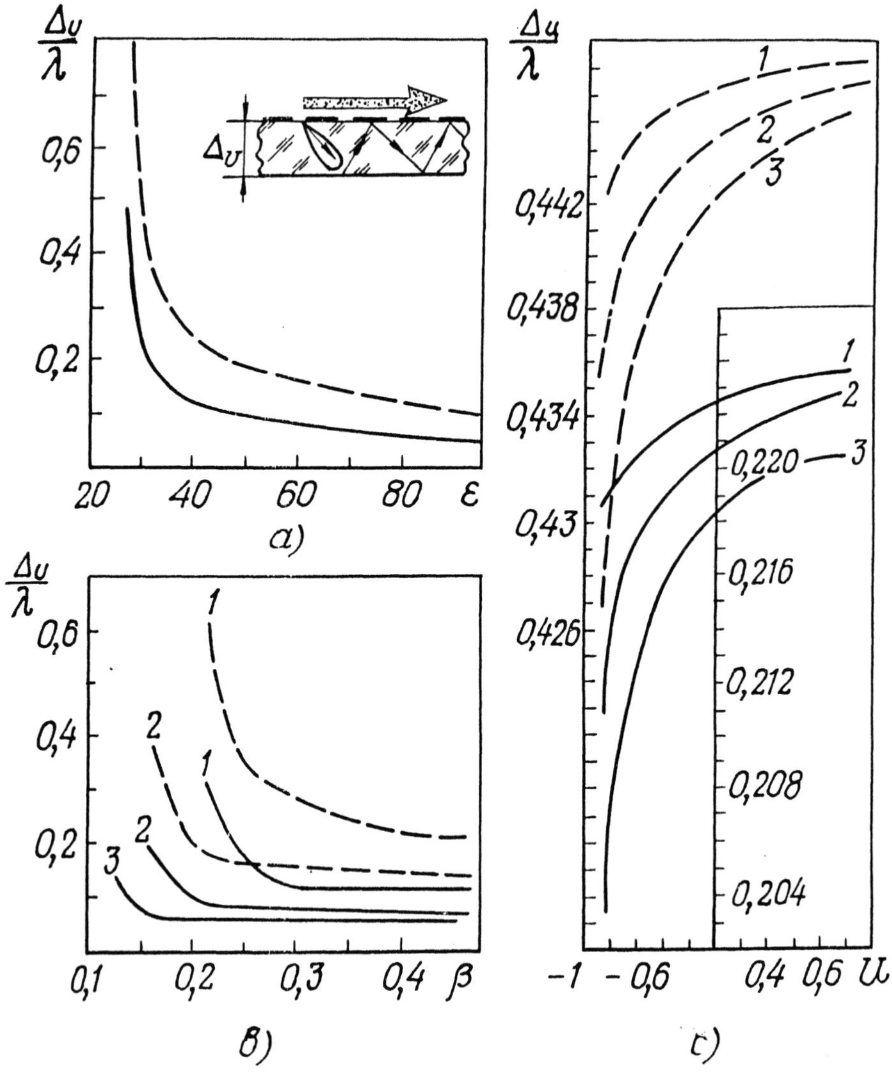

Fig. 43. Influence of the parameters of the grating and dielectric on the effective resonance thickness of dielectric plate:

a) $u = 0$; 1 – $\kappa = 0.5\epsilon^{-0.5}$; $\beta_e = 0.2$;

b) $u = 0$; $\kappa = 0.5\epsilon^{-0.5}$; 1 – $\epsilon = 25$; $\beta_e = 0.2$; 2 – $\epsilon = 50$; $\beta_e = 0.2$; 3 – $\epsilon = 100$; $\beta_e = 0.2$;

c) 1 – $\epsilon = 105$; $\beta_e = 0.1$; 2 – $\epsilon = 49$; $\beta_e = 0.15$; 3 – $\epsilon = 30$; $\beta_e = 0.2$; $k = 1$ – bold curve, $k = 2$ – dot curve.

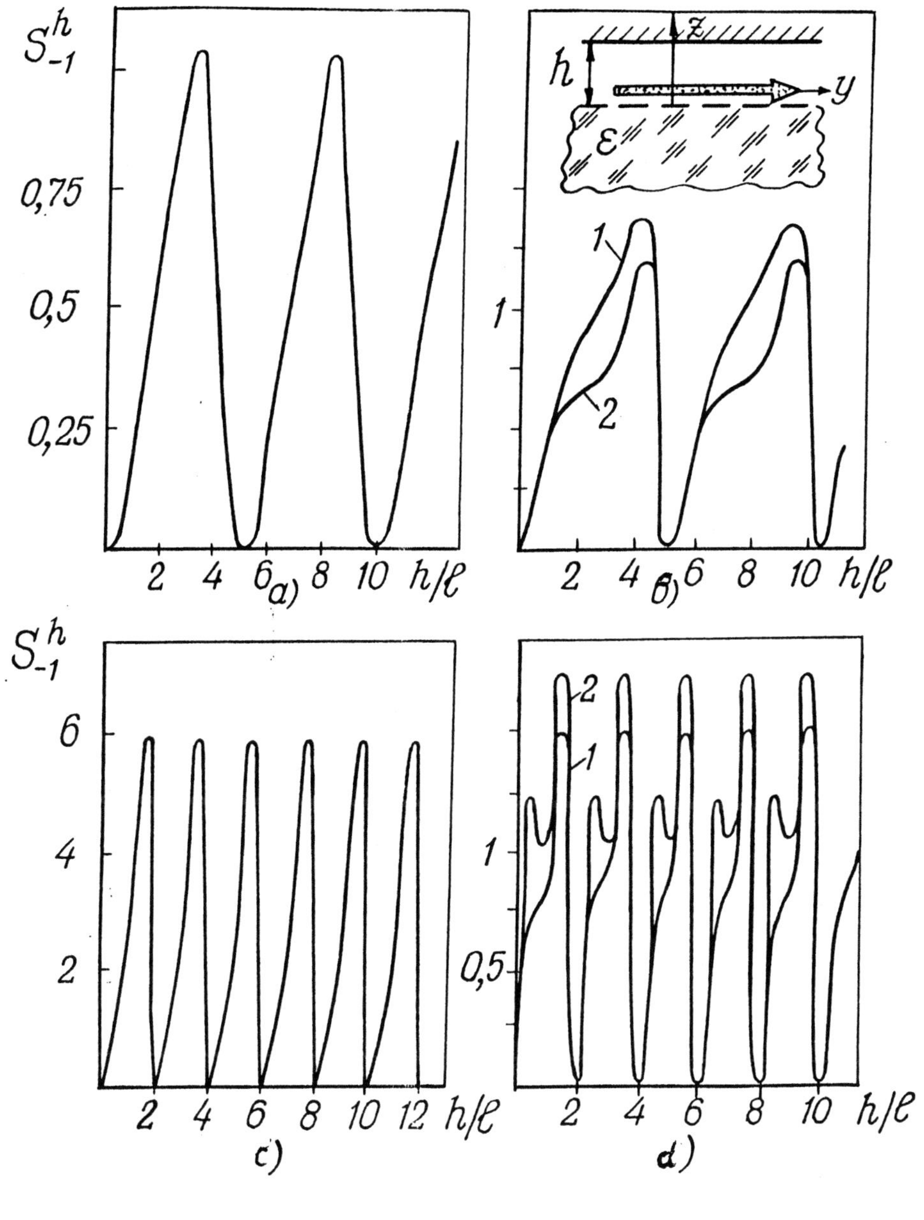

Fig. 44. Energy of diffractional radiation out of resonance system versus distance between reflection screen and grating: a) – $\kappa = \beta_e = 0.1$; $\epsilon = 1$; $u = 0.8$; b) – $\kappa = \beta_e = 0.1$; $\epsilon = 2\,(1),\ 3\,(2)$; c) – $\kappa = \beta_e = 0.25$; $\epsilon = 1$; d) – $\kappa = \beta_e = 0.25$; $\epsilon = 2\,(1),\ 3\,(2)$.

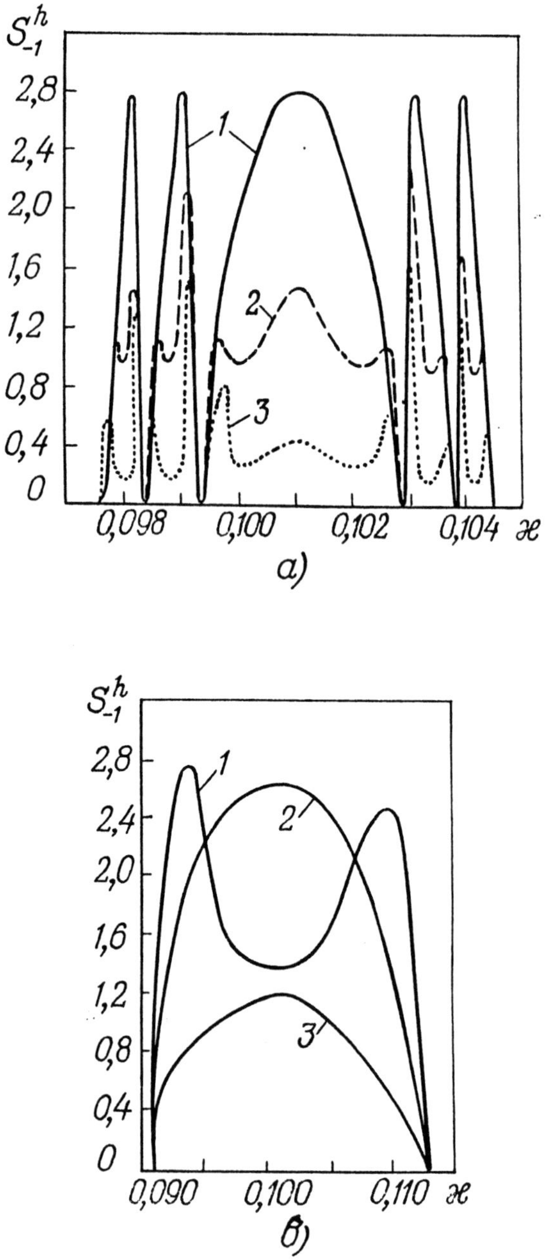

Fig. 45. Energy of Smith-Purcell radiation out of open resonance system versus parameter κ: a) – $h = 0.25l$; $\beta_e = 0.1$; $u = -0.8$; $\epsilon = 2\,(1)$, $4\,(2)$, $8\,(3)$; b) – $h = 3.5l$; $\beta_e = 0.1$; $u = -0.8$; $\epsilon = 1\,(1)$, $2\,(2)$, $4\,(3)$;

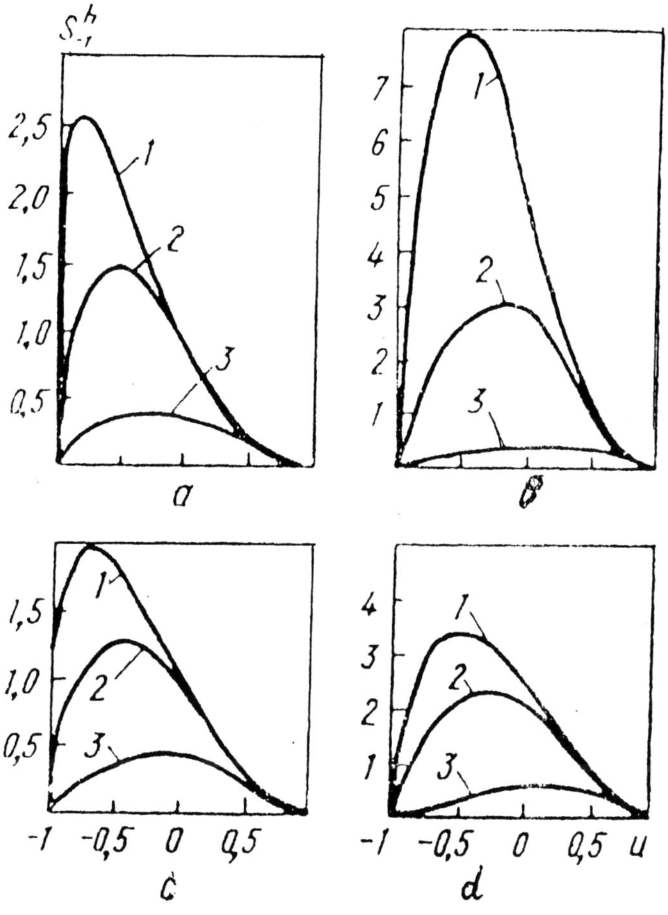

Fig. 46. Influence of the grating's filling coefficient on Smith-Purcell radiation out of resonance system for different values of ϵ: a) – $\kappa = \beta_e = 0.1$; $h = 3.5l$; b) – $\kappa = \beta_e = 0.25$; $h = 3.5l$; c) – $\kappa = \beta_e = 0.1$; $h = 252l$; $\epsilon = 2\,(1),\ 4\,(2),\ 10\,(3)$.

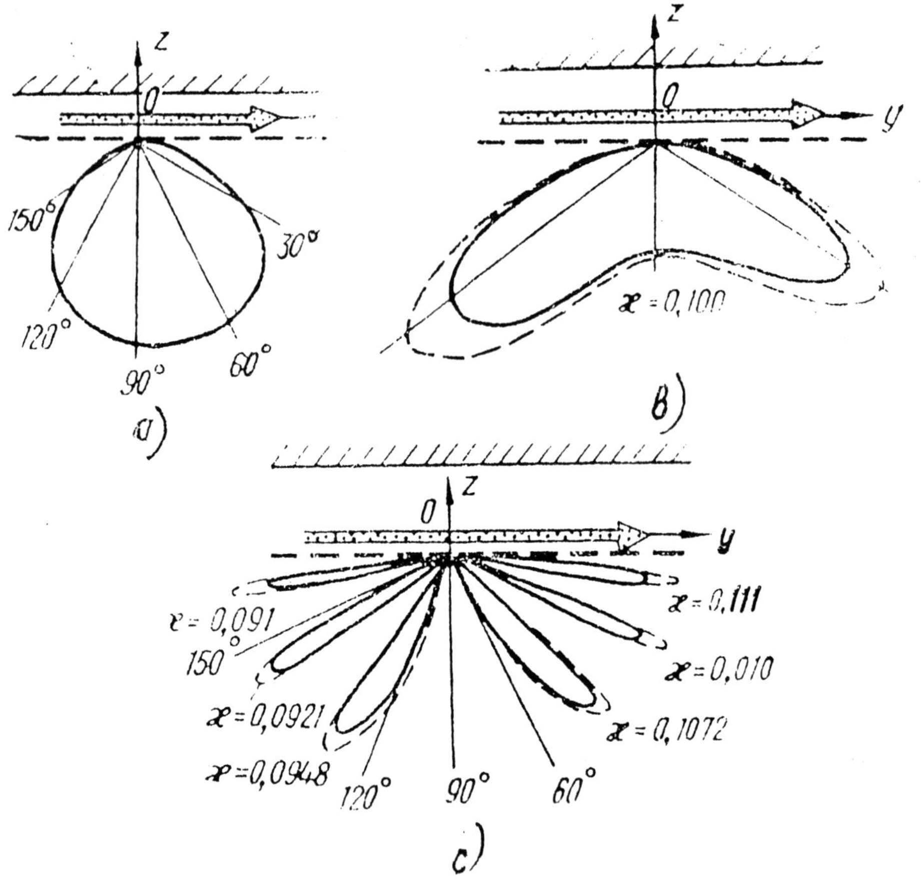

Fig. 47. Influence of the distance between reflection screen and grating on the radiation far field pattern of diffractional radiation: a) – $h = 2.2l$; $\beta_e = 0.1$; $u = -0.8$;

b) – $h = 3.5l$; $\beta_e = 0.1$; $u = -0.8$;

c) – $h = 15l$; $\beta_e = 0.1$; $u = -0.8$;

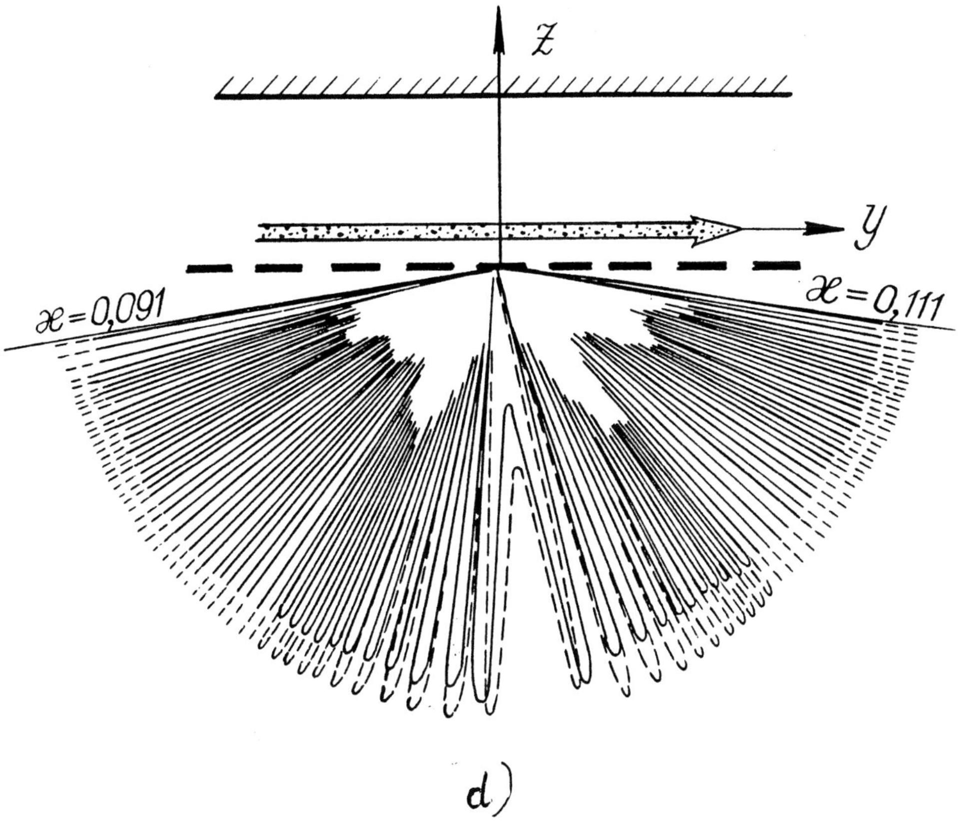

Fig. 47. (continued) d) multi-petal fan-type far field pattern of diffractional radiation, $h = 252l$; $\beta_e = 0.1$; dot lines correspond to infinitely conducting screen, bold lines – to the screen with finite conductivity.

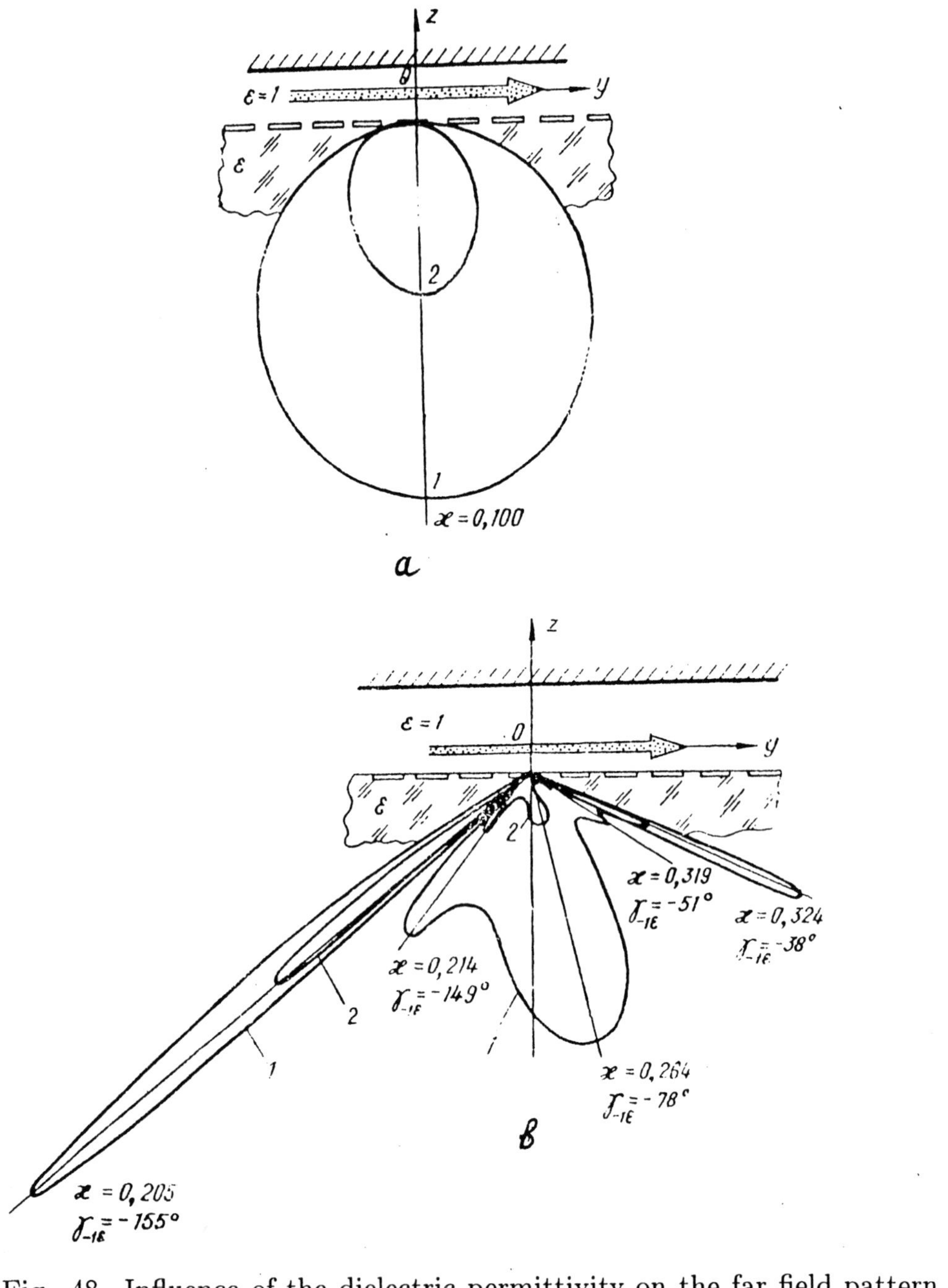

Fig. 48. Influence of the dielectric permittivity on the far field pattern of Smith-Purcell radiation out of resonance system: a) – $\beta_e = 0.1$; $u = -0.8$; $\epsilon = 2\,(1),\ 4\,(2)$; b) – $\beta_e = 0.1$; $u = -0.8$; $\epsilon = 6\,(1),\ 8\,(2)$.

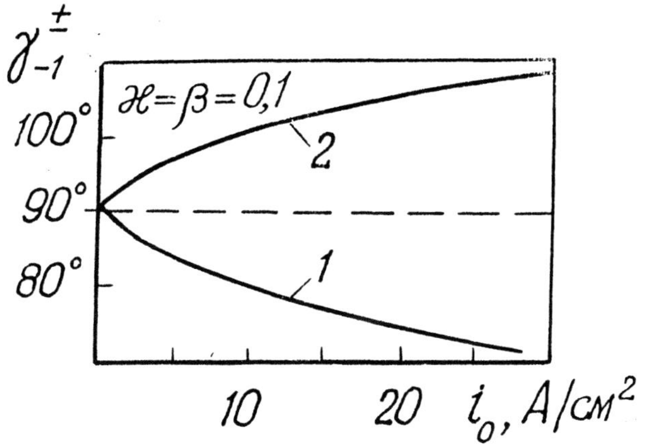

Fig. 49. Influence of the current's density on the angle of Smith-Purcell radiation: SWSC (1), FWSC (2) ($\kappa = \beta_e = 0.1$).

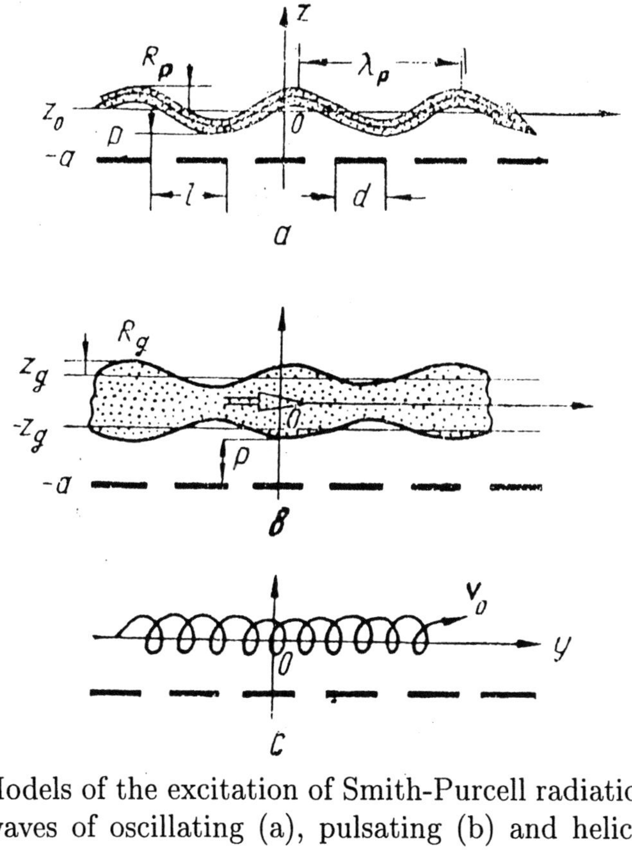

Fig. 50. Models of the excitation of Smith-Purcell radiation by transversal electron waves of oscillating (a), pulsating (b) and helical (c) electron flows.

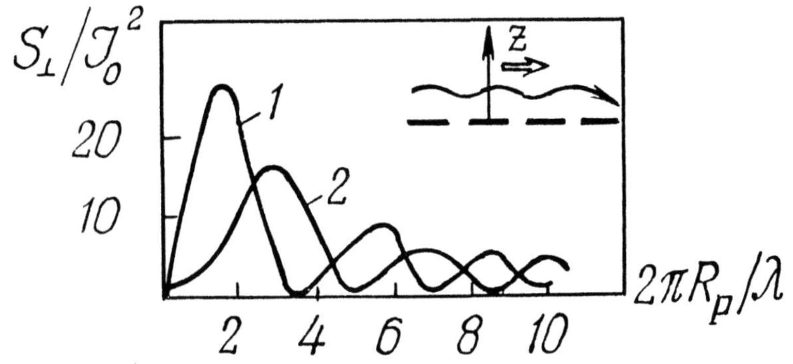

Fig. 51. Energy of "eigenoscillation" of oscillating electron flow in the free space versus pulsations' amplitude for the harmonics with indices $m = -1\,(1),\ -2\,(2)$ ($\lambda_P = -m\beta_0\lambda$, radiation angle 90°).

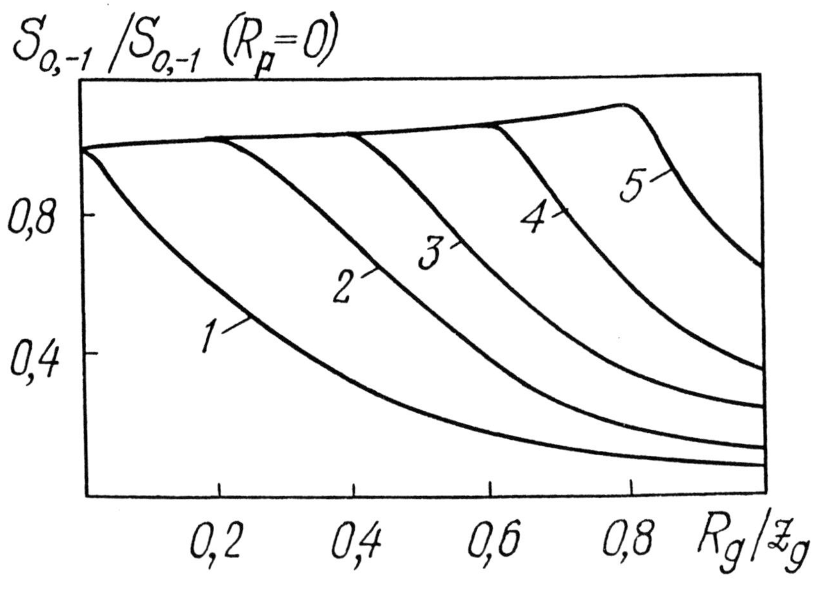

Fig. 52. Influence of the pulsations' amplitude of boundary electrons on the energy of diffractional radiation of SEW: 1 – $qp = 0$; 2 – $qp = 0.1\pi$; 3 – $qp = 0.2\pi$; 4 – $qp = 0.3\pi$; 5 – $qp = 0.4\pi$.

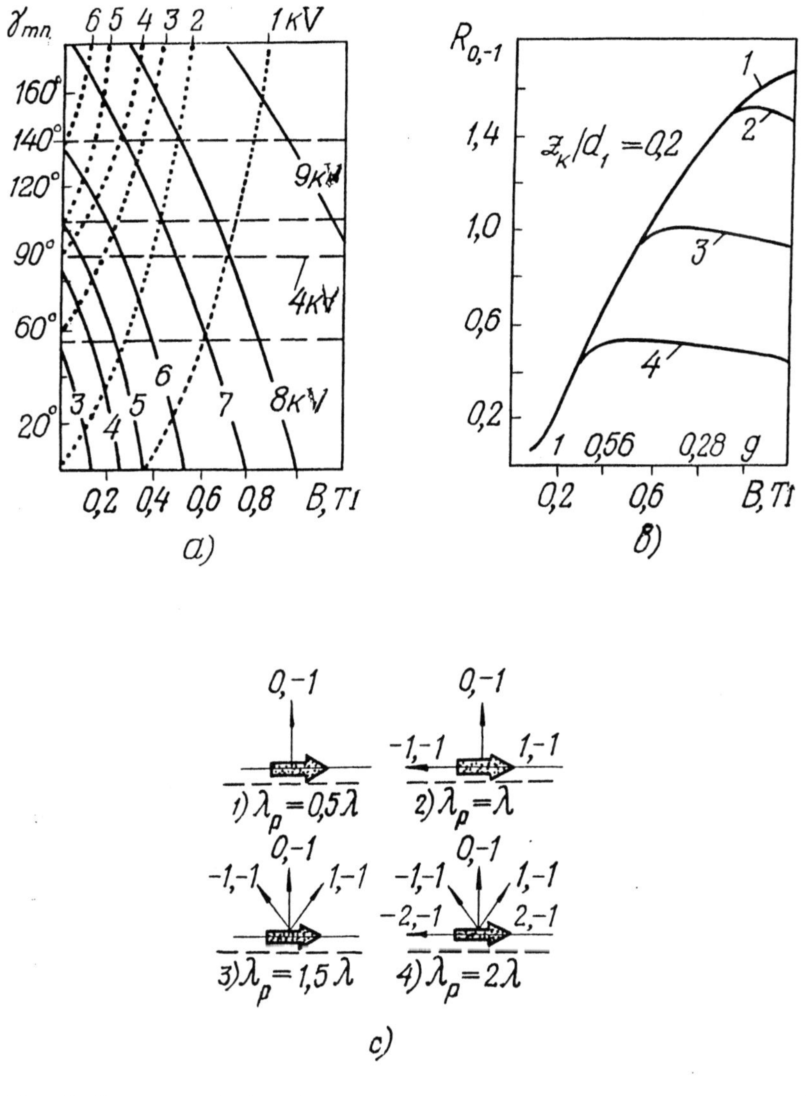

Fig. 53. Influence of the focusing magnetic field on the angle (a), energy (b) and splitting (c) of the far field pattern of Smith-Purcell radiation of transversal waves: a) – curves are marked by the values of accelerating voltage: b) – $z_C d^{-1} = 0.2$; 1 – $qp = 0$; 2 – $qp = 0.1\pi$; 3 – $qp = 0.2\pi$; 4 – $qp = 0.3\pi$.

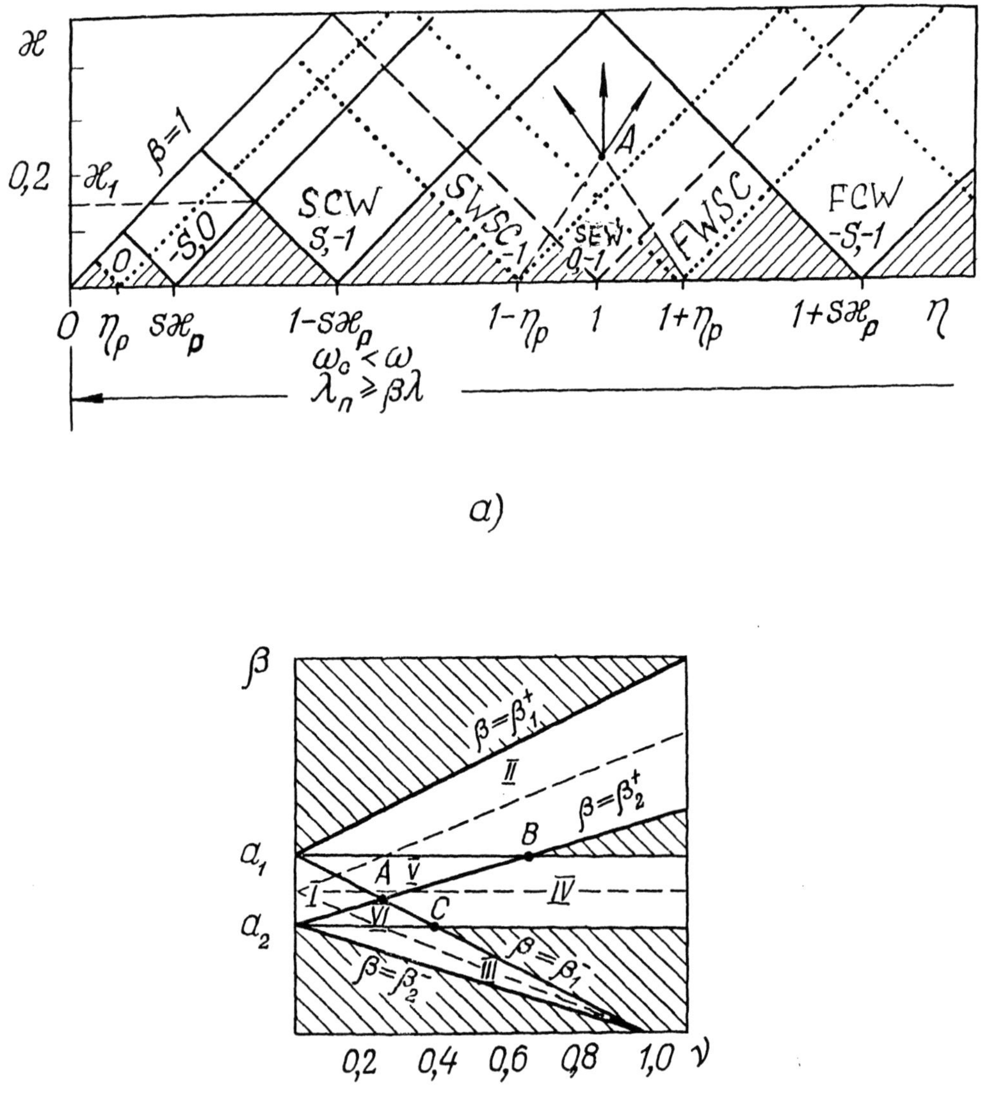

Fig. 54. Brillouine diagrams in the cases of excitation of of Smith-Purcell radiation by transversal (a) and longitudinal (b) electron waves.

Table 2. Parameters of the excitation of differactional radiation by longitudinal and transversal (cyclotron) electron waves.

Domain	Exciting Wave	Electrons' velocity β_0	Parameter $\nu = W_{w,c}/W$	Magnetic Field B, [Tl]
I	SWSC, FWSC, SCW, FCW, SSEW, FSEW	$\beta^+_2 \le \beta_0 \le \beta^+_1$	$\nu \le x/\|n\|$	$\beta \le \beta_A$
II	SWSC, SCW	$\beta^+_2 \ge \beta_0 \ge \alpha_1$	$\nu \le 2\alpha_1$	$\beta \le \beta_B$
		$\beta^+_1 \ge \beta_0 \ge \beta^+_2$	$\nu > 2\alpha_2$	$\beta > \beta_B$
III	FWSC, FCW	$\beta^-_2 \le \beta_0 \le \alpha_2$	$\nu \le 2\alpha_2$	$\beta < \beta_C$
		$\beta^-_2 \le \beta_0 \le \beta^-_1$	$2\alpha_2 \le \nu-1$	$\beta_A \le \beta \le \beta_C$
IV	SSEW,FSEW	$\beta^+_2 \ge \beta_0 \ge \beta^+_1$	$x/\|n\| \le \nu \le 2\alpha_2$	$\beta_A \le \beta \le \beta_C$
		$\beta^+_2 \ge \beta_0 \ge \alpha_2$	$2\alpha_2 \le \nu \le 2\alpha_1$	$\beta_C \le \beta \le \beta_B$
		$\alpha_2 \le \beta_0 \le \alpha_1$	$\nu \ge 2\alpha_1$	$\beta \ge \beta_B$
V	SWSC,SCW,SSEW, FSEW	$\beta^-_1 \le \beta_0 \le \alpha_1$	$\nu \le x/\|n\|$	$\beta < \beta_A$
		$\beta^+_2 \le \beta_0 \le \alpha_1$	$x/\|n\| \le \nu \le 2\alpha_1$	$\beta_A \le \beta \le \beta_C$
VI	FWSC, FCW, SSEW, FSEW	$\alpha_2 \le \beta_0 \le \beta^+_2$	$\nu \le x/\|n\|$	$\beta < \beta_A$
		$\alpha_2 \le \beta_0 \le \beta^-_1$	$x/\|n\| < \nu \le 2\alpha_2$	$\beta_A \le \beta \le \beta_C$

f – radiation frequency, [GHz]; $\beta_A = 3.6 * 10^{-2}\, x/|n|$; $\beta_B = 7.1 * 10^{-2}\, \alpha_1 f$; $\beta_C = 7.1 * 10^{-2}\, \alpha_2 f$;

CHAPTER THREE

Wave modeling of Smith-Parcell radiation

Various properties of charged particles are discovered owing to the electromagnetic field generated in the process of their movement. These electromagnetic waves of different ranges give the data not only about the properties of moving particles but also about those material objects which cause the radiation or its mechanisms.

Radiation of waves by a particle occurs when it non-uniformly moves in vacuum (brake, cyclotron and other radiation phenomena), uniformly shifts in the space filled by the definite medium (the Cherenkov effect) or in free space in the presence of obstacles (diffractional radiation). Observation of radiation phenomena occurring for real particles in natural environment or in the laboratory will entail great difficulties connected with organizing the experiment itself where first of all one must provide availability of the source of particles and indication of their radiation field.

The problem of pure wave modeling without participation of a particle in the process of forming the radiation arises in the studies and practical applications of the considered effects. It turns out that such statement of the problem is possible for the Cherenkov effect and SPR since these radiation phenomena have kinematic character; generated radiation depends on the value of a charge, its velocity and properties of the medium. Radiation effects of a uniformly moving particle may be treated as phenomena of diffraction and interference of electromagnetic waves. We consider radiation processes as transformation of particle's eigenfield to the field of radiation by representing the moving charged particle as a source of surface electromagnetic field and deflecting our attention from the particle itself. Particle's eigenfield is created artificially by various electrodynamical structures which can form surface fields specified in advance.

Microwave modeling enables one to state and to solve a number of physical problems ("freezing" in time of a transient act of radiation; modeling of radiation processes not only for real but also for hypothetical particles, etc) together with problems of diffractional electronics: wave modeling of the effect of diffractional radiation and constructing on this basis the "cold" purely wave model of GDR. It opens new perspectives for the studies of SPE and in diffractional electronics itself because one can "test" the phenomena occurring in generators (amplifiers) using appropriate models.

Numerous theoretical works dealing with transformation of the surface waves of dielectric waveguide to volume waves with the help of diffractional grating are mainly based on verifying characteristic equations for such open electrodynamical structures and calculating their roots. By this approach, based on the solution to homogeneous Maxwell equations leading to dispersion characteristics, one can analyze eigenregimes of the considered structures, namely, the types of waves (modes) together with phase velocities and to classify oscillations from the point of view of the character of their radiation to free space or of their propagation in waveguide systems. The roots of corresponding dispersion equations enable one to determine phase velocity of a waveguide mode in a section containing inhomogeneity.

This approach does not give the possibility to determine energetic characteristics of electromagnetic oscillations existing in the considered structures together with many other physical properties of the phenomena arising in the process of propagation of non-homogeneous waves of dielectric waveguides in the presence of various inhomogeneities.

Information about the behaviour of inhomogeneous electromagnetic waves interacting with inhomogeneities has of principle character in many cases for creating functional units and elements containing the source of inhomogeneous wave and a scatterer.

The rigorous mathematical methods of the diffraction theory are applied in this chapter for studying the interaction of inhomogeneous electromagnetic fields with normal and beveled strip grating; the scattered fields, in particular, those fields whose wavelengths are commensurable with structures' dimensions [4], are analyzed on the basis of obtained solutions.

Statement of electrodynamical problem is based on the diffractional method: the field of inhomogeneous wave propagating in free space is assumed to be given and obstacles brought into this field yield diffraction (scattering) of the given wave by inhomogeneity; the field scattered by obstacles should be found.

Unlike the method of dispersion equations the diffraction problems give interrelations which are determined by the given incident field of inhomogeneous plane wave and grating's parameters. The use of diffractional approach allows to effectively apply rigorous mathematical apparatus developed in the theory of diffraction of usual plane waves [61]. Mathematically correct solution to corresponding boundary value problems enables one to take into account variation of the parameters in wide range without limitations.

The present diffractional approach envisages both in the statement of electrodynamical problem and in practical use of the considered structures the presence of an external scattering obstacle in the field of inhomogeneous wave. In this case the waveguide structure of surface wave remains regular and existing distance from the source of inhomogeneous wave to the obstacle (aimed distance) enables one to vary the degree of connection in the system source–scatterer. Obvious advantage of the functional units with external scatterers in comparison with analogous elements of integrated optics, where the source of incident field and a scatterer is a single whole, is their manufacturing simplicity.

It is essential that rigorous method of the theory of diffraction of usual plane waves (the method of Riemann-Hilbert problem [61]) is used in the solution of stated problem. The latter is reformulated as the boundary value problem for a certain analytical function of complex variable which is reduced then to the infinite system of linear algebraic equations (SLAE) of the second kind with respect to unknown Fourier coefficients of the sought for scattered field.

Explicit analytical solutions will be given for a number of particular cases important in practical applications. Numerical data, which allows to choose optimal parameters, will be presented for arbitrary parameters of electrodynamical structure.

The main goal here is the study of the dynamics of transformation of surface waves to volume waves by the plane strip metallic gratings whose elements are placed perpendicular to the axis of a plane waveguide together with investigation of the physical properties of volume waves.

As it is shown in §§1–6, the energy of SPR does not depend on the mass of moving particle; the energy is determined by its velocity and charge and by the parameters of periodical structure. The number of radiated harmonics and their radiation angles at a given wavelength does not depend on a specific type of diffractional grating and they are constant for all periodical structures. The character of radiation is determined by the type of inhomogeneity.

Such properties of SPR formed the background for the construction of its wave model. Let us present some proofs of the possibility of creating this model.

The eigenfield of the plane monochromatic infinitely thin unbounded electron flow moving uniformly and straightforwardly with the velocity v has the form of inhomogeneous plane wave (1.4) given in §1. Similar wave may be formed by the specially chosen transmission lines supporting surface waves [3,4], for example, by the plane infinite dielectric waveguide with

$$H_x^{gb} = \frac{\epsilon}{uw_2} \sin \beta' e^{-a'\left(\frac{z}{a'}-1\right)} e^{-i\gamma y' - i\omega t} \tag{3.1}$$

where $w_2 = \sqrt{\frac{\mu_2}{\epsilon_2}} w_0$ is the wave resistance of external medium, $w_0 = 120\pi$ – of the vacuum; $u = c(v_\Phi \sqrt{\epsilon_2 \mu_2})^{-1}$ is the deceleration coefficient; α', β' are the external and internal wavenumbers, $R^2 = \alpha'^2 + \beta'^2$, $R = \pi d' \sqrt{\epsilon' \mu' - \epsilon_2 \mu_2}$ is the absolute waveguide's size, $d' = 2a'\lambda^{-1}$ is the relative waveguide's size.

If to express from (3.1) the field's tensity with respect to longitudinal, y, and transversal, z, coordinates versus relative waves' phase velocity $\beta = v_\Phi c^{-1}$ and to compare them with waves' eigenfields of the flow (1.4), one can conclude that they coincide as functions of y and have some divergence as functions of z caused by the finite transversal dimension of dielectric waveguide. Practical realization of the plane dielectric waveguide with infinite thickness is rather difficult. Hence, one should use bounded dielectric waveguides, waveguides having rectangular or elliptic cross-section. Such waveguides have greater differences of their transversal fields' distributions in comparison with the differences occurring for plane dielectric waveguides. However, they are not so important for modeling diffractional radiation, since here only the value of radiated power versus aimed distance may change. Phase characteristics of SPR for the wide class of excitators of plane waves remain invariable.

§8. Wave model of Smith-Purcell effect (scattering of waves of dielectric waveguide by strip diffractional gratings placed perpendicular to the waveguide)

1. Let the strip grating (with the period l and stripwidth d) is placed in the domain $x < 0$ perpendicular to the axis Oz at the distance b from the planar dielectric waveguide (the angle between every strip and the axis is 90°). The origin of the coordinate system x, y, z is situated in the arbitrary point on waveguide's axis and the axis Ox goes through the middle of the strip (Fig. 55). Decelerated wave propagates along the waveguide with the width $2a$ and dielectric permittivity ϵ_1; its electromagnetic field has the form

for E-wave

$$H_y^0 = ikg\epsilon_1 \left\{ \begin{array}{l} \cos ga \\ \sin ga \end{array} \right. e^{p(x+a)} e^{i\gamma z}, \tag{3.2}$$

for H-wave

$$E_y^0 = -ikg \left\{ \begin{array}{l} \cos ga \\ \sin ga \end{array} \right. e^{p(x+a)} e^{i\gamma z},$$

Let us consider the problem of diffraction of inhomogeneous wave (3.2) on a plane strip grating.

The given electromagnetic wave is scattered by a periodical inhomogeneity while propagating along the planar waveguide and the scattered field arises on both sides of grating's surface. The problem is to determine this field. Let us indicate the domain $-a - b < x < -a$ by the symbol I and the domain $x < -a - b$ – by the symbol II. Linearity of Maxwell equations allows to look for the solution in the form of total field for both domains: $\vec{E} = \vec{E}^0 + \vec{E}^s$, $\vec{H} = \vec{H}^0 + \vec{H}^s$ for the domain I, $\vec{E} = \vec{E}^s$, $\vec{H} = \vec{H}^s$ for the domain II, where $\vec{E}^0, \vec{H}^0$ is the known incident field, $\vec{E}^s, \vec{H}^s$ is the sought for scattered field. The obtained solution is completely determined by the parameters of incident field of inhomogeneous plane wave and by the properties of infinite periodical grating. Here it is not taken into account inverse influence of the scattered field on the source's eigenwave.

Let us consider the case when the total field is determined by the single non-zero E_y-component from interrelations

$$E_y = -ikg \left\{ \begin{array}{l} \cos ga \\ \sin ga \end{array} \right. e^{p(x+a)} e^{i\gamma z} + E_y^s; \;\; E_x = E_z = 0. \tag{3.3}$$

Here and in what follows $p = \sqrt{\gamma^2 - k^2}$; $g = \sqrt{k^2\epsilon_1 - \gamma^2}$ are the propagation constants along the axes Ox and Oz, respectively. The sought for function

E_y^s must satisfy homogeneous Helmholtz equation $\Delta v + k^2 v = 0$ (function v indicates either E_y^s or H_y^s) in the domains I, II and the boundary value condition $E_y = 0$ on metallic strips. The field is continuous out of the strips. Hence, the problem of scattering the inhomogeneous plane wave of magnetic type propagating along the planar dielectric waveguide on a strip grating is reduced to the Dirichlet boundary value problem for Helmholtz equation.

When the E-type waves propagate along the waveguide, the total field is described by the single non-zero H_y-component

$$H_y = ikg\epsilon_1 \begin{Bmatrix} \cos ga \\ \sin ga \end{Bmatrix} e^{p(x+a)} e^{i\gamma z} + H_y^s; \quad H_x = H_z = 0. \tag{3.4}$$

The sought for function H_y^s must satisfy homogeneous Helmholtz equation and the Nuemann boundary value condition $\frac{\partial H_y}{\partial x} = 0$ on metallic strips; the total field must be continuous out of the strips. In this case the scattering problem is reduced to the Nuemann boundary value problem for Helmholtz equation.

It is known that these boundary value problems may have several mathematically correct solutions. One can show that necessary and sufficient conditions providing uniqueness of their solutions are radiation conditions (absence of waves coming from infinity in the scattered field) and finiteness of field's energy in every bounded volume. The latter define the field's behaviour close to sharp edges of grating's strips. All these conditions together with periodicity of the considered obstacle along the axis Oz enable us to obtain constructive solution to the problems formulated above and to prove their existence and uniqueness.

Taking into account the expression (3.1) for incident field we represent the sought for scattered field in the form of Fourier series

$$E_y^{s(\mathrm{I})} = -ikg \begin{Bmatrix} \cos ga \\ \sin ga \end{Bmatrix} e^{-pb} \sum_{n=-\infty}^{\infty} a_n e^{ip_n(x+a+b)} e^{i\gamma_n z};$$

$$E_y^{s(\mathrm{II})} = -ikg \begin{Bmatrix} \cos ga \\ \sin ga \end{Bmatrix} e^{-pb} \sum_{n=-\infty}^{\infty} b_n e^{-ip_n(x+a+b)} e^{i\gamma_n z}; \tag{3.5}$$

$$H_y^{s(\mathrm{I})} = -ikg\epsilon_1 \begin{Bmatrix} \cos ga \\ \sin ga \end{Bmatrix} e^{-pb} \sum_{n=-\infty}^{\infty} \tilde{a}_n e^{ip_n(x+a+b)} e^{i\gamma_n z};$$

$$H_y^{s(\mathrm{I})} = -ikg\epsilon_1 \left\{ \begin{matrix} \cos ga \\ \sin ga \end{matrix} \right. e^{-pb} \sum_{n=-\infty}^{\infty} \tilde{b}_n e^{-ip_n(x+a+b)} e^{i\gamma_n z}; \qquad (3.6)$$

where $a_n, b_n, \tilde{a}_n, \tilde{b}_n$ are the sought for complex Fourier coefficients; $\gamma_n = \gamma + 2\pi n l^{-1}$ are the propagation constants of the Fourier-components of scattered field along the axis Oz; $p_n = \sqrt{k^2 - \gamma_n^2}$ are the propagation constants of the scattered field along the axis Ox. Here and in what follows $\sqrt{A}$ indicates the branch satisfying the condition $\Im\sqrt{A} > 0$ (if $\sqrt{A} = 0$, then $\Re\sqrt{A} > 0$). It is easy to show that representations (3.5), (3.6) satisfy all the abovementioned uniqueness conditions. Other components of scattered field for both types of the waves of incident field are found from the Maxwell equations. The time factor $e^{-i\omega t}$ is omitted. We will not prove here the uniqueness theorem because it is analogous to the proof given in [61].

The fields (3.5), (3.6) are the infinite sets of plane homogeneous and inhomogeneous waves going off the grating at certain angles in the domains $x > 0$, $x < 0$. In accord with the choice of root's branch one can separate these infinite sets into two types. If $\Im\sqrt{A} = 0$, p_n is real and for corresponding parameters of the scatterer and harmonic's number n such field's harmonics are the superposition of homogeneous plane waves. Let us determine the conditions of their existence.

It is easy to see that $\Im\sqrt{A} = 0$ when $(\gamma + 2\pi n l^{-1}) < k^2$, i. e., when radiation condition holds. Since the problem of diffraction of the given inhomogeneous wave with $\gamma > k$ is considered, it is obvious that only the harmonics with $n < 0$ satisfy this condition. Their propagation angles are found from the interrelation

$$|\cos\theta_n| = \frac{1}{k}\left(\gamma - \frac{2\pi n}{l}\right) \leq 1. \qquad (3.7)$$

which is the required radiation condition. For all other non-negative n one has $|\cos\theta_n| > 1$ and these waves form the set of inhomogeneous plane waves existing close to the grating and propagating along the axis Oz with phase velocities $v_\Phi < c$. One can conclude that when inhomogeneous plane wave of a planar waveguide is scattered by periodical scatterer in the form of a strip grating, the field of diffraction is a superposition of plane waves; a part of them goes to the free space at the angles determined by (3.7) in the form of a spectrum of volume waves and others are localized close to the scatterer as a spectrum of slow harmonics. One can guarantee the prevalence of the definite type of waves by choosing the parameters of electrodynamical structure and the surface wave propagating along dielectric waveguide (l, γ, k, n). Hence , the considered structure may act as a transformator of inhomogeneous waves of planar waveguides localized in the vicinity of a waveguide into homogeneous plane waves going to free space and it will be a wave analogue of SPR. Here the sector of the angles of radiation of volume plane waves is $0^\circ \leq \theta_n \leq 180^\circ$ ($0^\circ \leq \theta_n \leq 90^\circ$ for the dielectric half-space). This structure has good perspectives for practical applications.

2. Let us determine now the unknown complex amplitudes $a_n, b_n, \tilde{a}_n, \tilde{b}_n$.

Electrical field's component vanishes on perfectly conducting metallic strips and it is continuous in the slots what yields for $x = -a-b$ the equalities holding for all z

for H-wave

$$-ikg\left\{\begin{matrix}\cos ga\\ \sin ga\end{matrix}\right. e^{-pb}e^{i\gamma z} - ikg\left\{\begin{matrix}\cos ga\\ \sin ga\end{matrix}\right. e^{-pb}\sum_{n=-\infty}^{\infty} a_n e^{i\gamma_n z} =$$

$$-ikg\left\{\begin{matrix}\cos ga\\ \sin ga\end{matrix}\right. e^{-pb}\sum_{n=-\infty}^{\infty} b_n e^{i\gamma_n z};$$

for E-wave

$$-ip\epsilon_1\left\{\begin{matrix}\cos ga\\ \sin ga\end{matrix}\right. e^{-pb}e^{i\gamma z} - ig\epsilon_1\left\{\begin{matrix}\cos ga\\ \sin ga\end{matrix}\right. e^{-pb}\sum_{n=-\infty}^{\infty} p_n\tilde{a}_n e^{i\gamma_n z} =$$

$$ig\epsilon_1\left\{\begin{matrix}\cos ga\\ \sin ga\end{matrix}\right. e^{-pb}\sum_{n=-\infty}^{\infty} p_n\tilde{b}_n e^{i\gamma_n z}$$

and these equalities yield

$$a_0 + 1 = b_0; \qquad a_n = b_n;$$

$$1 - \tilde{a}_0 = \tilde{b}_0; \qquad \tilde{a}_n = \tilde{b}_n; \;\; n \neq 0. \tag{3.8}$$

Continuity of magnetic field's components in the slots for $x = -a - b$ gives the following interrelations

for H-wave

$$-pe^{i\gamma z} - i\sum_{n=-\infty}^{\infty} p_n a_n e^{i\gamma_n z} = i\sum_{n=-\infty}^{\infty} p_n b_n e^{i\gamma_n z};$$

for E-wave

$$e^{i\gamma z} + \sum_{n=-\infty}^{\infty} \tilde{a}_n e^{i\gamma_n z} = \sum_{n=-\infty}^{\infty} \tilde{b}_n e^{i\gamma_n z}. \tag{3.9}$$

Taking into account (3.8), (3.9) and the condition of vanishing the electrical field's components on metallic strips we obtain two infinite systems of coupled summation-type equations of the first kind having trigonometrical kernels with respect to unknowns b_n, $\tilde{a}_n$:

for H-wave

$$\begin{cases} \sum\limits_{n=-\infty}^{\infty} b_n e^{i\gamma_n z} = 0; & |z| < \dfrac{d}{2}; \\ \sum\limits_{n=-\infty}^{\infty} p_n b_n e^{i\gamma_n z} = ipe^{i\gamma z}; & \dfrac{d}{2} < |z| < \dfrac{l}{2}; \end{cases} \tag{3.10}$$

for E-wave

$$\begin{cases} \sum\limits_{n=-\infty}^{\infty} \tilde{a}_n e^{i\gamma_n z} = 0; & \dfrac{d}{2} < |z| < \dfrac{l}{2}; \\ \sum\limits_{n=-\infty}^{\infty} p_n \tilde{a}_n e^{i\gamma_n z} = ipe^{i\gamma z}; & |z| < \dfrac{d}{2}. \end{cases} \tag{3.11}$$

It follows from the systems of equations (3.10), (3.11) that one should treat in details only one of them since the solution is analogous in both cases; in addition to this, each solution may be obtained from another one if to replace slots by strips or vice versa due to the reciprocity principle. We will consider then only the system (3.11).

The obtained infinite system of functional equations with respect to the unknown Fourier coefficients of scattered field is of little use for numerical analysis and analytical investigations because of its low conditionality. However, one can regularize this system by applying the procedure of semi-inversion developed and justified in [61] for the problem of diffraction of usual plane waves. In this way let us introduce the following parameters

$$\nu = \frac{\gamma l}{2\pi}; \quad \kappa = \frac{kl}{2\pi}; \quad u = \cos\frac{\pi d}{l}; \quad \zeta = \frac{2\pi z}{l}.$$

Then

$$p_n = \frac{2\pi}{l}\sqrt{\kappa^2 - (n+\nu)^2} = \frac{2\pi i}{l}(1 - \epsilon_n)|n + \nu|, \tag{3.12}$$

where

$$\epsilon_n = 1 + i\sqrt{\frac{\kappa^2}{(n+\nu)^2} - 1}.$$

Representation of p_n in the form (3.12) enables one to separate the statical (singular) part of the operator of system (3.11) and its completely continuous part, i. e., to carry out the first stage of regularization. The system (3.11) takes the form

$$\begin{cases} \sum_{n=-\infty}^{\infty} \tilde{a}_n |n+\nu| e^{i(n+\nu)\zeta} = \frac{pl}{2\pi} e^{i\gamma z} + \sum_{n=-\infty}^{\infty} \tilde{a}_n |n+\nu| \epsilon_n e^{i(n+\nu)\zeta}; & |\zeta| < \frac{\pi d}{l}; \\ \sum_{n=-\infty}^{\infty} \tilde{a}_n e^{i(n+\nu)\zeta} = 0; & \frac{\pi d}{l} < |\zeta| < \pi. \end{cases} \tag{3.13}$$

Let us differentiate the second equation of (3.13) with respect to ζ and introduce the new unknown

$$x_n = \tilde{a}_n(n+\nu). \tag{3.14}$$

Then the system (3.13) will take the form

$$\begin{cases} \sum_{n=-\infty}^{\infty} x_n e^{in\zeta} = 0; & \frac{\pi d}{l} < |\zeta| < \pi; \\ \sum_{n=-\infty}^{\infty} x_n \frac{|n+\nu|}{n+\nu} e^{in\zeta} = \frac{pl}{2\pi} + \sum_{n=-\infty}^{\infty} x_n \frac{|n+\nu|}{n+\nu} \epsilon_n e^{in\zeta}; & |\zeta| < \frac{\pi d}{l}. \end{cases} \tag{3.15}$$

System (3.15) will be equivalent to the initial one if to add to it the second equality from (3.13) for $\zeta = \pi$:

$$\sum_{n=-\infty}^{\infty} (-1)^n \frac{x_n}{n+\nu} = 0. \tag{3.16}$$

The obtained system may be represented in the operator form

$$A_1 x + A_2 x = b,$$

where A_1, A_2 are the statical and completely continuous parts, respectively, of the initial summation-type operator. Operator A_1 containing singularity may be explicitly inverted with the help of the problem of conjugation of function depending on the complex variable z (the Riemann-Hilbert problem [61]). With this purpose let us introduce the functions of complex variable

$$x^+(z) = \sum_{n\geq 0} x_n z^n; \qquad x^-(z) = -\sum_{n<0} x_n z^n,$$

which are holomorphic inside and outside the unit disk, respectively. Then the first equation of the system (3.15) yields

$$x^+(z) - x^-(z) = 0; \quad \frac{\pi d}{l} < |\zeta| < \pi.$$

Now let us consider the equation of the system (3.15) which holds for $|\zeta| < \frac{\pi d}{l}$. Let introduce the whole number j satisfying inequality $0 < j < \nu < j+1$ and represent the infinite summation interval as $n[-\infty : -j-1] < 0$ and $n[-j : +\infty] > 0$. Then the considered equation will take the form

$$x^+(z) + x^-(z) = \frac{\pi l}{2\pi} - 2\sum_{n=-j}^{-1} x_n e^{in\zeta} + \sum_{n=-\infty}^{\infty} x_n \frac{|n+\nu|}{n+\nu} \epsilon_n e^{in\zeta}; \; |\zeta| < \frac{\pi d}{l}.$$

The latter differs from the analogous equation obtained for the problem of diffraction of usual plane waves coming at a certain angle to grating's plane [61] since the summing after the factor 2 is carried out by the indices belonging to the range which is less than the one in the case of usual plane waves and the difference is equal to unit. The physical reasons of this fact are as follows: scattering of inhomogeneous waves yields radiation of zero harmonic to free space and the number of harmonics in the case of diffraction of inhomogeneous waves is always less by one harmonic than in the case of diffraction of homogeneous plane waves.

Hence, the problem is reduced to the one of reconstructing analytical function by the sum of its limiting values on an arc L of the unit disc, which is usually called inhomogeneous conjugation problem or the Riemann-Hilbert problem [61]. In accord with [93], its solution is represented in the form of the Cauchy integral

$$\Phi(z) = \frac{X(z)}{2\pi i} \int_L \frac{g(t)dt}{X^+(z)(t-z)} + X(z)P(z),$$

where $P(z)$ is a polynomial; $X(z)$ is the solution to corresponding homogeneous conjugation problem ($g(t) = 0$); $X^+(z)$ is the boundary value of $X(z)$.

The Sohotsky formulas [93], [110], which determine boundary values of a function given by the Cauchy integral, yield infinite system of linear inhomogeneous algebraic equations with respect to the Fourier coefficients x_n of the functions x^+, x^-

$$\begin{cases} x_n = QV_n^0 - 2\sum_{s=-j}^{-1} x_s V_n^s + \sum_{s=-\infty}^{\infty} x_s \frac{|s+\nu|}{s+\nu} \epsilon_s V_n^s + 2CR_n; \\ 0 = QV_\sigma^0 - 2\sum_{s=-j}^{-1} x_s V_\sigma^s + \sum_{s=-\infty}^{\infty} x_s \frac{|s+\nu|}{s+\nu} \epsilon_s V_\sigma^s + 2CR_\sigma, \end{cases} \tag{3.17}$$

where $Q = \frac{pl}{2\pi}$; $V_n^s, V_\sigma^s, R_n, R_\sigma$ are the combinations of Legendre functions and polynomials P_n, P_ν of the argument $u = \cos\frac{\pi d}{l}$ defined by the formulas from [61].

The system (3.17) giver rigorous solution to the considered problem of diffraction of inhomogeneous E-waves of the planar dielectric waveguide on a strip grating which is treated as the wave model of SPE. One obtains the sought for diffractional field for arbitrary parameters of the diffractional structure by solving this system with respect to x_n and using (3.14), (3.6). It is easy also to obtain from (3.17) the solution in the case of scattering of waveguide's H-waves by the strip grating if to replace the argument u by $-u$ and $\tilde{a}_n$ by b_n in (3.14). We find then the required diffractional fields by using (3.8) and (3.5).

Before to analyze the system (3.17), let us represent it in the canonical form

$$x_n = \sum_{s=-\infty}^{\infty} A_{ns} x_s + B_s, \tag{3.18}$$

where

$$A_{ns} = \begin{cases} \frac{|s+\nu|}{s+\nu} \epsilon_s W_\sigma^{-s}, & -\infty < s < -j-1; \\ \frac{|s+\nu|}{s+\nu} \epsilon_s W_\sigma^s, & 0 < s < \infty; \\ W_\sigma^s(\frac{|s+\nu|}{s+\nu} \epsilon_s - 2), & -j < s < -1; \end{cases}$$

$$B_n = QW_\sigma^0;\ W_\sigma^s = V_n^s - V_\sigma^s \frac{R_n}{R_\sigma}.$$

The estimates verified in [61] for matrix elements A_{ns} show that the norm of system's matrix is bounded: $\sum_{n,s} |A_{ns}|^2 < \infty$ and the column of right hand sides satisfies the condition $\sum_n |B_n|^2 < \infty$. Hence, the system's matrix defines in the Hilbert space l_2 of the sequences x_n completely continuous operator and (3.18) is the Fredholm infinite system of the second kind. One can show [62] that corresponding homogeneous system has no trivial solution what yields existence and uniqueness of the solution to inhomogeneous system (3.18) in l_2. Fast decreasing of matrix elements with growing index guarantees efficiency of the method of reduction for numerical solution of the system in wide range of the parameters' variation. In some particular cases the condition $\|A_{ns}\| < 1$ holds for $\kappa < 1$ and $|u| \sim 1$, i. e., when strips and slots are sufficiently narrow. In this case one can use the method of successive approximations for obtaining explicit representations of the amplitudes of diffractional harmonics.

3. Let us consider at first the diffractional field of SPR for the small scatterer's period (in comparison with the wavelength) when $\kappa < 1$.The values p_n may vanish for certain κ. It yields the growth of small parameter ϵ_n up to unit and the norm of matrix becomes greater than unit what violates the possibility to apply the method of successive approximations. However, one can exclude this case by carrying out some special equivalent transformations of the system (3.15), (3.16).

Let us introduce the following notation: $\delta = \frac{2\pi d}{l}$, $\nu = \frac{\gamma l}{2\pi} = m' + \xi$ where m' is the digit closest to ν. It is obvious that $-0.5 < \xi < 0.5$. Transferring in (3.15), (3.16) to the new summation index defined by the equality $m = m'+n$ we rewrite the system in the following form

$$\begin{cases} \sum\limits_{m=-\infty}^{\infty} x_m^{m'} e^{im\xi} = e^{im'\xi}; & \delta < |\xi| < \pi; \\ \sum_{m=-\infty}^{\infty} x_m^{m'} \frac{|m|}{m} \frac{x_m^{m'}}{m+\xi} = e^{im\xi} = \sum_{m=-\infty}^{\infty} x_m^{m'} \frac{|m|}{m} \epsilon_m e^{im\xi}; & |\xi| < \delta; \\ \sum_{m=-\infty}^{\infty} (-1)^m \frac{x_m^{m'}}{m+\xi} = (-1)^{m'} \frac{1}{m'+\xi}; & \end{cases} \tag{3.19}$$

here

$$x_m^{m'} = \tilde{b}_{m-m'} \frac{m+\xi}{m'+\xi}; \qquad \epsilon_m = 1 + \frac{i\frac{l}{2\pi} p_m^{m'}}{\frac{|m|}{m}(m+\xi)};$$

$$p_m^{m'} = i\frac{2\pi(m+\xi)}{l}\frac{|m|}{m}(1-\epsilon_m) = \frac{2\pi}{l}\sqrt{\kappa^2 - (m+\xi)^2}. \tag{3.20}$$

It is obvious that $p_m^{m'}$ does not vanish for he considered values of κ.

The solution to inhomogeneous conjugation problem for the system (3.19) yields infinite system of linear algebraic equations of the second kind in the canonical form with respect to coefficients $x_m^{m'}$:

$$x_m^{m'} = \frac{1}{1 - \epsilon_m \frac{|m|}{m} V_{m-1}^{m-1}} \Big\{ \sum_{n \neq 0, m} x_n^m \epsilon_n \frac{|n|}{n} V_{m-1}^{n-1} - \frac{|m'|}{m'} V_{m-1}^{m'-1} + \delta_{mm'} -$$

$$P_m(u)\delta_{0m'} + x_0^{m'}[P_m(u) + \epsilon_0 V_{m-1}^{-1}]\Big\}; \tag{3.21}$$

$$x_0^{m'} = \frac{1}{P_{\xi-1}(u) + \epsilon_0 V_{-\xi-1}^{-1}}\Big\{-\sum_{n\neq 0} x_m^{m'}\epsilon_n \frac{|n|}{n} V_{-\xi-1}^{n-1} + \frac{|m'|}{m'} V_{-\xi-1}^{m'-1} + P_{\xi-1}(u)\delta_{0m'}\Big\};$$

$$V_{-\xi}^{n} = \frac{\xi-1}{2(n+\xi)}[P_{\xi-1}(u)P_{n+1}(u) - P_{\xi-2}(u)P_n(u)].$$

The norm of the matrix of the system (3.21) admits the estimate [61]

$$\|A_{mn}\| = \sum_{n\neq 0,m} \frac{|\epsilon_n||V_{m-1}^{n-1}|}{|1-\epsilon_m \frac{|m|}{m} V_{m-1}^{m'}|} < \frac{C_1\kappa^2\sqrt{1-u^2}}{\sqrt{1-4\kappa^2}}.$$

It yields applicability of the method of successive approximations for (3.21) when the scatterer's period is small and $u \approx 1$. Zero approximation gives the expression $(m \neq 0)$

$$x_m^{m'} = \frac{1}{1-\epsilon_m \frac{|m|}{m} V_{m-1}^{m-1}}\Big\{-\frac{|m'|}{m'} V_{m-1}^{m'-1} + \delta_{mm'} - P_m(u)\delta_{0m'}$$

$$+ x_0^{m'}[P_m(u) + \epsilon_0 V_{m-1}^{-1}]\Big\}(1+\omega_s); \tag{3.22}$$

$$x_0^{m'} = \frac{\frac{|m'|}{m'} V_{-\xi-1}^{m-1} + P_{\xi-1}(u)\delta_{0m'}}{P_{\xi-1}(u) + \epsilon_0 V_{-\xi-1}^{-1}}, \tag{3.23}$$

where

$$\omega_s < \frac{q}{1-q}.$$

Substitution of $x_0^{m'}$ to the expression for $x_m^{m'}$ yields representation for $x_m^{m'}$, $m \neq 0$ in the case of small period with error's estimation

$$x_m^{m'} = -\frac{|m'|}{m'} V_{m-1}^{m'-1} + \delta_{mm'} - P_m(u)\delta_{0m'} + \omega_m^{m'} + x_0^{m'}[P_m(u) + \epsilon_0 V_{m-1}^{-1} + \omega_m].$$

The values $\omega_m^{m'}, \omega_m$ are calculated in [61].

As it follows from (3.7), the diffraction field is a set of inhomogeneous plane waves localized close to the grating for $\kappa < 0.5$, $|\xi| > \kappa$, $m' = 0$. In this case the amplitudes of principal harmonic with index $n = 0$ for transmitted and reflected fields, respectively, are

$$\tilde{b}_0 = \frac{1}{1 - \sqrt{\xi^2 - \kappa^2} \ln \frac{1+u}{2}}; \tag{3.24}$$

$$\tilde{a}_0 = -\frac{\sqrt{\xi^2 - \kappa^2} \ln \frac{1+u}{2}}{1 - \sqrt{\xi^2 - \kappa^2} \ln \frac{1+u}{2}}; \tag{3.25}$$

When $\kappa < 0.5$, $|\xi| < \kappa$ or¡ in other words, when only one harmonic radiating to the free space, for example, with the index $n = -1$, exists in the spectrum of scattered field, then $m' = 1$ and $m = 0$ and we obtain from (3.23) for the principal non-radiating harmonic

$$\tilde{b}_0 = \frac{1 + \xi}{1 + u + (1 - u)\sqrt{(1 + \xi)^2 - \kappa^2}} \left\{ 1 + u + \right.$$

$$\left. \frac{[u - 1 + (1 + u)\xi \ln \frac{1+u}{2}][\xi(1 + u) + (1 - u)i\sqrt{\kappa^2 - \xi^2}]}{2(1 + \xi)((1 + i\sqrt{\kappa^2 - \xi^2} \ln \frac{1+u}{2})} \right\};$$

$$\tilde{a}_0 = \frac{U_0 - \frac{1}{2(1+\xi)}[u - 1 + (1 + u)\xi \ln \frac{1+u}{2}][\xi(1 + u) + (1 - u)i\sqrt{\kappa^2 - \xi^2}]}{(1 + u + U_0)((1 + i\sqrt{\kappa^2 - \xi^2} \ln \frac{1+u}{2})} \tag{3.27}$$

and for the first radiated harmonic

$$\tilde{b}_{-1} = -\tilde{a}_{-1} = \frac{1}{2}\frac{1 - u - (1 + u)\xi \ln \frac{1+u}{2}}{1 + i\sqrt{\kappa^2 - \xi^2} \ln \frac{1+u}{2}}, \tag{3.28}$$

where

$$U_0 = (1 - u)\sqrt{1 - \frac{\kappa^2}{(1 + \xi)^2}}.$$

Let us point out that (3.24)-(3.28) are obtained with taking into account interrelation

$$P_\xi(u) \approx 1 + \xi \ln \frac{1 + u}{2}$$

which holds for $\xi < \frac{1}{2}$. It is obvious that $\tilde{a}_0$ and $\tilde{b}_0$ defined by the formulas (3.25) and (3.24) are the reflection and transmission coefficients of inhomogeneous plane wave. Unlike analogous expressions given in [61] for the problem of inclined scattering by the grating of homogeneous plane wave, these representations are purely real. Hence, the phase of the reflection and transmission coefficients of inhomogeneous plane waves is equal to $n\pi$, $n = 0, \pm 1, \pm 2, \ldots$,; it does not depend on the scatterer's parameters and characteristics of incident surface wave. Similar data may be obtained as a result of the solution to the problem of scattering of inhomogeneous waves by the dielectric half-space in the case of absence of radiation.

4. The situation is quite different when radiation condition (3.7) holds for the harmonic with index $n = -1$. The expressions for $\tilde{a}_0$ and $\tilde{b}_0$ (see (3.27) and (3.26)) become complex-valued. Hence, the phase of reflection and transmission coefficients depends here on the parameters of incident inhomogeneous wave and on the shape of scattering obstacle.

As it follows from the equality $\nu = \frac{\gamma l}{2\pi} = m' + \xi$, characteristics of the source of inhomogeneous plane wave (dielectric permittivity ϵ_1 and width $2a$) are contained in the parameter ξ. Asymptotic solution to the dispersion equations for sufficiently thick waveguides makes it possible to represent the amplitude of radiating harmonic as an explicit function of ϵ_1:

$$\tilde{b}_{-1} = -\tilde{a}_{-1} = \frac{1}{2}\frac{1 - u - (1+u)(\kappa\sqrt{\epsilon_1} - 1)\ln\frac{1+u}{2}}{2 + i\sqrt{\kappa^2(1-\epsilon_1) - 2\kappa\sqrt{\epsilon_1} - 1}\ln\frac{1+u}{2}}; \tag{3.29}$$

But here one should take into account the multi-mode character of the planar dielectric waveguide and the fact that the values of its fundamental frequencies lie close to each other. It puts natural restrictions on practical application of the formula (3.29).

The given expressions correctly describe the behaviour of scattered field. If the filling coefficient $u = -1$, i. e., when the grating becomes the perfectly conducting plane, the reflection coefficient $\tilde{a}_0 = 1$ and transmission coefficient $\tilde{b}_0 = 0$. For $u = 1$, when there is no scatterer, $\tilde{a}_0 = 0$, $\tilde{b}_0 = 1$ and the field reflected from the perfectly conducting plane is an inhomogeneous plane wave propagating along the axis Oz with the phase velocity equal to the one of incident inhomogeneous wave. This results is directly connected with statement of the problem within the frame of the model of given field of dielectric waveguide. It follows from (3.29) that the amplitude together with electromagnetic field of the single radiating harmonic vanishes when $u \to \pm 1$, what corresponds to the absence of periodical structure.

The only one non-zero component of the magnetic field of radiating harmonic has the form

$$H_y^{\mathrm{II}} = \frac{ikg\epsilon_1 \begin{Bmatrix} \sin ga \\ \cos ga \end{Bmatrix} e^{-pb}[u + (1+u)\ln\frac{1+u}{2} - 1]}{2(1 + i\sqrt{\kappa^2 - \xi^2}\ln\frac{1+u}{2} - 1)}. \tag{3.30}$$

which holds for $\kappa < \frac{1}{2}$, i. e., when the scatterer's period is small in comparison with the wavelength λ_0 in free space. Using reciprocity principle it is easy to obtain from (3.25)-(3.28) the solution in the case of scattering of inhomogeneous H-waves of the planar waveguide by the strip grating.

One can verify explicit formulas from the system (3.21) in two physically important cases: narrow slots and narrow strips of the scatterer for arbitrary parameter κ.

Since here κ takes arbitrary values, it may happen that $\kappa = m + \nu$. Separating in the initial system the terms with x_0 and x_{m_ξ} and solving the obtained system by the method of successive approximations, we finally obtain

$$x_0^{m'} = \frac{\frac{(1-\delta_{n\xi})t_n}{1-t_n}[\frac{|m'|}{m}(V_{n_\xi-1}^{m'-1} + V_{-\xi-1}^{m'-1}) + \delta_{n_\xi m'} + \delta_{0m'} + P_{n_\xi}(u) + \delta_{0m'}P_{\xi-1}(u)]}{\frac{t_n}{1-t_n}\{[\epsilon_0 V_{n_\xi-1}^{-1} + P_{n_\xi}(u)](1-\delta_{n_\xi 0}) + \epsilon_0 V_{\xi-1}^{-1} - P_{\xi-1}(u)\}}, \tag{3.31}$$

where

$$t_n = \frac{|n_\xi|}{n_\xi}\epsilon_{n_\xi}V_{n_\xi-1}^{n_\xi-1}.$$

Let us determine from (3.31) the amplitude of radiating harmonic with $n = -1$, $m' = 1$, $s = 0$:

$$\tilde{b}_{-1} = \frac{m'+\xi}{s+\xi}x_s^{m'} = \frac{1+\xi}{\xi}x_0^1.$$

Since κ is an arbitrary number, then, in particular, it may be chosen such that $\kappa > \xi$. Taking into account that $s = 0$ and $n_\xi = 0$ we have for $m' = 1$

$$x_0^1 = \frac{V_{-\xi-1}^0}{P_{\xi-1}(u) + \epsilon_0 V_{-\xi-1}^0}. \tag{3.32}$$

Then in the case of narrow strips ($u \to 1$) complex amplitude of the first diffractional harmonic radiating to the free space has the form

$$\tilde{b}_{-1} = \frac{1}{2}\frac{u - 1 + \frac{1}{2}(1-u)\xi[\xi(1-u) + u - 1]}{1 - (1-u)(\frac{1}{2}\xi^2 + i\sqrt{\kappa^2 - \xi^2})}. \tag{3.33}$$

In order to verify explicit formula for the Fourier amplitude of the minus first space harmonic in the case of narrow slots we will use the Hobson expansion for the Legendre functions [110]

$$P_\xi(u) = \frac{\sin \pi\xi}{\pi}\left[\ln\frac{1+u}{1-u} + \psi(\xi) - \psi(-\xi-1) - 2\psi(0)\right]F_{21}(\xi,u) -$$

$$2\sum_{n=0}^{\infty}\left\{\sum_{\tau=1}^{n}\frac{1}{\tau}\frac{(-\xi)^n(1+\xi)^n}{(n!)^2}\left(\frac{1+u}{2}\right)^n\right\},$$

where $\psi(\xi)$ is the logarithmic derivative of Gamma-function; $F_{21}(\xi,u) = {}_2F_1(-\xi,\xi+1,1,\frac{1+u}{2})$ is the hypergeometrical Gauss function.

Using the properties of $\psi(\xi)$ and omitting the second order terms (for $u \to -1$) we obtain

$$\tilde{b}_{-1} = \frac{1}{2}\frac{\xi[(\ln\frac{1+u}{2} + \pi\cot\pi\xi)(u+1) + \frac{2u\xi}{1+\xi} + \frac{2(\xi-1)}{\xi}]}{\frac{1}{1+\xi} + i\sqrt{\kappa^2-\xi^2}[\ln\frac{1+u}{2} + \pi\cot\pi\xi + \frac{2\xi^2-1}{\xi(1+\xi)}]}. \tag{3.34}$$

Asymptotic expressions (3.33), (3.34) for the amplitude of space harmonic radiating to the free space yield rather sharp decrease of $\tilde{b}_{-1}$ in the case of narrow slots due to the logarithmic singularity of the denominator in (3.34). For the narrow strips ($u \to +1$) this decrease is more regular since the numerator in (3.34) tends to zero as $O(1-u)$. From the physical point of view one can give the following explanation of such difference in the behaviour of the harmonic with index $n = -1$. In the case of narrow slots the densities of shift currents arising in the slots between grating's strips grow while they approach each other and amplitude of the wave radiating to the free space must increase. For $u \approx -1$, when there are practically no slots, this amplitude vanishes. Hence, the function $\tilde{b}_{-1}(u)$ reaches its maximum and vanishes in a rather small interval of variation of the grating's filling coefficient what leads to sharp changes of the complex amplitude $\tilde{b}_{-1}$. For $u \to 1$ the density of shift currents is small because the distance between strips is large and the density tends to zero with its further growth. This effect is reflected in smooth fall of $\tilde{b}_{-1}(u)$ given by the expression (3.33) when the filling coefficient approaches the limiting value $u = 1$.

5. The field scattered by the strip grating consists of a set of space harmonics such that one part of them is localized close to periodical inhomogeneity and does not carry the energy to the free space and the other

part exists as homogeneous plane waves of SPR going to the free space. Let us analyze in this section energetic characteristics of the electromagnetic field of Smith-Purcell radiation. We will establish relationship between complex Fourier-amplitudes of the scattered field in the case of diffraction of inhomogeneous plane waves using the complex power theorem. As a closed volume we will choose, for example, the volume $V = \{(x, y, z) : -0.5 < x < 0.5, -0.5 < y < 0.5, -0.5l < z < 0.5l\}$ containing one scatterer's period. Then

$$\Re\vec{S} = \frac{c}{8\pi}\Re\int_S [\vec{E}, \vec{H}^*]\vec{n}\, ds = 0, \tag{3.35}$$

where $\vec{n}$ is the unit normal vector of the surface S bounding the volume V; $\vec{H}^*$ denotes complex conjugation of $\vec{H}$. Substituting to (3.35) the expressions of fields' components taken from (3.3), (3.4)-(3.6) for the case of scattering of E-waves of the planar waveguide we finally obtain

$$\frac{c}{8\pi} gk\epsilon_1 \cos gae^{-2pb}\left\{\sum_{\Im\sqrt[n]{A}=0} p_n|\tilde{a}_n|^2 + \sum_{\Im\sqrt[n]{A}=0} p_n|\tilde{b}_n|^2 + 2e^{pb}\Im\tilde{a}_0\right\} = 0. \tag{3.36}$$

This equality states the energy conservation law for the scattered field which has some differences if to compare with similar expressions given in §1 for the scattering of usual plane waves. This fact has been indicated for the first time in [73]. It is not difficult to obtain analogous result for the scattering of inhomogeneous magnetic waves. (3.36) is the necessary condition for the correctness of the obtained solution.

We will define the energy of diffractional field going off to the free space as the real part of the normal component of Pointing vector:

for E-wave

$$\Re\tilde{\vec{S}}_\perp = \frac{c}{8\pi} E_z H_y^* \vec{x}, \tag{3.37}$$

for H-wave

$$\Re\vec{S}_\perp = \frac{c}{8\pi} E_y H_z^* \vec{x}.$$

Substitution to (3.37) the expressions of scattered fields yields

for E-wave

$$\Re\tilde{S}_\perp = Q^2 e^{-2pb} \sum_{\Im\sqrt[n]{A}=0} p_n|\tilde{b}_n|^2, \tag{3.38}$$

for H-wave

$$\Re\vec{S}_\perp = Q_1^2 e^{-2pb} \sum_{\Im\sqrt[n]{A}=0} p_n|\tilde{a}_n|^2,$$

where $Q_1 = gk\epsilon_1 \cos ga$.

For the arbitrary parameters of inhomogeneous incident wave and periodical scatterer the complex amplitudes $a_n, \tilde{b}_n$ are determined from the infinite system of inhomogeneous algebraic equations (3.18). One can easily verify explicit formulas for the power flow's density of the radiated harmonic with $n = -1$ using explicit formulas for $\tilde{b}_{-1}$ in the case of narrow slots or strips. For example, substituting (3.29) to (3.38) we obtain for $\lambda >> l$

for E-wave

$$\tilde{S}_{-1} = Q^2 e^{-2pb} \frac{2\pi[(1-u) - (1+u)\xi \ln \frac{1+u}{2}]^2 \sqrt{\kappa^2 - (\xi - 1)^2}}{l[1 + (\kappa^2 - \xi^2) \ln^2 \frac{1+u}{2}]}; \qquad (3.39)$$

for H-wave

$$S_{-1} = Q_1^2 e^{-2pb} \frac{2\pi[(1+u) - (1-u)\xi \ln \frac{1+u}{2}]^2 \sqrt{\kappa^2 - (\xi - 1)^2}}{l[1 + (\kappa^2 - \xi^2) \ln^2 \frac{1-u}{2}]}. \qquad (3.40)$$

It directly follows from (3.39) and (3.40) that for sufficiently narrow slots ($u \approx -1$) electromagnetic energy of the radiated harmonic vanishes. But the law of decrease of the power radiated by the system planar waveguide + grating is not the same in the cases of scattering of electrical and magnetic waves. In the case of E-waves the field's decrease is sharper due to logarithmic singularity of $\tilde{b}_{-1}(u)$ for $u \to -1$. For H-wave radiation field decreases as $f(u+1)$ for $u \to -1$ and its fall has more smooth character. In the case of narrow strips ($u \to 1$) opposite situation takes place: the function describing radiation field of H-waves acquires logarithmic singularity. For arbitrary parameters κ one can use asymptotic formulas (3.33) and (3.34) for estimating radiation field when $u \to \pm 1$; these formulas yield vanishing of $S_{-1}, \tilde{S}_{-1}$ for every κ when $u \to \pm 1$.

Hence, it is established by means of (3.30), (3.34), (3.40) and (3.39) that on the edges of interval of variation of the grating's filling coefficient, for $u \approx \mp 1$, radiation field vanishes (both for $l >> \lambda_0$ and for arbitrary l).

It is natural to assume that for some arbitrary scatterer's parameters the density of radiation power reaches its maximal value. We will consider them as optimal parameters both of the planar waveguide and periodical scatterer.

Let us analyze the field of diffraction of inhomogeneous waves of the planar waveguide for arbitrary parameters of electrodynamical structure with the help of numerical solution to (3.18). It has been proved that this system of equations admitted the use of the method of reduction which transformed the infinite system to the finite one. The choice of the order of reduction provides obtaining of approximate solution with prescribed accuracy for arbitrary parameters of electrodynamical structure. It is accepted to call the results of such reduction procedure the rigorous solutions.

The density of power flow of the single radiated harmonic with the index $n = -1$ versus transversal size of the planar waveguide for E-waves (curve I) and H-waves (curve II) is presented on Fig. 56. Behaviour of the energy radiated to the free space has resonance character: the energy vanishes for the very small (but big in comparison with the wavelength λ_0) waveguide's thickness $2a$. The generated radiation is the result of transformation of inhomogeneous plane waves of the planar waveguide to volume waves in the process of scattering by the grating. Maximum of radiated energy of H-waves is much less than for E-waves because at the same wavelength and waveguide's thickness magnetic waves are more "pressed" to the waveguide than electric waves and, consequently, most part of the energy carried along the waveguide is concentrated inside it [111]. waveguide

It is studied also the density of power flow of the single radiated harmonic with the index $n = -1$ as a function of grating's filling coefficient for E- and H-waves and different scatterer's periods. There exist optimal values of the grating's filling coefficient providing maximal level of SPR (one-wave regime of SPR) in the case of scattering of electric and magnetic waves of dielectric layer.

Maximum of radiated power for the transformation of inhomogeneous electric waves lies in the domain of negative values of the filling coefficient, i. e., when the metallic surface covers most of the period. Maximum of radiated power for the magnetic waves is in the domain of narrow strips, where the filling coefficient is positive. This behaviour well corresponds to the conclusions made after the analysis of (3.39), (3.40) and its physical reasons are connected with the properties of currents generated on metallic strips [112].

In the case of H-polarization these currents have a break on the slots since generated current has the only one non-zero component j_z. In the domain of narrow slots tensity of electric field between the neighboring strips reaches its maximum under certain phase conditions connected with λ_0 and l. Radiation of electromagnetic wave in this case is maximal. Symmetry of Maxwell equations with respect to electric and magnetic currents yields maximal radiated energy in the domain of narrow slots for opposite polarization.

The multi-wave factor perturbing the minus first harmonic causes additional local maximums. It follows from the radiation condition (3.7) that variation of the filling coefficient does not influence emergence of new propagating harmonics but it can control redistribution of energy between the harmonics radiating to the free space in different directions.

The new harmonics radiating to the free space appear when the scatterer's period changes in such a way that the second derivative is discontinuous. Such features of the scattered field are observed also in the process of diffraction of usual homogeneous waves on the gratings [61] and they are analogous by their nature to the Wood anomalies known in optics.

For the further analysis of diffractional field of inhomogeneous waves of the planar dielectric waveguide scattered by periodical obstacle we will consider in brief one essential problem. It has been obtained in §1 within the model of given current approximation the rigorous solution to the problem of radiation of the plane density-modulated electron flow flying with constant velocity over the strip grating. Flow's eigenfield is an inhomogeneous wave and it coincides with the E-eigenfield of the planar waveguide as the function of x and z due to (3.4). We pointed out in the statement of the considered problem that it was solved within the model of given field of dielectric waveguide.

Qualitative coincidence of the results of §1 and the data presented here confirms the conclusion that eigenfield of the planar waveguide may well model an electron flow and electrodynamical structure planar waveguide + grating may be studied in the process of experimental investigations of SPR.

However, drawing an analogy between the phenomenon of SPR of electron flow and the effect of transformation of inhomogeneous waves of the planar waveguide to the volume waves one should indicate those characteristics of radiated field which enable one to identify these physical phenomena. Our considerations allow to conclude that the following magnitudes may play the role of such characteristics: dependence of radiation angle on the scatterer's period and phase velocity of inhomogeneous wave, radiated power versus grating's filling coefficient and its period for the fixed wavelength λ_0. It is essential here that comparison of these characteristics holds when one considers scattering of electric waves of the planar waveguide by the strip grating. Only in this case polarizations of incident fields for electron flow and planar dielectric waveguide coincide. Let us go back now to the analysis of diffractional field of the planar waveguide.

In the case of scattering of magnetic waves the points corresponding to generation of radiated harmonics are shifted towards smaller grating's periods in comparison with the scattering of E-waves since phase velocity of H-waves is less than phase velocity of E-waves for equal parameters of the waveguide layer [111].

It is obvious that if both electric and magnetic waves would simultaneously propagate along the planar waveguide, one can control radiation to the free space electromagnetic waves of the required polarization by choosing the grating's period.

Distribution of energy between the harmonics of E- and H-waves radiated to the free space are presented on Fig. 57, a, b. In both cases the density of power flow of the minus first harmonic is maximal; the energy falls with increasing number. One can clearly observe the domain of variation of grating's period corresponding to the single-mode radiation regime. The given curves are characterized by anomalous growth of S_{-1} corresponding to the emergence of even harmonic while its anomalous fall is accompanied by the emergence of odd one what confirms the results presented in [64].

The densities of power flow of the first four radiated harmonics versus wavelength of the E- or H-wave propagating along the dielectric layer are shown on Fig. 58, a, b. The general resonance behaviour of the curve for S_{-1} reflects dispersion properties of the planar waveguide. In fact, phase velocity of the scattering wave is small for small wavelengths and most part of the energy propagating along the waveguide is concentrated inside it. With growing wavelength a part of the energy supported by the external field (with respect to the layer) increases what is caused by growth of the curve in the interval between $\kappa_2 = 1.8$ and $\kappa_1 = 3$. With further growth of wavelength the curve tends to zero since effective size of the waveguide sharply increases and the energy density of electromagnetic field in external space falls.

It is clearly indicated the critical wavelength, when radiation of the minus first space harmonic vanishes (at the points $\kappa = 2.85$ for E-waves and $\kappa_{\text{cr}} = 2.6$ for H-waves). Main part of the energy irradiated to the free space falls on the harmonic, which is nearest to the minus first, for those parameters of the open structure when the latter is transformed to the non-radiated harmonic. In the considered case the harmonic with number $n = -2$ has the maximum. Let us note that there are points with non-differentiable singularity on all curves (see Fig. 58). For these parameters the energy sharply redistributes between all diffractional harmonics as well as in the case of variation of grating's period (Fig. 57). We conclude after the analysis of (3.7) the angle of radiation of the emerged harmonic is $\theta_n = 180°$. That is why such anomalous points are called sliding points. It follows from the character of the curves that there exists optimal value of the frequency of incident wave for fixed geometrical parameters of the considered electrodynamical structure.

6. Let us analyze properties of the field scattered by the strip grating by its amplitude-phase distributions close to inhomogeneity. It is given in [62] the pictures of amplitude-phase distributions in the case of diffraction of homogeneous plane waves on the gratings having different profiles as well as for SPR of monochromatic electron flow flying close to strip grating [5]. They illustrate the main processes occurring in the considered problems: going of electromagnetic energy through the grating and emergence (or absence) of SPR. However, as it will be shown below, amplitude-phase field's distributions contains information about the features of the scattered field which play basic role for the choice of optimal parameters of the structure. One can thus effectively control characteristics of the scattered field with the purpose to obtain desired distribution in the far zone. In addition to this, it is obviously demonstrated the space picture of transformation of the surface waves of planar waveguides to the volume waves.

As it follows from (3.6) and (3.5), the scattered field consists of infinite number of space harmonics with complex amplitudes and for certain parameters of inhomogeneous wave and periodical structure some harmonics radiate to the free space. Appropriate choice of these parameters enables one to separate a part of diffractional harmonics out of infinite spectrum which go off to the free space in the form of volume waves (SPR). The structure of scatterer's near field is determined by interaction of space harmonics localized close to the grating with a volume wave (when radiation condition holds).

Amplitude-phase distributions of the near field together with lines of equal phases (bold lines) and equal amplitudes (dot lines) at one scatterer's period are presented on Fig. 59-61. Values of phases and amplitudes are given in line breaks. Let us follow space dynamics of transformation of a surface wave of planar waveguides localized in its vicinity to the wave SPR in the form of volume waves going off to the free space. Fig. 59 corresponds to the case when radiation condition (3.7) gives imaginary value of radiation angle and, hence, diffractional spectrum consists only of surface harmonics. Really, one can clearly see on these figures that the lines of equal phases near the grating and far from it are perpendicular to its surface and field's amplitude noticeably falls with growing distance. As it is known for the two-dimensional problems, the lines of time average of energy flow carried by electromagnetic wave must be perpendicular to the lines of equal phases in every point. It directly follows from the equation for the lines of energy flow $\frac{dx}{dz} = \frac{S_x}{S_z}$ where S_x, S_z are corresponding components of Pointing vector. One can see after the proper drawing of the lines of energy flow that electromagnetic energy far from the grating (at the distance of about $2\lambda_0 - 3\lambda_0$) is carried along it and is not radiated to the free space. In the domains where the lines of equal phases intersect (approach each other) and phase's absolute value is not determined, it is formed so-called vortices of electromagnetic energy (the lines of equal phases meet in radial direction). The energy flow "circulates" in this domain and does not "escape" to the free space. The energy flow close to the grating and far from it rounds such domains proceeding along the grating. The domains of existence of the vortices of electromagnetic energy correspond to the minimal field's amplitude. At big distances from the grating the lines of equal phases are perpendicular to the lines of equal amplitudes and it is confirmed by the fact that diffractional spectrum consists of surface harmonics localized close to the grating; absolute values of field's amplitude show that the main part of energy is concentrated at the distance of about λ_0 from the grating. One can see also that the domain occupied by the field's vortex approaches grating's surface with growing period.

Fig. 60 illustrates the fulfillment of radiation condition (3.7) for the values of the parameters $a = 0.5$, $b = 0.02$mm, $u = 0.5$, $l = 1.93$mm. One radiated harmonic with $n = -1$ emerges in the spectrum of diffractional field. At the big distances of the order $3\lambda_0 - 4\lambda_0$ where one can omit attenuating harmonics we will represent the approximate field, in accord with (3.6), in the form

$$H_y = \tilde{b}_{-1} e^{-i\sqrt{k^2 - \left(\gamma - \frac{2\pi}{l}\right)^2} x} e^{i\left(\gamma - \frac{2\pi}{l}\right) z}. \tag{3.41}$$

Configuration of the lines of equal phases in the domain of existence of the vortices of electromagnetic field has sharply changed and become strongly asymmetrical with respect to the plane going through the middle of a vortex while the domain where these lines met has considerably decreased. Hence, in the domain of the space where earlier there was the boundary of circulation of electromagnetic field without radiation, it was established the lines of energy flow indicating that energy went off the grating to the free space. At big distances from the grating the lines of equal amplitudes coincide with the lines of equal phases what is evidence of the existence of plane electromagnetic wave in this domain. Let us note that asymmetrical domain of localization of the lines of equal phases emerged as a result of radiation is less than λ_0 distant from the grating.

If radiation is absent, this distance may be arbitrarily big and, in particular, it is equal to $1.5\lambda_0$ for the grating with parameters given on Fig. 59, a. Measurement of radiation angle by the lines of equal phases in the domain of existence of plane wave and its comparison with the value obtained from (3.7) give sufficiently good coincidence. In the considered case deviation is between $0.5^\circ - 1^\circ$ what is fully explained by the degree of graphical accuracy.

The case when radiated space harmonic is almost perpendicular to the grating's surface is presented on Fig. 61, a. It is distinctly indicated the domain where surface waves are transformed to volume waves together with the domain of existence of plane electromagnetic wave. Amplitude-phase distribution of electromagnetic field of the transformator of surface waves to volume waves for simultaneous radiation of the harmonics with $n = -1$ is presented on Fig. 61, b. In this case electromagnetic field admits the following approximate representation far from the grating

$$H_y^{\mathrm{II}} = i\left\{\tilde{b}_{-1}e^{-i\sqrt{k^2-\left(\gamma-\frac{2\pi}{l}\right)^2}x}e^{i\left(\gamma-\frac{2\pi}{l}\right)z} + \tilde{b}_{-2}e^{-i\sqrt{k^2-\left(\gamma-\frac{4\pi}{l}\right)^2}x}e^{i\left(\gamma-\frac{4\pi}{l}\right)z}\right\}. \tag{3.42}$$

Greater contribution to diffractional field is made by the minus first harmonic on the lines $z = \frac{ml}{\nu} - 1$, $m = 0, \pm 1, \pm 2, \ldots$, $\nu = \frac{\gamma l}{2\pi}$ and for $z = \frac{ml}{\nu} - 2$ the minus second harmonic is the principal one. Adding in the space, these two harmonics form amplitude "hills" and "cavities" separated by the lines which are definitely determined by the obtained conditions of minimum for corresponding harmonics. This figure is evidence of the much more complicated field's picture. Simultaneous existence of two radiated harmonics directed at different angles strongly bends the phase lines, therefore, it is rather difficult to determine the boundaries of establishing of the plane wave.

Systematic calculations and analysis of the pictures of amplitude-phase distributions show that the domain of existence of vortex plays the role of the transformator of electromagnetic energy of surface waves to the energy of waves going off the grating. While approaching the fulfillment of radiation condition (3.7) (vanishing of the imaginary part of radiation angle) they come from infinity to the grating's surface and violates electromagnetic field in the definite way what enables the field to tear off the grating in the form of SPR or plane waves going off the grating. The amplitude of volume wave grows while the vertex approaches the grating, since it finds itself in the domain of big amplitudes.

Amplitude-phase field's distribution in the domain I is shown on Fig. 61, b. Weak deviation of the lines of equal phases from the direction perpendicular to the grating to the left close to the first strip and to the right close to the second strip indicates that the flow of electromagnetic energy propagates between the waveguide and the grating almost by the sinusoidal trajectory proceeding along the axis Oz. Only in the closest vicinity of the grating a part of the flow goes off the gap to the free space. Finally, this data may be used for estimating the accuracy of the solution to infinite system obtained by the method of reduction in the following way: one should check the fulfillment of boundary value conditions on the grating, in particular, the continuity of field's components in the slot and their discontinuity on the metal. Our estimates (the system with complex matrix having the order up to 27 was solved numerically) give the relative error $10^{-4} - 10^{-5}$ for $\kappa \geq 1$, $\lambda_0 = 4.0$ mm.

The present method of the analysis of diffractional fields of planar structures may be developed for structures with bounded cross-section.

Transformation of surface waves to volume waves in the open structure dielectric waveguide + grating occurs either in the space close to structure or in the gap between the waveguide and strip grating. Here gradually form point to point a surface wave is transformed to volume wave with real wavenumber propagating perpendicular to the grating, while Brilluine waves go on propagating inside the waveguide due to total internal reflection from the waveguide's boundaries. The dynamical picture of emergence of volume waves close to the considered system may be constructed by the lines of equal phases, their ratio and phase difference between electric components of the scattered field perpendicular to waveguide's boundary and tangential to it. The latter is defined by the ellipticity coefficient and the angle of inclination of the big axis of ellipse describing the vector of scattered electric field in the plane perpendicular to the waveguide and going through the line along which the wave of dielectric waveguide propagates.

As it is known, the surface waves are characterized by the property of the end of the vector of electric field $\vec{E}$ (of the E-wave) to draw ellipse in the given plane [111]. For homogeneous plane waves propagating in the free space this vector has linear polarization. Phase portrait of the transformation of surface wave to volume wave for dielectric waveguide with the strip grating placed at the distance b over it is presented on Fig. 62. There it is also constructed ellipses determining polarization degree of the wave which is transformed by the grating from waveguide's surface wave to volume wave of the free space. The field has distinct elliptic polarization close to the surface of grating and waveguide. Hence, it has surface character. Going away from the structure, the character of polarization changes and ellipses become line's segments. At the distance of about $3\lambda_0$ the wave emerges with almost linear polarization whose electromagnetic field is already transformed from surface field to volume field (in this case to the plane wave propagating in the direction perpendicular to the system dielectric waveguide + grating).

§9. Theoretical study of the wave Smith-Purcell radiation on the dielectric waveguide and beveled strip grating.

As it is shown in §8, surface wave is transformed to volume waves in the open structure waveguide + grating close to their surfaces. Properties of scattered waves are controlled by varying the parameters of a waveguide (dielectric permittivity, size and shape), of a grating (period, stripwidth) and the distance between them. It is essential that strips of the grating are placed close to the waveguide strictly perpendicular to its axis. Such periodical structure is called *the normal grating.* New possibilities arise for the control of waves' transformation if to apply the slanted grating as a scatterer. Strip slanted grating may be formed if to place its strips at an angle which coincides with the one between grating's elements and direction of propagation of surface wave in the waveguide. One of the variants of such grating together with mathematically justified method of the solution to the problem of diffraction of plane wave by this structure is considered in [84].

In this section it is given rigorous solution to the problem and analysis of the transformation of surface waves propagating along the plane waveguide by the strip grating placed at arbitrary distance from the waveguide. Grating's axes form the angle α with the direction of propagation of surface wave. Diffractional field is completely determined for the given wavelength λ_0 and direction of propagation of surface wave by the position of grating's strips and the waveguide whose varying geometry leads to the possibility of effective control of scattered fields. Special attention is paid to their polarization and energy characteristics.

1. Let the plane dielectric waveguide with the thickness $2a$ and dielectric permittivity ϵ_1 is placed at the distance b from the grating whose metallic strips form an arbitrary angle α with the direction of propagation of surface wave. The coordinate system x, y, z is connected with the waveguide and x, s, τ – with the grating and we will call it *the slanted grating* (see Fig. 63). The axis Ox goes through the middle of grating's strip. The period $\tilde{l}$ and width of the gap between strips $\tilde{d}$ have the following form: $\tilde{l} = l \cos \alpha$, $\tilde{d} = d \cos \alpha$ where l is the period and d is the stripwidth of normal grating (the strip of normal grating is perpendicular to the axis Oz along which the surface wave propagates). The particular case of considered slanted grating is the so-called turning grating turned at the angle α without the change of period l and distance between the strips.

Surface wave of electric type propagates along the plane waveguide. Scattering of this wave by the normal grating is reduced to the problem of determining scattered field of the incident wave with $\vec{E}^0 = (E_x, 0, E_z)$, $\vec{H}^0 = (0, H_y, 0)$. As a result of its solution one can conclude that the scattered field repeats the structure of incident field (see §8). For the slanted grating incident field with the components $\vec{E}^0 = (E_x, E_s, E_\tau)$, $\vec{H}^0 = (0, H_s, H_\tau)$ excites in the scattered field the whole spectrum of electric and magnetic components and one should expect that the total diffraction field, being their superposition, will have elliptic polarization. The solution to the stated problem is to be sought in the form satisfying the system of Maxwell equations, boundary value, edge and periodicity conditions together with conditions in infinity. The space periodicity condition for the field yields representations

$$E_s^{(1)} = e^{i\eta s} \sum_{n=-\infty}^{\infty} c_n e^{ip_n(x+a+b)} e^{i\xi_n \tau}; \quad H_s^{(1)} = e^{i\eta s} \sum_{n=-\infty}^{\infty} a_n e^{ip_n(x+a+b)} e^{i\xi_n \tau};$$

$$E_s^{(2)} = e^{i\eta s} \sum_{n=-\infty}^{\infty} d_n e^{-ip_n(x+a+b)} e^{i\xi_n \tau}; \quad H_s^{(2)} = e^{i\eta s} \sum_{n=-\infty}^{\infty} b_n e^{-ip_n(x+a+b)} e^{i\xi_n \tau}; \tag{3.43}$$

where

$$\eta = \gamma \sin\alpha; \; \xi = \gamma\cos\alpha; \; \xi_n = \xi + \frac{2\pi n}{\tilde{l}}; \; p_n = \sqrt{k^2 - \eta^2 - \xi_n^2};$$

γ is the propagation constant of surface wave along the axis Oz; a_n, b_n, c_n, d_n are the complex amplitudes of scattered field; indices (1) and (2) denotes the domains $x > -a$, $x < -a - b$, respectively. The branch of $\sqrt{A}$ is chosen in such a way that $\Im\sqrt{A} > 0$ (if $\Im\sqrt{A} = 0$, then $\Re\sqrt{A} > 0$). The form of other components of scattered field are determined from Maxwell equations. The time factor $e^{-i\omega t}$ is omitted. The components of surface eigenwave of the plane waveguide are given by the following expressions in the coordinate system x, s, τ (in the domain $x < -a$):

$$E_s^0 = -pQe^{p(x+a)} \sin\alpha e^{i(\eta s + \xi\tau)}; \quad E_\tau^0 = -pQe^{p(x+a)} \cos\alpha e^{i(\eta s+\xi\tau)};$$

$$E_x^0 = i\sqrt{\eta^2 - \xi^2} Q e^{p(x+a)} e^{i(\eta s+\xi\tau)}; \quad H_s^0 = -ikQe^{p(x+a)} \cos\alpha e^{i(\eta s+\xi\tau)};$$

$$H_\tau^0 = -ikQe^{p(x+a)} \sin\alpha e^{i(\eta s+\xi\tau)},$$

where $p = ip_0$, $Q = g\epsilon_1 \cos\alpha$, g is the propagation constant of waveguide's surface wave for $|x| < a$.

The complete solution to the problem is sought in the form
for the domain 1
$\vec{E} = \vec{E}^0 + \vec{E}^{(1)}$, $\vec{H} = \vec{H}^0 + \vec{H}^{(1)}$;
for the domain 2
$\vec{E} = \vec{E}^{(2)}$, $\vec{H} = \vec{H}^{(2)}$.

Subordinating the fields to boundary value conditions in the plane of diffractional grating and carrying out transformations analogous to the ones performed in §8, we obtain infinite system of functional summation-type equations whose kernels are trigonometric functions with respect to unknowns $x_n = a_n$, $n \neq 0$; $x_0 = b_0 - \tilde{P}$, $n = 0$ and $y_n = b_n$:

$$\begin{cases} \sum_{n=-\infty}^{\infty} y_n e^{i\xi_n \tau} = 0; & |\tau| < \frac{d}{2} \\ \sum_{n=-\infty}^{\infty} (y_n p_n k + x_n \xi_n \eta) e^{i\xi_n \tau} = -\frac{1}{2}(\tilde{Q} + \tilde{P}\xi\tau)e^{i\xi\tau}; & \frac{d}{2} < |\tau| < \frac{l}{2}; \end{cases} \quad (3.44)$$

$$\begin{cases} \sum_{n=-\infty}^{\infty} x_n e^{i\xi_n \tau} = 0; & \frac{d}{2} < |\tau| < \frac{l}{2}; \\ \sum_{n=-\infty}^{\infty} (y_n \xi_n \eta - x_n p_n k) e^{i\xi_n \tau} = \frac{1}{2} ipk\tilde{P}e^{i\xi\tau}; & |\tau| < \frac{d}{2} \end{cases} \quad (3.45)$$

where

$$\tilde{Q} = p\eta\xi + kp_0 N + i(\eta^2 - k^2)kQ \sin\alpha e^{-pb}; \; Q = g\epsilon_1 \cos ga; \; \tilde{P} = 2ikQ\cos\alpha e^{-pb};$$

$$M = pQ(\eta^2 - k^2)\cos\alpha e^{-pb}; \; g = \sqrt{k^2\epsilon_1 - \gamma^2}; \; N = pQ\sin\alpha e^{-pb}$$

and the following conditions hold for coefficients: $c_n = d_n$, $n \neq 0$; $c_0 = d_0 + N$; $kp_n(c_n + b_n) = \eta\xi_n(d_n - c_n)$, $n \neq 0$; $kp_0(c_0 + b_0) = \eta\xi_0(d_0 - c_0) + M$.

These systems of functional equations determine the field scattered by slanted grating.

As it is shown in §8, the solution to the problem of complex analysis of reconstruction the function by its limiting values on the arc of unit disc which is called the Riemann-Hilbert problem [61] is the most effective way of reduction of the weak-convergent system (3.44), (3.45) to the canonical Fredholm infinite system of linear algebraic equations. Before to apply this method, one should fulfil a number of necessary transformations of initial systems and first of all to break it in two independent systems with respect to unknowns x_n and y_n defined on the whole grating's period.

Let us differentiate the first equations of (3.44), (3.45) with respect to τ and subtract the second equations from the obtained ones. After introducing the notation

$$\nu = \xi L; \; \mu = \eta L; \; \tilde{\kappa} = kL; \; \zeta = \tau L; \; \xi_n = \frac{\nu + n}{L}; p_n = iL|\nu + n|(1 - \tilde{\epsilon}_n);$$

$$w^2 = \tilde{\kappa}^2 - \mu^2; \ \epsilon_n = 1 + i\sqrt{\frac{w^2}{(\nu+n)^2} - 1}; \ L = \frac{\tilde{l}}{2\pi}$$

and taking into account additional equalities in the point $\tau = 0.5\tilde{l}$ one can reduce (3.44), (3.45) to the form acceptable for application of the method of Riemann-Hilbert problem [61]. Omitting intermediate calculations, let us write down infinite system of linear algebraic equations of the second kind

$$\tilde{x}_n + \sum_{s=-\infty}^{\infty} A^1_{ns}\tilde{x}_s = \tilde{B}^1_n; \tag{3.46}$$

$$\tilde{y}_n + \sum_{s=-\infty}^{\infty} A^2_{ns}\tilde{y}_s = \tilde{B}^2_n, \tag{3.47}$$

where $\tilde{x}_n = (n+\nu)x_n$, $\tilde{y}_n = (n+\nu)y_n$

$$\tilde{B}^{1,2} = B_{1,2}W^0_{\sigma 1,2}; \ W^s_{\sigma 1} = V^s_n(u) - V^s_\sigma(u)\frac{R_n(u)}{R_\sigma(u)};$$

$$W^s_{\sigma 2} = V^s_n(\tilde{u}) - V^s_\sigma(\tilde{u})\frac{R_n(\tilde{u})}{R_\sigma(\tilde{u})}; \ \ u = \cos\frac{\pi d}{l}; \ \tilde{u} = \cos\frac{\pi\tilde{d}}{l};$$

$$A_{ns1,2} = \begin{cases} q_{\nu s}W^s_{\sigma 1,2}, & -\infty < s \le -j-1; \\ q_{\nu s}W^s_{\sigma 1,2}, & 0 \le s < \infty; \\ (q_{\nu s} - 2)W^s_{\sigma 1,2}, & -j \le s \le -1; \end{cases}$$

here $q_{\nu s} = |\nu + s|(\nu + s)^{-1}$; j is the whole part of ν; $V^s_n(u)$, $V^s_\sigma(u)$, $R^s_n(u)$, $R^s_\sigma(u)$ are calculated in [61].

Now we determine from (3.46), (3.47) the unknowns $\tilde{x}_n$, $\tilde{y}_n$ and then obtain the solution to the problem of transformation of waveguide's surface waves to volume waves with the help of a grating transferring to corresponding formulas for expansion coefficients of diffractional fields.

2. Scattered fields given by the expression (3.47) have the form of an infinite spectrum of space harmonics. Depending on the parameters of surface wave and slanted grating these harmonics may either radiate to the free space as outgoing volume (plane) waves ($\Re\sqrt{A} > 0$) or propagate in the plane waveguide + grating as surface (slow) waves whose amplitude exponentially falls in the direction perpendicular to grating's surface ($\Re\sqrt{A} \le 0$).

Let us state the criterion of existence of outgoing plane waves in the diffractional field and write down the expression for radiation angle. In general the direction of propagation of radiating wave is given by the cosines of angles which radiation vector forms with coordinate axes Ox, Oy, Oz (in the considered case it is determined by propagation constants of diffractional harmonic along these axes). Let us introduce two angles $\tilde{\theta}$, $\tilde{\phi}$ (Fig. 63) formed by the projection of vector $\vec{k}$, the plane xOy and the axis Oz ($\tilde{\theta}$) and by the projection and vector $\vec{k}$ itself ($\tilde{\phi}$). The radiation direction is determined by the function

$$\sin\tilde{\phi}_n = -\frac{n}{\tilde{\kappa}}\sin\alpha; \; \tilde{\kappa} = \kappa_0\cos\alpha; \; \kappa_0 = \frac{kl}{2\pi};$$

$$\cos\tilde{\theta}_n = \frac{\frac{1}{\beta} + \frac{n}{\kappa}\cos\alpha}{\sqrt{1 - \frac{n^2}{\kappa^2}\sin^2\alpha}}. \tag{3.48}$$

It follows from (3.48) that variation of the angle of strips' inclination α_i of slanted grating yields scanning of radiation in the angles' sector determined by the parameters of surface wave and properties of normal grating ($\alpha = 0°$). When α reaches a certain critical value (α_n^c), radiation turns out to have the sliding character and propagates along the grating and for all other angles which are greater than this critical value radiating harmonic becomes a surface wave. We will determine the interval of variation of α for which there exists radiation to the free space by the condition $\Re\sqrt{A} > 0$ which yields

$$0 < \alpha_n < \arccos\frac{1}{2}\left(\frac{\kappa_0}{n\beta} + \frac{n\beta}{\kappa_0} - \frac{\kappa_0\beta}{n}\right)$$

$$0 < \tilde{\alpha}_n \leq \arccos\frac{n\beta}{\sqrt{\kappa^2(\beta^2-1) + 2\kappa_0\beta_n}}, \tag{3.49}$$

where $\beta = \frac{k}{\gamma}$; α and α_n are the angles of strips' inclination for the turning grating and slanted grating, respectively.

(3.49) yields high angle radiation sensitivity $\frac{\partial(\tilde{\phi},\tilde{\theta})}{\partial\alpha}$ in the case of diffraction of surface waves on the slanted grating ($\tilde{l} = f(\alpha)$). It is essential that, if there is only one propagating harmonic (for example, with $n = -1$) in the radiation spectrum for $\alpha = 0°$ (for normal grating), the new propagating harmonics do not emerge for all $\alpha \in [0°, 90°]$. And what is more, for $\alpha = \alpha_n$ the harmonic radiating to the free space becomes a surface one. In the case of multi-wave radiation with $\alpha = 0°$ and when the grating's turn angle varies the harmonics with big numbers are the first to disappear and to transfer to surface harmonics with big indices. Let us indicate that the angle sensitivity of radiation fields depends on harmonic's number: greater the number, higher the sensitivity. If there are no radiating waves for $\alpha = 0°$ in the diffractional spectrum, then they cannot emerge for any value of α.

For $\alpha > \alpha_n^c$ corresponding space harmonic is transformed to surface one and propagates at the angle δ_n to grating's strip where

$$\cos\delta_n = \frac{\sin\alpha}{\sqrt{1 + 2n\cos\alpha\frac{\beta}{\kappa} + \frac{n^2\beta^2}{\tilde{\kappa}^2}}}. \tag{3.50}$$

Phase velocities of these harmonics are given by corresponding propagation constants; they depend on the strips' inclination angle and harmonic's number n. Only zero space harmonic repeats the properties of incident surface wave and, as well as in the case of scattering of surface waves by the normal grating, it propagates along the axis Oz (for $n = 0$ $\beta_0 + \alpha = 90°$). It is obvious that surface wave of scattered field is reflected from the grating at the angle determined by the Bragg condition for those α which provide the fulfillment of this condition.

3. Let us determine energy characteristics of the transformation of surface waves to volume waves by the open structure dielectric waveguide + slanted grating. The calculated density of energy flow of the minus first space harmonic in the coordinates (α, u) for different values of initial period are presented on Fig. 64, 65 (the one-wave radiation regime for $\alpha = 0$ is on Fig. 64, the multi-wave regime is on Fig. 65). It follows from Fig. 64 that one can achieve the radiation of constant energy to the free space in the whole sector of variation of the angle α and, hence, in the whole interval of the angles of scanning by choosing the proper value of the grating's filling coefficient. The maximal density of energy flow is achieved for the narrow slots what is explained by strong currents generated by E-wave on the strips along the axis $O\tau$. For $u \to \pm 1$ as well as for $\tilde{u} \to \pm 1$ radiation vanishes and the degree of decrease in the domain of narrow slots grows due to the asymptotic properties of the density of energy flow which here has the order $\ln^{-1} 0.5(u+1)$.

Fig. 65 illustrates behaviour of the density of energy flow of the minus first harmonic when the harmonics with indices $n = -2$, $n = -3$ are present in the radiation spectrum. One can clearly observe the cut off points of higher order harmonics. Let us indicate that calculations are made for the slanted grating where both the period and filling coefficient are the functions of turning angle. One can easily obtain from this figure the essential information about the behaviour of radiating power for the turning grating, if to properly draw the cross-sections. Then the period and filling coefficient do not depend on α and required functions have gradual character. When grating's strips are turned at a certain angle to the direction of propagation of surface wave, the harmonics with big numbers are the first to disappear being transformed to surface waves.

As it has been already pointed out, the polarization of surface and scattered waves was not violated for $\alpha = 0°$ in the volume wave $\vec{E} = (E_x, E_s, E_\tau)$, $\vec{H} = (H_x, H_s, H_\tau)$ in the process of scattering of surface waves by a slanted grating. In the general case, when $\alpha \neq 0°$, the waveguide's electric field generates on the strips the currents with $j_s \neq 0$ such that H_τ corresponds to this current's component. Hence, the scattered field has linear polarization for $\alpha = 0°$ and for $\alpha \neq 0°$ it has elliptic polarization. Calculation results of the ellipticity coefficient of radiating harmonic show that for small angles of inclination radiation wave is almost linearly polarized. Analogous field's behaviour is observed for the grating with narrow slots or with narrow strips for all α. The ellipticity coefficient remains constant for arbitrary grating's filling coefficient beginning from certain values of angles of inclination. It is rather important for practical applications. With growing deceleration of surface wave the domain of big ellipticity coefficients shifts towards semi-filled grating.

4. The data presented in the previous section are obtained as a results of numerical solution to (3.46), (3.47). Properties of the matrix elements of these equations enable one to find in some particular cases analytical expressions for the amplitudes of scattered harmonics. Let us consider in details the regime of absence of radiation to the free space. It takes place when $\alpha > \alpha_n^c$ (α_n^c is determined in (3.50)) or when $\alpha = 0°$ (i. e., for such value of the period of normal grating l when the radiation condition does not hold).

The system of equations (3.46), (3.47) admits obtaining of analytical solution by the method of successive approximation when the norm of its matrix is less than unit. It is known [61] that the following estimate holds for this norm: $q < cw^2\sqrt{1-u^2}(1-4w^2)^{-0.5}$. It is clear that $q < 1$ for $u \to \pm 1$ or for $w < 1$. The interesting case is when u is arbitrary number and $w < 1$. It occurs when

$$\left| \tilde{\kappa}^2 - \frac{\kappa^2 \sin^2 \alpha}{\beta^2} \right| < 1. \tag{3.51}$$

Fulfillment of (3.51) gives possibility to obtain analytical solution to (3.46), (3.47). Let the parameters of open structure are chosen in such a way that the condition $\nu < 0.5$ holds what is equivalent to the demand $\beta > 2\tilde{\kappa}\cos\alpha$. Then approximate solution of (3.46), (3.47) yields the expressions for the principal harmonic of diffractional field

$$b_0 = \frac{ikQ\cos\alpha e^{-pb}}{1-U}; \; a_0 = \frac{-ikQ\cos\alpha e^{-pb}U}{1-U}; \tag{3.52}$$

$$d_0 = \frac{q\tilde{l}}{2\pi}\frac{p\sin\alpha e^{-pb}\tilde{U}}{1-\tilde{U}}; \; c_0 = \frac{pQ\sin\alpha e^{-pb}}{1-\tilde{U}},$$

where

$$U = \ln\frac{1+u}{2}; \; \tilde{U} = \ln\frac{1+\tilde{u}}{2}.$$

It follows from (3.52) that for $u(\tilde{u}) \to \pm 1$ (scattering of E-waves by the slanted grating with narrow slots) the H-component makes less contribution to the resulting field and the field of diffraction whose polarization almost coincides with the one of incident field, i. e., of the field of dielectric waveguide, has linear polarization. With growing slotwidth or the angle of strips' inclination the ellipticity coefficient increases. For $u = -1$ the field gone through the grating vanishes and for $u = +1$ it is identical to the field of surface wave of dielectric waveguide.

One can obtain from (3.52) the expressions for phase difference of E_s- and E_τ-components of electric field of the system dielectric waveguide + diffractional grating

$$\phi_{E_s} - \phi_{E_\tau} = \arctan\left(\frac{\tilde{\kappa}U(1-\tilde{U})}{(1-U)\tan\alpha}\right).$$

The formula for $\phi_{H_s} - \phi_{H_\tau}$ may be also obtained by simple calculations.

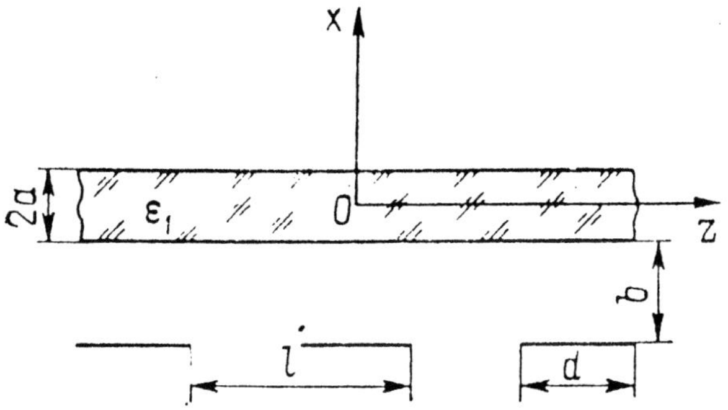

Fig. 55. Wave model of the Smith-Purcell effect.

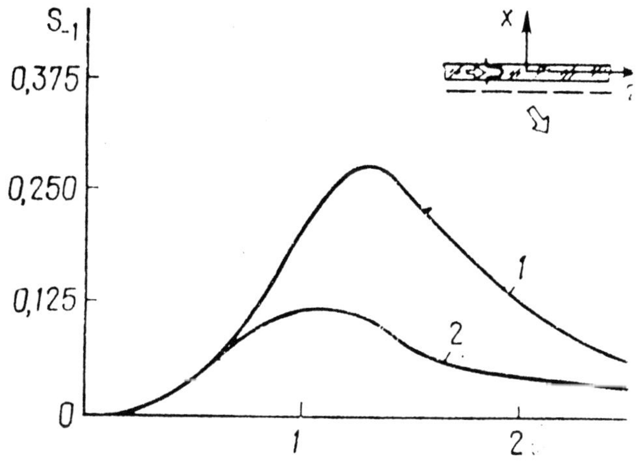

Fig. 56. One-wave power of of Smith-Purcell radiation.

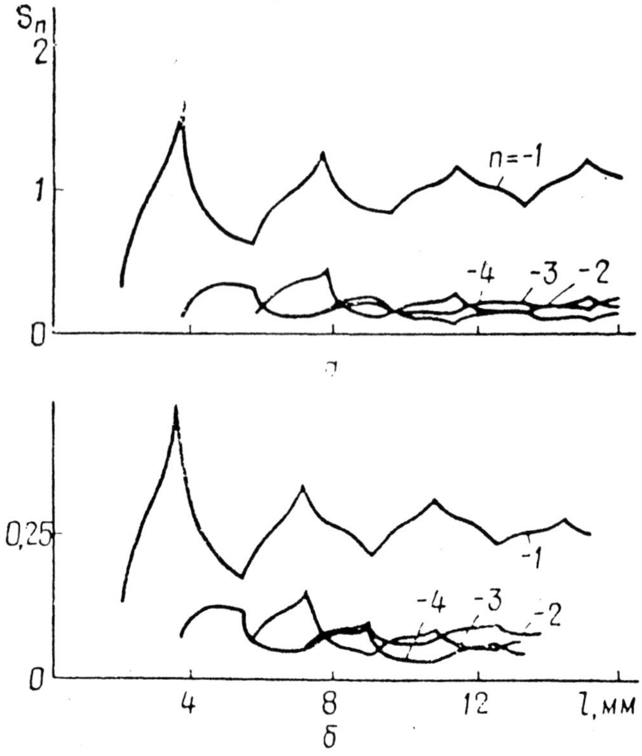

Fig. 57. Radiation power flow of Smith-Purcell radiation at the minus first harmonic.

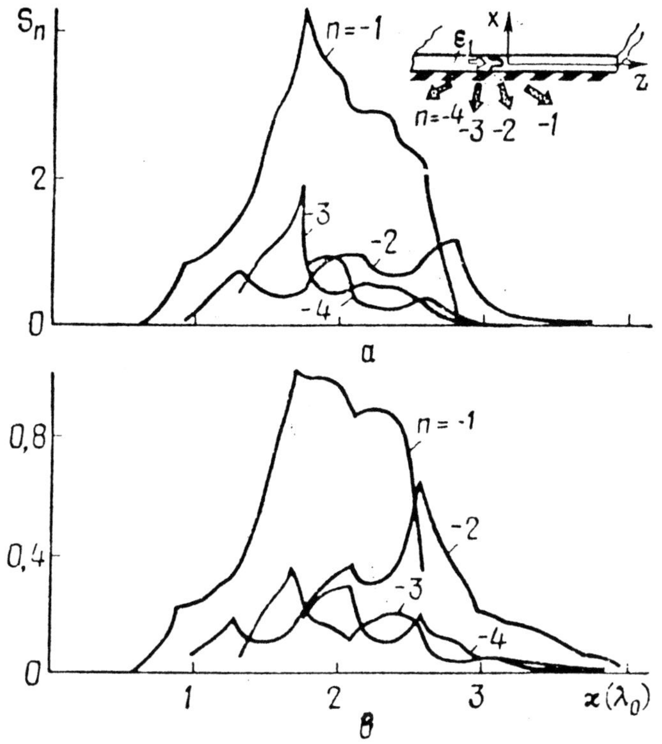

Fig. 58. Radiation power flow of Smith-Purcell radiation at the −1, −2, −3, −4 harmonics.

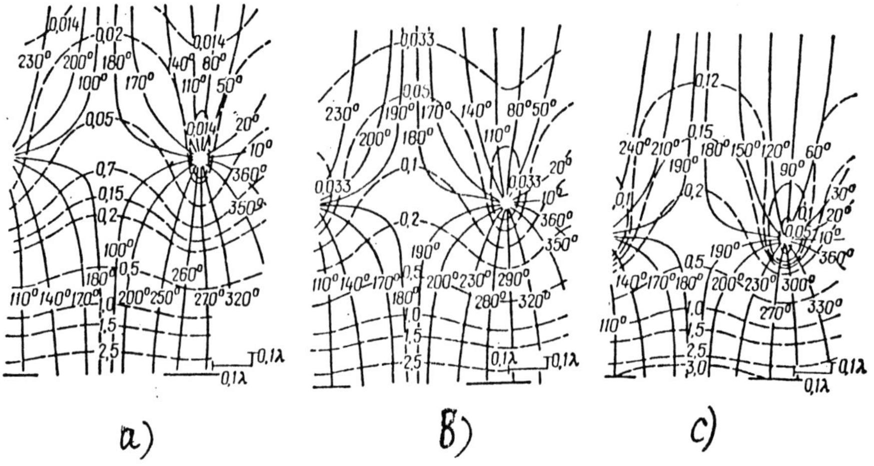

Fig. 59. Amplitude-phase distribution of the field of dielectric waveguide scattered by the grating when only the surface waves are formed ($a = 0.5$ mm, $b = 0.02$ mm, $u = 0.5$): a) – $l = 0.910$ mm; b) – $l = 0.915$ mm; c) – $l = 0.020$ mm.

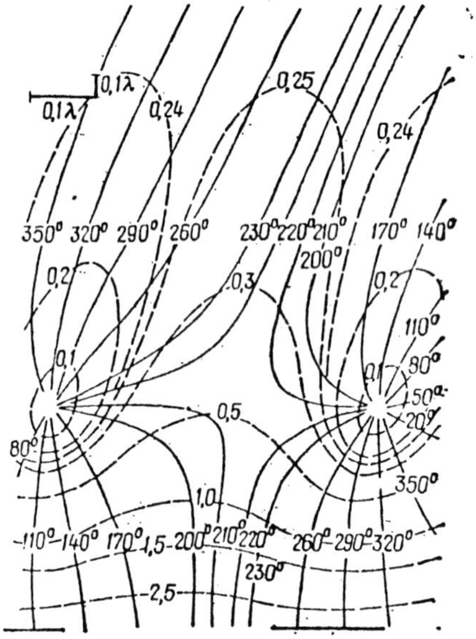

Fig. 60. Amplitude-phase distribution of the field of dielectric waveguide scattered by the grating when Smith-Purcell radiation is formed at the minus first harmonic ($l = 1.930$ mm).

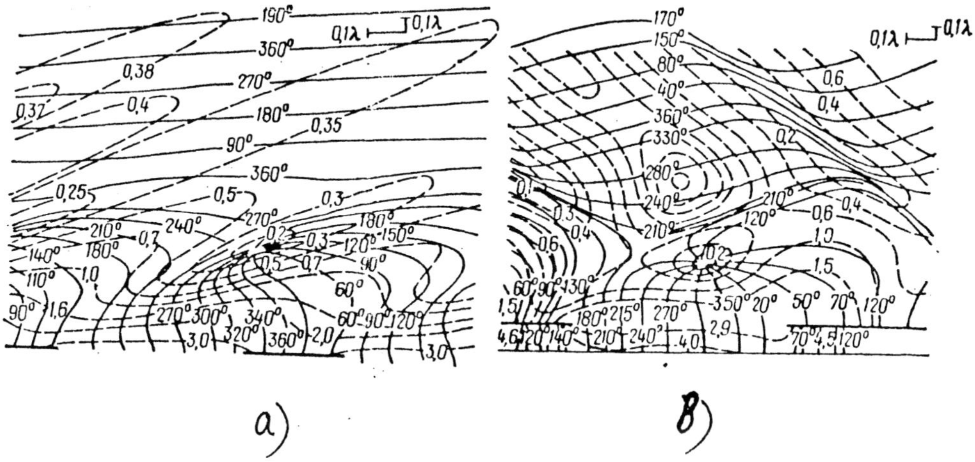

Fig. 61. Smith-Purcell radiation almost at the angle 90° to grating's surface: a) – $a = 0.5$ mm; $b = 0.02$ mm; $u = 0.5$; $l = 0.3$ mm; b) – radiation of two field's harmonics, $a = b = 0.5$ mm; $u = 0.4$; $l = 0.4$ mm.

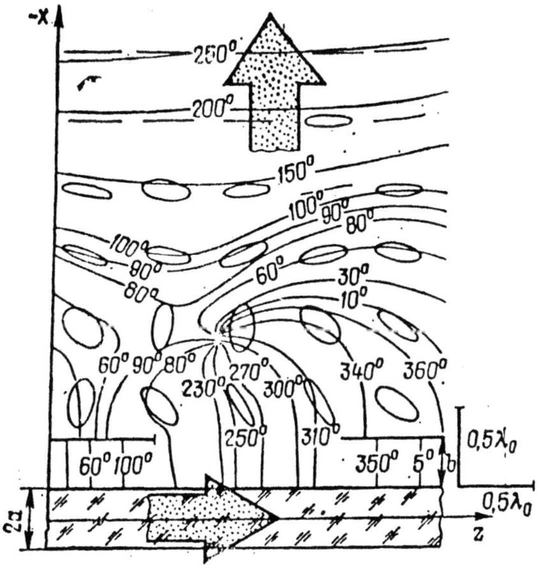

Fig. 62. Phase portrait amd polarization picture of Smith-Purcell radiation.

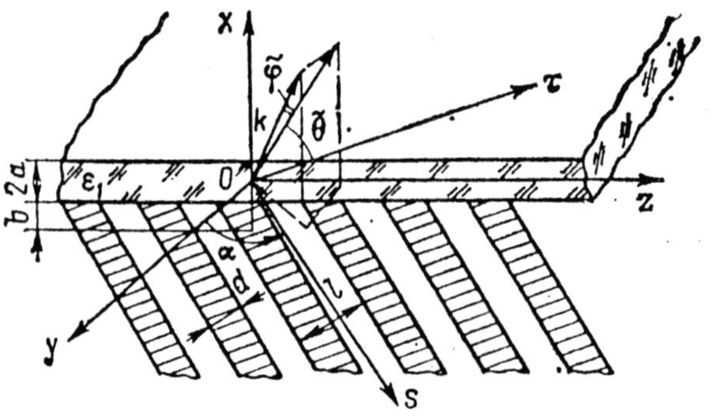

Fig. 63. Dielectric waveguide and beveled grating.

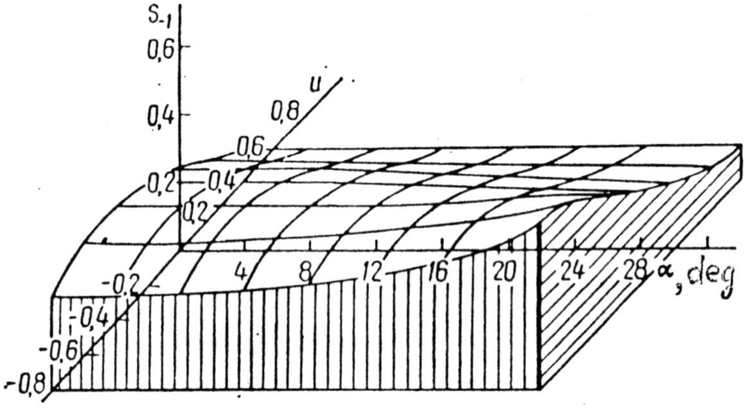

Fig. 64. One-wave regime of Smith-Purcell radiation for normal grating ($\alpha = 0$).

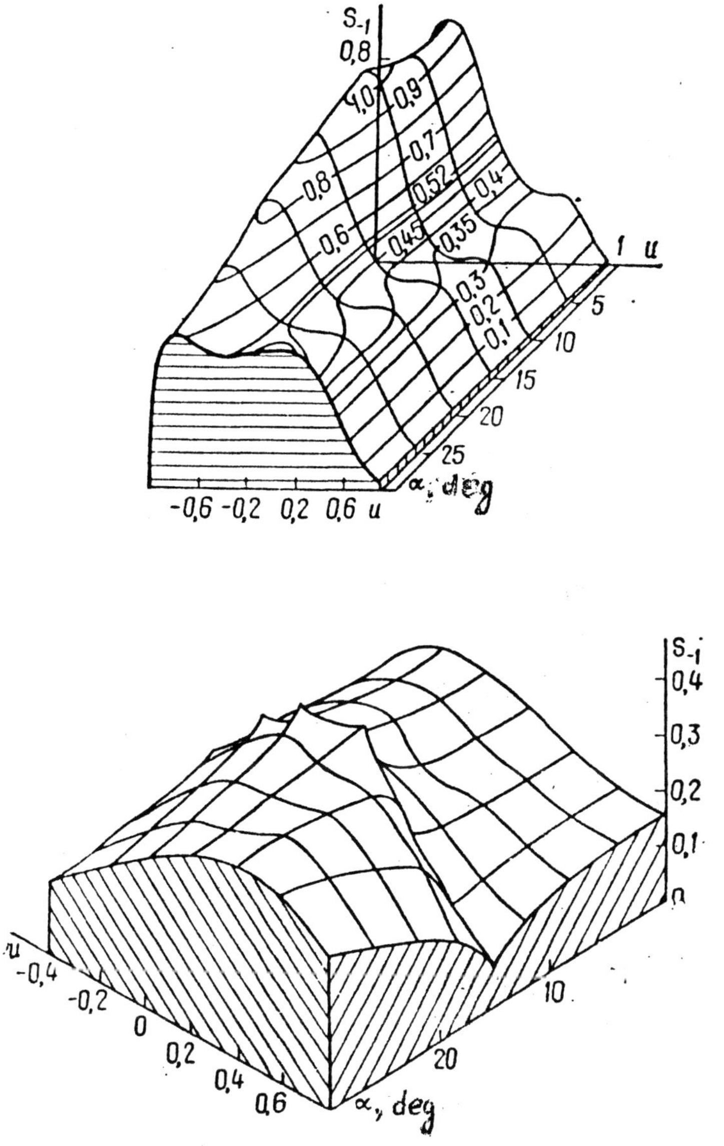

Fig. 65. Multi-wave regime of Smith-Purcell radiation for beveled grating ($\alpha \neq 0$).

Part 2

EXPERIMENTAL STUDY OF SMITH PURCELL EFFECT

Comparatively small number of works are devoted to the "pure" experimental study of Smith-Purcell effect. Here those measurements are assumed to be "pure" which are carried out only for electron flow moving close to a grating when there is practically no inverse influence of radiation on the movement of particles (except of the influence of grating's eigenfield).

Experimental investigations of SPR are mainly connected with applications of more complicated electrodynamical structures, mostly open resonators (OR) consisting of two mirrors such that one of them is a plane diffractional grating and the other is a part of spherical surface (or having other canonical form) [43, 113-120]. The presence of a plane mirror completely covered by a grating caused essential difficulties in the studies when radiation's intensity grows because of high non-controlled ohmical and radiation losses in a grating. Therefore, in early works the question concerning the possibility of application of SPE for generation (or amplifying) of electromagnetic oscillations in the millimeter and submillimeter ranges remained open (due to low powers). Naturally, it was also burdened by circumstances arising from the theory of blinking dipole (see Chapter I).

Of principle new ways for experimental investigations of SPE had been opened when we proposed OR whose lower mirror had the form of plane (or cylindrical) surface partially covered by a grating. OR with a grating occupying 15-75 per cent of resonator's mirror have been already considered in the first theoretical and experimental works [63-65, 122-127]. The results of these innovations in constructing the theory of SPE are analyzed in the second chapter of this book.

Carrying out experiments with OR having complicated mirrors required radical reconstruction of the whole experimental technique. First of all, the methods of visualization of electromagnetic fields in OR have been elaborated and optimal types of open structures have been constructed on this basis for their effective applications in the millimeter and submillimeter technology. Then experimental holographic studies of fields in these ranges have been developed. If to add to these approaches the methods of investigations of fields' polarization properties, it would practically complete the system of experimental tools for the study of SPE.

One should apply together with enumerated methods the broad set of modes for the study of SPE in its "cold" development, i. e., under the conditions of wave modeling. This qualitatively new situation in the consideration of radiation phenomena allows to carry out directly "on the table" rather complicated (and very expensive!) experiments which enable one to measure basic kinematic characteristics of SPE. Such experiments are of considerable independent significance, since they open the way for creating new family of antenna systems, as well as units and devices of millimeter and submillimeter technology.

These results as a whole formed the basis for the chapters 4,5 and 6. The "pure" SPE or the "hot" SPE in its natural development is considered in the fourth chapter; the "hot" SPE in OR is treated in the fifth chapter and the sixth chapter is devoted to the "cold" SPE, i. e., to its wave modeling.

CHAPTER FOUR

"Hot" experiments with "pure" Smith-Purcell effects.

Here we will consider first of all the experiments carried out by Smith and Purcell in 1953 analyzing them from the modern point of view. It is essential in our interpretation of these experiments that we make an assumption about close connection between electromagnetic fields accompanying electron flow, which moves uniformly and straightforwardly, and eigenfields of diffractional grating. This approach enables one to completely explain the results of Smith-Purcell experiments (unlike the blinking dipole theory).

It concerns even to a more considerable extent to late experiments described in [55] where, as well as in [1], the main conclusion may be formulated in the following way: with growing wavelength λ the intensity of SPR falls as λ^{-2}. Hence, one can expect more or less high power of SPE in the optical range. For longer waves (infrared, millimeter or submillimeter) this power is negligibly small.

It had been shown already in 1963-1965 in Kharkov that these results following from the blinking dipole theory turned out to be mistaken. All further experiments carried out in the millimeter and submillimeter ranges confirmed the main conclusion: SPE is perspective not only in optical but in other wave ranges.

§10. Smith-Purcell experiments and their analysis.

In Smith-Purcell experiments the relativistic flow of electrons well focused by the static and magnetic field was passed through close to the plane optical reflection diffractional grating with the period $l = 1.67$ μm [1,2]. As a result, radiation emerged at different angles visible by naked eye. The central radiation wavelength varied from 0.57 μm to 0.485 μm in the observation sector $\Delta\theta \approx 3.5°$ at the angle $\theta \lesssim 20°$ to the direction of pencil's movement when accelerating voltage grew from 309 kV to 340 kV. The obtained data, given on Fig. 66, where it is also indicated the phase raid per period ψ and wavelength λ, are rather simply interpreted by the model of modulated electron flow.

Before to analyze the Smith-Purcell experiments, one should point out that the modulation of electron flow is not mentioned in [1]. Besides, its density would be too small for the modulation in optical range ($i_0 < 0.028$ Acm^{-2}). Average distance between electrons was more than 50 μm what was in 100 times greater than the wavelength ($\lambda \approx 0.53$ μm). Only several (about eight) electrons could simultaneously move along the whole length of diffractional grating ($L = 48$ mm) in the interaction zone, i. e., at the distance less than $0.1l$ from the grating. The questions naturally arise: what is the reason of radiation, do the usual charges radiate or is there some other mechanism of SPE.

Mirror image of isolated charge flying close to a periodical structure oscillates with the frequency $\Omega = 2\pi v_e l^{-1}$ what yields $\Lambda = l\beta_e^{-1} \approx 2.1\mu$m, where $\beta_e = v_e c^{-1}$. If $\vec{k}$ is the wave vector in the direction of observation, Doppler frequency transformation has the form [105]: $\omega - \vec{k}\vec{v}_e = \Omega$ and, consequently, $\lambda = (1 - \beta_e \cos\theta)\Lambda \approx 0.48 - 3.8\mu$m. Such radiation is multi-directional and its spectrum is continuous. One can receive only one photon from every electron coming to interaction zone under the conditions of the experiment described in [1], i. e., calculated power of the registered radiation generated by the whole flow of equal electrons has the order of $2 \cdot 10^{-8}$ Wt. Unlikely to discover such power without special equipment.

More significant reason of the disparity of radiation of isolated electrons and the Smith-Purcell effect is the presence of certain anomalies occurring in the process of variation of frequency and direction of radiation. For example, in accord with the Huygens construction, $\lambda = l(\beta_e^{-1} - \cos\theta)$, what formally coincides with the Doppler frequency transformation, and $\Delta\lambda = 0.05\mu$m (if ϕ =const, $\beta_e = 0.78 - 0.80$) in place of the value $\Delta\lambda = 0.09\mu$m registered by Smith and Purcell. If to consider the true variation of wavelength, the radiation angle will vary in the sector $\Delta\theta \approx 3.5°$, i. e., from $\theta \approx 16.5°$ for $\lambda = 0,485\mu$m, $\beta_e = 0.8$ to $\theta \approx 20°$ for $\lambda = 0,57\mu$m, $\beta_e = 0.78$ in spite of the constant value of observation angle. Even the width of the line of observed radiation in the presence of a number of factors promoting its broadening is only $\delta\Lambda = 0.020\mu$m, i. e., it is narrower than the calculated value $\delta\Lambda \approx 0.035\mu$m for the sector $\Delta\theta \approx 3.5°$ and β_e =const.

One can observe in the optical radiation range not only the first ($m = 1$) but also the higher orders ($m \leq 5$) emerging for $\theta = 90°$ where there is almost no Doppler frequency shift (i. e., $\lambda = \Lambda \approx 2.1\mu$m). It completely excludes the possibility to explain SPR by the Doppler frequency shift in the radiation of isolated charges because such effect occurs only in the presence of higher order current's harmonics [1]. Hence, there exists the mechanism of phase sorting of isolated electrons in Smith-Purcell experiments in spite of the absence of inverse communication in the effect of diffractional radiation itself. The regime of eigenwaves of diffractional grating acts like this mechanism, since for it one cannot in principle eliminate inverse communication.

2. Eigenwave of diffractional grating is continuously fed by single electrons only at those frequencies for which $v_e \geq v_{\Phi n}$. Electrons most effectively communicate with eigenwave in the case of Bragg reflection when $|v_{\Phi 0}| \approx |v_{\Phi -1}|$ at the wavelengths $\lambda = 2l\beta_e^{-1} \approx 4.2\mu$m [3]. Surface wave propagating to meet electron flow carry out positive inverse communication by means of transversal HF field's component in the presence of strong longitudinal magnetic field and slightly inclined electrons' movement. It draws electrons with required phase from inside the flow ($\sim 2l$) into the space of effective interaction ($\sim 0.1l$) not only placing them close to the grating at the distance divisible by $2l$ but also increasing their density in the domain of interaction. Finally we obtain considerable growth of intensity and the degree of orientation of radiation.

Modulated electron flow always has a lot of higher order current's harmonics which generate diffractional radiation. Radiation frequency at every time harmonic with the index p is divisible by the frequency of modulation. At every harmonic the flow radiates its specific wavelength $\lambda_p = \lambda_w p^{-1} = lp^{-1}(\beta_e^{-1} + cv_{\Phi_0}^{-1})$ at specific angle, hence, the color of radiation depends on observation angle [1]. For example, one can simultaneously observe at different angles not only blue (or green) ($p = 8$) but also dark blue ($p = 9$), yellow (or orange) ($p = 7$) or even red ($p = 6$) radiation (see Fig. 66). Radiation at longer wavelengths ($\lambda = 0.8 - 2.1\mu$m) ($p = 5 - 2$) could not be registered by optical devices since it was in infrared part of the spectrum of electromagnetic waves.

We may conclude that positive inverse communication in the Smith-Purcell radiation is realized by eigenwaves of diffractional grating which provide phase correction and participation in interaction of noticeably greater number of electrons at minimal distances from the grating, promoting thus considerable (in ~ 1000 times) growth of radiation intensity in discrete directions up to the levels visible by naked eye. It completely agrees with physical principles of emerging diffractional radiation on periodical inhomogeneities by means of the transformation of eigenfield of uniformly moving charges localized in the space to homogeneous plane waves of radiation. It is obvious that one can consider as inhomogeneities not only bodies but also HF fields close to them.

3. If to carry out calculations by the formulas of the theory of blinking dipole (1.1), (1.2) and to compare them with the data obtained in Smith-

Purcell experiments [1] where it is established coherent luminescence of intensive pencil of electrons moving from Van-der-Graaph generator close to optical diffractional grating, the analysis of wavelength λ as a function of θ shows that radiation occurs at a certain angle θ_r and variation of the color (wavelength) depends on observation angle, accelerating voltage (β_e^2) and grating's inclination (variation of its period) what is confirmed by theoretical expression (1.1). Here the light of radiation was strongly polarized with electric vector perpendicular to grating's surface. At the same time, one cannot correctly calculate radiation's intensity, determine its structure and the properties of variation of the main characteristics versus grating's parameters and radiation angle θ by the interrelation (1.2).

§11.

Experiments of Ishigaro and Tako

1. Experiments of Ishigaro and Tako [55] were carried out ten years after Smith-Purcell experiments. They obtained for the first time the measurement data of SPR of non-relativistic electron flow (up to 10 kW). They verified in [55] more precise expression for the total radiation intensity than in [1,2] using the approximation of blinking dipole:

$$I_{\text{tot}} = 2.2 \cdot 10^{-4} \frac{d^2}{l^4} L U^{1.5} J \cos\theta, \tag{4.1}$$

where I, [Wt cm^{-1}]; U, [kV] is the voltage; J, [μA], is the pencil's current; d, l [μm]; L, [cm].

Formulas (1.2), (4.1) establish the following properties of Smith-Purcell effect: with growing wavelength radiation's intensity falls proportional to λ^{-2}; intensity of Smith-Purcell radiation increases proportional to the growth of pencil's current ($I \sim J$); with growing accelerating voltage radiation's intensity increases proportional to $V^{1.5}$ (unlike the result given by formula (1.2) where W decreases as $V^{-0.5}$); the flow generates maximal power in the direction of electrons' movement ($\theta = 0°$) and it does not radiate energy at the angle $\theta = 90°$ to the grating's surface.

Considerable fall of radiation's efficiency with growing wavelength and the obtained experimental data are evidence of the fact that the use of Smith-Purcell radiation in the long-wave part of optical range and, consequently, in millimeter and submillimeter ranges is not perspective. That is why on the first stage they studied this effect in the optical range of wavelengths.

The studies were carried out in the range $\lambda = 6 - 60\mu$m. General scheme of the experimental model of Ishigaro and Tako together with experimental data are presented on Fig. 67. Temperature effects of heating the cannon's cathode and model as a whole were the limiting factors which considerably influenced the measurements of radiation's characteristics with the help of thermopair. Nevertheless, the authors could register the radiation power up to 50μWt and obtain wavelength and radiation's intensity versus accelerating voltage and pencil's current J (Fig. 67, b-d). These characteristics yield qualitative agreement of experimental data with the properties of SPE established in accord with (4.1).

2. In spite of the obtained reliable experimental results (including discovery and investigation of some features of variation of SPR characteristics

in the optical range of wavelengths), physical interpretation of theoretical data within the frame of blinking dipole approximation has led to mistaken conclusions. It turned out to be especially negative in the attempt to use SPE in the millimeter and submillimeter technology. Really, in accord with (1.2) and (4.1), radiation intensity essentially falls with growing wavelength and it is not possible to realize the discovered effect at the angle $\theta = 90°$. It is known [3] that these very characteristics of SPR form the basis for development of the modern diffractional electronics. The latter required creating the theory of diffractional radiation within the frame of the model of given current and taking into account inverse influence of the field of diffractional grating on the movement of electron flow.

§12. Experimental methods for the study of Smith-Purcell radiation in the millimeter range and the results of experiments.

Theory of SPE which has been developed in the first two chapters enabled us to make fundamental conclusion about the possibility to generate powerful radiation in the millimeter and submillimeter wave ranges for non-relativistic electron flows. It opened unique ways for broad implementation of SPE in various areas of science and technology. In order to complete the solution to the problem one should confirm theoretical results by the experiment. In these sections we will develop from the physical point of view diverse system of experiments which will not only justify the theory but also reveal utterly new verges of SPE itself putting it in to the row of already known radiation phenomena and allowing to assert that this effect is characterized by original and distinctive display.

1. Experimental study of the "pure" SPE, which should be made only in the presence of electron flow and diffractional grating, requires appropriate methods and specially created plants. It is constructed SPR analyzer for the use in millimeter range, $\lambda = 3.8 - 5.6$ mm, whose scheme is presented on Fig. 68. Its detailed description is given in [5, 12, 128].

Analyzer contains electron cannon 2, which forms the strip electron flow 1 with cross-section 0.1×3.8 mm^2, diffractional grating 9, electrons' modulator 3, electrons' collector 8 and thin-wall vacuum tank 4. A flow is focused over the grating by constant magnetic field generated by the electromagnet. The grating with the period $l = 0.4$ mm is placed along the axis of cylindrical mirror with curvature radius 110 mm and aperture's size 40×40 mm^2. The specially constructed small-size GDR [21] is used as the modulator 3. Its OR is formed by the short-focus spherical mirror with curvature radius 15 mm (aperture's diameter is 18 mm) and by a part of cylindrical mirror with the grating L_1 of the length 16 mm placed from the side of anode of electron cannon.

The considered SPR 5 is excited by the modulated electron flow on the grating L_2 situated betweeen modulator and collector 8.

SPR-registering device has the mouthpiece receiving antenna 7, polarimeter 6 described in [4], SHF-modulator and detector section. SPE signal comes from detector to the measuring oscillograph. Receiving apparatus is placed on the scanning device which measures the main characteristics of SPR and analyzes its far field pattern in equatorial plane in the range of angles $\phi = 60° - 120°$ counted with respect to the direction of electrons' movement, and in meridional plane in the range of angles $\phi = -90° - 90°$ counted with respect to the plane perpendicular to the grating and going through the axis of electron flow. A part of the energy of SPR comes from polarimeter through calibrated coupler to the thermistor measuring bridge registering radiation's power. The measurement accuracy of SPR angles by the mouthpiece antenna is $\pm 30''$. Polarization of radiation field is controlled by the analyzer in the form of rotating segment of the waveguide junction joining circular and rectangular cross-sections. Rotation angle of polarization plane is counted with the accuracy $\pm 5''$. Frequency modulation of electron flow in the process of experiment is established by varying electrons' velocity v_0, i. e., by the regulation of anode voltage U_0 of electron cannon and by mechanical readjustment of the resonator of small GDR.

Angles of diffractional radiation is determined by the formulas (2.25), (2.46) and the width of far field pattern of the main petal (at the level -3 dB related to maximal intensity) is calculated by the interrelation known from antenna technology [3] and applied for synphase radiators, $\delta\phi° = 65\lambda(lN_l)^{-1}$, where N_l is the number of periods of diffractional grating L_2.

Of principle feature of the measurements of SPR made by the plant presented on Fig. 68 is not only the use of millimeter waves (and, consequently, the whole set of measuring facilities applying in this range) but also the specially created units such as small electrons' modulator constructed on the basis of GDR, polarimeter, *et. al.* These units themselves are the embodiment of SPE in the devices of diffractional electronics. Application of these experimental means enabled us to complete deep and comprehensive study of SPE.

In order to exclude side effects connected with radiation capabilities of modulator's resonator and various field's re-reflections by inhomogeneities, cold modeling of radiation's excitation in the analyzer has been preliminarily carried out.

2. Diffractional fields of quasi-optical complex have been studied in the cold regime of excitation of SPR in order to determine the sector of working angles where one could observe and investigate with high trustworthiness characteristics of diffractional radiation of electron flow.

The following effects may influence radiation far field pattern of analyzer's grating: external resonance field of OR itself caused by the diffraction of wave pencil on resonator's mirrors; shading of the registered radiation by modulator's mirror; radiation of excitator's eigenfield out of OR. In accord with these factors, cold experimental studies of the radiation far field pattern in the model analyzer are carried out under the conditions listed below:

a) excitation of the system by the field of H-polarized wave generated by the communication slot cut in the middle of the mirror of modulator's OR along the axis of the grating $L_2 = 18$ mm;

b) excitation of the model analyzer by dielectric waveguide placed over periodical structure.

In all cases the origin of rotation of the mouthpiece antenna with linear polarization lies on the axis going through the center of the grating L_2 perpendicular to its plane what corresponds to the real conditions of hot measurements of diffractional radiation ($\phi' = 90° - \phi$).

The considered cases of cold modeling of excitation of diffractional radiation are studied in details in [10]. Let us present the most interesting results indicating the features of the forming of far field patterns in the analyzer of SPR.

Dielectric waveguide placed over periodical structure was used as an excitator in order to bring the conditions of cold measurements closer to hot experiment with quasi-optical electrovacuum complex. If an eigenmode is excited in the dielectric waveguide, it provides interaction of the slow wave of dielectric waveguide with periodical structure and its transformation to volume wave. As it is known, bringing narrow dielectric waveguide into the volume of OR leads to perturbation of its resonance field because such waveguide acts as a strong phase inhomogeneity (similar to local reflection diffractional grating [3]). Therefore, we use for the modeling of diffractional radiation planar single-mode dielectric waveguide in the form of thin dielectric disk whose diameter corresponds to the size of plane mirror with diffractional grating. This waveguide is fed by SHF generator of millimeter range through the waveguide-dielectric junction surrounded by an absorber.

The grating's period is chosen for excitation of diffractional radiation in accord with the formula $\cos\phi = \beta_\epsilon^{-1} + n\kappa^{-1}$ where $\beta_\epsilon = v_\epsilon c^{-1}$, v_ϵ is phase velocity of the wave in dielectric waveguide, n is the number of diffractional harmonic. The number of periods of the gratings L_1, L_2 considerably decreases ($N_1 = 6$, $N_2 = 7$) due to high values of β_ϵ what may cause broadening of the far field pattern of excited diffractional radiation.

The angles of SPR (of maximal intensity) of electrodynamical system (plane mirror partially covered by periodical structure excited by dielectric waveguide) versus radiation frequency are given on Fig. 69. For the frequency of feeding generator $f = 70$ GHz the system radiates close to normal direction ($\phi' = 0$) and when the frequency f varies, radiation angle ϕ' obeys theoretical interrelation $\cos\phi = \beta_\epsilon^{-1} + n\kappa^{-1}$. One should point out good agreement of experimental and theoretical data modeling radiation of electron flow. Characteristic radiation patterns on the mirror with diffractional grating are presented on Fig. 70. For $f = 70$ GHz maximum of far field pattern is concentrated close the normal direction and has one petal in the dynamical range 20 dB. With varying frequency far field pattern keeps the one-petal shape, although one can observe emerging of additional interference petals with the level of intensity up to $0.4F_{\max}$ (Fig. 70, b).

Let us consider the forming of patterns of radiation from the analyzer in the presence of spherical short-focus mirror placed over dielectric waveguide in such a way that the edge of its aperture coincides with the edge of plane mirror from the side of waveguide -dielectric coupler. When a part of the grating is covered by metal foil, the far field pattern has complicated form (Fig. 71, a). It is connected with broadening of the pattern of radiation which feeds OR and with diffraction of radiation by resonator's spherical mirror; the radiation also interferes with scattered fields. The grating L_2 strongly influences radiation far field pattern (Fig. 71, b): the main petal emerges which corresponds to excitation of diffractional radiation on the grating L_2 at the angle given by the formula $\cos\phi = \beta_\epsilon^{-1} + n\kappa^{-1}$. Lateral petals in the interval $-30° < \phi' < 0°$ are mainly caused by external diffractional field of OR; for $\phi' > 18°$ they emerge due to the scattering on the boundary of plane mirror with the grating and by variation of the coefficient of communication between dielectric waveguide and periodical structure. If for a certain radiation frequency the maximal value of intensity of diffractional radiation lies in the interval $0° < \phi' < 30°$, the presence of spherical segment leads to the $4° - 5°$ shift of this maximum towards positive angles ϕ'. When the petals with maximal intensity belong to the sector of negative angles $\phi' < 0$, this shift is more essential and, for example, reaches 8° for $f = 68$ GHz.

Hence, appropriate choice of the parameters β_0, κ for the given size of spherical mirror of modulator's OR always allows to separate the petals of the far field pattern of radiation excited on the grating L_2. Reliability of such separation grows when the number of periods of diffractional grating increases, since here the width of radiation far field pattern decreases what is characteristic for the case of excitation of diffractional radiation by an electron flow.

In the process of cold modeling of excitation of the prototype analyzer of diffractional radiation the ratio of field's amplitudes in the maximums of far field patterns is 18 – 24 dB at the main component and cross-components, i. e., mainly the H-polarized wave is excited in such system. However, it is shown below that polarization characteristics considerably change when radiation is excited by an electron flow.

3. One can study with the help of the plant depicted on Fig. 68 not only usual characteristics of SPR but also the thin structure of diffractional radiation. It is established that, as well as in theory, the measured far field patterns of diffractional radiation are multi-petal what is caused by excitation of longitudinal and transversal waves in electron flow. Maximal intensity of some petals reaches $1 - 3$ MWt cm^2. The difference between angles of diffractional radiation of SWSC and FWSC is $15° - 25°$ while for SSEW and FSEW this difference is $6° - 8°$ when the current's density $i_0 \approx 30$ A cm^{-1} and focusing fields $B = 0.25 - 0.5$ Tl. In accord with radiation conditions (2.5), (2.25), one can excite radiation by one or by several electron waves simultaneously for the given electrons' velocity, if to perform mechanical readjustment of OR-modulator. In the first case radiation far field pattern has one (main) petal oriented at the angle whose value is close theoretical figure obtained by (2.5), (2.46) for a certain electron wave excited in a flow. In the second case one registers two- or three-petal radiation far field pattern. The two-petal pattern corresponds to radiation of SWSC and SSEW or of FWSC and FSEW. In the three-petal pattern (Fig. 72) central petal corresponds to diffractional radiation of SEW, right petal – of SWSC (smaller value of the angle ϕ), left petal – to FWSC (greater value of the angle ϕ). Variation of the density of pencil's current i_0 and electrons' velocity β_0 gives similar

results.

The data of measurements of radiation angles in the plane of Brilluin diagram us presented on Fig. 72. Circles indicate the parameters calculated for SEW, triangles – for SWSC and FWSC, squares – for SCW and FCW; experimental data is marked by crosses. They lie on the lines κ =const. One can see that for $U_0 = 1100$ V ($\kappa = \kappa_1$) radiation is excited by two electron waves, FWSC and SEW, in the working range of angles ϕ determined above; for $U_0 = 2555$ V ($\kappa = \kappa_2$) radiation is excited by one SWSC. For $U_0 = 2545$ V ($\kappa = \kappa_3$) three-petal far field pattern emerges which corresponds to excitation of diffractional radiation by SWSC ($\phi = 71°$), FWSC ($\phi = 94°$) and SEW ($\phi = 83°$); calculated radiation angles are depicted by bar lines on Fig. 72. The measured radiation angles deviates from their theoretical values obtained with the help of formulas (2.5), (2.46) without taking into account the reduction coefficient of plasma frequency. In the cases presented on Fig. 72 SCW and FCW do not excite diffractional radiation since here the radiation condition (2.45) is not valid (theoretical values of the parameters κ, $\beta_c^{\pm} = \beta_0(1 \pm \omega_c\omega^{-1})^{-1}$ in the domain of excitation of surface waves are marked by squares).

It has been established in the process of measurement of polarization characteristics of SPR that electron waves excite diffractional radiation with polarization of electromagnetic field which differs from the expected H-polarization predicted theoretically without taking into account spiral electrons' movement in the strip flow and also from the polarization obtained as a result of cold modeling of excitation of diffractional radiation. Far field patterns for H- and E-polarizations are shown on Fig. 73. One can see that theoretical patterns well agree with experiment and electron flow effectively excites diffractional radiation in both cases. Orientation angle of polarization ellipse γ_E (counted with respect to the positive direction of electrons' movement), ellipticity coefficient $k = ba^{-1}$ and radiation intensity in some petals essentially depends on current's density i_0, accelerating voltage which focuses magnetic field and the form of electron wave exciting this radiation.

It has been established experimentally that SWSC and FWSC excite diffractional radiation with linear polarization of electromagnetic field. For example, Fig. 74 shows the behaviour of intensities of SWSC diffractional radiation and the angle γ_E versus U_0 within the limits of generation zone of the small GDR-modulator for different values of pencil's current. Here the ellipticity coefficient of diffractional radiation takes the values $k \approx 0-0.15$ in the whole range of U_0. The strongest variation of angle occurs in the domain of smooth generation regime. With growing SHF field's amplitude in GDR or falling density of electron flow the range of variation of γ_E decreases. With growing U_0 the angle γ_E reaches certain minimal value ($12° - 15°$) which does not change until the breakdown of generation when the field of SPR has some deviation from the H-polarized field keeping nevertheless the domains of presumably linear polarization. The latter is caused by the influence of transversal electron waves on excitation of diffractional radiation by WSC and this effect is confirmed by experimental investigations of the influence of focusing magnetic field on the state of polarization of diffractional radiation: with growing magnetic field the angle γ_E of diffractional radiation excited by SWSC and FWSC increases and depends on the direction of force lines of B related to electrons' movement (Fig. 75). This data describing γ_E versus the magnitude and direction of B are measured in the domain of smooth regime of excitation of SWSC in the modulator. They are characterized by the presence of sharp turn of polarization plane of diffractional radiation with varying direction of focusing magnetic field.

Unlike longitudinal waves of space charge, SCEW and FCEW propagate in a flow with equal phase velocity which approximately coincides with the velocity of unperturbed electrons v_0. It is known [4] that this factor essentially impedes in classical SHF electronics the possibility of experimental separation of CEW. However, these waves are easily discovered by the characteristics of diffractional radiation (polarization and "splitting" of radiation far field pattern), since phase velocities of SCEW and FCEW negligibly differs under the influence of space charge [104].

In the "double" signals of diffractional radiation of synchronous waves presented on oscillograms (Fig. 76) the right signal corresponds to the radiation of SWSC ($\gamma_{0,-1} = 98°$), the left signal – to the radiation of FWSC ($\gamma_{0,-1} = 104°$). These results well agree with the data calculated by the formula (2.46); in accord with this formula, diffractional radiation of SEW must be observed at the angle $\gamma_{0,-1}$ close to $100°$. However, such conditions turn out to be very sensible to variations of i_0 and magnetic field B what is illustrated by oscillograms on Fig. 76, a, b. Decreasing current's density (for constant magnetic field, Fig. 76, a) or B (for i_0 =const, Fig. 76, b) yields narrowing of generation zone. The "splitting" effect in radiation pattern vanishes beginning from the certain moment, hence it is difficult to separate SWSC and FWSC by radiation angle.

Oscillograms of SPR showing the features of separating the radiation of left- and right-polarized SEW with the help of receiving polarization analyzer are given on Fig. 76, c. The rotation angle of polarization plane for these two waves are $\gamma_E = 45°$ and $\gamma_E = 48°$, respectively. One can conclude that strong variation of ellipticity coefficient of radiation electromagnetic field calculated as a function of flow's parameters is the characteristic property of diffractional radiation of SEW (in comparison with WSC). It is shown on Fig. 77, where intensity, rotation angle of polarization plane and ellipticity coefficient of the field of diffractional radiation are given as functions of accelerating voltage within the limits of generation zone of GDR-modulator. Here ellipticity coefficient varies in the interval $0.01 \leq k = ba^{-1} \leq 1$ what means that polarization of radiation field changes from the linear (H-polarization) to circular polarization ($k \approx 1$) and the latter is found in the domain of transition of SWSC radiation field to the field of FWSC. This domain is characterized by sharp change of orientation of polarization ellipse in diffractional radiation ($\Delta\gamma_E \approx 80°$).

The shape of generation zone of GDR-modulator changes together with coefficients k, γ_E when the diffractional radiation of SEW is excited (in comparison with the shape of generation zone in the case of excitation of radiation by SWSC, see Fig. 76). With growing U_0 at first SSEW is excited ($\gamma_E = 20° - 40°$) and then – FSEW for which $\gamma_E = 40°$. One should point out that the angle of orientation of polarization ellipse of the SEW field of diffractional radiation also does not depend on the value and direction of focusing magnetic field. However, unlike SWSC radiation, the angle γ_E of diffractional radiation of synchronous waves decreases with growing magnetic field B and the range of its variation is much greater than in the case of radiation of longitudinal WSC.

It has been established in the process of experimental studies that variation of characteristics of diffractional radiation in the system thin strip flow – reflection diffractional grating (RDG) is mainly connected with forming of longitudinal WSC and SEW in the modulated flow. At the same time it is rather difficult to investigate the properties of diffractional radiation excited by SCW and FCW in this analyzer due to complicated procedure of choosing appropriate values of flow's velocity, modulation frequency and focusing magnetic field. With this purpose one must change analyzer's design: in order to control the flow's velocity it is necessary to place additional electrode in the space of radiation behind the modulator and to modulate the flow by external HF signal coming to OR through communication slot in the spherical mirror. But even in this case one registers only the diffractional radiation of principal ($m = \pm 1$, $n = -1$) transversal cyclotron waves (slow and fast) what means weak efficiency of generating cyclotron space harmonics of convectional current in this system. It is more perspective to apply double periodical structures (as well as in classical cyclotron parameter devices [104]) or diffractional grating having the form of aperture waveguide with flight channel for cylindrical electron flow [3].

Hence, theoretical and experimental studies enable us to conclude that characteristics of diffractional radiation are essentially determined by electron-wave processes in the electron flow being modulated by HF field. Emerging in the flow of longitudinal WSC and transversal (cyclotron) waves leads to the splitting of radiation far field pattern and variation of polarization characteristics of excited field. One can change radiation's characteristics varying the flow's parameters.

The discovered effects are of great scientific and practical importance, since they allow to specify the physical processes of transformation of the energy of electron flow to diffractional radiation, stimulate further development of the theory of devices of diffractional electronics and point out to the possibility of creating of principle new devices, in particular, parameter devices of diffractional radiation, analyzers of charged particles *et. al.* Nowadays these effects are used for elaborating the methods of remote measurements of dynamical parameters of the electron flow in the devices of diffractional electronics of the millimeter and submillimeter ranges, i. e., in the frequency domain where the known classical methods [19, 20] do not work.

§13. Interaction of Smith-Purcell radiation with the field scattered by a grating (inverse Smith-Purcell effect).

The features of thin structure of diffractional radiation of the preliminarily modulated electron flow have been established in the second chapter. In the real devices, where electron flow interacts with the field scattered by a grating the problem arises about excitation of SPR by electron flow interacting with additional electromagnetic field scattered by a grating [13, 14, 18]. These effects we will also relate to "pure" SPR which is not connected with the possibility of amplifying (or absorbing) additional field falling on the grating. This phenomenon was called inverse Smith-Purcell effect (ISPE). In this section we will present some results of its experimental and theoretical study.

1. Analysis of the processes of excitation of diffractional radiation by the electron flow interacting with the field scattered by a grating within the approximation of given incident field is in successive solution to three interconnected particular problems: diffraction of electromagnetic waves by the grating (determination of amplitude and phase characteristics of volume and surface waves of the space harmonics of scattered field); interaction of electron flow with scattered field (determination of convectional current and power of interaction); excitation of SPR by convectional current generated as a result of interaction.

In order to simplify the analysis, let us assume in the beginning that the incident H-polarized wave has the angle 90° to the grating's surface [61] (see Fig. 78, a, b) and non-modulated thin electron flow moves close to the grating with constant velocity $\vec{v}_0 = \vec{j}\beta_0 c$ (incident field is depicted by bold lines with arrows, reflected and transmitted field – by dot lines). One solve the problem in similar way for other periodical structures: comb, grating made of bars, echelette type grating *et al* [62]. In order to find SPR we write expressions for the fields under and over the grating and determine by the method of Riemann-Hilbert problem [61] only the reflected (incident) field and the one of surface waves (in the case of non-relativistic electron flows only zero harmonic is reflected from the grating).

Electrons interact with longitudinal E-component of electromagnetic field which we will represent in the form

$$E_y^1 = A_0 e^{-ikz} - a_0 e^{ikz} - \sum_{m\neq 0} a_m \theta_m e^{-q_m|z|+ih_m y}, \tag{4.2}$$

where $k = \omega c^{-1}$; $q_m = \sqrt{h_m^2 - k^2} = k\theta_m$; a_0, b_0 are the amplitudes of reflected and transmitted waves; a_m are the amplitudes of surface diffractional harmonics; A_0, ω, λ are the amplitude, frequency and wavelength of incident field. For $\kappa \leq 0.5$ one has for the principal harmonics of scattered field $(m = 0, 1)$

$$a_0 = \frac{i\kappa}{\Delta} A_0(4U - \chi_1 p_1); \; a_1 = -\frac{2i\kappa}{\Delta} A_0(1-u); \; b_0 = \frac{1}{\Delta} A_0(4 - \chi_1 p_2), \quad (4.3)$$

where

$$\Delta = 4(1 + i\kappa U) - \chi_1(i\kappa p_1 + p_2); \; p_1 = 2(1-u^2) + (4u^3 - 5u^2 + 4u - 1)U;$$

$$p_2 = 4u^3 - 7u^2 + 1; \; \chi_1 = 1 - \sqrt{1-\kappa^2}; \; U = \ln \frac{1+u}{2},$$

$u = \cos \frac{\pi d}{l}$ is the grating's filling coefficient; χ_1 is the small parameter.

We assume that input non-modulated electron flow with constant velocity v_0 and current's density i_0 comes to the interaction space (of the length L); then equations of movement and continuity together with (4.2) and (4.3) yield the expressions for convectional current $\tilde{i}(y)$ and power of interaction P

$$\tilde{i}(y) = i\frac{h_e}{h_p}\frac{i_0}{2U_0} \int_0^y E_y(\zeta) \sin h_p(y-\zeta) e^{-ih_e(y-\zeta)} d\zeta; \quad (4.4)$$

$$P = \frac{1}{2} \int_V \tilde{i} E_y^* dV; \quad (4.5)$$

After integration we obtain

$$\tilde{i}(y) = \sum_{s=1}^{\zeta} i_s e^{i\Gamma_s y}; \quad (4.6)$$

$$P = P_a + iP_r = \mathrm{P}[F_a^+ + F_a^- + i(F_r^+ + F_r^- + F_1)], \quad (4.7)$$

where for the convenience of further analysis the following notation has been introduced:

$$i_s = -ih_e i_0 a_m \theta_m L^2 \frac{e^{-q_m|a|}}{2U_0\delta_s};$$

δ_s, Γ_s are the amplitude and propagation constant of the s-th electron wave; i_0, U_0 are the constant values of current and voltage; for $s = 1, 2, 3$ one has $\delta_1 = \delta^+\delta^-$, $\delta_2 = -2\phi_p\delta^+$, $\delta_3 = 2\phi_p\delta^-$, $\Gamma_1 = h_m$, $\Gamma_2 = h_2 + h_p$, $\Gamma_3 = h_e -$

h_p, $h_p = \omega_p v_0^{-1}$ are the wavenumbers; $\delta^{\pm} = -\Phi_{0m} \mp \phi_p$, $\Phi_{0m} = 2\pi N_l(\eta - m)$ is the parameter determining the degree of synchronization; $\phi_p = L\omega_p v_0^{-1}$ is the parameter of space charge; $N_l = Ll^{-1}$ is the number of grating's periods; $\mathrm{P}_1 = 0.25|a_m|^2\theta_m^2 L^3 h_e i_0 U_0^{-1} exp(-2q|a|)$; functions $F_a^{\pm}$, $F_r^{\pm}$, F_1 are calculated by formulas

$$F_a^{\pm} = \mp[1 - \cos(\Phi_{0m} \mp \phi_p)][2\phi_p(\Phi_{0m} \mp \phi_p)^2]^{-1};$$

$$F_r^{\pm} = \mp \sin(\Phi_{0m} \mp \phi_p)[2\phi_p(\Phi_{0m} \mp \phi_p)^2]^{-1}; \tag{4.8}$$

$$F_1 = (\Phi_{0m}^2 - \phi_p^2)^{-1}.$$

If electrons interact only with the principal field's space harmonic (4.2) ($m = 1$), the amplitudes i_s, $s = 1, 2, 3$ are given by interrelations

$$i_1 = -\frac{G_0}{\delta^+\delta^-}; \; i_2 = \frac{G_0}{2\phi_p\delta^+}; \; i_3 = -\frac{G_0}{2\phi_p\delta^-}, \tag{4.9}$$

where

$$G_0 = A_0\frac{i_0 h_e}{U_0}\left[\sqrt{1-\kappa^2}(1-u)e^{i\psi_0 - \frac{2\pi a}{l}\sqrt{1-\kappa^2}}\right]\left[\kappa^2(4U - \chi_1 p_1)^2 + (4 - \chi_1 p_2)^2\right]^{-0.5};$$

$$\psi_0 = \arctan\frac{4 - \chi_1 p_2}{4\kappa U - \kappa\chi_1 p_1}.$$

Now one can find energy density, electromagnetic field and other parameters of SPR by means of (4.9) after obtaining the solution to the problem with the help of given current approximation. For the current (4.9) energy density and direction of radiation of the principal space harmonic in the domain $z > 0$ are calculated by the formula

$$S_{s,-1} = \frac{\pi}{8c}|i_s|^2(1-\beta_s)^2\frac{u - 1 + U(\eta_s + 1)(u + 1)}{1 + iU\sqrt{\kappa^2 - (\eta_s - 1)^2}}e^{-2q_s|a|}, \tag{4.10}$$

$$\gamma_s = \arccos\frac{\eta_s + n}{\kappa}. \tag{4.11}$$

The obtained interrelations enable us to analyze excitation of diffractional radiation by electron waves (4.9) when the flow is modulated by the field falling at the angle 90° to grating's surface.

If the grating with electron flow is irradiated by the field falling at arbitrary angle $\alpha' = 90° + \alpha$ (for example, by the H-polarized wave $H_x = A_0 e^{i\psi}$ where $\psi = \vec{k}\vec{r} - \omega t = k(y \sin\alpha - z\cos\alpha)$ is the phase of incident field), the problem is solved in the way analogous to [18, 61]. In this case only zero ($m = 0$) space harmonic reflects at the angle $\alpha_0 = 90° - \alpha$ if the grating's period satisfies the condition $l < \lambda(1 + \sin\alpha)^{-1}$; the rest of the field is the superposition of surface diffractional harmonics propagating along the axis Oy with phase velocities

$$v_\Phi = \frac{\kappa c}{m + \kappa \sin\alpha}. \tag{4.12}$$

In accord with formulas (4.6), (4.7) for the power of interaction P (in the functions $F^{\pm}_{\alpha,r}$) and current's density one must replace flight angles Φ_{0m} by the parameter characterizing the degree of synchronization

$$\Phi_{\alpha m} = 2\pi N_l[\eta - (m + \sin\alpha)], \tag{4.13}$$

depending on $\eta \frac{\kappa}{\beta_0}$ and the angle α of incident field (for $\alpha = 0$ $\Phi_{\alpha n} = \Phi_{0m}$).

In the general case it follows from interrelations (4.6)-(4.9) that the power of interaction P (the functions $F^{\pm}_{\alpha,r}$) and convectional current $\tilde{i}(y)$ differs from the magnitudes given by the formulas for classical O-type devices with long interaction by the amplitude (G_0) and electrons' flight angle Φ_{0m} in the scattered field. The combined analysis of these interrelations together with expression (4.10) for the energy density and radiation conditions (4.11) yields a number of new properties and effects having considerable scientific and practical significance.

Effect of splitting of the far field patterns. The spectrum of electron waves propagating with different phase velocities is excited in an electron flow interacting with electromagnetic field scattered by a periodical structure when the velocities of electrons $v_0 = v_\Phi$. It leads to excitation of SPR in different directions symmetrical with respect to the direction of propagation of incident field reflected from the grating (the angle α_0), where SWSC excite diffractional radiation at the angle $\gamma^+ < \alpha_0$, FWSC excite it at the angle $\gamma^- > \alpha_0$ and unperturbed wave – at the angle $\gamma = \alpha_0$ (see Fig. 78, b, Fig. 79).

Effect of absorbing and amplifying of scattered field – ISPR. In the process of interaction of electron flow with the scattered field SWSC gives back and FWSC absorbs energy of the field of surface harmonic what causes absorbing and amplifying of scattered field by the flow. FWSC, while absorbing energy, partially spends it on radiation, i. e., unlike classical devices where FWSC only absorbs energy of the field of surface wave, here electron flow "cools itself down".

Maximal amplifying and absorbing of scattered field together with maximal energy density of diffractional radiation occur when SWSC and FWSC are synchronous with the field of space diffractional harmonic, i. e., when

$$\Phi^{\pm}_{\alpha m} = \mp\phi_p,\ \phi_p > \pi;\ \Phi^{\pm}_{\alpha m} \approx \mp\pi,\ \phi_p \leq \pi; \tag{4.14}$$

Hence, maximal degree of amplifying and absorption holds for the field having angle of incidence $\alpha = \alpha^+$ and $\alpha = \alpha^-$, respectively, where

$$\alpha^{\pm} = \arcsin \frac{\eta \pm \frac{\phi_p}{2\pi N_l} - m}{\kappa}. \tag{4.15}$$

For the given angle of incidence α maximal amplifying and absorption of incident field are provided by increasing and reducing of electrons' velocity up to the values

$$v_0^{\pm} = v_{\Phi}\left(1 \pm \frac{\omega_q}{\omega}\right). \tag{4.16}$$

Effect of the control of interaction process. It follows from interrelations (4.13)-(4.15) that varying the angle α and phase ϕ one can control the process of interaction of electron field with scattered field and thus to amplify or to weaken as much as possible the diffractional radiation of electron flow. This effect yields the possibility of flow's phase focusing in the devices of diffractional electronics if to provide appropriate configuration of reflection mirror of an open electrodynamical system.

In addition to the effects indicated above one should underline that under the optimal conditions (4.14) the radiation angle of synchronous SWSC (or FWSC) coincides with the angle α_0 of incident field reflected from the grating ($\gamma^+ = \alpha_0$ or $\gamma^- = \alpha_0$). (4.10) yields maximal energy density of diffractional radiation for $u \approx 0.7 - 0.3$ what corresponds to the stripwidth (or to the width of comb's lamellae) $d \approx (0.75 - 0.6)l$. Since $S_{s,-1}$ is proportional to $|i_0|^2$ and inversely proportional to the parameters δ^+, δ^-, the energy density of diffractional radiation grows with increasing current's density and it is optimal when SWSC is syncrhonized with the field of surface diffractional harmonic. OPtimal grating's period is determined by the condition $\Phi^{\pm}_{\alpha m} =$

$-\phi_p$ and is calculated with the help of formula

$$l = \lambda\beta_0\left(m + \sin\alpha - \frac{\phi_p}{2\pi N_l}\right), \tag{4.17}$$

which coincides with (2.27) obtained within the frame of the model of given current for $m = 1$, $\alpha = 0$.

Theoretical values of radiation angles versus current's density i_0 of the active component of interaction power $P_a^{\pm}P_{\max}^{-1}$ of SWSC and FWSC with principal ($m = 1$) space diffractional harmonic together with angles of diffractional radiation versus accelerating voltage are presented on Fig. 78, 80. Dot vertical lines on Fig. 80 indicate the voltages $U_0, U_0^{\pm}$ for which $v_0^{\pm} = v_\Phi$. This data are evidence of the fact that when SWSC and FWSC are synchronized with the field of surface diffractional harmonic for given current's density i_0 the radiation of electron waves is always directed at the angle of reflected SHF signal, while optimal values of $U_0^{\pm}$ essentially depends on i_0 and the angle of incident field α'. The powers $P_0^+ < 0$, $P_a^- > 0$, hence, SWSC amplifies and FWSC absorbs SHF energy. Variation of α' changes phase velocity of electron waves and, consequently, interaction power of electron flow, what points out to the control of the process of interaction of electrons with the running volume wave of scattered field.

2. Interaction of electron flow with scattered field has been studied experimentally in the following way: the grating was irradiated by electromagnetic field perpendicular (the case of normal irradiation) or at a certain angle (oblique irradiation) to its surface [14, 18]. As a result, theoretically predicted effects have been confirmed and the new data have been obtained: cyclotron-diffractional radiation excited by transversal (slow and fast) cyclotron electron waves have been discovered.

Experimental study of the interaction of electron flow with scattered field in the case of normal irradiation was carried out in the millimeter range ($\lambda = 3.8$ mm -5.6 mm) by the plant shown on Fig. 81, a. The plant contains diffractional grating 1, diode electron cannon 3 which forms electron flow 2, electrons' collector 11, receiving-transmitting mouthpiece-type antenna 4 with the aperture 12 mm^2 $\times$ 15 mm^2, SHF source 8 and measuring section. The grating with the period $l = 0.4$ mm, width 10 mm and length 40 mm was placed along the axis of reflection cylindrical mirror with curvature radius 110 mm and the aperture 40 mm^2 $\times$ 50 mm^2.

Electron flow is focused in the interaction space by constant magnetic field. Electrons are modulated by the H-polarized electromagnetic wave which is formed by the mouthpiece-type antenna placed at the angle $\alpha' = 90^\circ$ ($\alpha = 0$). The control of electrons' interaction with the field of space diffractional harmonic ($m = 1$) is provided by the possibility to vary the distance between mouthpiece and grating within the limits $d_1 \approx \lambda - 0.15\lambda$ by means of the special device 4. Calibrated SHF signal having the given frequency f is sent from generator 8 to mouthpiece-type antenna through directional coupler 6 whose straight shoulder is adjusted with coordinated load 10 and inverse shoulder – with the second coupler 9 connected with power measuring device W, wavemeter λ and detector section with indicator (oscillograph).

It has been established that for the given angle α', distance d_1 and frequency f one could guarantee the effects of amplifying and absorption of SHF energy by electron flow if to vary electrons' velocity in the interval (2.25). These effects are discovered for small distances between mouthpiece and reflection mirror (of about one or several wavelengths) and for big d_1 (in the experiment the maximal distance $d_1 \approx 15\lambda$). In particular, oscillograms of the excited SPR signals on Fig. 81, b illustrate how the power (intensity) of diffractional radiation behaves itself with respect to electrons' velocity when d_1 gradually grows. One can control interaction within small limits of variation of d_1 ($\Delta d_1 \leq 0.25\lambda$) by separating the cases of smooth transfer from the process of amplifying (absorption) to the process of absorption (amplifying) of SHF energy by the flow (oscillograms 1, 3) and also of explicit absorption of FWSC (oscillogram 2) or of amplifying of SWSC (oscillogram 4), i. e., to observe ISPE experimentally.

Experimental curves of SPR intensity versus accelerating voltage measured in the constant regime for different values of current's density are presented on Fig. 81, c. Here and in what follows it is used the electron flow with cross-section 3.7 mm^2 $\times$ 0.1 mm^2. It is clear that these curves qualitatively coincide with theoretical data (see Fig. 80). However, in the experiment $U_0^{\pm}$ non-uniformly depends on current's density i_0 what indicates the distinctive features of the dynamical parameters of real electron flow in comparison with its theoretical model where it is assumed that reduction coefficient $R = 1$. These features are applied for constructing the method of measurement of the dynamical parameters of electron flow in the devices of diffractional electronics.

It has been created the special model-analyzer [14] (see Fig. 81, a) for experimental studies of the electron-wave processes of energy exchange between electron flow and scattered field when the grating is irradiated at different angles. This device is characterized by the following property: one uses in place of the mouthpiece (see Fig. 81, a). the special receiving-transmitting antenna in the form of diffractional grating 2 with planar dielectric waveguide 3 coordinated at the edges with standard metal waveguide 1. Dielectric waveguide is fed by SHF generator at fixed frequency in constant regime. When SHF signal P_{in} is sent to the input of dielectric waveguide, the waves of dielectric waveguide are transformed to volume wave and antenna excites H-polarized field P_0 falling to the mirror with diffractional grating 5 at the angle $\phi : \cos\phi = \beta_e^{-1} + n\kappa^{-1}$. If there is no electron flow 6 generated by electron cannon 7, then only a part of non-transformed input signal comes to the output P_{out} of dielectric waveguide. In the presence of electron flow in the interaction space and under the conditions (4.11) diffractional radiation emerges which is excited by SWSC and FWSC symmetrical to the direction of propagation of the field reflected by the grating. In this case, when the grating is irradiated at the angle 90° ($\kappa = \beta_0$), radiation of longitudinal (or transversal) electron waves is transformed along two separated channels (Fig. 82, a, dot lines, $s = 2, 3$). This radiation is received by antenna and is registered at input and output edges of the planar waveguide as isolated SHF signals $P_{s=2,3}$ due to inverse transformation of the volume wave to the wave of dielectric waveguide.

Oscillograms of SHF radiation signals on Fig. 82, b show amplitudes of these signals as functions of anode voltage (electrons' velocity v_0) for different distances $d_1 = c + b$ between grating's surface and lower mirror. Variation of d_1 (for given input frequency) leads to the cases when interaction power (and SPR) of SWSC or FWSC have opposite phases ($b = 4.92$ mm) or equal phases ($b = 5.76$ mm).

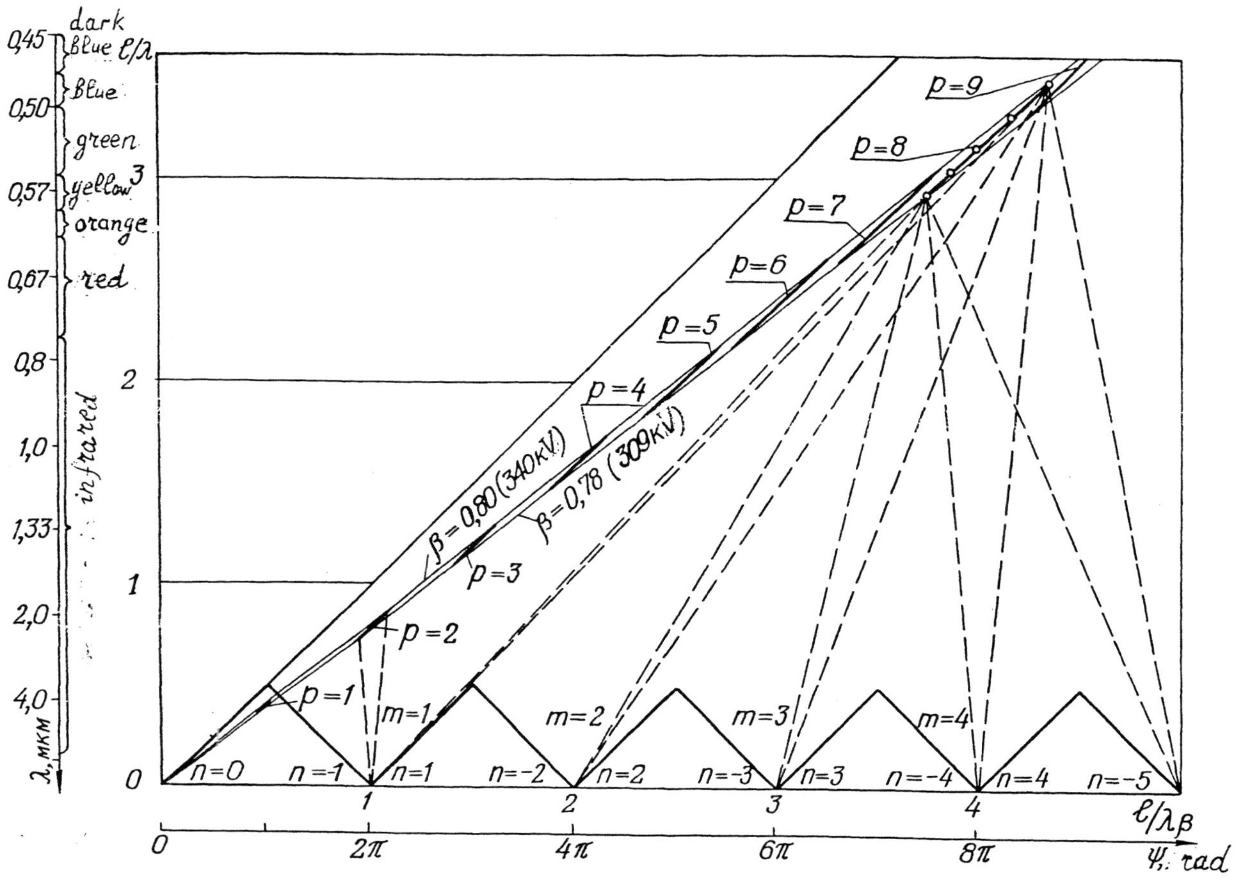

Fig. 66. Brillouine diagrams for Smith-Purcell radiation.

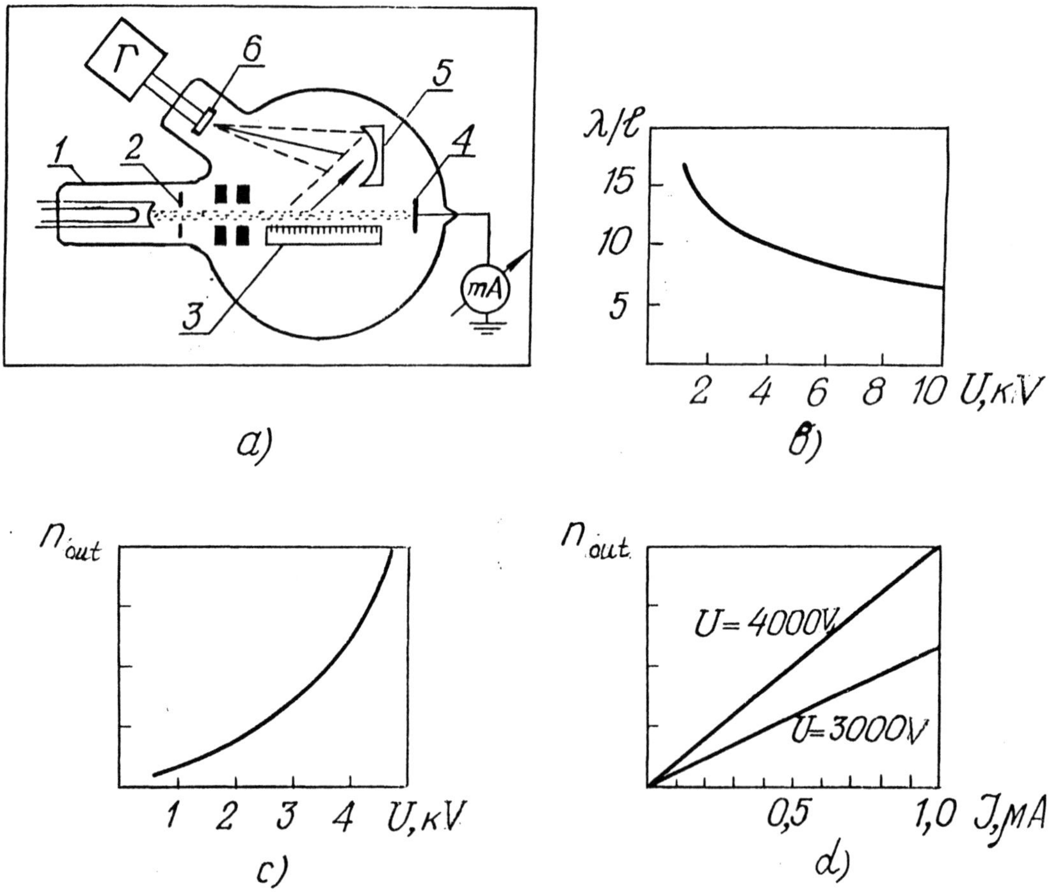

Fig. 67. Discovering of the Smith-Purcell effect for non-relativistic electron flow: experimental model of the Ishiguro and Tako analyzer; a: 1 – vacuum tube (diamater 30 cm, height 15 cm); 2 – electron cannon; 3 – optical diffractional grating ($l = 1\mu$m or $l = 10\mu$m in further experiments); 4 – mirror reflecting the radiation to thermopair 3 connected with galvanometer Γ; 5 – electrons' collector; 6 – diffractional grating.

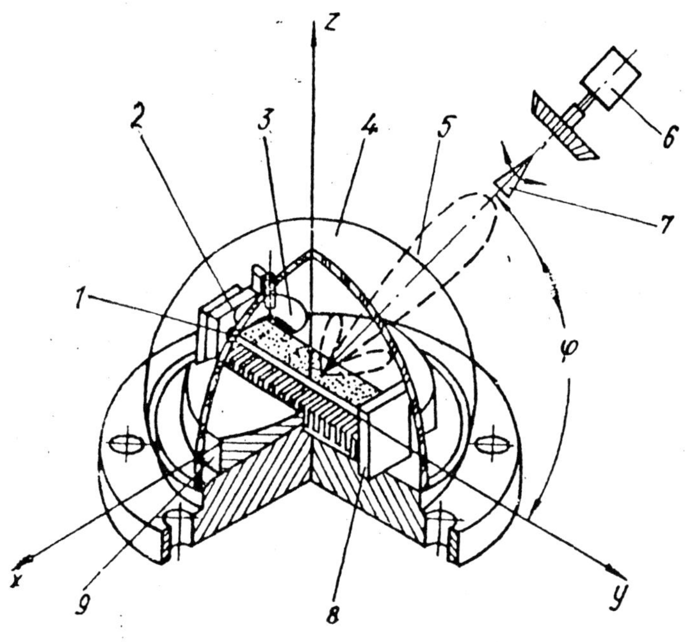

Fig. 68. Analyzer of Smith-Purcell radiation in the millimeter wave range.

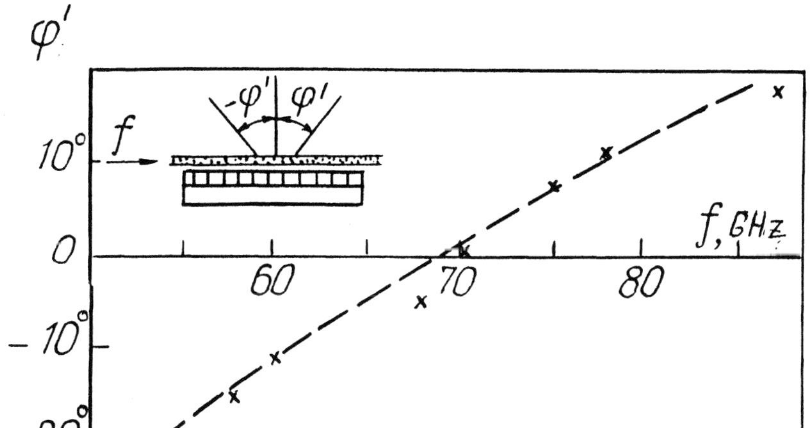

Fig. 69. Angle of Smith-Purcell radiation versus modulation frequency of electron flow (wave model: planar dielectric waveguide + grating): crosses indicate experimental data, dot line – theoretical data.

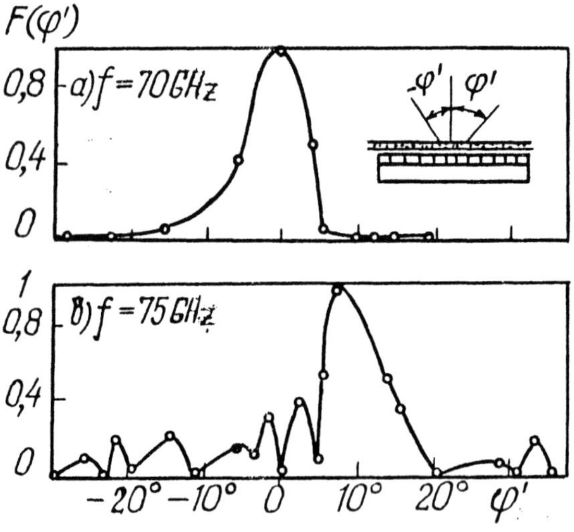

Fig. 70. "Cold" modeling of radiation far field pattern for the diffractional radiation of the mirror with analyzer's grating: a) – $f_0 = 70$ GHz; b) – $f_0 = 75$ GHz.

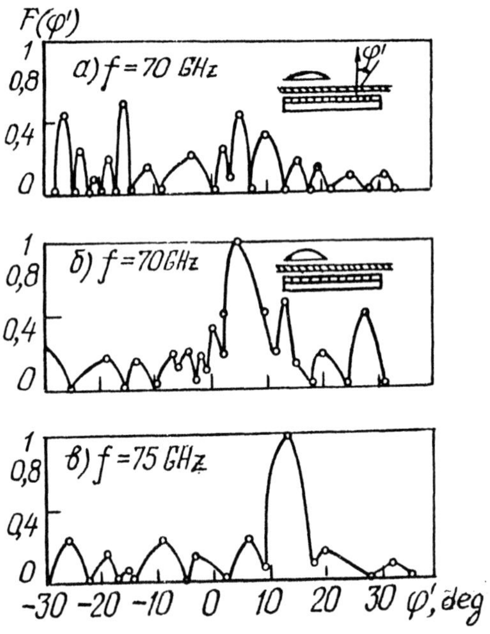

Fig. 71. Modeling of radiation capabilities of the analyzer of diffractional radiation: a), b) – $f_0 = 70$ GHz; c) – $f_0 = 75$ GHz.

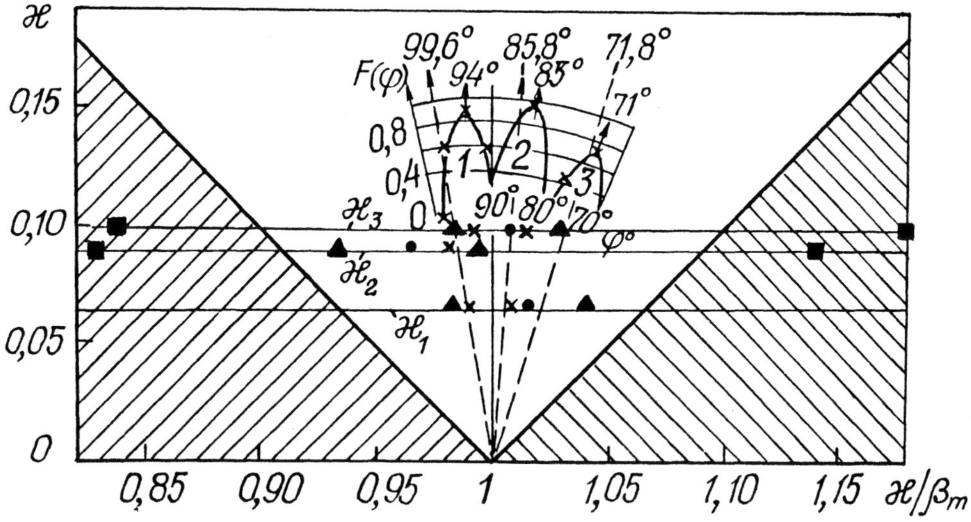

Fig. 72. Three-petal radiation far field pattern excited by SWSC (beam 1), SEW (beam 2) and FWSC (beam 3). It is shown also experimental (crosses) and theoretical (black triangles and squares) values of κ, β_m and radiation angles of electron waves in the plane of Brillouine diagram.

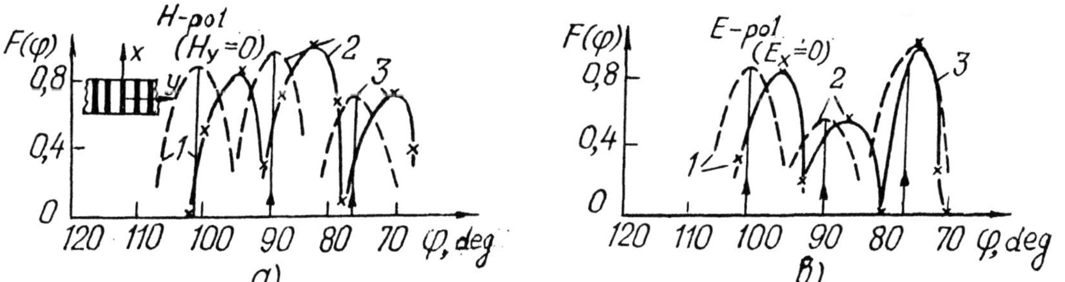

Fig. 73. Experimental (bold line) and theoretical (dot line) radiation far field patterns of electron flow measured in the planes H- (a) and E- (b) polarizations: 1 – FWSC; 2 – SEW; 3 – SWSC.

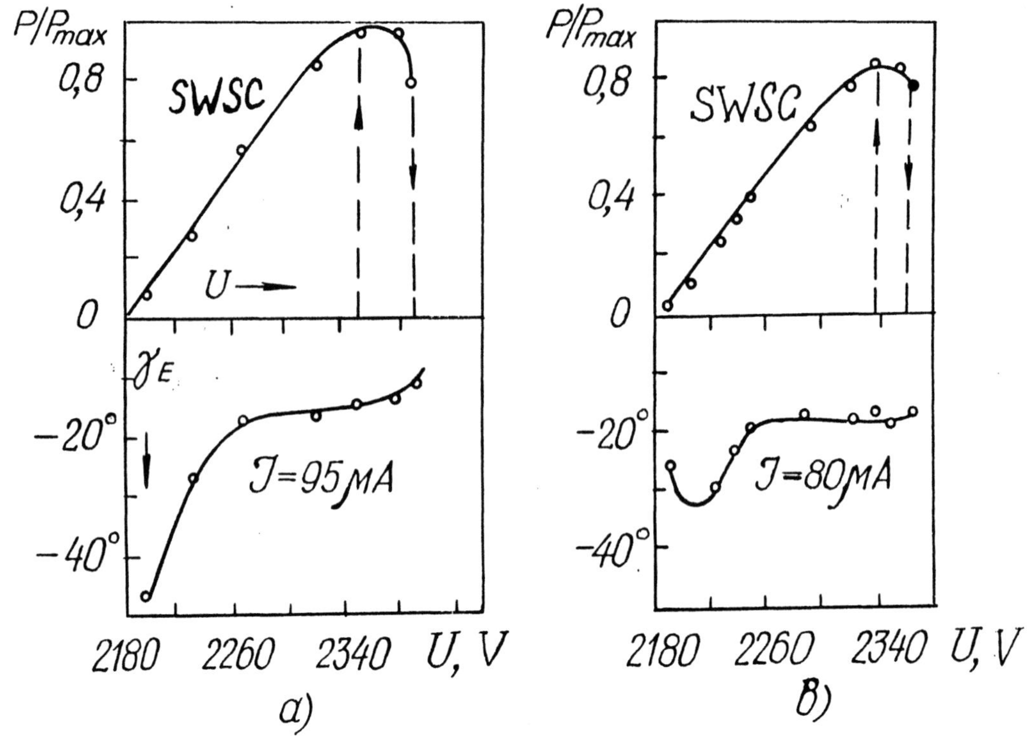

Fig. 74. Variation of Smith-Purcell radiation of SWSC and field's polarization angle in modulator's generation zone, $J = 95\mu$A (a), $J = 80\mu$A (b).

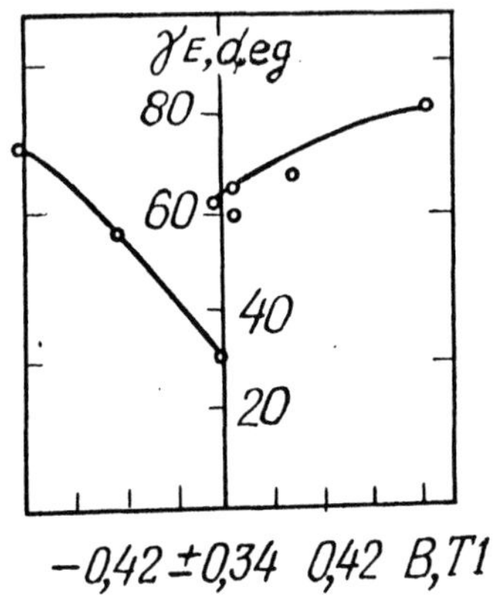

Fig. 75. Measured polarization angle of the Smith-Purcell radiation field versus value and direction of focusing magnetic field.

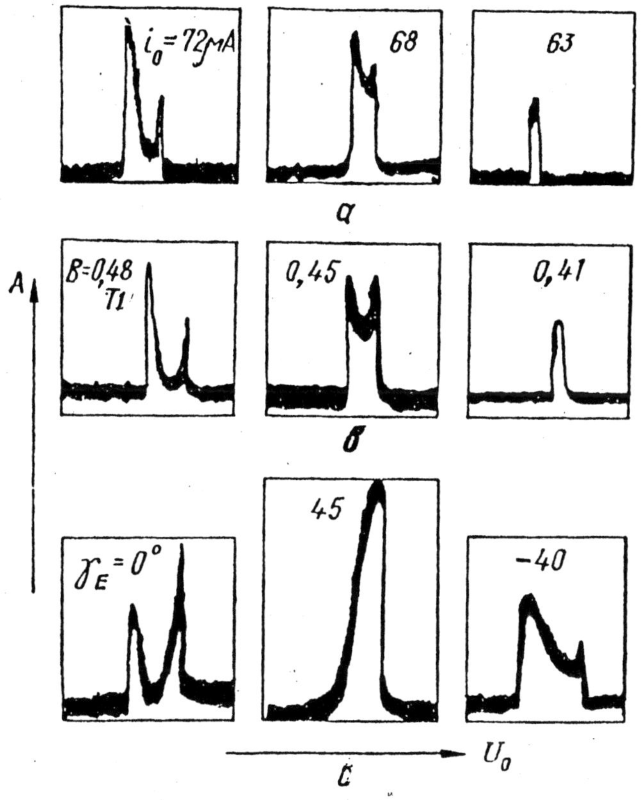

Fig. 76. Oscillograms of SEW radiation signals (a – c) registered for varying current's density and focusing magnetic field ($U_0 = 2700$ V; $\lambda = 3.968$ mm).

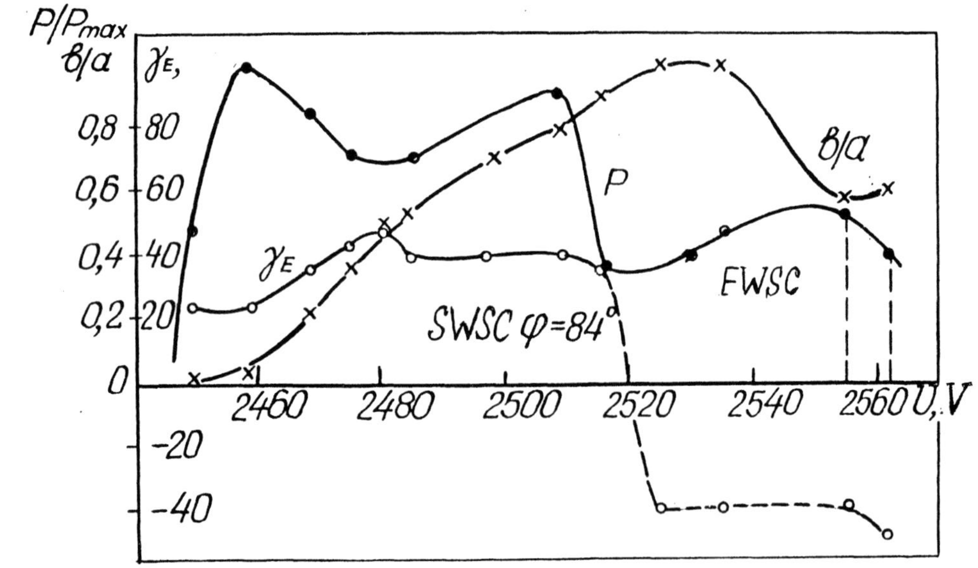

Fig. 77. Experimental curves of the intensity, polarization angle and ellipticity coefficient for the Smith-Purcell radiation field of SEW within the limits of modulator's generation zone.

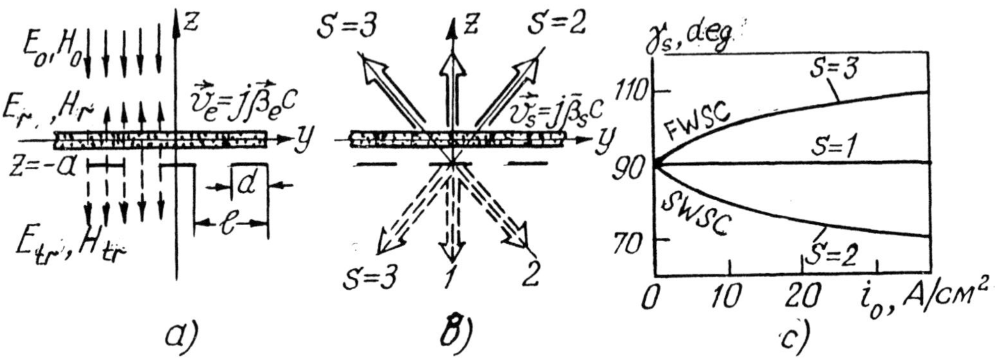

Fig. 78. Theoretical model of the excitation of of Smith-Purcell radiation by electron flow interacting with scattered field when the grating is radiated at the angle 90°: a) – problem's statement; Smith-Purcell radiation of electron waves; b) – theoretical data for radiation angles versus current's density for electron waves, $s = 1, 2, 3$.

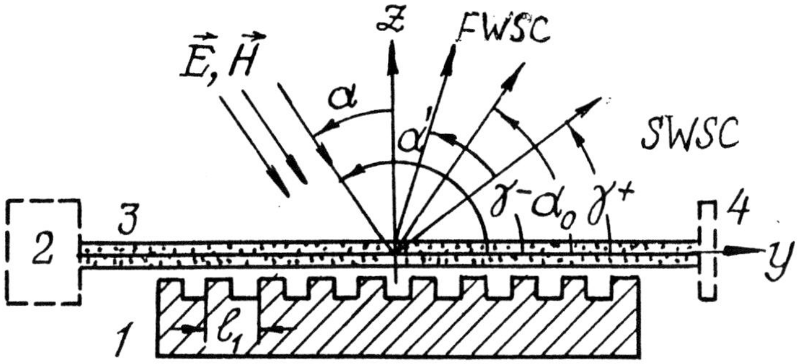

Fig. 79. Formation of the radiation far field pattern of Smith-Purcell radiation when the angle of incident field varies; 1 – diffractional grating; 2 – electron cannon; 3 – electron flow; 4 – collector.

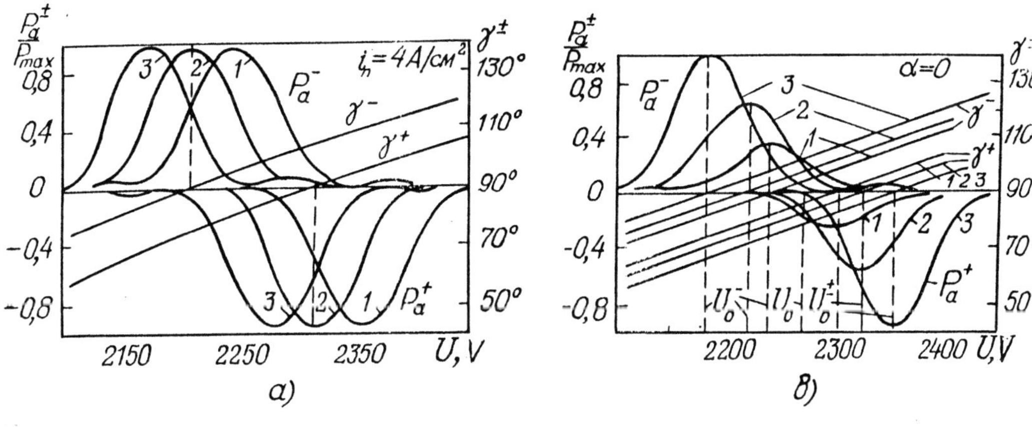

Fig. 80. Theoretical data for the radiation angle, SWSC ($1' - 3'$) and FWSC ($1 - 3$) interaction power versus accelerating voltage: a) – $i_0 = 4\text{A}\cdot\text{cm}^{-2}$; $\alpha' = 95°$ (1), $90°$ (2), $85°$ (3);
b) – $\alpha' = 90°$; $i_0 = 2\text{A}\cdot\text{cm}^{-2}$ (1), $4\text{A}\cdot\text{cm}^{-2}$ (2), $10\text{A}\cdot\text{cm}^{-2}$ (3).

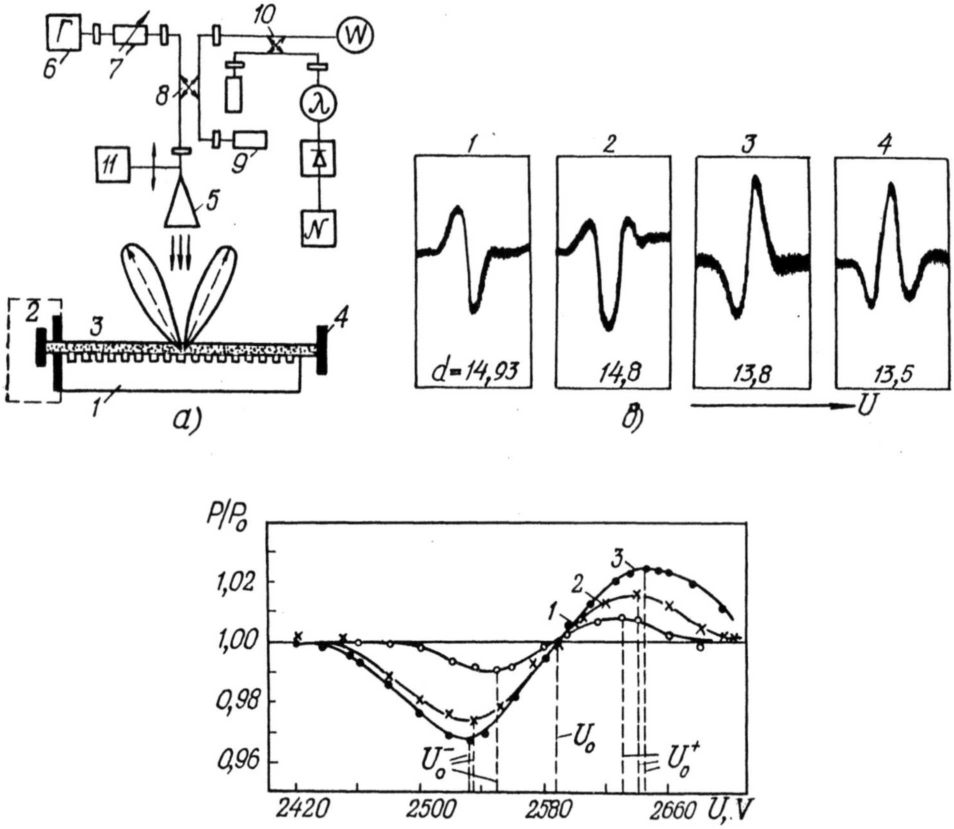

Fig. 81. Experimental plant (a) for the study of the processes of energy exchange between electron flow and the field scattered by a grating ($\alpha' = 90°$); b) – oscillograms of the discovered effects of amplifying and absorption of electromagnetic field; c) – effects of amplifying and absorption of scattered field by the electron flow in the continuous regime ($f_0 = 74.61$ GHz; $\lambda = 4.018$ mm): $i_0 = 15.6\text{A}\cdot\text{cm}^{-2}$ (1), $11.6\text{A}\cdot\text{cm}^{-2}$ (2), $4.8\text{A}\cdot\text{cm}^{-2}$ (3).

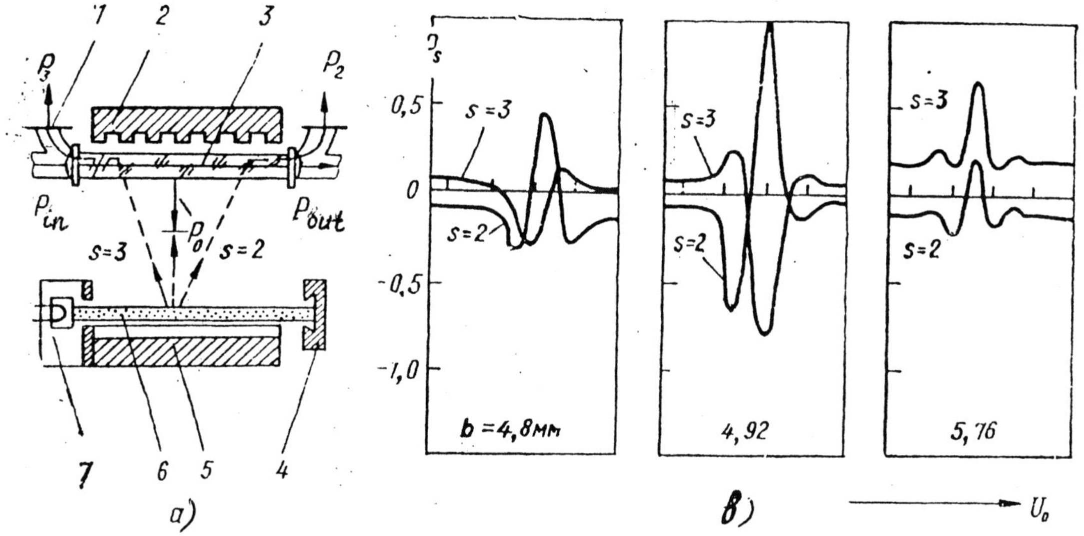

Fig. 82. Divisor-analyzer of the WSC radiation of electron flow interacting with scattered field (a); oscillograms of the separated signals: in SWSC channels ($s = 2$) and in FWSC channels ($s = 3$) of the analyzer (b).

CHAPTER FIVE

Excitation of open resonators by Smith-Purcell radiation

In the experimental treatment of SPR with the use of OR one must choose two-mirror OR with a grating covering a part of one of the mirrors. Such complicated open structures have been studied in details (namely, their excitation by concentrated sources) in [3, 4].

The discovered properties of thin structure of SPR (see chapter two) stimulated the development of the deeper investigation of complicated OR. One of the most important tasks here is to study the features of variation of amplitude-phase behaviour of the fields excited in OR and the possibilities to control them by varying the grating's parameters and choosing the shape of resonator's mirrors.

§14. Open resonators with local reflection diffractional grating

Let us consider the properties of OR with reflection diffractional grating [3, 4] paying a special attention to polarization characteristics and the possibilities to control the resonance field what is important for the study of thin structure of SPR in open electrodynamical structures.

1. Let the two-mirror OR consists of spherical mirror and of the plane mirror with grating on it (Fig. 83). Spectral characteristics and Q-factor of such resonators with bounded gratings are studied in [62], in particular, the influence of grating's width d, its period l, relative slotwidth and depth h on the Q-factor and amplitude of the field of resonance oscillations. The main result is in establishing non-monotonous dependence of the properties of OR on grating's parameters. The measurements are evidence of existence in this OR of rather many modes having field's minimum over periodical structure in addition to high Q-factor oscillations with maximal field's amplitude on the grating. Theoretical analysis of OR with local rectangular grating on one of its mirrors is a complicated boundary value problem.

Under certain conditions it is sufficient to construct physical model of OR based on known rigorous solution to the problem of diffraction of plane wave by a two-dimensional grating [62]. Let us limit ourselves only to the one-wave scattering regime considering excitation of infinite grating with narrow slots ($\hat{\chi} << 1$) by the plane H-polarized wave with angle of incidence θ. Then the phase of reflected wave is determined by the formula

$$\mathrm{Arg}A_0 = 2\arctan\frac{\hat{\chi}\tan kh}{(1+\kappa\ln\sin\pi\hat{\chi})\cos\theta} + \mathrm{O}(k^2\theta^2), \qquad (5.1)$$

where $\hat{\chi} = pl^{-1}$ is the relative slotwidth; $k = 2\pi\lambda^{-1}$; h is the grating's depth; $\kappa = l\lambda^{-1}$. From the point of view of an observer in the far zone such grating is equivalent to a reflection plane whose position is determined by the phase of reflected wave. The distance z_0 between grating's surface and reflection plane is given by the expression

$$z_0 = \frac{\mathrm{Arg} A_0}{2k \cos\theta}. \tag{5.2}$$

Hence, periodical structure covering a part of resonator's mirror may be treated as equivalent reflection plane placed at the depth z_0 under the mirror's surface. Sharp perturbation of the wave front of incident wave yields, firstly, additional scattering out of OR leading to spectrum's selection and, secondly, taking down degeneration of axial modes as well as after bringing phase filter in OR, i. e., the grating acts like phase inhomogeneity. Essential feature of such inhomogeneity in comparison with phase filters is in the fact that reflection grating possesses anisotropic properties what yields considerable difference of OR characteristics for E- and H-polarizations. Strong perturbation caused by such inhomogeneity excites oscillations whose field's amplitude in the domain of the grating is less than over the smooth mirror's surface. The field of H-polarized mode is not already described by the Hermit-Gauss function. However, as it follows from (5.1), (5.2), perturbation caused by the finite grating may be reduced, if to combine equivalent reflection surface with mirror's surface (what is called compensation of phase inhomogeneity).

Experimental studies of OR with phase inhomogeneity in the form of reflection grating having the parameters $d_0 = 10$ mm, $l = 0.4$ mm, $\lambda = 4.01$ mm, $h = 0.92$ mm, $\hat{\chi} = 0.35$ are carried out on the plant described in [3]. The mirrors' dimensions are $2a_s = 2a_p = 60$ mm, the curvature radius of spherical mirror is $R = 150$ mm. Amplitude field's distribution is measured along the diameter of plane mirror with grating parallel to lamellae. Fields' measurements show that many oscillations with amplitude's "gap" close to the grating are excited in this OR. Lifting or lowering of the grating leads in a number of cases to frequency degeneration [128] of these oscillations having maximal amplitude over the grating, since each of them has different perturbation degree determined by z_0 and by the amplitude of resonance field in the domain of the grating [129]. In this sense influence of the grating on OR is equivalent to the perturbation caused by phase filter. It is confirmed by the model experiment where the grating is replaced by a moving metal piston of the width d in the plane mirror. Comparison of typical resonance oscillations in OR containing a piston lowered at the depth $z_0 = 0.75$ mm with the ones of OR having the grating placed on the level of mirror's surface is evidence of their identical structure. However, the Q-factor of oscillations excited in OR with a metal piston is higher than the one of the oscillations in OR with the grating, since the grating increases resonator's ohmical losses. Investigation of the amplitude distribution of TEM_{20q} oscillation allows to control the decrease of field's amplitude over the grating when piston is being lowered at the depth z_0 (grating's edges are marked on Fig. 84 by vertical dot lines).

Let us study compensation of phase inhomogeneity brought in OR by diffractional grating. The degree of compensation is determined by phase difference of the waves reflected from the plane mirror and the grating. Decrease of this difference is achieved by moving the grating along the mirror's surface. As an example let us consider degeneration of two oscillations having maximum and minimum of the field in the center of plane mirror (Fig. 85, a). When lifting the comb, i. e., in the process of "smoothing" phase inhomogeneity, the phase raid for oscillation with maximal field's value over the comb sharply changes, while phase raid for the field with the gap in the center remains practically constant, what is illustrated by the dependence of resonance distances for these modes versus z_0 (Fig. 85, b). Degeneration occurs when the mirror's surface coincides with equivalent grating's reflection surface. Deviation of corresponding value z_0 from the theoretical figure $z_0 = 0.75$ mm is caused by the fact that z_0 is calculated in the case of normal incidence without taking into account angle spectrum of plane waves in OR. Q-factor $Q - 15 \cdot 10^3$ of degenerated oscillation is in one and a half time higher than Q-factor of the mode with maximum in the center.

Let us consider properties of OR with the grating for the principal TEM_{20q} oscillation [130]. Study of influence of the grating on the structure and Q-factor of this oscillation shows that in OR with the grating perturbation caused by the comb is negligible, if the minimum in field's distribution is on the boundary between the grating and the mirror. It takes place when grating's width is greater or equal than the width of the main petal of the field of TEM-oscillation. If there exists the field on the boundary grating-mirror, oscillation close to the grating is destroyed and minimum in the field's distribution over the grating is formed. Thus, there are strict limitations for the choice of comb's width and the range of distances between OR reflectors which provide effective interaction of the field of OR with electron flow. Here one must take into account that the field in the center of plane mirror decreases for narrow gratings, while ohmical losses grow with increasing grating's width.

2. Results of experimental investigations of the compensation of phase inhomogeneity allow to assert that perturbing influence of a grating on the fields' structure in OR may be reduced. Field's distribution of oscillation having the structure similar to TEM_{20q}-mode at the distance 10λ from the plane mirror is presented on Fig. 86, a, b; fields in the domain close to the grating, where $Oz = 0.25\lambda$, are shown on Fig. 86, c, d. It is clear that perturbing influence of the grating is developed close to the plane mirror. When the grating is placed at the level of plane mirror (Fig. 86, c), the Q-factor of this oscillation is not high and equal to $6 \cdot 10^3$. Lifting of the grating at the height $z_0 = 0.6$ mm ($z_{0q} = 0.75$ mm) change the field's structure (Fig. 86, d).

Redistribution of the field in the process of compensation of phase inhomogeneity causes not only the growth of amplitude of H-polarized field close to the grating, but also the growth of Q-factor due to field's fall along the mirror's edge. In the given case field's amplitude near the grating increased in 4.5 times and Q-factor raised in 3 times up to the value $18 \cdot 10^3$. It is easy to see that improvement of field's characteristics in OR containing a grating by means of compensation of phase inhomogeneity with the help of the finite diffractional grating is possible at fixed wavelength.

Dependence of z_0 versus frequency calculated by (5.2) is shown on Fig. 87, a. General variation of the properties of OR with respect to the position of reflection grating may be followed with the help of the frequency curve of spectrum's density as a function of z_0 (Fig. 87, b). Spectrum's density grows with increasing frequency and for every fixed frequency it has extreme value corresponding to such level of grating's lifting when resonator has minimal perturbation caused by the local grating. $z_{0\max}$ determined by extreme values ρ and shown on Fig. 87, b well agree with $z_0(f)$. Maximal spectrum's density $\rho(z_0)$ in OR is developed in the case of compensation of phase inhomogeneity. Sharper fall of these curves for $z_0 > z_{0\max}$ is evidence of spectrum's selection due to the growth of scattering by the grating; here equivalent plane runs out over the surface of plane mirror.

Compensation of phase inhomogeneity may be provided by the application of a mirror (on which the grating is drifted) having a special profile in such a way that its continuation coincides with equivalent grating's surface. Partial compensation of phase inhomogeneity is thus realized in the spherical-cylindrical OR [3] and in OR with changing angle of inclination of mirrors with respect to the grating [130].

3. Local objects in the form of transparent single and double diffractional gratings brought into OR play the role both of phase inhomogeneity (for the wave of given polarization) and of the filter selecting angle spectrum. The may initiate strong perturbations of fields' structure violating, in particular, rectangular modes' symmetry. The strongest perturbations may be caused by a resonance object. Similar situation occurs when the double diffractional grating of finite width is placed on one of resonator's mirrors (Fig. 88, a). It is expediently to use this grating for raising the efficiency of interaction of electron flow with resonance field.

For experimental determination of the characteristics of OR with double diffractional gratings it was used the grating having the following parameters: $d_0 = 8$ mm; $l = 0.4$ mm; $p = 0.13$ mm; $h = 1.0$ mm. Reflection grating was placed under diffractional grating made of bars with the parameters $d_0 = 10$ mm; $l = 0.4$ mm; $p = 0.13$ mm; $h = 1.7$ mm. Measurements were made at the wavelength $\lambda = 4$ mm (the curvature radius of spherical mirror $R_{\text{sp}} = 150$ mm). Fields' distributions were registered near the plane mirror with double grating along its diameter and parallel to grating's lamellae. Transparent grating was placed at the level of plane mirror. Reflection grating was fixed on moving piston which could be shifted along the axis of OR. Distribution of the field excited in such OR is presented on Fig. 88, b (distance between the grating $c_0 = 1.52$ mm). The presence of two diffractional gratings leads to excitation of the field with complicated structure what is not favorable for interaction of electron flow with resonator's field. High field's amplitude on the edge yields low Q-factor of such oscillation ($Q \approx 10^3$). It is practically impossible to separate predominating mechanisms responsible for worsening the properties of OR due to the complexity of considered system.

Amplitude field's distribution of eigenoscillation in OR with the grating lifted 0.1 mm higher its primary position is shown on Fig. 88, c. Field's amplitude near the gratings increased in four times. Amplitude field's distribution were registered for different types of eigenoscillations. It was established that Q-factor of OR on these modes (3600 – 4000) was in three-four times higher than Q-factor of the same resonator but without grating's correction.

Mutual positions of mirrors and gratings, for which resonance oscillations are excited, are given on Fig. 88, d. Crosses indicate high Q-factor oscillations when field's distribution on the grating provides intensive interaction with electron flow.

Although OR with two diffractional gratings is a complicated objects difficult for theoretical analysis, the consideration of such objects is of great interest for creating both effective oscillating systems of the generators of millimeter waves and two-frequency resonance cells of the radiospectrometers of magnetic resonance [131, 132], as well as systems for pumping polarized nuclear targets in the millimeter wave range [4, 48].

4. Let us study waves' scattering in OR with local diffractional grating (having the parameters indicated above and curvature radius 150 mm). Presence of local inhomogeneity in the volume of OR (or on the surface of one of the mirrors) leads to additional lighting of electromagnetic energy out of the system what causes lowering of the Q-factor of resonance oscillations as well as change of intensity and structure of diffractional field. The latter may considerably perturb spectral characteristics of OR shielded by a metal screen. Measurements at the frequency 65 GHz show that the spectrum of OR which has a grating on a screen strongly thickens, if to compare it with the spectrum of non-shielded resonator [53]. Eight oscillations of cylindrical resonator are excited in shielded OR when $d(2a)^{-1} = 1.13$ (d is the diameter of shielding cylinder) due to the waves excited out of OR; their amplitude is comparable with the one of OR containing the grating and resonance wavelength is in two times greater.

The number of oscillations of cylindrical resonator is determined by relative amplitudes of the fields E_s and $\hat{E}_1$. Level of the field scattered by resonator's inhomogeneity depends on its parameters and on the way how it influences the characteristics of OR. One can reduce the losses caused by the scattering of resonance field on the grating and, consequently, to lower $\hat{E}_1$ and communication with external space, if to apply compensation of inhomogeneity.

5. Control of the field out of OR with the help of compensation of inhomogeneity on the mirror is illustrated by the far field pattern in H-plane of the H-polarized TEM_{20q} oscillation excited in the resonator for the cases of non-compensated (bold curve) and compensated (dot curve) grating. Resonator is excited by the concentrated communication element from the side of spherical mirror. Local inhomogeneity essentially perturbs the far field pat-

tern of diffractional field, if to compare with classical two-mirror OR without inhomogeneity. Field near the axis has a gap for the main polarization. It means that the main contribution to resonator's radiation is made not by the field scattered on mirrors but by the diffraction on inhomogeneity (oscillations with rectangular symmetry are excited in OR with a local grating). Presented far field patterns enable one to understand the process of spectrum's selection. If the grating is compensated, maximum in the diffraction far field pattern comes closer to the axis (Fig. 89, dot line) and the level of those petals falls, which are radiated at big angles to the axis and determine intensity of radiated energy. Replacement of the plane mirror by cylindrical one allows to reduce interaction of OR with external space due to partial compensation of the focusing properties of cylindrical mirror by phase inhomogeneity (at least, in one plane).

6. Let us consider polarization characteristics of external and internal fields of OR with a grating. Consider the conditions leading to excitation of oscillations with degenerated polarization. In one-wave range for $\hat{\chi} << 1$ one can represent the matrix of phase anisotropy caused by infinite diffractional grating using the plane waves approximation [134]:

$$M = \begin{pmatrix} e^{2i\arctan a_0^E} & 0 \\ 0 & e^{2i\arctan\frac{\hat{\chi}\tan kh}{1-\kappa\ln\sin\pi\hat{\chi}}} \end{pmatrix}, \tag{5.3}$$

where $\arctan a_0^E$ is the phase of E-polarized field's component reflected from the grating.

For the values $\kappa = 0.2; \hat{\chi} = 0.2; n \sim l$ applied in practice the phase difference between orthogonal components of reflected wave is $15° - 20°$. But for the grating with finite width it is rather difficult to obtain even approximate estimate for phase anisotropy, since, as it has been established, the grating influences field in OR not only as a polarizer but as a phase filter; it also selects angle spectrum. Taking into account all these factors considerably complicates solutions to the problems in electrodynamics of open systems. Therefore, experimental studies take on special significance for the analysis of such systems. They enable researchers to underline the main physical properties of OR for further use in order to create OR with given characteristics. Here we will use compensation of inhomogeneity for the forming of oscillations with degenerated polarization in OR with a grating by applying the levelling of phases of waves reflected from the grating and mirror's surface together with differential shift of the frequencies of oscillations with different space structure. Since z_0 may not vanish also for the E-polarized wave, compensation occurs for two orthogonal field's components.

The measured data for the E-polarized TEM_{40}^{E} mode and H-polarized TEM_{20}^{H} mode versus z_0 are presented on Fig. 90. Compensation of inhomogeneity in the case of E-polarized wave takes place for much lower z_0 ($\kappa = 0.2; \hat{\chi} = 0.2; h \sim l; \lambda = 4$ mm), what well agrees with the results of rigorous solution to the problem of diffraction of plane wave on infinite diffractional grating [62]. Compensation in the case of H-wave yields strong perturbation of H-polarized field up to the breakdown of resonance.

If to use cylindrical mirror in place of plane mirror, one can increase the number of E- and H-oscillations with close frequencies (due to partial compensation of inhomogeneity). The parts of spectrums for OR formed by spherical ($R_s = 150$ mm) or cylindrical ($R_c = 110$ mm) mirrors with the grating of the length 40 mm having the parameters $l = 0.4$ mm; $h = 0.92$ mm; $\hat{\chi} = 0.35$ are presented on Fig. 91. Amplitudes of H-polarized waves are much lower than amplitudes of E-polarized waves because of additional losses caused by diffractional grating. It is confirmed by the fact that H-oscillations whose field are mostly concentrated on the smooth mirror's surface have smaller losses and, hence, greater amplitude. For example, TEM_{80}^{H} oscillation has maximal amplitude in the spectrum of H-polarized waves. Distributions of E-polarized fields are close to the fields described by Gauss-Hermit functions. There are distortions in the central part of distributions caused by perturbing influence of the grating. Crosses on Fig. 91 indicate oscillations having close frequencies.

Let us consider creation of frequency degeneration in OR with local grating taking as an example oscillations with orthogonal polarization marked by the index 4 on Fig. 91: TEM_{02}^{H} and TEM_{100}^{E}. Analysis of the oscillogram of resonance characteristic near the point where their frequencies coincide with respect to the height of lifting of diffractional grating shows that the growth of z_0 leads to sharp perturbation of TEM_{100}^{E}-mode. This example is evidence of the possibility to compensate local reflection grating playing the role of polarization inhomogeneity.

Study of the diffractional far field pattern for two polarizations in the plane parallel to lamellae show that preservation of the polarization of resonance mode is the general property of the fields radiated out of OR in lateral directions (Fig. 92). Some deviation from this polarization in lateral petals (ellipticity and a small turn of polarization plane) is caused by radiation of edge waves. Diffractional far field patterns in the case of degeneration of TEM_{02}^{H} and TEM_{100}^{E} oscillations are presented on Fig. 93, a. The field has elliptic polarization between their maximums (the ratio of half-axes is ~ 0.1).

Polarization patterns for $\theta = 63°$ are shown on Fig. 93, c. Here resonance distances between the mirrors of OR for two orthogonal polarization have the difference equal to 0.001 mm. One can see that readjustment of one oscillations with respect to another yields negligible variation of polarization of diffractional field. Readjustment of OR in wider limits along the resonance curve yields strong change of polarization pattern. Variation of excitation conditions in the point of degeneration (the turn of polarization plane of exciting field) may be registered in the diffractional far field pattern; polarization angles in diffractional field and in OR have the difference of not more than $1.5° - 2°$ for small turning angles ($\sim 10°$).

We may conclude that determination of the state of polarization of exciting field in quasi-optical resonators with a grating is possible by the control of polarization in diffractional field.

§15. Interaction of electron flow with the field of Smith-Purcell radiation in open resonators

The key problem in the theory of SPR excitation by OR with a grating is the study of thin structure of the process of transformation of the energy of electron flow to diffractional radiation in OR. Recently it has been obtained a number of important results which enabled the researchers to determine the limits of applicability of the created theory and stimulated further development of the technology of diffractional electronics [3, 4, 5]. It has been established that the discovered effects of thin structure of diffractional radiation (see Chapter 2) were characteristic also for excitation of oscillations in OR.

1. In the general case it follows from the solution to the problem obtained within the frame of the model of given current that convectional current $\tilde{i}(y)$ and the power of interaction P of the flow with the field of OR and the grating admits representation in the form [3]

$$\tilde{i}(y) = \frac{1}{4}\frac{i_r}{U_0}\frac{h_e}{h_r}\left[e^{-ih^- y}\int_0^y E_y(\zeta)e^{-ih^-\zeta}d\zeta - e^{-ih^+ y}\int_0^y E_y(\zeta)e^{-ih^+\zeta}d\zeta\right];$$

$$P = P_a + iP_r, \tag{5.4}$$

where $E_y(\zeta)$ is the field's distribution of surface diffractional harmonic in the space of interaction (on the grating); $h^{\pm} = h_e \pm h_r$ are the propagation constants of electron WSC; $h_e = \omega v_0^{-1}$, $h_r = \omega_r v_0^{-1}$.

Analysis of (5.4) shows that in the process of energy exchange before generation of oscillations electron WSC are formed in the flow of electrons of OR with a grating, which propagate with phase velocities (2.22). SWSC give energy to resonator's field and FWSC absorbs the energy. Maximal interaction intensity occurs when the condition (4.14) holds, i. e., when SWSC (or FWSC in the case of absorption of energy) is synchronous with the field of space diffractional harmonic. With growing current's density the power P_a increases and when it reaches the value of W equal to the power stored in OR self-excitation of oscillations begin (value of the current J_S, for which this process begins, is called starting current).

Considered electron-wave mechanism of transforming the energy of electron flow in OR with a grating is confirmed by experimental investigations carried out in the millimeter range on the plant shown on Fig. 94. Let us describe the method of experimental studies. Given type of oscillation is excited in OR 10 by external HF generator 1 (GDR); signal goes through preliminary attenuator 2, resonance wave-meter 3 and directional coupler 4. A part of the power of this oscillation comes through the second shoulder of the coupler to measuring HF section containing directional coupler 5, power-meter 6, control wave-meter 7, detector section 8 and oscillograph 9. Electron flow is let through near the grating; it is modulated in the process of electrons' interaction with resonator's field (electron WSC are formed) what yields absorption and amplifying of the field in OR by the flow. These phenomena are observed on the screen of oscillograph, if electron flow amplified by accelerating (anode) voltage comes to the input of OR (diode electron cannon is fed by an oscillating high-voltage source having the frequency 50 Hz).

Analysis of oscillograms confirms electron-wave mechanism of excitation of diffractional radiation in OR. For fixed distance between the mirrors with growing velocity of electron flow it is observed at first absorption of energy by FWSC and then amplifying of energy by SWSC. With growing pencil's current the power of interaction of SWSC and FWSC increases; auto-generation of diffractional radiation in OR is indicated at amplified signal what points out to the transformation of energy of SWSC to HF oscillation. Experiments enable one to discover several new phenomena which are not studied theoretically. Let us analyze some of them.

a) Changing distance between the mirrors of OR (or the phase of incident field) influences the considered processes of energy exchange. In particular, gradual variation of this distance $D+\zeta$ within small limits, ($\zeta << 0.25\lambda$) provides explicit absorption (amplifying) of resonator's field by the flow or separates anomalous energy characteristic: with growing U_0 (the parameter of synchronization) electromagnetic field at first is amplified and then absorbed. In this case the signal in OR emerges for the value of accelerating voltage which is smaller than in the considered case of "normal" energy characteristic.

b) With growing J_0 auto-generation starts in the process of absorption of energy at fixed frequency but when the parameter of synchronization takes two close values placed to the left and to the right of energy characteristic, where the strong absorption of signal is observed. With growing pencil's current the energy absorbed by electrons gradually decreases and for a certain value of i_0 both "emerging" signals of diffractional radiation join together to one oscillation of OR capturing the signal of absorption. Such initial stage of excitation of auto-oscillation is characteristic for interaction of electron flow with higher order types of oscillations where suppression of accelerated electrons may be connected with regrouping of electron flow, after which most of accelerated electrons come to the hampering field's phase.

Let us point out that weak pulsations appear in the initial moment of "emerging" the signal in OR with a grating and electron flow which indicate either that frequency of auto-oscillations is close to the one of pumping signal or that synchronization takes place in the presence of signal from external source.

2. Experimental studies show that longitudinal electron waves (SWSC and FWSC) are mainly excited in the process of interaction of electron flow with the field scattered by the grating in OR containing semi-spherical or spherical-cylindrical mirrors and a grating placed in the central part of the "lower" mirror. However, as well as in the analyzer of diffractional radiation with preliminary modulation of electron flow (see Chapter 2), it is difficult to separate transversal (cyclotron-diffractional) electron waves in such generator because of weak power of flow's interaction with transversal component of electric field of reflection diffractional grating.

Efficiency of forming in the flow of cyclotron-diffractional space harmonics of electron current essentially raises if the flow in OR is let through between two diffractional gratings (as in the classical cyclotron parameter devices [104]). It has been established experimentally that longitudinal and transversal electron waves were effectively formed in OR with two gratings in the process of interaction of electron flow with the field of OR. These waves are easily separated by the character of variation of interaction power via focusing magnetic field. Oscillograms enable one to determine the domains of excitation of longitudinal and transversal electron waves when accelerating voltage grows for different values of focusing magnetic field. They are also evidence of such characteristic feature of the interaction power of WSC as gradual transition from absorption of electromagnetic field of OR by FWSC to its amplifying by SWSC with growing accelerating voltage.

Energy exchange of cyclotron-diffractional electron waves is characterized either by absorption of the (fast) wave or by amplifying of the (slow) wave by the electron waves of the field of OR. In this processes the growth of focusing magnetic field for WSC negligibly influences the optimal value of U_0 (when the powers of absorption and amplifying are maximal), while for FCW this value shifts with increasing focusing magnetic field towards smaller accelerating voltages and for SCW it shifts towards greater values of U_0.

The complete experimentally discovered spectrum of longitudinal and transversal electron waves in OR with two gratings is presented on Fig. 96 in the plane of modified Brilluin diagram (in the coordinate system of voltage U_0 and focusing magnetic field B connected with the parameters β, ν given on Fig. 54, b). Eleven electron waves are separated including six WSC (SWSC and FWSC with indices $m = 0$; $n = -1, -2, -3$), the principal SCW ($m, n = -1, -1$) and four higher order cyclotron-diffractional waves ($m, n = \pm 1, -2; \pm 1, -3$). One may conclude that theoretical interrelations (2.25), (2.26) for the radiation conditions allow to determine in a sufficiently accurate way with the help of Brilluin diagrams the domains on the plane of parameters U_0, B where the given longitudinal or transversal electron wave is excited in the electron flow of OR with two gratings. Unlike WSC, excitation of cyclotron-diffractional electron wave depends on the level of accelerating voltage and focusing magnetic field; these waves have rather big steepness of the graph of variation of magnetic field B and accelerating voltage (big $\frac{dU_0}{dB}$) while for WSC these quantity is practically constant.

Hence, excitation of electromagnetic oscillations of diffractional radiation in OR is characterized, as well as energy exchange between the flow and the field scattered on the grating, by the electron-wave mechanism of transformation of electrons' energy to radiation. Signals of HF oscillations of OR with a grating are "produced" by the slow (longitudinal and transversal) electron wave propagating near the grating with constant phase velocity. Fast electron waves (or accelerated particles) absorb the energy of the field of OR; but, since the condition of diffractional radiation holds for these waves, one may assume that accelerated particles in such open structures are formed not as effectively as in the devices of surface waves what points out to raised efficiency and higher quality of the spectrum of HF signal. Besides, these phenomena are to be observed also for the diffractional radiation excited in OR because electron waves are characterized by the effect of different polarization of convectional current.

§16. Polarization effects of Smith-Purcell radiation in open resonators

Polarization regimes of the excitation of OR by diffractional radiation are studied by in their diffractional field. One can determine the state of field's polarization, i. e., polarization of the field of electron flow flying over diffractional grating, if to take into account existence of the domains of polarization degeneration in the spectrum of OR [134].

1. Polarization characteristics of the fields in OR with a grating and a flow of electrons have been investigated by means of the polarimeter presented on Fig. 97, a. The signal measured in the plane ξ, η comes through quartz window (which is not shown) to mouthpiece antenna 10 and then to polarization's divisor 9 connected with mixers 4. Phase shifter 3 in the supporting channel connected with waveguide output of energy 2 in the spherical mirror 1 of GDR provides registering of the signals at intermediate frequency for E_ξ- and E_η-components of diffractional field in OR. In oscillating regime (for accelerating voltage) polarizations in diffractional field are measured by the waveguide analyzer connected to mouthpiece antenna 10. Or has the following parameters: $2a_s = 60$ mm; $2a_c = 40$ mm; $R_s = 150$ mm; $R_c = 110$ mm; the grating is lifted on the level 0.12 mm. Reflection diffractional grating has the parameters $d_0 = 10$ mm; $l = 0.4$ mm; $\frac{d}{l} = 0.35$; $h = 0.92$.

It is discovered two oscillations with strong dependence of the orientation of polarization plane on the focusing magnetic field B in the whole range of combined readjustment of the device. With growing B the amplitude of SHF field grows, what well agrees with the results of [62], and orientation of the electric vector changes. Relative variation of orientation angle $\tilde{\beta}$ of polarization ellipse in the diffractional field for oscillation 1 is given on Fig. 97, b. The turn of polarization plane is caused by faster growth of E_η-component perpendicular to grating's lamellae. Of course, one can estimate only relative variation of the orientation of electric vector of SHF field within the frames of this experiment, since initial polarization of the diffractional field in OR may be influenced by the hermetic cover of the device, construction of the window, etc. However, indirect proofs have been experimentally obtained which were evidence of the fact that electric component of SHF field was oriented at a certain angle to grating's lamellae even for rather high tensity of magnetic field. This phenomenon is a consequence of the rotation of polarization plane in diffractional field occurring in the process of distance resonator's readjustment. It takes place only for some oscillations excited in OR with weak phase anisotropy in the presence of both components of electric field in the exciting field. For other types of oscillations E_τ-component was mainly observed (15 modes were analyzed in total), what agreed with the data characterizing polarization degeneration in the spectrum of OR.

Discovered effect of the magnetic-optical rotation of polarization plane in the diffractional field of OR enables one to conclude that in some cases idealized models of excitation of oscillations do not completely describe physical processes of interaction for a real device. The curve of magnetic-optical rotation for such oscillations has more complicated character. Relative variation of orientation angle of polarization ellipse versus focusing magnetic field is presented on Fig. 98 ($\tilde{\beta}_{\min}$ denotes orientation of the small half-axis). The plant works in the regime of frequency modulation what allows to measure the curves of magnetic-optical rotation in two generation zones excited for growing (curve 1) and decreasing (curve 2) accelerating voltages. One can see that the regimes of model's operation are different in these zones. With growing magnetic field the axis of polarization ellipses rotates in opposite directions (Fig. 98, a).

Let us compare polarization characteristics of the generation signals and the signals radiated to free space (through the window). In this case the model is adjusted to the generation of H-polarized wave (polarization is controlled through the circular diaphragm in spherical mirror loaded by the circular waveguide). OR is excited at the same wavelength ($\lambda = 4.05$ mm) by an external generator in "cold" regime, i. e., without electron flow. Diffractional far field pattern is measured through the quartz window by the angle θ (see Fig. 92) and it is determined the domain of the minimum of diffractional field for the oscillation in OR. Then polarization analyzer is placed to one of these minimums and the model is again initiated at the given mode. Main contribution to the signal received by analyzer is made by radiating electron flow what is confirmed by the far field patterns measured with respect to the angle $\tilde{\phi}$ for fixed θ. The patterns registered through the window for two values of accelerating voltage and the directions of diffractional radiation calculated for OR with the grating having the same parameters are shown on Fig. 99. Calculation and experimental data (Fig. 99, a) are in good correspondence. With growing voltage the main (calculated) petal comes closer to the direction perpendicular to the grating while lateral (experimental) petal approaches the position given by the angle $\tilde{\phi} = 0$. The window of finite size undoubtedly influences the measured far field patterns; therefore, although the presented data have illustrative character, they correctly determine qualitative characteristics of the field directly radiated by electron flow.

In order to compare the signals radiated from the window and the ones registered through the communication element in spherical mirror of OR in the regime of frequency modulation, the latter are sent to the pair of inputs of two-beam oscillograph. One may conclude then that these signals differ both by power's variation versus accelerating voltage U_0 and by polarization properties of radiated field. Obtained oscillograms illustrate the change of signals' shape with respect to varying magnetic field.

2. Measurements of the polarization state in the zone of electron readjustment of OR revealed the rotation of polarization plane from the side of smooth excitation regime. After the splitting domain orientation of polarization plane has a sharp leap. Amplitude of generated signal in the splitting domain (curve 1) and ellipticity coefficient (curve 2) for one of the signals excited in OR are presented on Fig. 100, a and variation of the angle of rotation of polarization plane is shown on Fig. 100, b. In the low-voltage peak the field has polarization close to H-wave and in the high-voltage peak polarization plane is oriented along the direction of grating's lamellae. There exist critical values of the current and magnetic field for which the registered zone of generation of diffractional radiation has no obvious anomalies. It is important that, unlike magnetic-optical rotation observed in two-three points of OR, the described anomalous splitting of generation zone are met in more than the half of generated oscillations and, apparently, it is connected with the forming of SWSC and FWSC in the flow of electrons.

The shape of divided zone in diffractional field is determined not only by orientation angle $\tilde{\beta}_a$ of polarization analyzer but also by the angles θ and $\tilde{\phi}$. It confirms the assumption that the radiation field of the flow is registered through the window and well agrees with the data describing internal and diffractional fields of OR. In particular, considerable turns of polarization plane occurring in the vacuum experiment for the waves radiated out of OR far from the points of polarization degeneration cannot be the result of edge effects, since in this domain of the field of OR the influence of edge waves on polarization is negligible.

3. Thus, we conclude that interaction of electron field with the field of SPR excited in OR yields generation of transversal and longitudinal electron waves in the flow, which influence the structure of the field of diffractional radiation. In OR with a grating (situated in the central part of plane or cylin-

drical mirror [3]) diffractional radiation is mainly excited by SWSC having almost H-polarized field (some deviations are observed in the domains of smooth and hard generation regimes). At the same time, there exist several oscillations essentially differing from the H-polarized field what indicates excitation of transversal (cyclotron-diffractional) electron waves in the flow.

Efficiency of excitation of transversal and longitudinal electron waves raises for OR with double diffractional gratings (electron flow moves between two periodical structures). The domain of variation of the parameters of diffractional grating, velocity of electron flow (accelerating voltage) and focusing magnetic field, which provide the specific case of excitation of diffractional radiation, are determined in a sufficiently accurate way by the proposed Brilluin diagrams.

Discovered effects of the thin structure of the processes of SPR excitation and interaction of electron flow with the field of OR are of considerable scientific and practical interest. The theory of GDR created in [3-5] gives sufficiently good description of the processes of energy exchange between longitudinal electron waves and the field of OR but does not contain the models of longitudinal-transversal interaction where one must take into account variations of polarization characteristics of diffractional radiation.

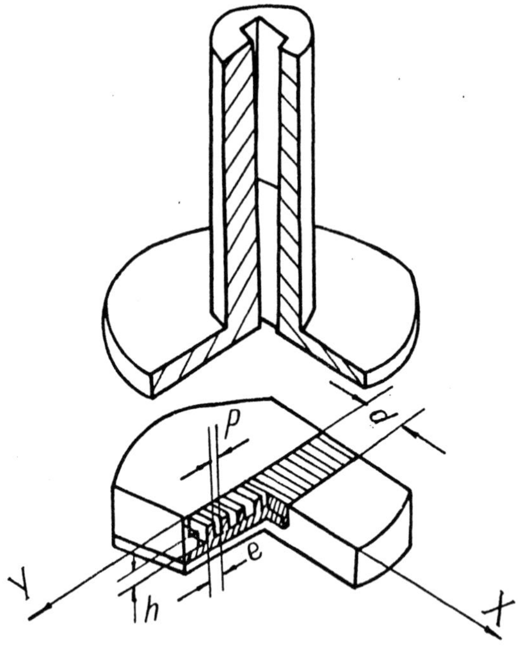

Fig. 83. Semi-symmetrical open resonator with reflection diffractional grating.

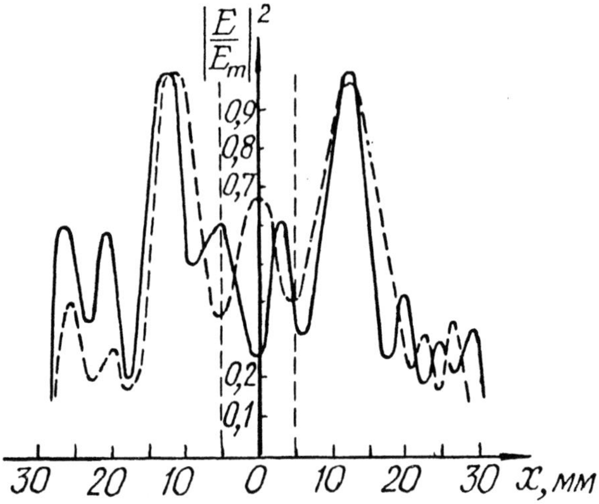

Fig. 84. Field's distribution in the open resonator with metal plate: dot line – slat is placed at the distance z_0; dot line – slat is combined with plane mirror.

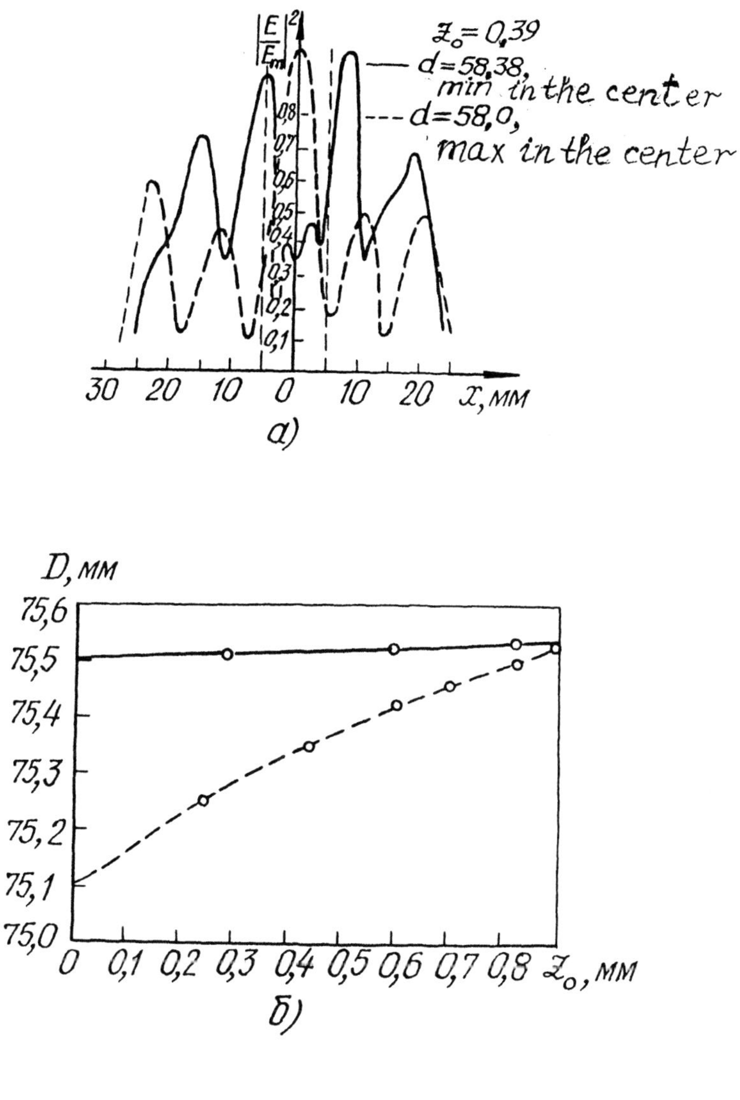

Fig. 85. Degeneration of oscillations in the open resonator with diffractional grating: a) – fields' structure; $z_0 = 0.39$ mm, $d = 58.38$ mm; b) resonance distance versus z_0; bold line – calculation data; dot line – experimental data.

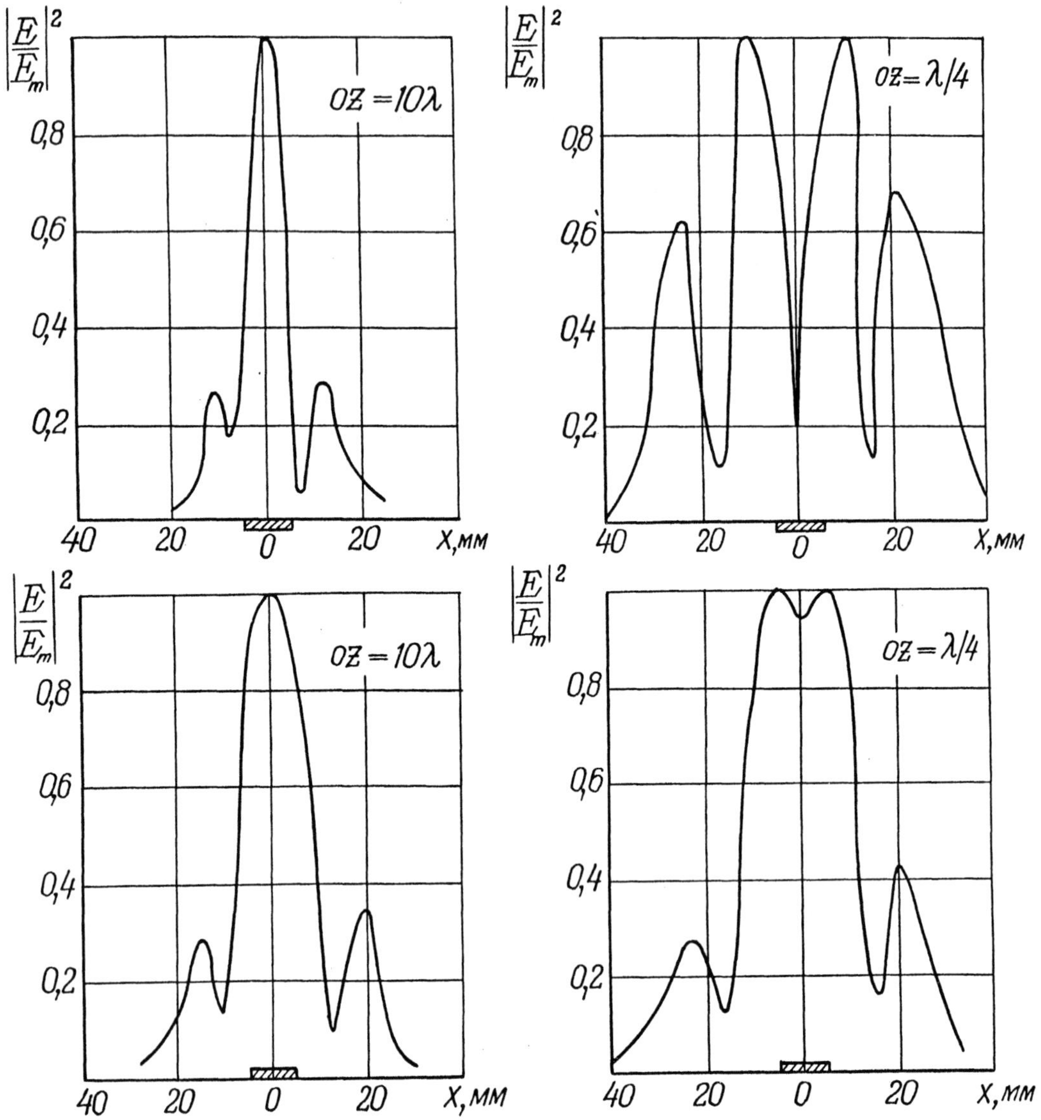

Fig. 86. Field's distribution in the open resonator with a grating placed in its center (a, c) and over the plane mirror (b, d).

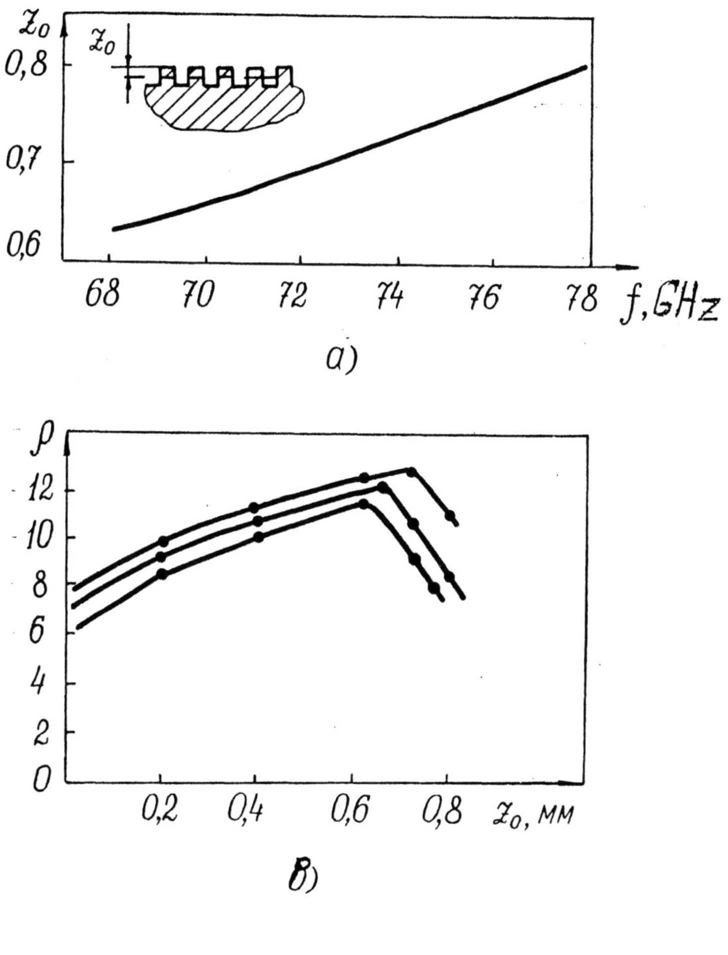

Fig. 87. Dependence of z_0 (a) and spectral density (b) of the open resonator on frequency: b) 1 – $f_0 = 74$ GHz; 2 – $f_0 = 71$ GHz; 3 – $f_0 = 68$ GHz.

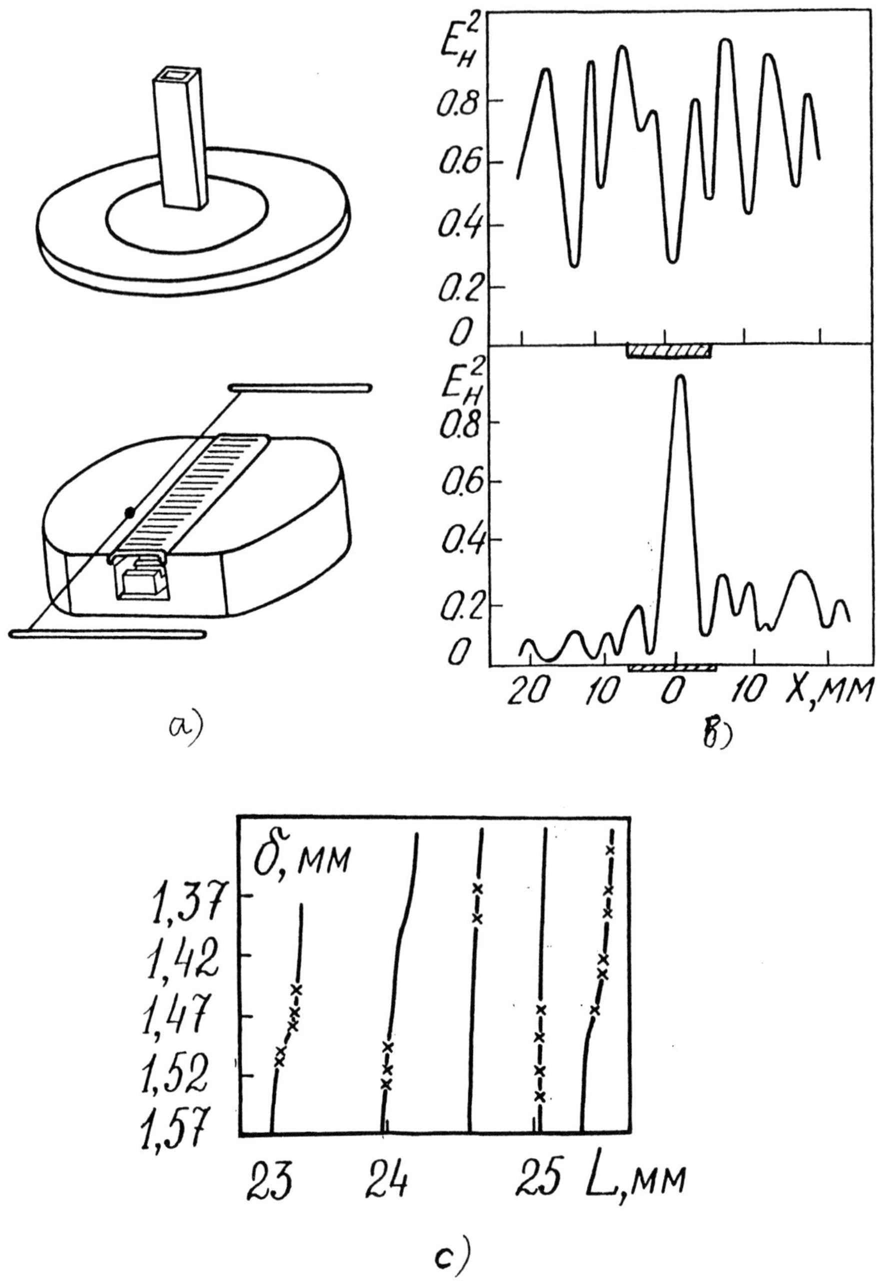

Fig. 88. Open resonator with double grating (a); field's distribution in the resonator (b); resonator's readjustment diagram (c).

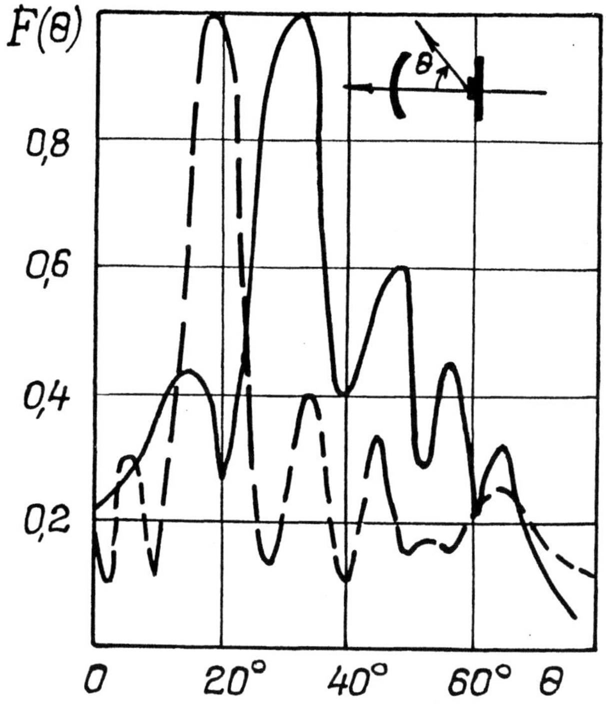

Fig. 89. Radiation far field pattern of the scattered field of the open resonator with a grating.

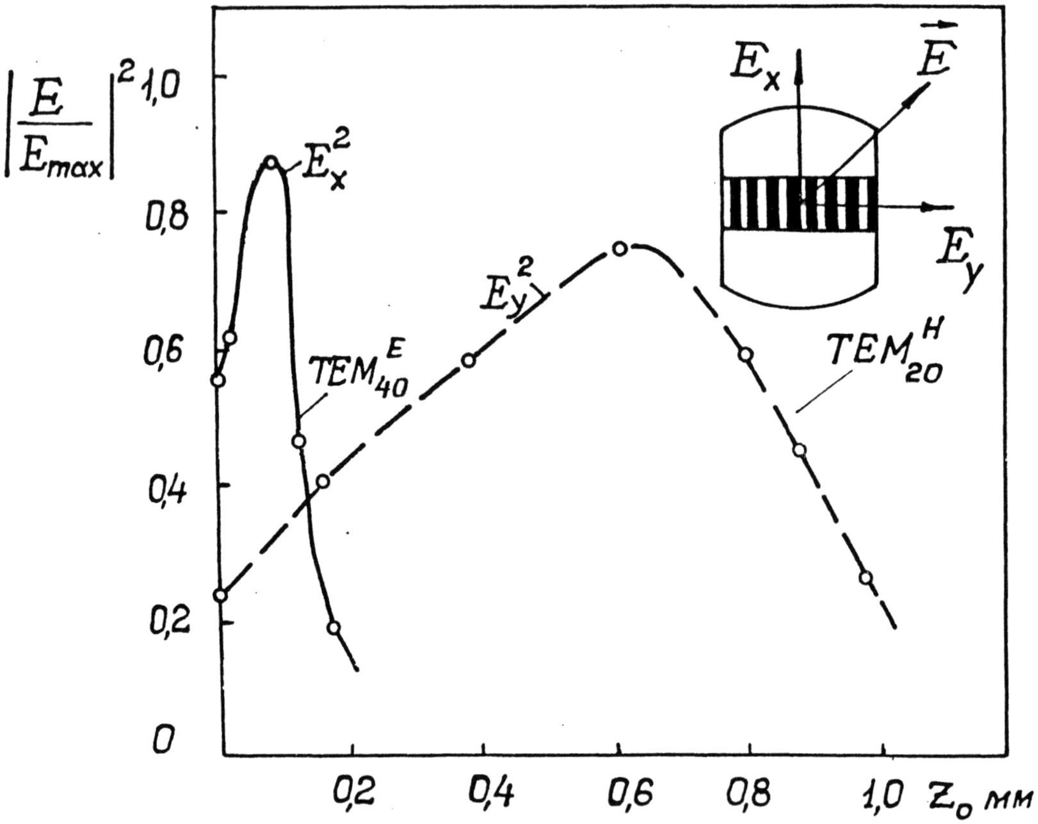

Fig. 90. Relative variation of the amplitudes of TEM_{20}^H and TEM_{40}^E oscillations versus grating's position.

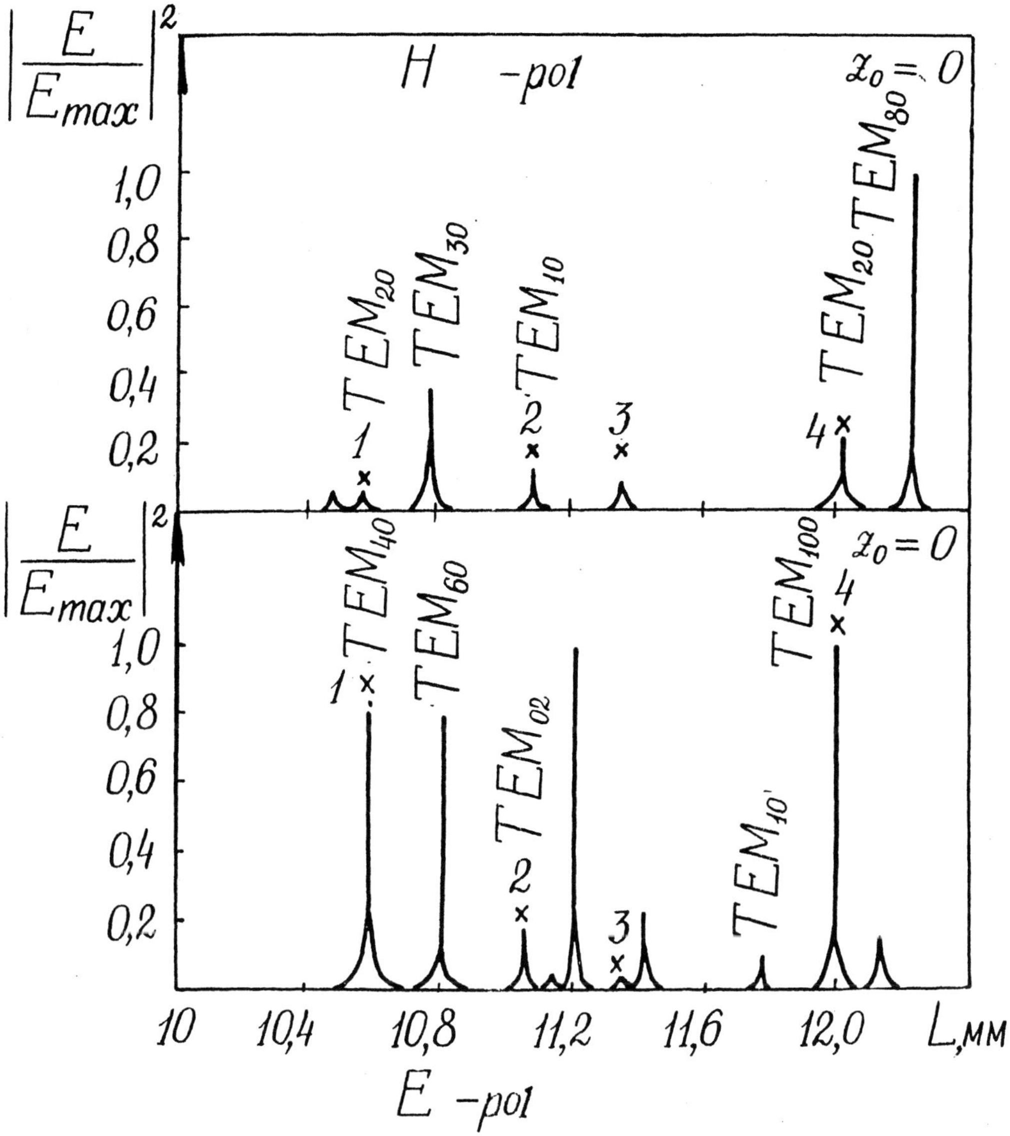

Fig. 91. Part of the spectrum of the open resonator with a grating at the frequency 68 GHz for H- (a) and E- (b) polarizations.

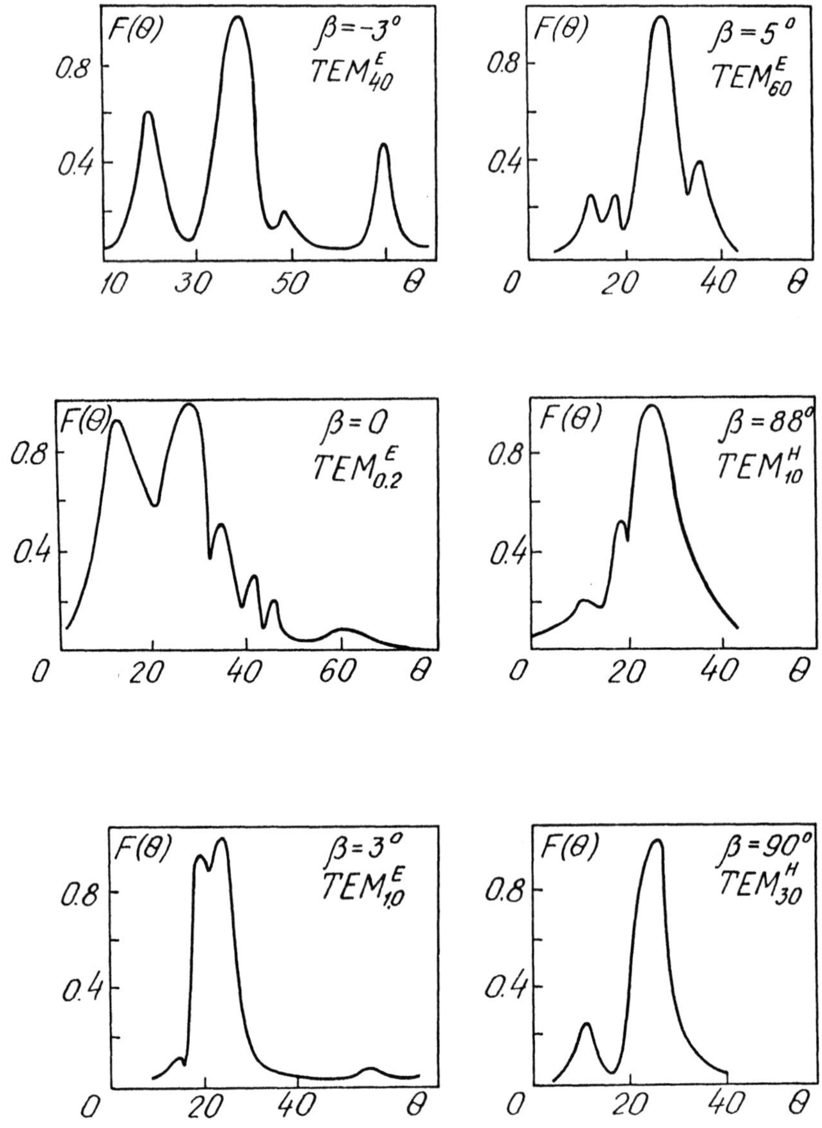

Fig. 92. Radiation far field pattern of the Smith-Purcell radiation for an open resonator of GDR.

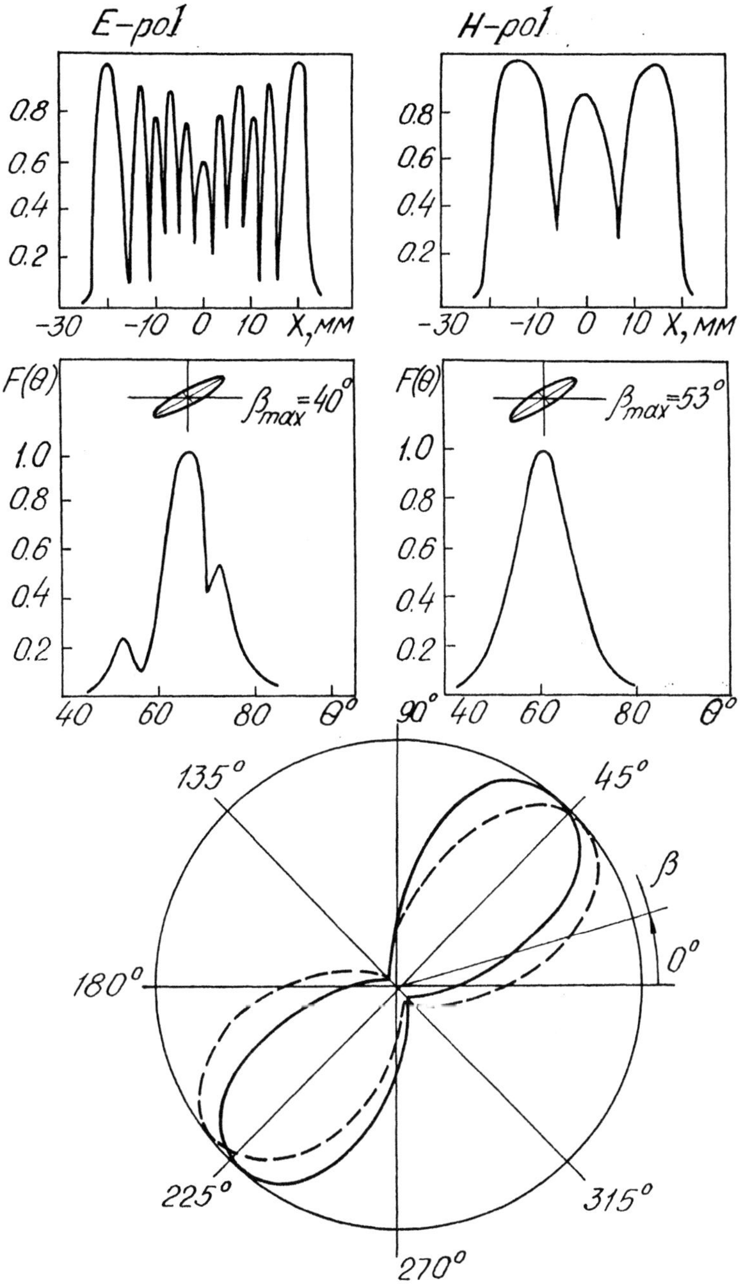

Fig. 93. Field's distribution (a), far field patterns (b) and polarization diagrams (c) of TEM^{E}_{100} (bold curve) and TEM^{H}_{20} (dot curve) oscillations close to the domain of degeneration.

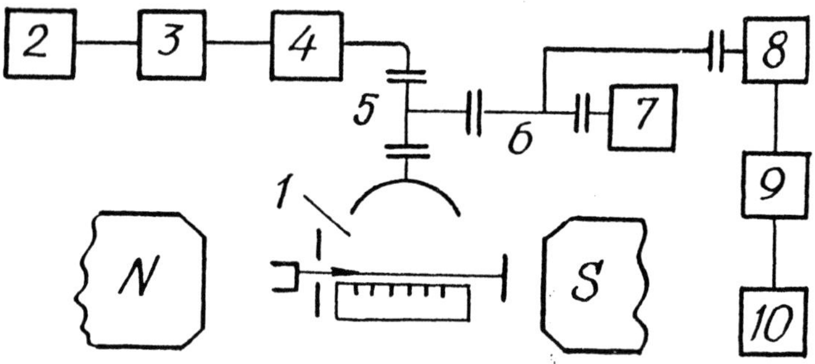

Fig. 94. Experimental plant for the study of the processes of interaction of electron flow with the field of an open resonator of GDR.

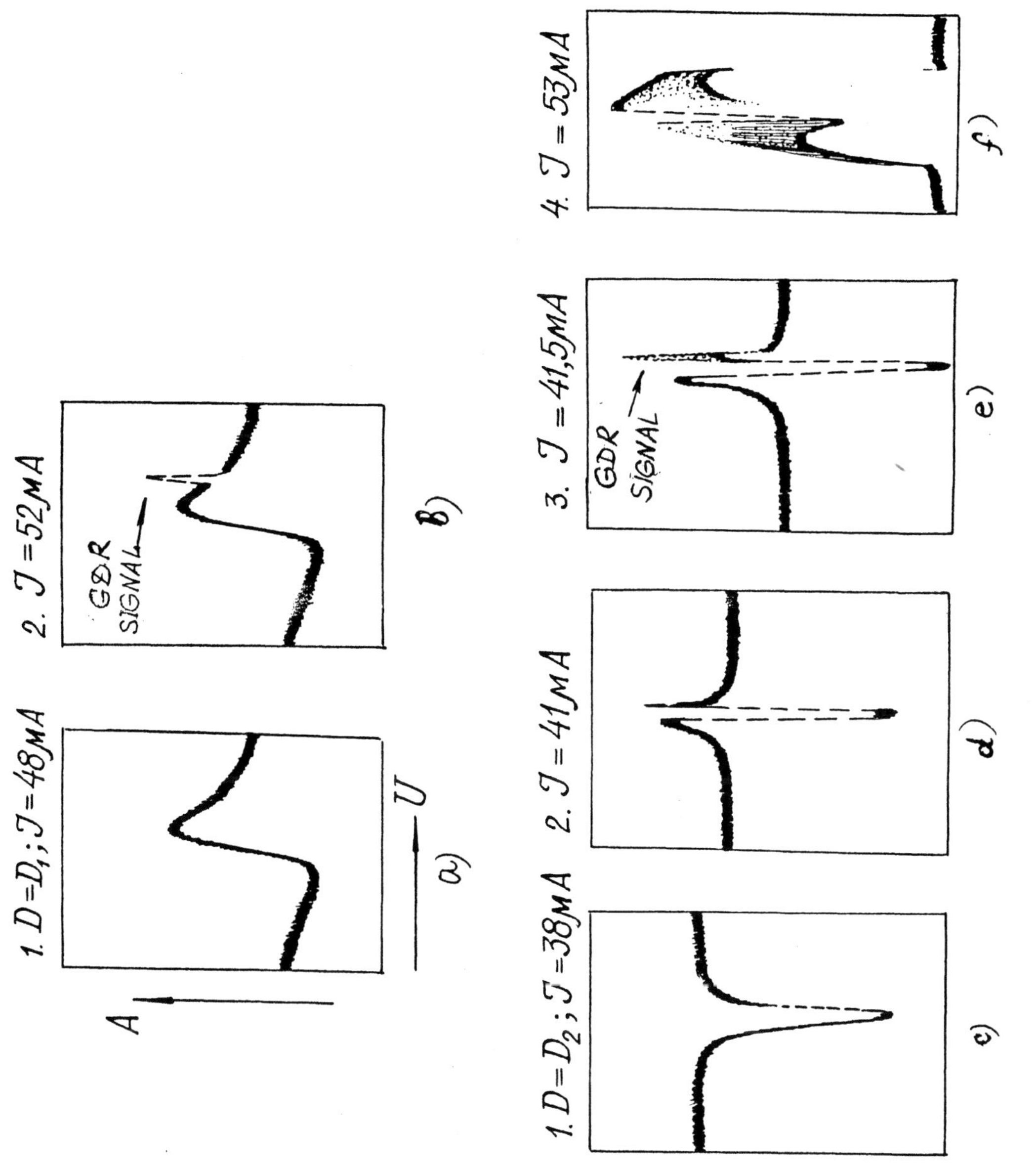

Fig. 95. "Emergence" of GDR signals in the cases of normal interaction power (a, b) and strong absorption of the field of an open resonator by the electron flow (c–f); oscillograms (e, f) indicates the phenomenon of synchronizing the GDR oscillations by external signal.

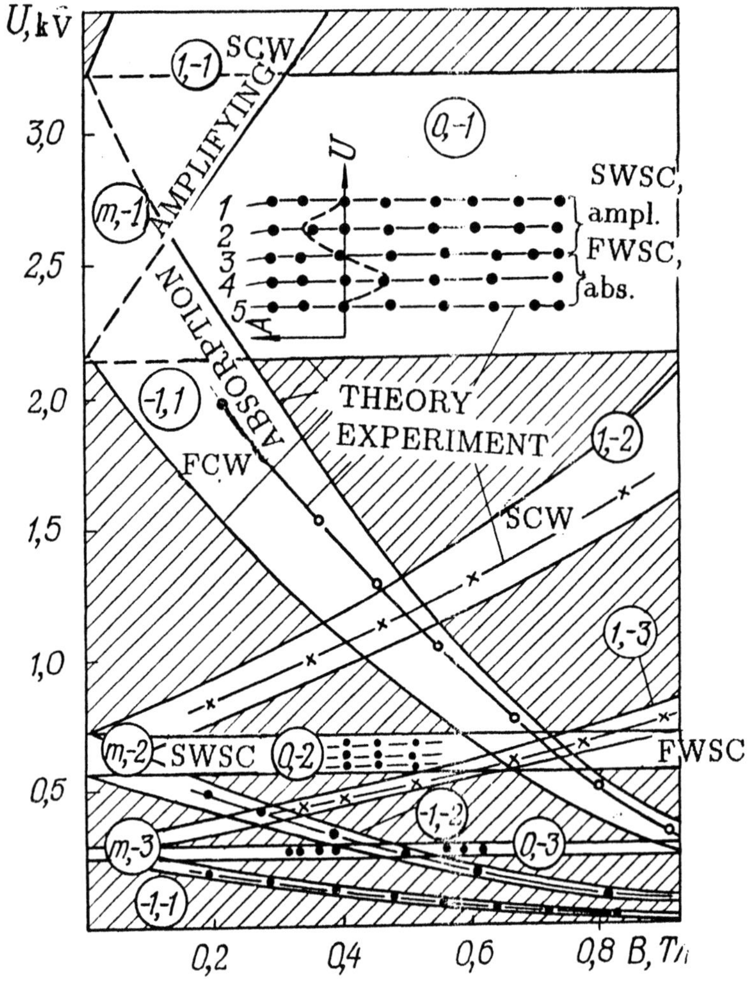

Fig. 96. Theoretically predicted domains of radiation of WSC and cyclotron-diffractional electron waves in GDR with double grating on the plane of modified Brillouine diagram and experimental data.

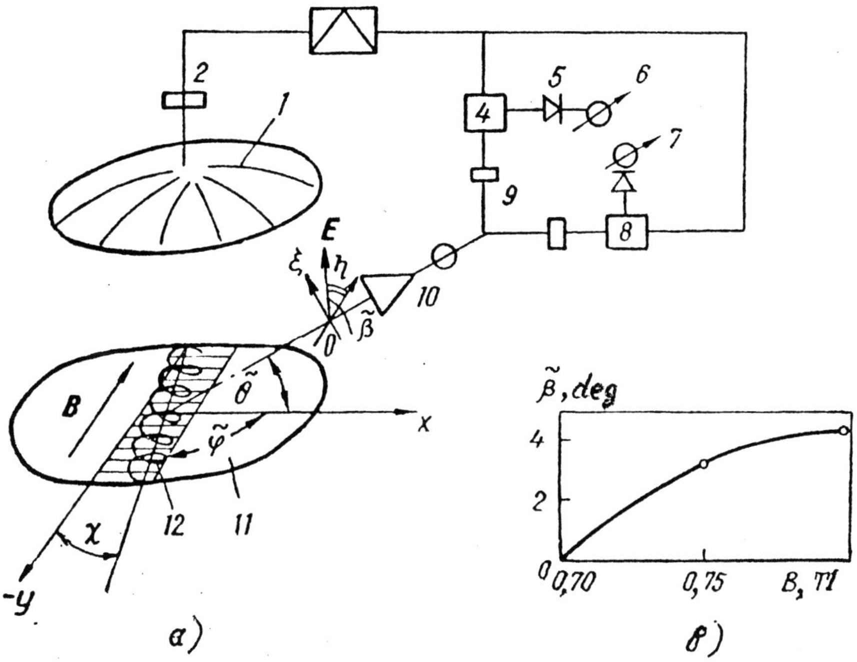

Fig. 97. The plant for the measurement of field's polarization in GDR (a) and variation of angle β versus focusing magnetic field (b): 1 – spherical mirror; 2 – waveguide output of energy; 3 – frequency shifter; 4, 8 – T-joint; 5 – detector; 6, 7 – indicating registering devices; 9 – divisor of polarization; 10 – mouthpiece antenna; 11 – mirror with a grating; 12 – trajectory of electron's movement in focusing magnetic field.

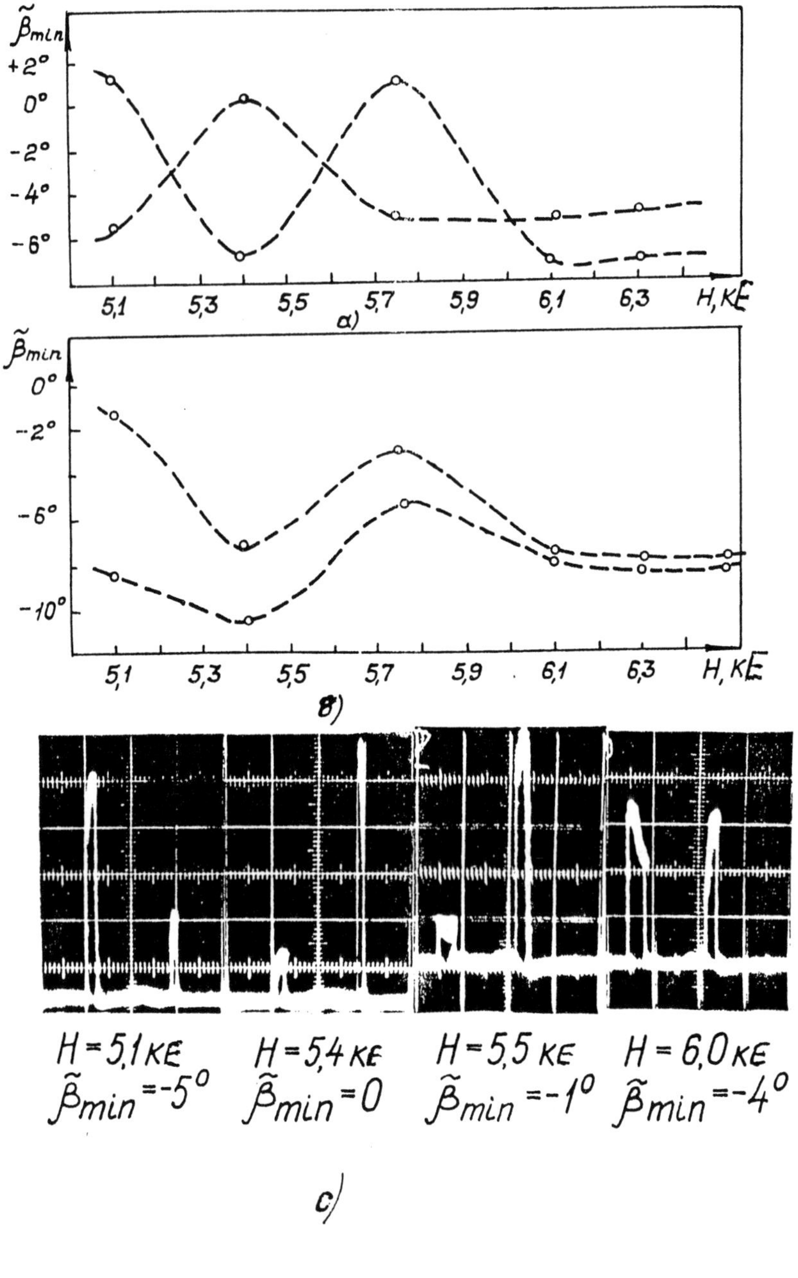

Fig. 98. Polarization characteristics of scattered field in GDR versus focusing magnetic field (a–c).

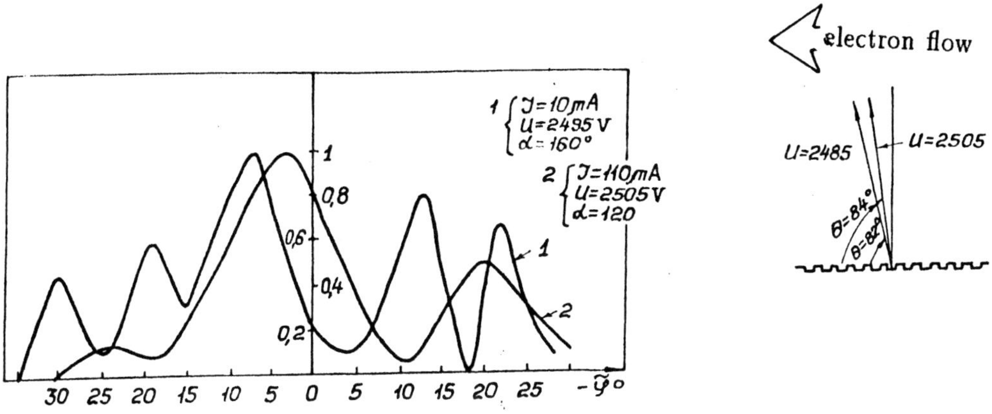

Fig. 99. Far field patterns of the field radiated out of the window of GDR: 1 – $J = 10\mu$A, $U_0 = 2495$ V, $\alpha = 160°$; 2 – $J = 110\mu$A, $U_0 = 2505$ V, $\alpha = 120°$.

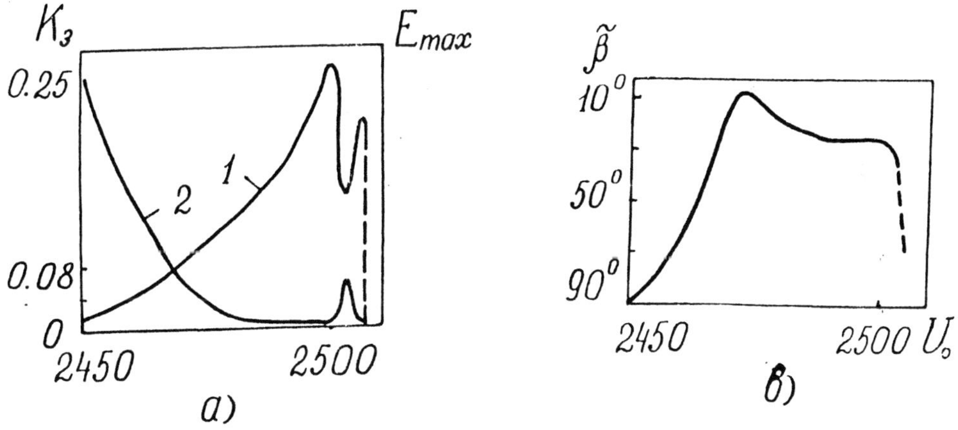

Fig. 100. Ellipticity (a) and angle $\tilde{\beta}_a$ in splitted zone (b) versus accelerating voltage.

CHAPTER SIX

Experimental wave modeling of Smith-Purcell effect

Rigorous theoretical model of the wave SPE is created in the second chapter. There the "hot" transformation of surface eigenwaves of electron flow is replaced by transformation of "cold" surface waves of the planar dielectric waveguide to the volume waves of SPR and one can almost completely study kinematical properties of SPR. The required measuring devices may be essentially simplified and are mainly connected with the use of antenna measurements which are well developed in the millimeter and submillimeter wave technology. Wave modeling of surface waves of electron flow by means of various decelerating devices has no limitations what yields the possibility of experimental observation of not only SPE but also the Cherenkov effect and their combinations together with other radiation effects, especially in the relativistic case and even for hypothetical particles. Besides, wave modeling enables one to "freeze" in time transient radiation acts and to spread this approach for "cold" modeling of the devices of diffractional electronics.

We have already demonstrated some experiments of wave modeling for SPR (determination of radiation angle at which SPR is observed for certain, §12). In this chapter we will pay special attention to the arrangement and realization of experiments on transformation of surface waves of different decelerating structures to volume waves of radiation. The solution to this problem is connected not only with deeper understanding of various radiation effects, but also with dissemination of these ideas in radiophysics, electronics, radio engineering, plasma and solid state physics. These studies are of special interest for antenna technology, cybernetics, fiber optics, *et al.*

§17. Justification of the method of wave modeling of transformation of surface waves to volume waves by diffractional gratings

Of principle question in the wave modeling of SPE (in more general case, in the modeling of transformation of arbitrary surface field to volume waves by metal diffractional gratings) is the creation of such experimental conditions which could at most approach experiments with "hot" SPE (in the presence of charges moving near the grating) to the ones where "cold" measurements are applied. It lays special demands upon not only the methods of measurements but upon the measuring devices used for "cold" experiments.

1. Let us consider diffractional grating (reflection grating or the one made of rectangular bars) over which at the distance a it is placed the waveguide of surface waves (see Fig. 101). For the given size and dielectric permittivity

of the waveguide the slow surface wave propagates with phase velocity $\beta = v_\Phi c^{-1}$ along the axis Oy (v_Φ is the wave's phase velocity in the waveguide and c is the velocity of light).

A part of the power propagating along the waveguide is concentrated in external domain. This very surface field provides diffraction of surface waves on the elements of periodical structure and enables one to model SPR only by means of wave fields. Finally, the energy of surface waves is transformed to the energy of fast space harmonics radiating to surrounding space. Experimental investigation of diffractional radiation is reduced to determination of the number of space harmonics, direction of their maximal radiation, dependence of power versus wavelength and polarization versus aimed distance [52].

It follows from the physical properties of SPR that in order to carry out experimental modeling of the diffraction of surface waves on periodical structures the measuring device must provide transformation of slow waves to volume waves, measurement of radiation angles of space harmonics in the sector $0° - 180°$ with absolute error not greater than $1°$ together with the wavelength and power of surface waves propagating in the waveguide; registering of that part of power of the waveguide of surface waves which forms radiation losses; determination of deceleration coefficient of the waveguide of surface waves; measurement of the distance between the structure and this waveguide. The plant must also provide absence of radiation of the waveguide of surface waves in the places of inhomogeneities (bends, windings) [135, 136].

The scheme of experimental plant [52] operating in the four-millimeter range is presented on Fig. 101. Clystron generator 4 feeds waveguide of surface waves 2 through coupling 5 and measuring 6 attenuators, directional coupler 7, measuring section 10 and excitation unit 3. In order to work in the regime of running wave (providing minimal reflection coefficient in the transmission line) the opposite end of waveguide 2 is connected with coordinated load 12 through unit 3 and directional coupler 11. Frequency and power measurement in the section are controlled by coupler 7, wavemeter 8 and indicating device 9. Directional coupler 11 registers the part of the power of the waveguide of surface waves which is lost in the process of diffractional radiation. Coefficient of standing wave is determined with the help of measuring section 10. Elliptic dielectric waveguides with the ratio of axes

$\frac{b}{a} = 0.5$ and dielectric permittivity $\epsilon = 2.5$ are used as waveguides of surface waves. Deceleration coefficient of dielectric waveguide is changed by varying the size of cross-section and its position in the space is provided by the use of hard foam plastic templet. Curvature radius R of the templet is chosen in such a way that $R > 40\lambda$ to avert radiation close to bends.

Surface field of the waveguide scattered by periodical structure radiates to the free space. In order to obtain diffractional far field patterns one must construct the dependence of field's directions on angle coordinates. With this purpose receiving mouthpiece 13 is placed on rotating bar (according to the method of moving antenna). In order to reduce errors of measurements of angle coordinates the axis of rotation of receiving mouthpiece must lie in the plane of diffractional grating. The signal from the mouthpiece comes after detecting to amplifier 14 and registering (self-writing) device 15. The obtained far field patterns enable one to find the number of radiating space harmonics and their radiation angles.

Radiation far field patterns are determined by measuring radiation power for varying frequency. Unlike the case of field's excitation by electron flow moving near diffractional grating, the described method does not allow to obtain radiation far field patterns experimentally for $\beta =$const due to dispersion of the waveguides of surface waves. Polarization pattern may be found by the method of rotation of receiving linearly polarized antenna around the axis coinciding with the direction of maximal radiation.

2. In spite of broad applications of periodical structures in many fields of modern physics, the phenomenon of diffraction of electromagnetic waves by different periodical obstacles were mainly considered within the frames of two approaches: investigation of volume fields emerging as a result of diffraction of plane incident waves (foundations of optical spectroscopy, antenna measurement technology, etc.); study of propagating surface fields localized near electrodynamical structure (such problems have considerable significance for design and creation of decelerating systems for the use in electron devices of classical electronics).

Experimental investigations of the diffraction of surface waves on periodical structures were carried out in the four-millimeter range of wavelengths in accord with the methods presented above for semi-transparent gratings (made of rectangular bars) and reflection gratings (combs, echelette with rectangular lug). The former was assembled in a packet out of copper ground elements and the latter were made of copper on milling machines and planers.

Characteristics of these gratings are brought together in Table 2.

3. The gratings applied for experimental studies of diffractional fields are bounded along the axes Ox and Oy (forming square apertures with an opening $15 - 20\ \lambda$ where λ is the wavelength). Bounded size of a grating along the axis Oy provides finite width of the far field pattern of emerging SPR (in the plane containing the direction of propagation of surface wave). Registering self-writing device whose strip moves synchronously with the turn of receiving antenna (Fig. 101, b) enables one to obtain far field pattern caused by the diffraction of surface waves on periodical structure having finite size along the direction of their propagation. One of such records is shown on Fig. 102. Length of the evolvent corresponds to the sector of space angles ($0° - 180°$) counted by measuring the distance between the coordinate of pattern's maximum and the origin ($0°$ corresponds to the last stroke on the right of the strip) with further evaluation of angle's value. In the case presented on Fig. 102 ($\kappa = 0.66$; $\beta = 0.842$) one-wave regime is realized occurring for those l, λ, β which satisfy radiation condition (1.6) only for one space harmonic. This regime is the most acceptable for practical purposes. Here the whole energy of diffractional radiation is concentrated in one beam going off reflection grating or in two beams spreading on both sides of transparent grating.

Radiation condition ties together radiation angle and characteristics of the source of electromagnetic field: relative velocity β, wavelength and grating's period. In accord with (1.6), the aimed parameter a (distance between the source of surface waves and the grating) does not influence the value of radiation angle. But in experimental studies of the diffraction of surface waves of dielectric waveguide on periodical structures this dependence takes place due to communication of the waveguide of surface waves and the grating.

Communication coefficient of such electrodynamical system depends on aimed parameter and characteristics of the waveguide of surface waves: relative dimensions, deceleration and type of propagating mode. They determine also effective cross-section of this waveguide where the most part of propagating energy is concentrated. Hence, variation of aimed parameter yields corresponding changes of the communication coefficient of electrodynamical system waveguide-grating. If the grating is out of active waveguide's cross-section or on its boundary, the case of weak communication occurs. Strong communication holds when the grating is within the limits of active cross-section.

In case of strong communication the grating perturbs field's distribution and, hence, the waveguide's propagation constant. Its deceleration also changes (if to compare with weak communication). Besides, in this case the wave propagating along the waveguide and the wave scattered by diffractional grating interfere what may cause emergence of several modes in the waveguide and, consequently, space harmonics of diffractional radiation.

Qualitative variations of the far field pattern registered for strong communication is shown on Fig. 103. Strong communication (aimed parameter $\Delta = 0$) and interference perturbed far field pattern in such a way that it transformed to the one of multi-wave regime, while the conditions of experiment ($\kappa = 0.66$; $\beta = 0.943$; $\lambda = 4.21$ mm) required one-wave regime.

Experimental data of radiation angle of the minus first harmonic versus aimed distance (for the system diffractional grating-reflection comb, $\kappa = 0.658$; $\lambda = 4.22$ mm) are presented on Fig. 104. Experimentally determined angles are marked by crosses and bold lines denote the curves calculated by (1.7) without taking into account communication of the grating and dielectric waveguide. These results give the possibility to quantitatively estimate radiation angle of the considered system with and without communication. In a number of practical measurements radiation angle must be found with very high accuracy. Therefore, investigations of the dependence of radiation angle on aimed distance acquires independent significance. They enable one to find the value of aimed distance (for the given waveguide of surface waves) which provides minimal error of determination of radiation's direction and to establish measurements' accuracy at small distances. The necessity of such measurements may be caused by low sensibility of registering devices (radiation intensity decreases exponentially with growing aimed distance).

After determination of radiation angle as a function of aimed distance and its values which correspond to the minimal error of measurements, series of experiments were carried out in order to calculate radiation angle of various periodical structures versus relative phase velocity of elliptic dielectric waveguide at several frequencies (measurements were made for $\lambda = 3.89$ mm, $\lambda = 4.15$ mm; relative phase velocity of the waveguide of surface waves varied in the interval $\beta = 0.66 - 0.94$). Radiation angles versus parameters of the waveguide of surface waves were obtained at two frequencies by the constructed far field patterns. Experimental data marked by crosses are presented on Fig. 105 together with the curves calculated by (1.7) for infinite gratings. One can compare the theoretical curves characterizing radiation conditions for the structure infinite along the axis Oy with experimental data obtained for the finite structure.

4. Radiated power is the most important characteristic of the process of transformation of surface waves to volume waves. In particular, raising of the efficiency of electro-vacuum devices operating with the use of diffractional radiation is connected with providing such a regime of radiation when maximal power of the source of electromagnetic field is taken away. The value and behaviour of radiated power with respect to the parameters of the source, waveguide of surface wave and the grating predetermine the possibility of creation various devices and units (directional couplers, power divisors, antennas, etc.) whose operation principles are based on the phenomenon of diffraction of surface waves by finite periodical structures.

Power of diffractional radiation (1.49) is concentrated in the petals of far field pattern. Two methods may be applied for its measurement. The first one, which is frequently used in antenna measurements, consists in the measuring of a far field pattern in at least two perpendicular planes and in further integrating of the obtained results over the whole space of observation. This technique is too cumbersome and laborious, so one applies another method based on the measurement of power in the waveguide of surface waves 2 (Fig. 101, b) which goes to coordinated load 12. If there is no diffractional grating, the power P_0 propagates in the dielectric waveguide. When the grating is brought into the field of surface wave, a part of the power P_0 (P_{rad}) is radiated to the free space because of transformation of slow wave to volume waves. The power equal to $P_0 - P_{\text{rad}}$ gets into the load 12 and is measured by directional coupler 11. One can recount $P_0 - P_{\text{rad}}$ to P_{rad} with the help of simple calculations according to the following procedure: to determine a part of the total power of dielectric waveguide which coincides with the power of diffractional radiation, to draw experimental curves of P_{rad} versus aimed parameter, to compare the characteristics of radiated power of different finite diffractional gratings.

One should carefully fix and move periodical structures relative to immovable dielectric waveguide taking into account exponential decrease of radiated power versus aimed parameter. With this purpose we place the gratings on a subject table shifted by a micrometer screw and equipped with a stop. Grating is tightly pressed to the stop by a holder. These measures guarantee equal values of aimed parameter, what is especially important for the comparative analysis of different periodical structures.

Radiation power was measured at the fixed wavelength $\lambda = 4.15$ mm by indicating the dependence of $P_0 - P_{\text{rad}}$ on the aimed parameter, calibrating the curve by graduated attenuator and calculating P_{rad} for each of periodical structures: reflection grating with the period 2 mm, grating made of bars with the period 2.04 mm and echelette with the period 2.345 mm with two elliptic dielectric waveguides having different relative phase velocities $\beta_1 = 0.768$, $\beta_2 = 0.940$. The results are presented on Fig. 106 for the case of diffraction of slow waves of the waveguide with $\beta = 0.768$ (and also for $\beta = 0.94$); $P_{\text{rad}}P_0^{-1}$ is marked on the axis of ordinates, 1 corresponds to the grating made of bars, 2 – to the echelette and 3 – to the reflection grating made of rectangular bars.

Analysis of the curves allows to assert that diffraction on periodical structures may provide such a regime when practically all power propagating along the waveguide is radiated out of it (Fig. 106, curve 1). This property of diffractional radiation has basic character for the design and creation of antenna devices of millimeter range.

The observed deviation of exponential decrease of radiated power in the domain of small aimed distances is caused by big value of the coefficient of communication between the grating and the waveguide as well as by corresponding variation of relative phase velocity. Monotonous decrease of radiated power with growing aimed distance (Fig. 106) may be used for creating directional couplers with oscillating transfer attenuation.

In addition to this, the curves on Fig. 106, giving a comparative characteristic of radiation power of different periodical structures, enable one to choose the type of diffractional grating (and its parameters) for calculation and manufacturing of the devices, which radiate maximal power. The present experimental plant and methods may be applied for all kinds of diffractional gratings. It is of special interest for the study of diffraction on those structures which do not admit application of rigorous mathematical technique. In this case experimental methods form the only possible approach to the solution of such problems.

The curves on Fig. 106 give power's distribution of dielectric waveguides along the direction of variation of aimed parameter, hence, they may be used for determining effective waveguide's cross-section. Error of calculation of power's distribution grows with decreasing aimed distance but it is not essential for finding effective cross-section [137].

5. Interrelation (1.49) for the power of diffractional radiation is evidence

of its functional dependence on the parameters $\kappa, \beta, \theta, \delta$. Variation of relative period κ and relative phase velocity β leads to the scanning of the maximum of far field pattern in free space. This maximum as a function of κ, β, λ is given by the formula (1.7).

For practical purposes it is important to carry out experimental measurements of radiation power for varying wavelength λ and, hence, κ when other parameters β, θ, δ are fixed. Graphic description of such dependence is considered as radiation diagram. Here we will study radiation diagrams of (reflection) diffractional grating made of rectangular bars (the comb). By definition, radiation diagram does not depend on the aimed parameter. Therefore, explicit interrelation for radiation diagram coincides with squared value of the amplitude of space harmonic.

Radiation diagrams were measured by means of the plant presented on Fig. 101, b. Let us describe the scheme of experiment in the one-wave regime. One varies generator's frequency by supporting constant level of power in the feeding waveguide by attenuator 6 with the help of indicating device 9 and directional coupler 7. Receiving mouthpiece is placed in the direction of maximal radiation at every wavelength determined by wavemeter 8. Wavelength is varied within the limits of the range of clystron's readjustment. The work in one-wave regime and within the limits of frequency range allows to assume that diffractional far field pattern has equal width. Hence, the method of moving antenna enables one to measure relative power's variations in the process of generator's readjustment. Changes of amplitude are calibrated by graduated attenuator.

Elliptic dielectric waveguide served as a source of electromagnetic field scattered by the grating. Its inevitable dispersion causes of deceleration and, hence, of the relative phase velocity $\beta =$ in the considered frequency range and by this the radiation far field pattern (radiation diagram), being a result of diffraction of surface waves on a grating, differs from the far field pattern of "classical" diffractional radiation where $\beta =$const. This condition is connected with the support of constant accelerating voltage. However, the indicated variation of relative phase velocity may be determined by the formulas from [138]. It enables one to use the obtained results for β in the whole frequency range as the data for computer codes for calculating the required radiation diagram.

Calculated diagrams for the comb with the period $2l = 2$ mm are presented on Fig. 107 where experimental far field pattern is marked by crosses.

Relative phase velocity of the wave of dielectric waveguide β belongs to the interval $0.74 - 0.80$. Deviation of the obtained curve form calculated one is caused both by the influence of finite size of the grating and by the difficulty to support constant feeding power in the whole frequency range. The latter may be explained by the presence of varying communication coefficient of the measuring units in the frequency range. Nevertheless, experimental and calculated data qualitatively coincide.

Such qualitative and quantitative correspondence of the obtained results confirms the basic idea that the fields excited by electron flows and by surface fields moving close to periodical structures have common nature – the diffractional radiation, i. e., SPR, what in principle proves the possibility to create wave model of the scattering of surface waves on periodical structures.

Successful application of diffractional gratings in the devices of millimeter and submillimeter range is connected with the right choice of dimensions of periodical structures. In diffractional electronics determination of such minimal permissible size of a grating, when radiated power and radiation angle are practically the same as for almost infinite gratings, has of principle importance for creating miniature devices, reducing operation gaps and, hence, for decreasing the feeding power of magnetic focusing systems. In measurement technique one must take into account the influence of axis grating's length on radiation angle in the devices based on the phenomenon of diffraction of surface waves on periodical structures (wavemeters, directional couplers, measuring devices of the deceleration coefficient of the waveguides of surface waves), since the measured radiation angle determines operation accuracy of such devices. In antenna technology characteristics of finite arrays and, in particular, propagation constant of the antenna of running wave are calculated [139] with the help of results obtained for similar infinite structures. Such methods put certain restrictions on the considered object (grating's size is much greater than the wavelength, excitation amplitude on edges gradually vanishes) what is not always practicable.

Thus, study of the influence of the length of periodical structure on diffractional fields is an independent and rather important problem of the modern radiophysics. Here we will present some results describing the influence of grating's size on the fields scattered by periodical structures. We will determine such their characteristics as radiated power, radiation angle and the width of far field pattern.

Radiation angle (direction of maximal radiation) and the width of far field pattern may be obtained by the analysis of the far field pattern itself (characteristics of the scattered fields are found in accord with the methods presented in the first section of this paragraph and radiated power as a function of the length of a structure is calculated in accord with the procedure stated in the fourth section).

Originality of the diffraction of surface waves on periodical structures is in the necessity of specific choice of the value of aimed parameter, which should be less than the wavelength due to the surface character of the waves propagating along the waveguide of slow waves. In this way, one cannot vary grating's length by means of the diaphragms (apertures) in the millimeter range because it is necessary to manufacture them of absorbing sheets with the thickness less than one millimeter. An attempt to make such an absorbing aperture (of dense paper covered by the layer of absorber) has failed, since the applied substance had low attenuation ($\sim$ 3 dB) and the edges of aperture made strong contribution to diffractional field whose amplitude was comparable with the one of the field scattered by the grating. Therefore, one had to decline this way of measurement. Diffractional grating with different number of periods were used for creating the grating of varying length. Echelette of varying length with rectangular tooth and the period $2l = 2.345$ was used in experimental measurements of the far field patterns and radiated power caused by the diffraction of surface waves of elliptic waveguide.

Such far field patterns related to the amplitude of radiation field of the grating with maximal number of periods ($N = 31$) and measured for the waveguide with relative phase velocity $\beta = 0.98$ are presented on Fig. 108 for two values of the communication coefficient of the system waveguide-grating. The value of this coefficient was controlled by varying aimed parameter. Radiation far field pattern in the case of strong communication (aimed parameter is equal to zero) is also given there. Radiation far field pattern for the case when the aimed parameter coincides with half of the wavelength is shown on Fig. 109. Number of periods is the varying parameter for these patterns.

Presented curves reflect basic features of radiation characteristics (direction of radiation, width of far field pattern) for varying length of diffractional grating. One can see that radiation angle increases with growing number of periods. For small number of periods the speed of variation of radiation angle is big. With further growth of the number of periods radiation angle asymptotically tends to the value calculated by (1.7) which is valid for infinite grating.

Quantitative dependence of radiation angle of the minus first harmonic on the number of periods for different values of communication coefficient is depicted on Fig. 110. The character of these curves may be explained with the help of the method of far field patterns multiplication [140]. Far field pattern of antenna grating is mainly determined by grating's factor in the case of big number of radiating elements. With decreasing number of elements individual properties of every radiator begin to count and the obtained curve (Fig. 110) may be divided in two parts: the domain with small number of elements where radiation angle is determined by individual scattering characteristics of a separate element and the domain with many periods where the direction of radiation is given by (1.7). It is natural that the dependence $\phi_{-1} = f(N)$, where N is the number of periods, must be different in different domains what is confirmed by Fig. 110.

Far field patterns on Fig. 108, 109 allow to follow the behaviour of their width on the level 3 dB as a function of the number of periods. Corresponding curves for two values of the aimed parameter are shown on Fig. 111 where the dot line presents theoretical data calculated in [140] for the plane grating made of elements with equal phases by interrelation $2\Delta\phi_0 = 65\lambda(Nd)^{-1}$, where N is the number of elements, d is the distance between them. Satisfactory coincidence of experimental and calculated data is observed only for sufficiently big value of N ($N \geq 20$). For $N = 6-8$ the speed of decreasing of the width of far field pattern sharply falls with growing length of the grating. This very number of periods divides the curves on Fig. 110 in two domains differing by the speed of variation.

Hence, sharp growth of the width of far field pattern in the case of diffraction of surface waves on the grating with the number of periods $N \leq 8$ is caused by the influence of directional properties of a single radiating element. Growth of communication coefficient is accompanied by the increasing of the width of far field pattern. This property of diffractional field, which is caused by transformation of surface fields to volume waves as a result of scattering on periodical structures, is explained by the decrease of grating's effective length due to the influence of strong communication. Results presented on Fig. 110, 111 allow to conclude that the system waveguide-diffractional grating may be considered as a structure of leaky waves. It is known [136], that the latter is the running wave whose energy continuously falls because of radiation. Hence, end elements of the grating are less excited than its "input" elements. In case of strong communication between the source and diffractional grating excitation efficiency of peripheral elements falls what yields decrease of the effective length of periodical structure and, hence, growth of the width of far field pattern.

The present electrodynamical system may be also considered as a structure of running (fast) waves. For the fixed length of diffractional grating determined by the degree of communication the regime of satiation of energy of diffractional radiation must come. In the case of strong communication this regime occurs when the number of periods is less than in the case of weak communication. Experimental curves characterizing radiation power versus the number of periods are given on Fig. 112. Axis of ordinates corresponds to the value $P_r P_0^{-1}$ where P_r is the radiated power and P_0 is the power propagating along dielectric waveguide in the absence of grating. Qualitatively the obtained data coincide with the predicted behaviour of this system considered as the radiating structure of running (fast) leaky waves.

Remarkable feature of the results of investigation of the influence of grating's length on diffractional field (see Fig. 110-112) is one common property of all curves describing different characteristics of diffractional fields (direction of radiation, width of the far field pattern, radiated power), namely, all these curves indicate minimal length of the diffractional grating whose radiation characteristics are determined by interrelations which are valid for infinite grating. In particular, in the problem of scattering of surface waves by periodical structures the length of such "minimal" grating is $L_{\text{gr}} \approx 26$ mm. This value well agrees with the parameters of operating GDR where the minimal length of reflection grating in the four-millimeter range is 25 mm. One could not excite GDR with shorter grating.

We may conclude that the results of experimental studies of the influence of grating's longitudinal dimensions on characteristics of diffractional field confirm the possibility of working out devices for the use in millimeter and submillimeter ranges applying the phenomenon of diffraction of surface waves on periodical structures having finite length ($L \leq 10\lambda$) [141].

§18. Experimental modeling of Smith-Purcell and Cherenkov radiations

Let us formulate the statement of basic problem [142, 143] describing excitation of Smith-Purcell and Cherenkov radiations. Assume that two-dimensional transversal-magnetic wave having field's components $E_{0x} = 0$, $H_{0y} = 0$, $H_{0z} = 0$ propagates with phase velocity v_B along the axis Oy of planar dielectric waveguide (Fig. 113, domain II) placed at the distance $z = -a$ from strip diffractional grating, which is drifted on the surface of dielectric medium (Fig. 113, domain IV).

Phase velocity v_B is determined by waveguide's dielectric permittivity, its transversal size and the properties of surrounding medium. In this way let us introduce effective dielectric permittivity $\epsilon_B^* = c^2 v_B^{-2}$, then $\beta_B^* = v_B c^{-1} = (\epsilon_B^*)^{-0.5}$.

It follows from the field's representations in separate domains of dielectric waveguide (Fig. 113) that transformation of its surface waves to radiation is possible only for phase velocities satisfying inequalities

$$\frac{\kappa c}{|n| + \kappa} \leq v_B \leq \frac{\kappa c}{|n| - \kappa}, \quad z > -a \tag{6.1}$$

in vacuum and

$$\frac{\kappa c}{|n| + \kappa\sqrt{\epsilon_c}} \leq v_B \leq \frac{\kappa c}{|n| - \kappa\sqrt{\epsilon_c}}, \quad -\delta < z < -a;$$

in dielectric; direction of radiation is determined by the angles

$$\gamma_{n1} = \arccos\left(\sqrt{\epsilon_B^*} + \frac{n}{\kappa}\right); \tag{6.2}$$

$$\gamma_{n\epsilon} = -\arccos\frac{\sqrt{\epsilon_B^*} + \frac{n}{\kappa}}{\sqrt{\epsilon_c}}. \tag{6.3}$$

Formulas (6.1)-(6.3) are similar to interrelations (1.7)-(1.9) obtained for monochromatic electron flow, if to assume that phase velocity of the wave of dielectric waveguide coincides with electrons' velocity. Hence, here, as well as for diffractional and Cherenkov radiations, one can separate different cases of transformation of the waves of dielectric waveguide to volume waves by means of corresponding Brilluin diagram (see §2). Such equivalence of radiations' excitation makes possible to model excitation with the help of considered electrodynamical system [100, 101, 142, 143].

Solution to the present problem in long-wave approximation enables one to obtain equivalent amplitudes of diffractional and Cherenkov radiations of monochromatic electron flow (1.3), if to assume that dielectric waveguide is "transparent" for the excited radiation. For $\delta \to \infty$ the amplitudes of principal space harmonics have the form [4, 145]

$$A_{-1} = K_{-1}\frac{1}{\epsilon_c}\sqrt{\frac{\epsilon_c\kappa - k_B}{\kappa^2 - k_B}};$$

$$K_{-1} = -\frac{1}{2}ia_\epsilon\frac{(1+\sqrt{\epsilon_c})\sqrt{(\epsilon_c(\epsilon_B^*)^{-1}-1)(\kappa^2-k_B)}}{\sqrt{(\epsilon_c\kappa^2 - k_B + \epsilon_c)(\kappa^2 - k_B)}}\Phi_{-1}^1; \tag{6.4}$$

$$K_0 = \frac{a_\epsilon}{1+G_0\Phi_G}; \quad a_\epsilon = -2i\frac{hpA}{\sqrt{\epsilon_B^*}}\frac{i\epsilon_c\sqrt{\epsilon_B^*-1}}{\sqrt{\epsilon_c-\epsilon_B}+i\epsilon_c\sqrt{\epsilon_B^*-1}}; \quad k_B = (\kappa\sqrt{\epsilon_B^*}-1)^2.$$

Formulas (6.4) coincide with expressions (2.6), (2.7) for diffractional and Cherenkov radiations of monochromatic electron flow, if to make change of variables $\epsilon_B^* = \beta_e^{-2}$, $(ihpA) \to 2\pi\rho_0$. Then planar dielectric waveguide will be equivalent to the electron flow (1.3) with fictitious charge's density $\rho_0 = 2\pi\lambda^{-2}A\sqrt{\epsilon_B^*(\epsilon_B^*-1)}$ moving with constant velocity $v_e = v_B$ [100, 101].

Let us consider particular cases of waves' transformation (see Fig. 37).

(1) The domain I_0. If $\epsilon_c > \epsilon_B^*$ and $u = 1$ (grating is absent), the energy goes out of dielectric waveguide at the angle

$$\cos\gamma_\epsilon = \sqrt{\frac{\epsilon_B^*}{\epsilon_c}}, \tag{6.5}$$

which coincides with the angle $\cos\gamma_{0\epsilon}$ of Cherenkov radiation. In this case the energy density of leaky waves equivalent to Cherenkov radiation (2.10) is determined by interrelation

$$S_\epsilon = \frac{c}{2\pi\epsilon_B^*}(hpA)^2\frac{\epsilon_c^{1.5}(\epsilon_B^*-1)}{\epsilon_c-\epsilon_B^*+\epsilon_c^2(\epsilon_B^*-1)}e^{\frac{-4\pi}{\lambda}\sqrt{\epsilon_B^*-1}|a|}. \tag{6.6}$$

The presence of periodical structure on the surface of dielectric half-space allows to form other channels of leaky waves. For example, when $n = 0$ the angle $\gamma_\epsilon = \gamma_{0\epsilon}$ and the direction of radiation along diffractional channel of zero space harmonic coincides with the channel where total internal reflection (equivalent to Cherenkov radiation in the domain I_0 on Brilluin diagram, Fig. 37) is violated. Its radiation density S_0 essentially depends on grating's parameters:

$$S_0 = S_\epsilon \frac{\epsilon_c^2(\epsilon_B^* - 1) + (\epsilon_c - \epsilon_B^*)}{(\epsilon_c - \epsilon_B^*)[1 - (1+\sqrt{\epsilon_c})\sqrt{1 - (\epsilon_B^*)^{-1}\Phi_G}]^2 + \epsilon_c^2(\epsilon_B^* - 1)}. \quad (6.7)$$

Here influence of strip grating may be also considered from the point of view of the effect of shielding waveguide's communication with dielectric half-space in the process of radiation caused by violation of total internal reflection. One control the intensity of S_0 by variation of grating's filling coefficient.

(2). The domain 2. If parameters of the system satisfy the condition $\sqrt{\epsilon_c} <| \sqrt{\epsilon_B^* - n|\kappa|^{-1}} |$, then there is no transformation of electromagnetic field of dielectric waveguide to volume waves [111].

(2). The domain 3. If $\epsilon_c < \epsilon_B^*$, the waveguide wave is transformed to volume wave (going to dielectric) which propagates along diffractional channel. For $n = -1$, $\kappa < (2\sqrt{\epsilon_c})^{-1}$ there exists one such channel and energy density in it may be determined by the formula (6.7) if to replace S_ϵ by $\hat{S}_\epsilon$ where

$$\hat{S}_\epsilon = \frac{c}{2\pi\epsilon_B^*}(hpA)^2 \frac{\epsilon_c^{1.5}(\epsilon_B^* - 1)}{(\sqrt{\epsilon_B^*} - \epsilon_c + \epsilon_c\sqrt{\epsilon_B^* - 1})^2} e^{\frac{-4\pi}{\lambda}\sqrt{\epsilon_B^* - 1}|a|}. \quad (6.8)$$

(4). The domain $4_{0,-1}^{-1}$. The conditions $\epsilon_c > \epsilon_B^*$, $\sqrt{\epsilon_B^*} \le |1 - |n|\sqrt{\epsilon_B^*}||\kappa|^{-1}$ provide existence of the channels of propagation of leaky waves only to dielectric medium. If $n = -1$ and $\kappa < (1+\sqrt{\epsilon_c})^{-1}$, there are two main channels of transformation: one is diffractional (equivalent to diffractional radiation) and the other (equivalent to Cherenkov radiation) corresponds to the case of violation of total internal reflection. Energy density in the channel of total internal reflection is calculated by the formula (6.6) and for $n = -1$

$$S_{-1} = \frac{1}{4} S_\epsilon \frac{(1+\sqrt{\epsilon_c})^2(\epsilon_c - \epsilon_B^*)[(\sqrt{\epsilon_B^*} - \lambda l^{-1})^2 - 1]}{\epsilon_c^2[(\sqrt{\epsilon_B^*} - \lambda l^{-1})^2 - 1] + \epsilon_c - (\sqrt{\epsilon_B^*} - \lambda l^{-1})^2} \mid \Phi_{-1} \mid^2. \quad (6.9)$$

Under the conditions $\kappa > |\kappa\sqrt{\epsilon_B^*} - |n||$, $\epsilon_c > \epsilon_B^*$ there exist the channel of transformation to radiation of electromagnetic field of dielectric waveguide simultaneously to vacuum and to dielectric medium (the domain $4_{0,-1}^{-1}$).

(5). The domain 5_{-1}^{-1}. This channel of transformation is realized when $\epsilon_c < \epsilon_B^*$. For $n = -1$ there are two such channels: one is directed to vacuum and the other to dielectric. In this case at least two-petal diagram of diffractional radiation is formed.

Presented theoretical analysis opens the way of modeling radiation effects occurring when electron flow moves close to periodical structure.

2. Let us consider experimental treatment of excitation of diffractional and Cherenkov radiation of the millimeter wave range [142]. With this purpose we will describe experimental plant for "cold" modeling of these excitation processes (Fig. 114). Rectangular tephlon dielectric prism of the size $60 \times 100 \times 300$ mm^3, whose ribs are much longer than wavelength, models dielectric space. Strip grating made of aluminium foil having the thickness 0.02 mm is drifted on the side 60×300 mm^2.

HF signal of SHF generator excites dielectric waveguide. Then the signal before to go over excited scatterer is sent through the channel in the absorber A in order to reduce the influence of additional scattered field caused by coordinating device CD which tights together rectangular and dielectric waveguides. Dielectric substrata is surrounded by the blocks made of absorber which is coordinated with free space (these blocks are marked by dot lines on Fig. 114). Absorber is removed from the prism's side when electromagnetic radiation is studied.

Axis of rectangular dielectric waveguide is placed perpendicular to the strips of periodical structure in the central part of prism's side with the grating. Length of the strips is 300 mm. One can investigate for $n = -1$ all theoretically considered cases of transformation of the wave propagating in dielectric waveguide to the volume waves, if dielectric permittivity of the prism is fixed ($\epsilon_p = 2$), by varying the period l, working wavelength in the range 4–9 mm and wave's velocity v_w.

Pencil of electromagnetic radiation emerging on dielectric's boundary $z = -a$ and propagating at the angle γ_ϵ or $\gamma_{n\epsilon}$ to the direction of wave's propagation in the waveguide, refracts on the prism's sides which are parallel to the plane xOz after covering a certain distance in dielectric medium and then goes off to free space at the refraction angle ϕ_ϵ or $\phi_{n\epsilon}$ (depending on the type of the channel of leaky waves). Intensity distribution in free space is registered by the receiver having mouthpiece antenna with the aperture 40×50 mm^2 moving on the circle of the radius 500 mm in the plane yOz which contains all indicated angles ϕ_ϵ, $\phi_{n\epsilon}$, γ_ϵ.

Radiation of leaky wave on the side perpendicular to waveguide's axis along the channel of total internal reflection is observed in free space at the angle ϕ_ϵ such that

$$\sin^2 \phi_\epsilon = \epsilon_p - \epsilon_B^*. \tag{6.10}$$

The radiation in this channel going through the side parallel to waveguide's axis is not observed due to the total internal reflection on this side. The last interrelation reflects sufficiently complete information about the features of radiation along the channel of total internal reflection where $\epsilon_p > \epsilon_B^*$ and the condition of total internal reflection on the output side: $\epsilon_p - \epsilon_B^* < 1$. Hence, the parameters of dielectric waveguide were chosen to satisfy these two demands: $\epsilon_p > \epsilon_B^* > \epsilon_p - 1$.

Conditions of transformation of electromagnetic waves of dielectric waveguide in the cases 3 and 4 along the channel of total internal reflection and diffractional channel are studied by measuring the radiation going through the side of prism which is perpendicular to waveguide's axis. On other sides the regime of total internal reflection takes place for Cherenkov radiation, as it follows from interrelations (6.6) and (6.3). Radiation angle $\phi_{n\epsilon}$ of the wave radiating from the considered prism along diffractional channel is determined by the formula

$$\sin^2 \phi_{n\epsilon} = \epsilon_p - \left(\sqrt{\epsilon_B^*} - |n|\frac{\lambda}{l}\right)^2. \tag{6.11}$$

Analysis of (6.3) enables one to conclude that in the cases 4 and 5 radiation along diffractional channel may be discovered only by registering the signals through the side parallel to xOy. Total internal reflection occurs for this radiation channel on the side perpendicular to waveguide's axis. Pencil's refraction on parallel sides yields $\gamma_{n\epsilon} = \gamma_n$ where $\gamma_{n\epsilon}$ is radiation angle. Then the velocity of electromagnetic wave in dielectric waveguide is determined directly by the experimental data [3, 144]. Inevitable perturbations caused by periodical system and dielectric prism are taken into account automatically. In other cases wave's velocity v_B is determined by means of additional experiments proposed and described in [111, 144, 145].

3. let us analyze the results of experiment. Distribution of radiation field in the free space with respect to angle ϕ is shown on Fig. 115 in the form of curves for $PP_{\max}^{-1}$ as functions of ϕ measured in the radiation channel of total internal reflection on the smooth prism's side (marked by 1 and 2) and on the side covered by grating (marked by 3 and 4). The curve 3 corresponds

to the first case of excitation of surface waves when there are no radiation channels ($\kappa = 0.232, \lambda = 8.62$ mm, $l = 2$ mm) and the curve 4 corresponds to the case of one radiation channel ($\kappa = 0.429, \lambda = 4.66$ mm, $l = 2$ mm). Radiation is excited by the tephlon waveguide ($\epsilon_B^* = 1.52$) having the cross-section 6×3 mm^2 ($\epsilon_p = 2.05$). Shift of radiation's maximums with varying working wavelength is caused by dispersion properties of the waveguide.

Periodical system on dielectric surface yields 20–25 per cent reduction of radiation's intensity due to the shielding effect of metal grating without noticeable change of direction of radiation maximum and width of far field pattern. These results confirm theoretical conclusions. The width of far field pattern is determined by the relative length of communication section between waveguide and scatterer (see Fig. 115) and it gets narrow with shortening the wavelength.

Radiation far field patterns measured at the wavelength $\lambda = 5.14$ mm ($\kappa = 0.989$) in the half-space with tephlon prism for the cases 3 and 4 are given on Fig. 116. In order to provide the required velocity v_B it is used rectangular tephlon waveguide with the cross-section 6×3 mm^2 for the curve 2 and polystyrole waveguide with the cross-section 4×7.2 mm^2 for the curve 1. Radiation intensity in the channel of total internal reflection turns out to be one order greater than the one in diffractional channel. Therefore, for the measurement of radiation in the channel of total internal reflection one introduces 10-dB slackening and radiation intensity presented in the first quadrant is related to this value. Vertical lines indicate radiation angles $\phi_{n\epsilon}$ and ϕ_n calculated by the formulas (6.10), (6.11). Variation of $\phi_{n\epsilon}$ caused by the change of waveguide, varying wavelength λ or period l leads to corresponding shifts of maximums. In addition to the main maximums, a number of "parasitic" resonances emerge on the curves (having lower intensity) connected with finite size of the prism and manufacturing defects of periodical system. Radiation developing near the directions $\phi = 0°$ and $\phi = 180°$ is caused by parasitic radiation going from the edges of dielectric waveguide. Radiation maximum at the angle $180° - \phi_\epsilon$ is connected with overreflection of the radiation of total internal reflection on the prism's sides. Existence of two waves in dielectric waveguide yields bifurcation of the main maximum on the curve 2 (here and on other figures it is presented qualitative description of radiation channels corresponding to the domains on Brilluin diagrams).

Excitation in different domains is illustrated by Fig. 117. Relative variation of radiation intensity $PP_{\max}^{-1}$ versus ϕ in the half-plane with dielectric prism is given in the upper part of the figure. Behaviour of this parameter for the case when radiation is excited in diffractional channel and goes off to free space is shown in the lower part of the figure. The curves marked by 1 are obtained in the presence of channel of radiation caused by total internal reflection (the domain $4_{0,-1}^{-1}$). Excitation of radiation for the cases 4 and 5 (curves 2) is provided at the same wavelengths by those two dielectric waveguides which were described above. Relative radiation intensity I_R in diffractional channel in these cases is comparable with radiation intensity in the channel of total internal reflection I_T and in maximums I_R is equal to 40–50 per cent of I_T (unlike the regimes 3 and 4 on Figs. 41, 47 where the difference between radiation intensities is much greater). It is connected with the dependence of intensity on the angle ϕ where maximal level is observed when $\phi \sim \frac{\pi}{2}$. Difference between the angles γ_n and $\phi_{n\epsilon}$ of maximal intensity is caused by small deviations from parallelism of those prism's sides which are parallel to waveguide's axis.

The considered experimental method allows to model with high accuracy all channels of transformation of the energy of electron flow to diffractional, Cherenkov or diffractional-Cherenkov radiation which are predicted by the theory of given current. In addition to this, the presented method enables one to deeper understand the main appropriateness of Smith-Purcell effect.

These theoretical and experimental approaches have been further developed for the study of thin structure of diffractional radiation. It stimulated working out the technique of remote measurements of dynamical parameters of electron flow and new devices of diffractional electronics.

One has to point out that in order to carry out inverse communication emerging in Smith-Purcell effect due to properties of the fields formed by diffractional grating, it is necessary to apply in theoretical and experimental models additional assumptions characterizing self-coordinated and non-linear interaction of electron flow with the fields of various electrodynamical systems.

§19. Antenna properties of Smith-Purcell radiation for beveled gratings

It has been shown in §18 that position in the space of radiating part of scattered field is determined in accord with (3.4) by two angles $\tilde{\phi}_n$ and $\tilde{\theta}_n$ (Fig. 119) for dielectric waveguide and the grating placed at the distance b from it at the angle α to waveguide's axis excited by inhomogeneous plane wave of dielectric waveguide. When $\alpha = 0$, the volume waves tear off the grating and propagate only in the plane crossing waveguide's axis which is perpendicular to the plane occupied by waveguide. In the case of beveled grating ($\alpha \neq 0$) space distribution of radiation has two-dimensional character. There exists critical value $\alpha = \alpha_{\rm cr}$, $\alpha_{\rm cr} = \arccos 0.5(\kappa\beta n^{-1} - n\beta\kappa^{-1} - \kappa(\beta n)^{-1})$ when the scattered field has the form of inhomogeneous plane wave propagating along grating's surface. Hence, for $\alpha = \alpha_{\rm cr}$ the considered open structure is characterized by the regime of surface waves and further increasing of α leads to Bragg diffraction when $\alpha_{\rm Br} = \arccos\left[n\beta(2\kappa)^{-1}\right]$, $n = 1, 2, \ldots$ is the order of Bragg diffraction. In this regime inhomogeneous plane wave is transformed to surface wave which propagates at the angle $\pi - \alpha$ to the direction of propagation of initial inhomogeneous wave. The degree of transformation is determined by longitudinal grating's dimensions and the level of its communication with the waveguide. Knowledge of polarization characteristics of scattered fields is essential for modeling the beveled grating. Corresponding experimental investigations are carried out in [146–148].

1. Methods of experimental studies of transformation of waveguide's surface waves to volume waves on beveled grating are similar to the ones applied for modeling such processes on usual grating. One must only take into account that depending on the angle determining the turn of the grating related to the direction of propagation of surface wave in the planar waveguide it may occur either total transformation of surface wave to volume wave (for $\alpha = 0 - \alpha_{\rm cr}$) or its transformation to surface wave for a certain value of α.

It is known that there simultaneously exist two principal E- and H-waves having different deceleration coefficients. Interaction of these waves with beveled grating depends both on the properties of the waveguide and grating. One can effectively excite E-wave in the waveguide by means of the volume H-polarized wave, if the grating is of reflection type (a comb) with resonance narrow slots; the comb with wide slots excites H-wave in the waveguide.

Polarization effects are brightly developed for beveled grating with $\alpha \neq 0$.

Since dielectrics which are used for manufacturing the waveguides have permittivity $\epsilon = 2 - 3$, E- and H-waves have comparatively low deceleration and small difference ($\approx 2^\circ - 3^\circ$) between scattered beams. Independent measurements of scattered fields is possible only for such grating's apertures which provide high degree of separation in the space of the beams corresponding to E- and H-waves. For cosine field's distribution the aperture should have the dimensions $50\lambda_0$. It is rather difficult to apply such apertures in experimental plants. Real radiating apertures have the size of about $10\lambda_0$. While processing the results, one assumes here that mutual interaction of E- and H-waves in planar waveguides may be described theoretically.

Characteristics of scattered fields should be determined when it is provided plane phase front and single-mode regime for surface waves propagating in planar waveguide with the help of special way of excitation of transmission line along the narrow and wide waveguide's side. It is required also excitation of the minus first space harmonic by the beveled grating what yields one-petal far field pattern of the scattered field. Dimensions of interacting aperture should not change when the grating turns (rotates) at arbitrary angle with respect to its initial position.

Experimental studies of transformation of waveguide's surface waves to volume waves on beveled grating are reduced to determination of the following characteristics of scattered fields:

a) the number of radiated space harmonics for E- and H surface waves propagating in the planar dielectric waveguide;

b) the direction of propagation of the scattered volume or surface waves (the angles $\tilde{\theta}_n$, $\tilde{\phi}_n$, $\alpha_{\rm cr}$, $\alpha_{\rm Br}$ *et. al*) as functions of angle α and aimed distance;

c) dependence of the radiated power of E- and H waves for different values of α;

d) ellipticity coefficient and the angle of orientation of polarization ellipse of scattered fields versus α and b.

In connection with these requirements experimental plant must provide

– accurate measurements of scattered fields in the regimes of transformation of surface waves to volume waves for $\alpha = \pm 30^\circ$ and of surface waves to surface waves with absolute error not greater than 1°;

– determination of the amplitude and phase field's distribution in every point of radiating aperture;

– measurement of the wavelength and deceleration coefficient of surface wave;

– gradual regulation of the turn angle α and aimed distance b;

– measurement of the ellipticity coefficient and the angle of orientation of polarization ellipse of scattered waves in the sector $\pm 180°$;

– determination of the mode composition, attenuation in the waveguide together with the efficiency of transformation of surface waves to volume waves.

Experimental plane on Fig. 120 enables one to carry out all necessary measurements in the four-millimeter range for determining characteristics of the field emerged as a result of transformation of surface waves by a beveled grating. The plant is used in accord with the following scheme. GDR 21 feeds mouthpiece antenna 14, which provides linearly polarized radiating field, with modulated oscillations through directional coupler 20, unbinding 19 and measurement 16 attenuators. Limb 7, which registers polarization characteristics, is stiffly connected with antenna 14 and attenuator 16 and rotates in the sector $0° - 180°$ around immovable axis 15 transferring to supporting upright 13 (angle positions of mouthpiece antenna are measured with the accuracy 3°). Junctions between empty metal waveguides 17 and linear waveguide 18 provide communication with GDR. Frequency in the section is controlled by frequency-meter 22 through coupler 20 and indicating device 6. Homogeneous linearly polarized wave radiated by antenna 14 from the far zone is received by electrodynamical structure consisting of planar waveguide 2 and grating 1 (where homogeneous plane wave is transformed to inhomogeneous surface wave). Surface wave comes to detector section 4 through mouthpiece excitator 3. Modulation frequency of GDR is separated after detecting and goes through amplifier 5 to registering device 6.

Transversal field's picture along the planar waveguide is close to cosine distribution with plane phase front; transversal size of the waveguide made of fluorine plastic, $\epsilon = 2.05$, $\tan\delta = 2 \cdot 10^{-4}$, is $10\lambda_0$ and it supports single E-mode, $cv_{\Phi}^{-1} = 1.84$. Finally, one can obtain far field pattern of the width $5° - 6°$ in the H-plane what is sufficient to carry out all necessary measurements. Polystyrole cartridges 10 on rotating platform 11 stiffly hold waveguide 2. Limb 7 whose turn angles are counted equally to 14 is stiffly fastened in the central part of the platform. The considered beveled grating 1 in the form of circular disc (providing constant radiation aperture in the whole sector of rotation $0° - 90°$) is supported by the axis 7 freely rotating around the limb and going through the middle of planar waveguide 2 along the axis Oz' (the accuracy of mounting the grating with respect to planar waveguide is 0.01 mm).

Platform 11 is fastened in the middle of frame 12 by rotation device 8. Frame's orientation is determined by the axis Oz' which coincides with Oz for $\alpha = 0°$ and rotation device 9 is placed on its left side. Rotation devices 8,9 (see Fig. 119) which indicate the values of angles $\tilde{\theta}_{-1}$ and $\tilde{\phi}_{-1}$, respectively, are combined with potential-meter primary devices and enable one to register far field patterns of the considered apertures in E- and H-planes with the accuracy $\pm 0.5°$. Initial position of the plane containing the aperture of mouthpiece antenna 14 is parallel to the plane going through the frame 12 (the axis Ox'' is always perpendicular to the aperture's plane of antenna 14 for different angles of grating's turn and goes through its phase center).

In the process of experiment ($\alpha = 0$ for the grating) we achieve maximal signal at the output of detector 4 by rotating the platform 11 and determine $\tilde{\theta}_{-1}$ with the help of devices 8, while $\tilde{\phi}_{-1} = 0$ (the axes Oy and Oy' coincide and collinear with the axis O_0y_0' of the mouthpiece antenna 14; the axis Ox' coincides with Ox'' and cavities of the grating 1 are directed along the axes Oz, Oz'). Wave vector of the incident linearly polarized field lies in the plane xOz (big axis of field's polarization ellipse is oriented along the axis Oz' of the mouthpiece antenna, since ellipse's orientation angle is counted from the axis O_0y_0 and it is equal to $90°$).

If $\alpha \neq 0°$, we use the devices 8, 9 in order to provide the following condition: wave vector $\vec{k}$ of the radiating harmonic is perpendicular to the aperture's plane of antenna 14 with taking into account maximal signal of the detector 4. In this case the axes Oz', O_0z_0 and Oy', O_0y_0 are collinear. Grating 1 and planar waveguide lie in the plane xOy. Angle divergences between Oy', Oy and Oz', Oz measured by the device 9 determine the values of $\tilde{\phi}_{-1}$ and $\tilde{\theta}_{-1} - 90°$, respectively.

Orientation of linearly polarized radiating field is given by the position of antenna 14 rotating in the plane perpendicular to the wave vector of receiving radiation. Dumb-bells curve determining ellipticity coefficient and ellipse's orientation angle is registered in the sector of antenna's rotation angles $(0° - 180°)$ with the help of amplifier 5 and primary device 6. Hence, the plant provides determination of polarization characteristics of the considered aperture in the receiving regime.

Mode composition with respect to waveguide's width and transversal field's distribution are registered by the sensor in the form of narrowed end of a planar metal waveguide. Longitudinal field's distribution is determined by the sensor moving along waveguide's axis Oz.

Losses in mouthpiece-lens junction 3, efficiency of transformation of surface waves to volume waves and coefficient of standing wave are indicated by a measurement section combined with measurement and unbinding attenuators which joins the junction 3 and GDR. Opposite end of planar waveguide 2 is connected with metal screen. Characteristics of radiating and non-radiating fields are determined with the help of mouthpiece junction which replaces screen and enables one to register E- and H-waves passed through the planar waveguide on the equal energy level. Here it is measured the "image" picture of scattered fields, i. e., the part of energy which remains in planar waveguide after the transformation occurring when the waveguide turns in the sector $0° - 90°$.

2. Beveled and reflection gratings with different filling coefficients have been studied by the experimental plant described above.

Transversal field's distribution of the principal E-wave of the planar waveguide measured along the axis Oy (see Fig. 119) is presented on Fig. 121 where 1 indicates calculated curve of the cosine-square distribution, 2 – experimental data without the grating. Divergence between E- and H-waves is 42 dB. Coefficient of standing wave for $\alpha = 0$ is equal to 1.12 and the losses in mouthpiece-lens junction are 1.15 dB. Deviation of the lines of equal phases from straight lines along the axis Oz measured in different parts of planar waveguide does not exceed 57°. Hence, under the condition of absence of communication between the waveguide and the grating field's distribution is characterized by the high purity of the principal waveguide's wave.

In order to determine the influence of gratings on the field of principal E-wave it is measured the part of energy which remains in planar waveguide after the act of transformation. The picture of E-wave at the edge of planar waveguide in the case of reflection grating ($u = 0.5$) for different values of aimed distance b and angle $\alpha = 0° - 90°$ is shown on Fig. 122 (figures 1, 2, 3 correspond to the following values of b: 1 – $b = 0$; 2 – $b = 0.1\lambda_0$; 3 – $b = 0.2\lambda_0$). Sector of angles $\alpha = 0° - 45°$ corresponds to the regime of transformation of surface wave to volume wave; the calculated value $\alpha_{cr} = 44.3°$. Difference between calculated and experimental values of α_{cr} and radiated power occurring when b varies may be explained by strong influence of the scatterer on phase velocity of surface wave.

When $\alpha = \alpha_{cr}$, diffractional grating transforms the surface wave of planar waveguide to surface wave propagating at the angle δ_n counted from the axis Oy such that $\tan\delta_n = -\kappa(n\beta\sin\alpha)^{-1} = -\mathrm{cotan}\alpha$ (see Fig. 119). In the process of measurements resonance emergence the surface wave of planar waveguide occurs for $\alpha = \alpha_{\rm cr}$.

Further increasing of $\alpha > \alpha_{\rm cr}$ leads to the ceasing of transformation of surface wave to other waves by the beveled grating while for $\alpha = \alpha_{\rm Br}$ the energy is taken away from planar waveguide and the latter has a resonance character (the calculated values are $\alpha_{\rm Br} = 55.3°$ for E-wave and $\alpha_{\rm Br} = 57.2°$ for H-wave). The value of $\alpha_{\rm Br}$ and the level of energy which is taken away are determined by the degree of communication between dielectric waveguide and the grating.

Fig. 122 is evidence of two operation regimes of the system waveguide+grating: transformation of surface wave to volume wave ($0 \le \alpha \le \alpha_{\rm cr}$) and transformation of surface wave to surface wave ($\alpha_{\rm cr} \le \alpha \le 90°$). If $\alpha = \alpha_{\rm cr}$ or $\alpha = \alpha_{\rm Br}$, then resonance transformation of surface wave to volume wave takes place.

Emergence of H-waves in the planar waveguide is discovered for varying α (see Fig. 123 where b takes the same values as on Fig. 122). Let us note that the principal H-wave is the result of transformation of E-wave by the turn grating and this transformation is effective in the case of strong communication between planar waveguide and diffractional grating. For $\alpha = 0°, 90°$ intensity of H-wave is negligible. The minimums on Fig. 122 for H-wave correspond to the values $\alpha = 45°, 57°$, i. e., to the angles $\alpha = \alpha_{\rm crE}$ or $\alpha = \alpha_{\rm BrE}$. For $\alpha = 66°$ E- and H-waves have equal levels.

It is of special interest to consider the properties of E- and H-waves in the system waveguide+grating when reflection gratings with the filling coefficient $u = 0.87$ and varying depth of cavities are used. The behaviour of energy of E-wave at the end of planar waveguide and detailed analysis of the near field show that the energy is effectively taken away when the degree of communication between planar waveguide and diffractional grating is low ($b > 2\,$mm) and there is no transformation of surface wave to volume wave for other angles. Polarization of radiated wave corresponds here to the direction of vector $\vec{E}$ perpendicular to grating's axis with the degree of linearity 30 dB. Maximum of the energy which is taken away shifts with decreasing h to the values $\alpha \approx 0$. Field's distribution and angle dependence of radiated wave versus grating's parameters enable one to conclude that H_{11}-wave of the circular waveguide with wide slot 17 is excited in grating's cavities. Phase of this wave is synchronized with the E-wave of planar waveguide (for example, they are synchronized for $\alpha = 17°$ when $h = 2$ mm and these calculation results well agree with experiment).

The behaviour of transmitted E-wave in the planar waveguide depending on the angle α may be investigated by measuring the coefficient of standing wave. For the beveled grating with $u = -0.5$ and wide planar waveguide (having the size $\sim 30\lambda_0$ and dielectric permittivity $\epsilon = 2.25$) the calculated values $\alpha_{\rm cr} = 45.4°$, $\alpha_{\rm Br} = 56.6°$ what agrees with experimental data.

High degree of coordination in a section for different values of aimed distance b (when α varies from $0°$ to $44°$ and $\alpha = \alpha_{\rm Br} = 57°$) confirms the described properties of E-wave propagating in planar waveguide when $\alpha = \alpha_{\rm cr}$ and $\alpha = \alpha_{\rm Br}$.

Hence, it has been established that planar waveguide may be effectively fed by E-wave. It has been also revealed the influence of grating's rotation (of the angle α) on the characteristics of propagating E- and H-waves. Now one can carry out with the help of the plant shown on Fig. 120 further studies of volume waves, in particular, measurements of their far field patterns in E and H planes for different values of α.

Such far field patterns in E (a, c) and H (b, d) planes for the reflection grating with $u = 0.5$, $b = 0.4$ mm for $\alpha = 0°$ (a, b) and $\alpha = 20°$ (c, d) are presented on Fig. 124. The width of far field pattern in E plane at the level of 3 dB is equal to 3° and in H plane it is equal to 6° (deviation is connected with different aperture's width in E and H planes).

Calculated values of $\tilde{\theta}_{-1}$ (bold curves) and $\tilde{\phi}_{-1}$ (dots) are given on Fig. 124. Deviation between experimental and theoretical data does not exceed 5° for $\tilde{\theta}_{-1}$ and 3° for $\tilde{\phi}_{-1}$ (the smallest deviation is observed for strip and reflection gratings with over-limiting size of cavities).

Experimental (1,2) and theoretical (3–5) values of ellipticity coefficient and orientation angle of polarization ellipse are practically the same. Hence, the vector $\vec{E}$ of received field is always perpendicular to grating's strips.

The following assertions may be also formulated after the present theoretical and experimental analysis:

– calculated and experimental polarization characteristics for strip gratings with $u = -0.5$ have considerable difference;

– if cavities of such gratings do not support any propagating mode, polarization properties of the grating must coincide with theoretical characteristics of the strip grating with narrow slots.

3. Let us consider in details resonance phenomena occurring in the process of diffraction of inhomogeneous plane waves propagating in dielectric waveguides by reflection grating (its cavities may be either of waveguide or of over-limiting type for the cross-component of excited E-wave). In this section we have already discussed resonance effects in the open structure dielectric waveguides+reflection grating which become apparent in the high degree of ellipticity of the scattered resonance field and in the coincidence of orientation of big axis of polarization ellipse with grating's turn angle (all these effects are connected with orientation of $\vec{E}$ of the E-wave excited in the waveguide). In resonance domains one observes excitation of waveguide and slot waves for the gratings differing by the filling coefficient and cavities' depth. It has been established that waveguide resonance takes place only for the gratings having waveguide-type cavities while the slot resonance occurs for all reflection gratings.

It is necessary also to clear up the role of interaction of E- and H-waves in these resonance processes, to determine energy and polarization characteristics of surface-volume waves' transformation by means of beveled reflection gratings.

Let us consider at first reflection gratings with over-limiting width of cavities, for which $u = 0.5$ and whose depth of cavities is equal to 1 mm (this grating has already been described above). Here the level of transformation of E-wave to H-wave is rather strong. H-wave contributes to the scattered homogeneous field and when $b = 0.3\lambda\,(\alpha = 30°)$ the Bragg diffraction is essential (for $b = 0\,\alpha_{\mathrm{Br}} = 60°$ and with decreasing communication of the grating with planar waveguide α_{Br} tends to the calculated value $\alpha_{\mathrm{Br}} = 55.3°$). Analysis of experimental curves for $E_x(\alpha)$, $E_y(\alpha)$ shows that slot resonance occurs when $\alpha - 45°- = 47°$. With growing degree of communication slot resonance is observed for smaller values of α because phase velocity of slot wave falls when waveguide's dielectric surface comes closer to grating's slot what prevails over the decrease of phase velocity of E-wave caused by bringing metal objects in waveguide's volume.

Let us consider now the behaviour of scattered fields for reflection gratings with waveguide width of cavities. Pictures of $E_x(\alpha)$, $E_y(\alpha)$ for the gratings with $u = 0.87$ and different depths of cavities 0.75 mm, 1.0 mm ($\lambda = 4$ mm) are evidence of slot resonance as well as waveguide and Bragg diffraction.

Growth of cavities' depth leads to increase of the angle of phase synchronization for the slot wave, since phase velocity in the cavity falls. Maximal level of the slot resonance occurs for E-wave of the planar waveguide when cavities have resonance depth.

Bragg diffraction for the gratings with waveguide-type cavities is very distinct for the E-wave (for the H-waves scattered by the grating) when cavities have resonance depth. The angle of waveguide resonance tends to the value $\alpha = 0°$: when $\alpha = 0°$, phase synchronization of the excitation of waveguide wave occurs already for $h = 1$ mm. Further decrease of the depth up to $h = 0.75$ mm does not change this angle; efficiency of taking the energy away from the waveguide falls and resonance curve broadens. Weak transformation of the surface E-waves to volume waves is characteristic for all gratings except of the case when parameters vary in the domain of waveguide resonance where surface waves are transformed to volume waves by finite number of grating's periods. In the considered gratings waveguide resonance is observed in the sector of angles corresponding to this transformation (while space characteristics in the domain of waveguide resonance remain the same).

Ellipticity coefficient in the domain of waveguide resonance is close to zero and big ellipse's axis is oriented along $\vec{E}$ perpendicular to grating's lamellae. Far field patterns noticeably broadens. The degree of transformation of E-waves to H-waves is determined by the difference between cavity's depth and its resonance value for the field's cross-component.

Trustworthiness of theoretical conclusions concerning resonance phenomena in the gratings with resonance width of cavities is controlled by the study of scattering on one cavity having appropriate geometric parameters. Experimental measurements of $E_x(\alpha)$ and $E_y(\alpha)$ confirm the possibility of excitation of waveguide and slot wave in one cavity: the former is excited for $\alpha = 20°$ when the width and the depth of the cavity are both equal to 2.5 mm, the latter is excited for $\alpha = 47° - 50°$ depending on the degree of communication. These synchronization angles are very close to the ones for the grating having the same cavities.

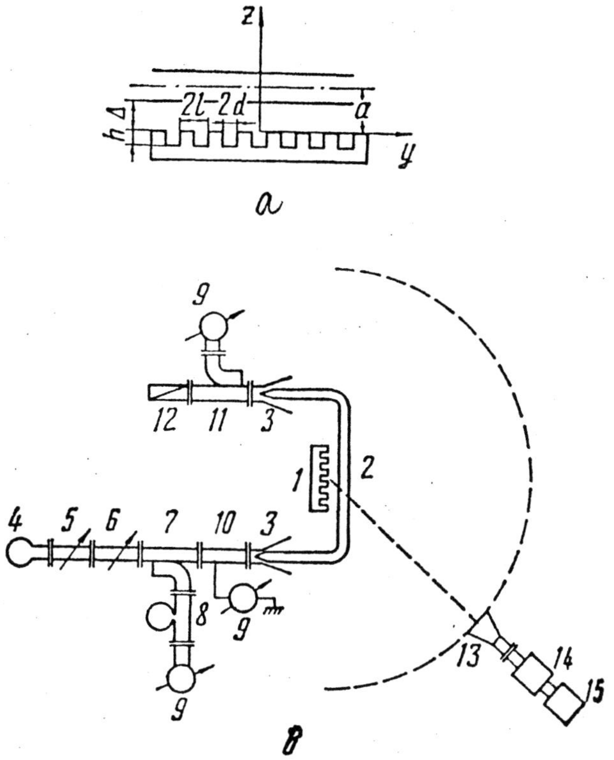

Fig. 101. Experimental plant for the study of wave model of Smith-Purcell effect.

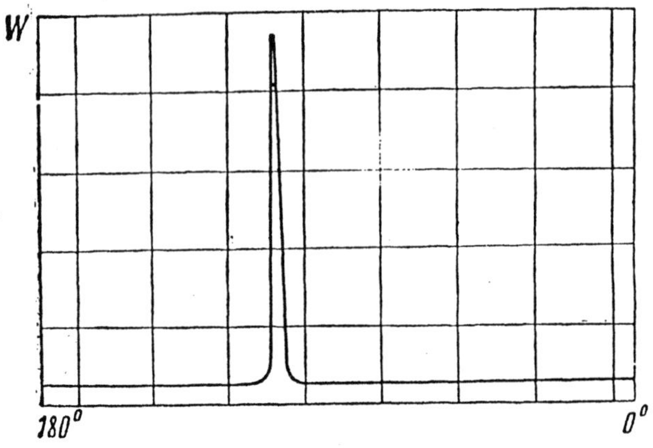

Fig. 102. Far field pattern of the radiation of dielectric waveguide when its surface wave is transformed to volume wave.

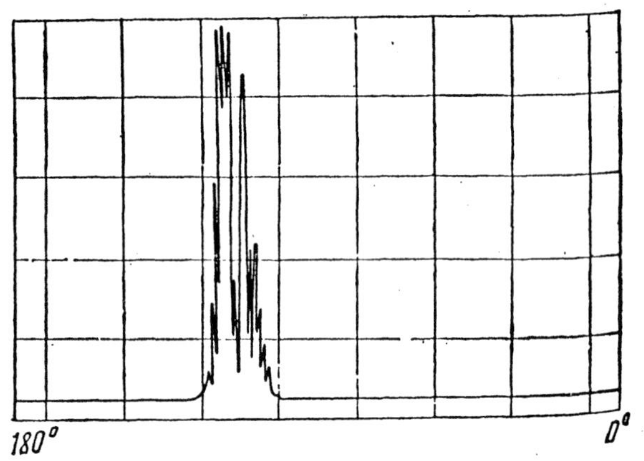

Fig. 103. Far field pattern of the transformed waveguide wave in the case of strong communication waveguide–grating.

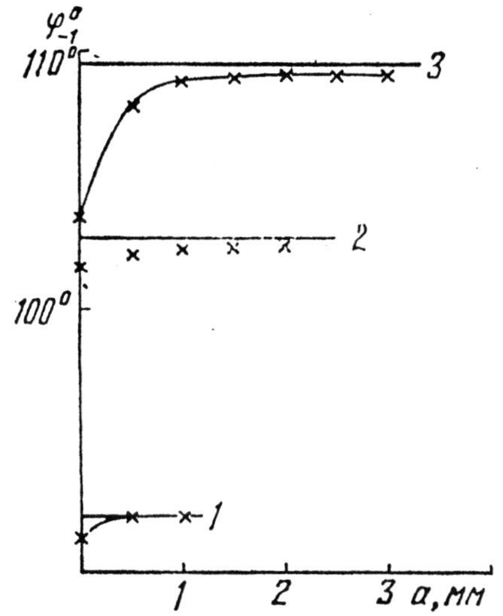

Fig. 104. Radiation angle of the minus first harmonic versus aimed distance: 1 – $\beta = 0.672$; 2 – $\beta = 0.774$; 3 – $\beta = 0.846$.

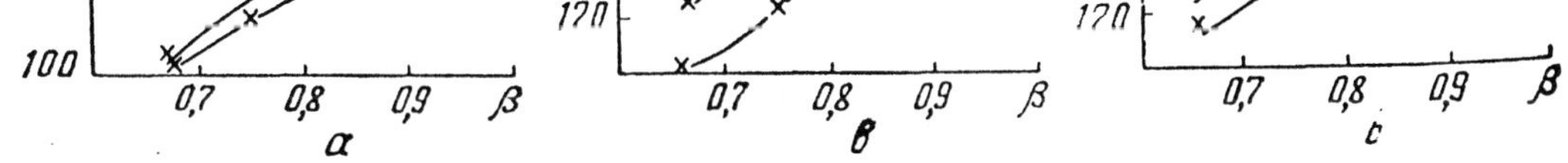

Fig. 105. Radiation angle of the surface waves of dielectric waveguide versus κ taking the values: a) 1 – 0.606; 2 – 0.568; b) 1 – 0.528; 2 – 0.495; c) 1 – 0.518; 2 – 0.485.

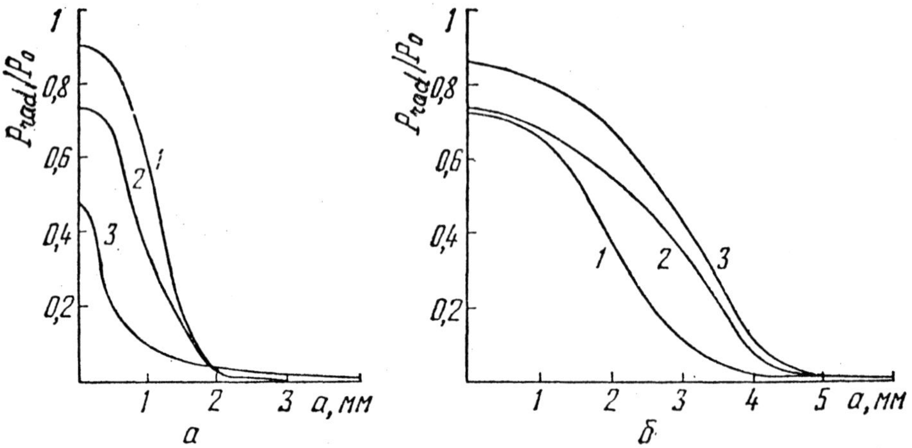

Fig. 106. Radiated power versus aimed distance.

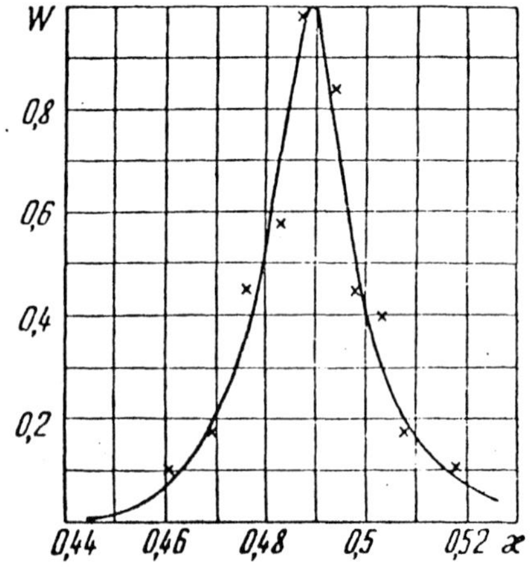

Fig. 107. Radiation far field pattern of the reflection grating made of rectangular bars.

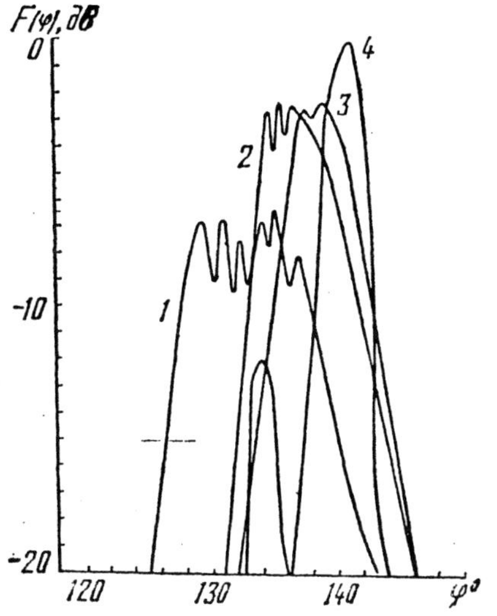

Fig. 108. Far field pattern of the waves scattered by the finite echelette (1 – $N = 6$, 2 – $N = 12$; 3 – $N = 18$; 4 – $N = 31$) in the case of strong communication between dielectric waveguide and diffractional grating, $a = 2$ mm.

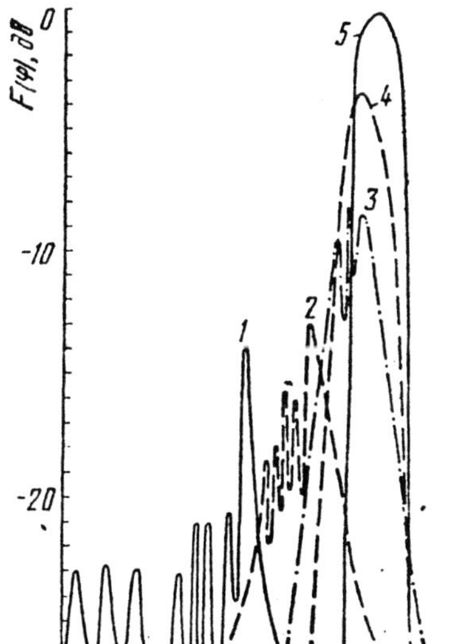

Fig. 109. The same data (as on Fig. 108) but in the case of strong communication of N: 1 – $N = 2$, 2 – $N = 6$; 3 – $N = 12$; 4 – $N = 22$; 5 – $N = 31$.

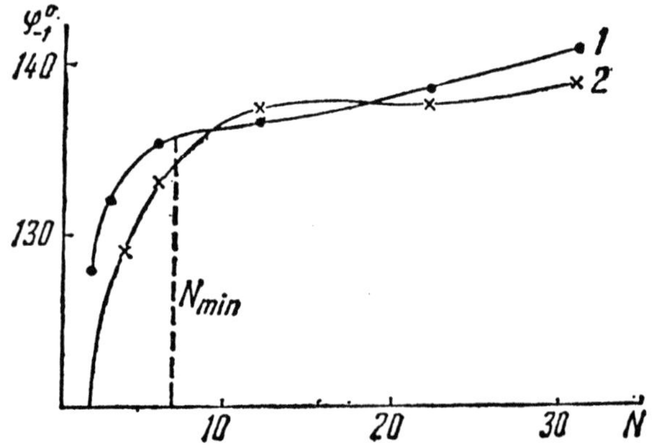

Fig. 110. Radiation angle of the minus first harmonic (dielectric waveguide–echelette) versus the number of grating's periods for different values of a, [mm]: 1 – $a = 0$; 2 – $a = 2$.

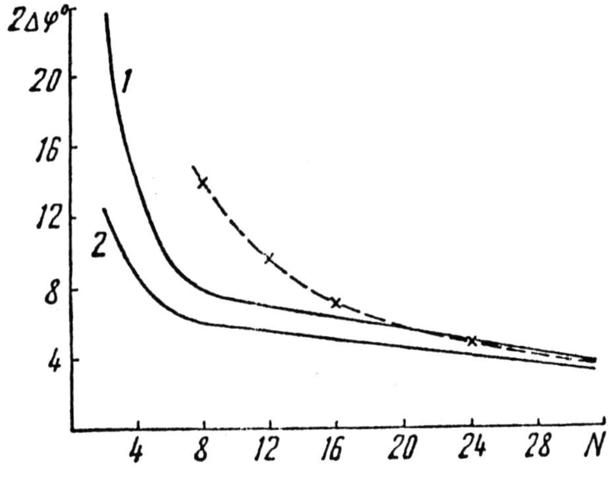

Fig. 111. Width of the far field pattern versus the number of grating's elements, a, [mm]: 1 – $a = 0$; 2 – $a = 2$.

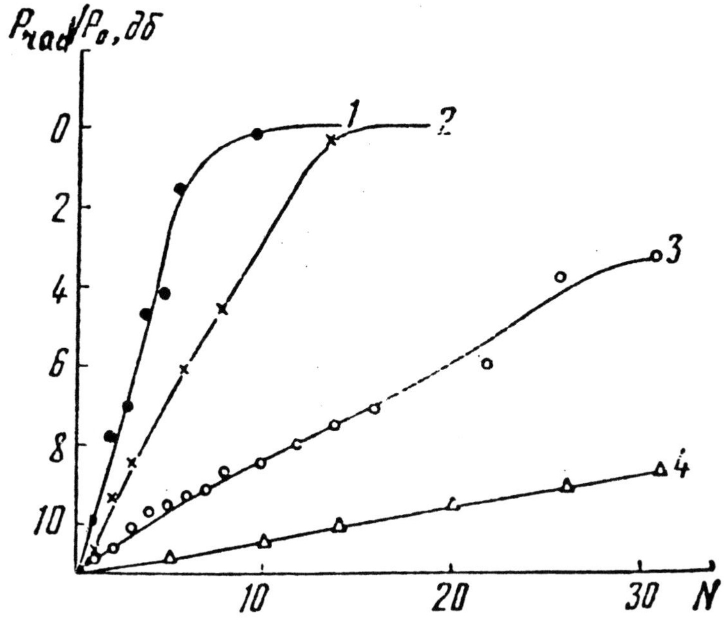

Fig. 112. Radiation power versus grating's period, a, [mm]: 1 – $a = 0$; 2 – $a = 1$; 3 – $a = 2$; 4 – $a = 3$.

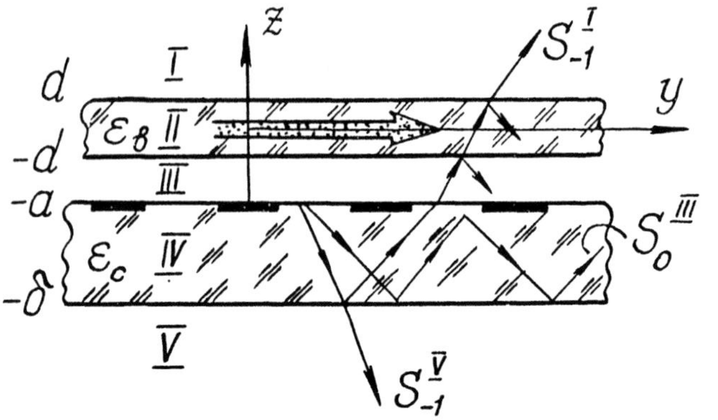

Fig. 113. Theoretical modeling of the excitation of of Smith-Purcell and Cherenkov radiation.

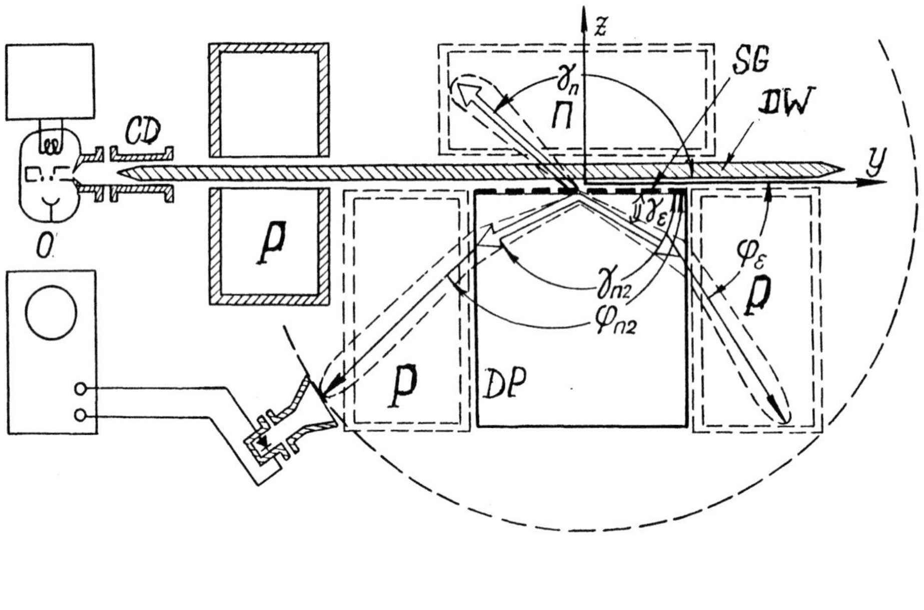

Fig. 114. Experimental plant for the modeling of Smith-Purcell and Cherenkov radiation.

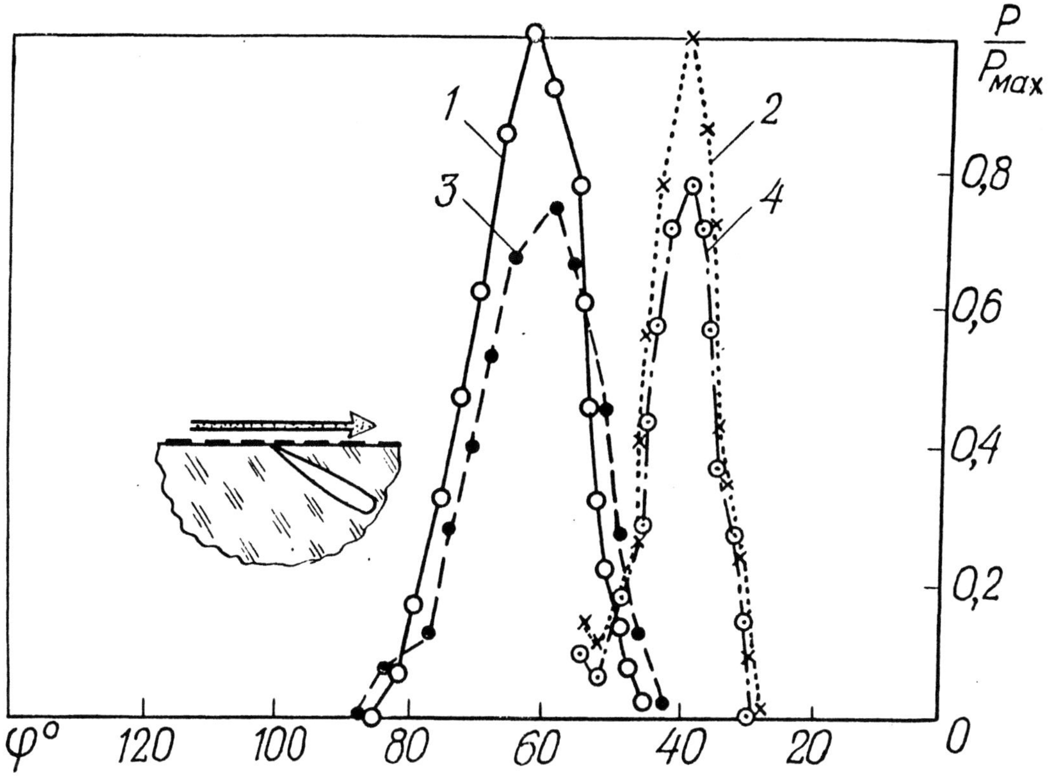

Fig. 115. Far field pattern of the Cherenkov radiation in the channel of the violation of total internal reflection: 1,2 – without a grating, $\lambda_1 = 8.62$ mm, λ_2 – 4.66 mm; 3,4 – with a grating, $l = 2$ mm, $d = 1$ mm.

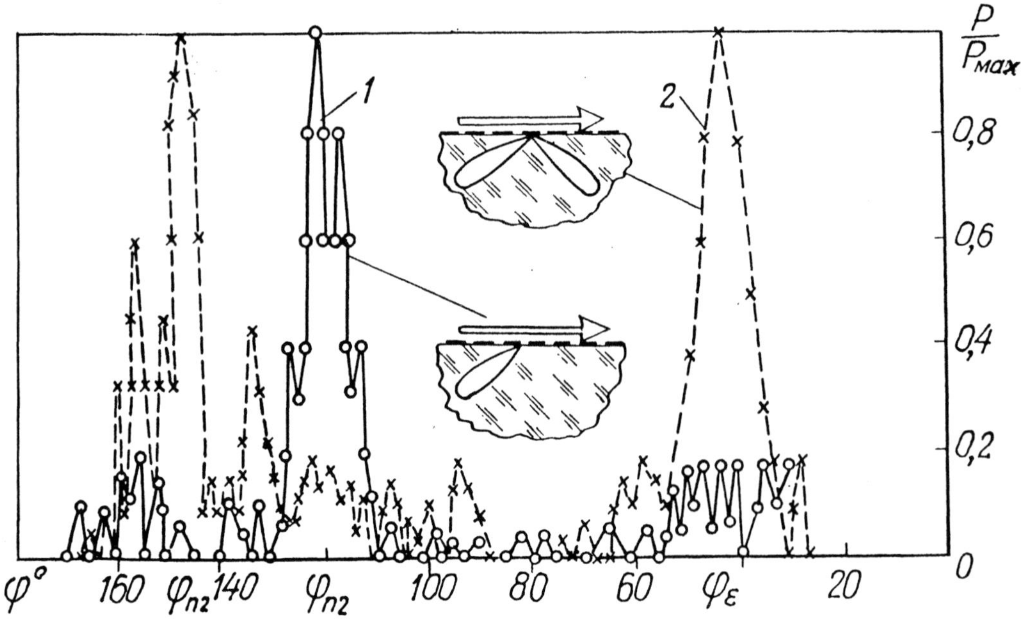

Fig. 116. Far field pattern to the dielectric: 1 – excitation of Smith-Purcell radiation (domain 3^a_{-1}; $\epsilon = 2.1$); 2 – simultaneous excitation of diffractional and Cherenkov radiation (domain $4_{0,-1}$; $\epsilon_b = 1.52$; $n = -1$).

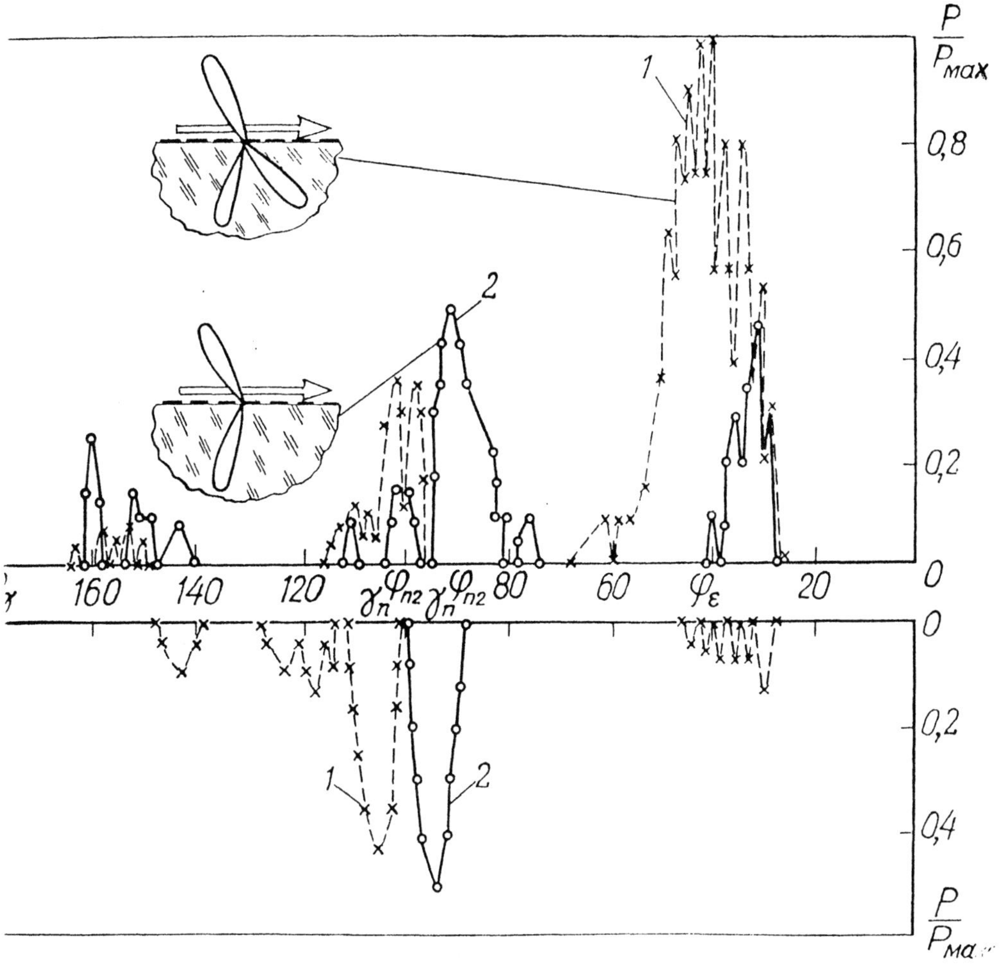

Fig. 117. Far field patterns to the vacuum and dielectric: 1 – simultaneous excitation of of diffractional and Cherenkov radiation (three-petal far field pattern, domain $4^{-1}_{0,-1}$); 2 – excitation of Smith-Purcell radiation (domain 5^{-1}_{-1}).

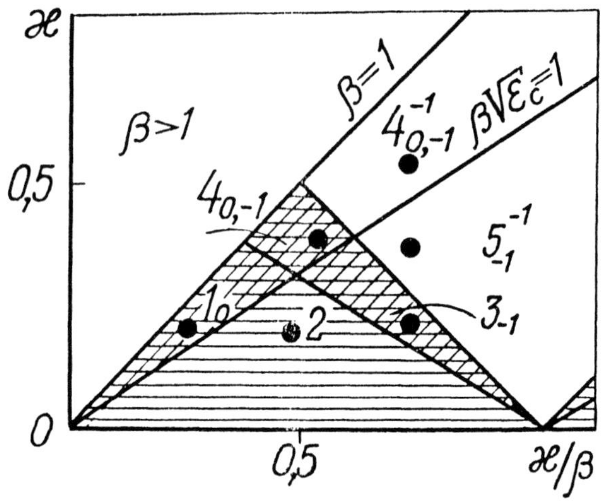

Fig. 118. The cases of excitation of the diffractional and Cherenkov radiation in the plane of modified Brillouine diagram realized in experiment (experimental and theoretical data coincide within the frame of the accuracy of applied methods).

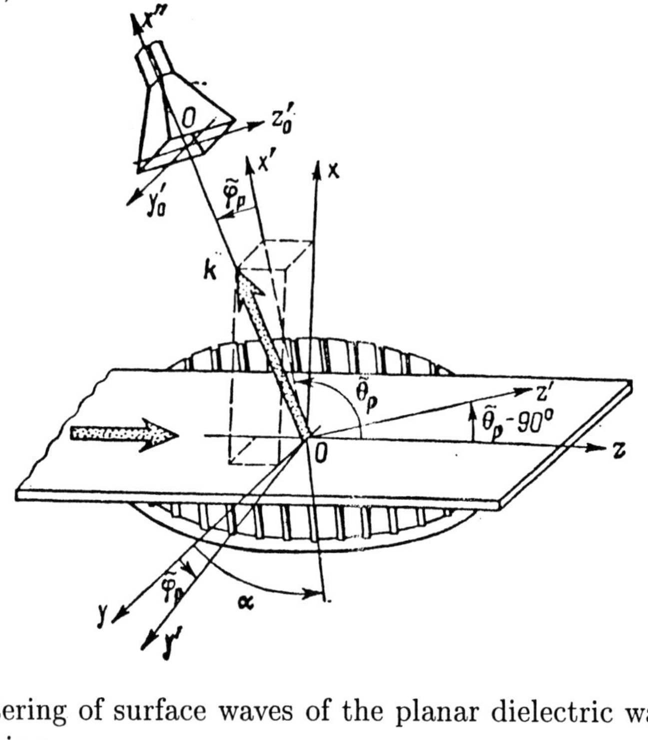

Fig. 119. Scattering of surface waves of the planar dielectric waveguide by the beveled grating.

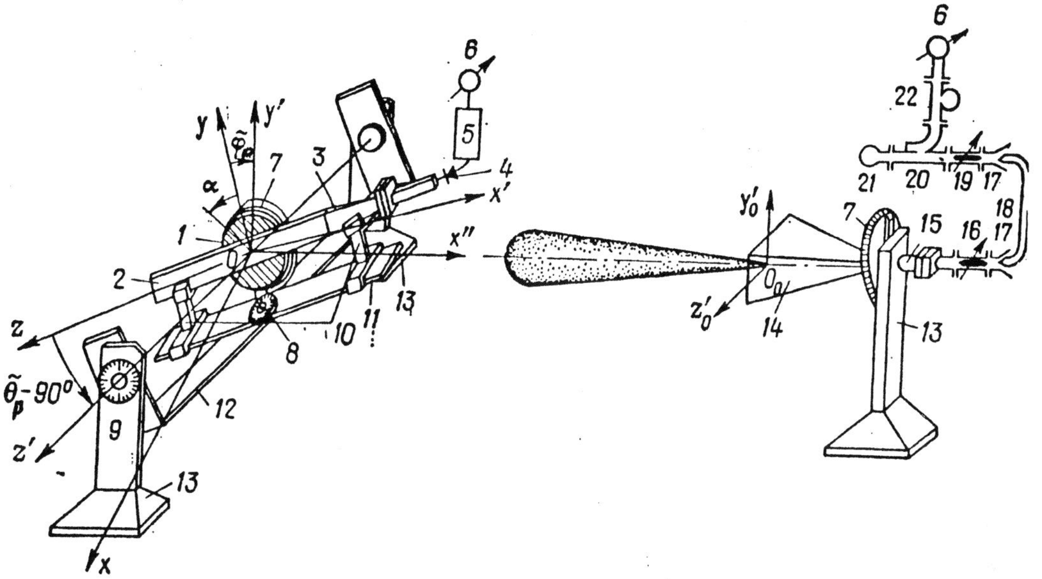

Fig. 120. Experimental plant for the antenna measurements of the wave Smith-Purcell radiation scattered by the beveled grating.

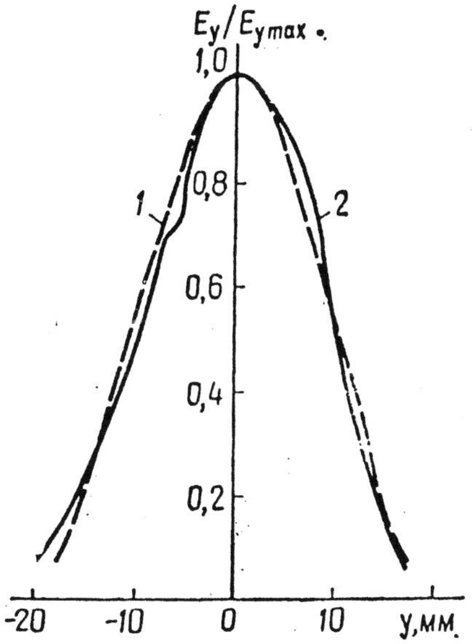

Fig. 121. Transversal distribution of the field of E-wave of planar dielectric waveguide.

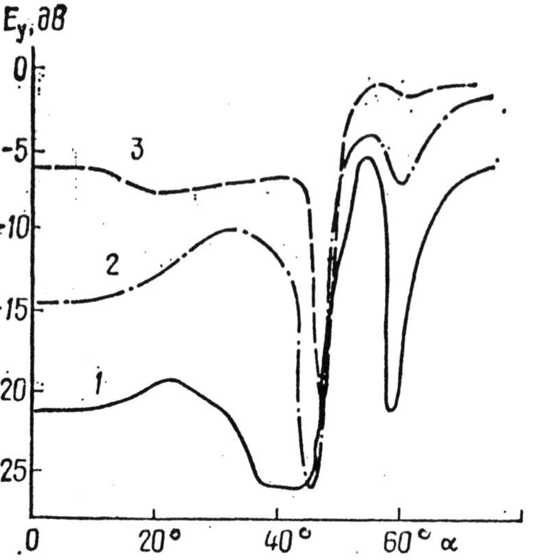

Fig. 122. Scattering of E-wave of planar dielectric waveguide by the beveled grating.

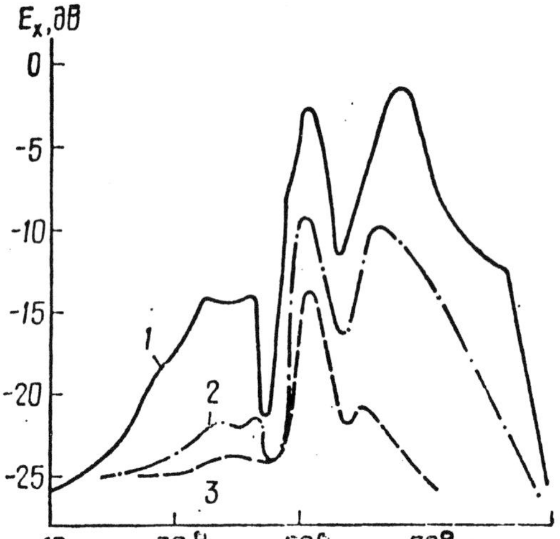

Fig. 123. Scattering of H-wave of planar dielectric waveguide by the beveled grating.

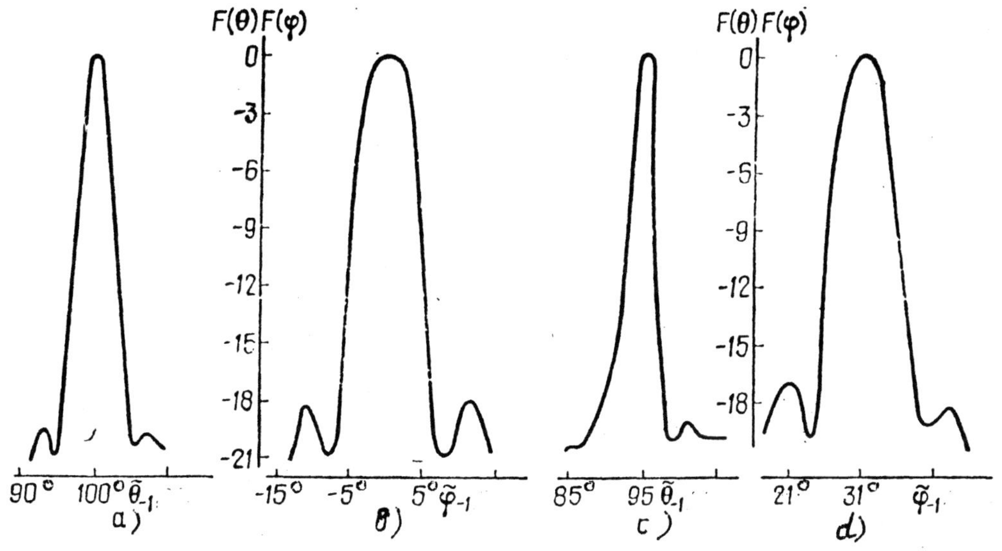

Fig. 124. Far field patterns of Smith-Purcell radiation for beveled gratings.

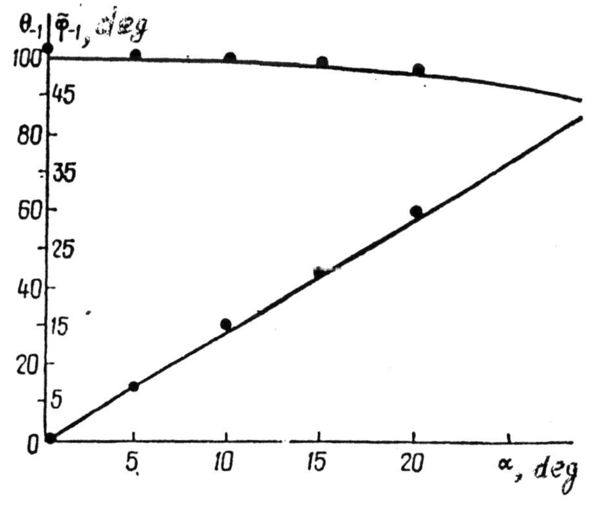

Fig. 125. Radiation angles of Smith-Purcell radiation for beveled gratings.

Part 3

APPLICATIONS OF SMITH-PURCELL EFFECT

More than three decades of investigations of Smith-Purcell effect enable us to formulate the conception which determines its place among different radiation phenomena and establish the ways of application in various domains of science and technology. We discussed it in the introduction. Now let us point out to other aspects of this problem.

First of all let us underline that the main results stated in the previous chapters lead to the conclusion that together with Cherenkov effect which lies in the background of modern physics Smith-Purcell effect does not yield to the Cherenkov effect in its internal content and varied displays. These two allied phenomena occupying everyone its own specific niche in the series of fundamental phenomena of nature and supplementing each other may be described on the common basis. The general problem of transformation of surface waves to volume waves (and vice versa) by different material objects (including dielectric bodies, diffractional gratings *et. al.*) causes here the necessity to use and develop the known approaches elaborated in the diffraction theory.

The of principal idea is to study Smith-Purcell effect in the millimeter and submillimeter ranges. Evidently, the nature itself prepared this effect for this wave range. Although originally in the experiments of Smith and Purcell this radiation phenomenon was observed in the optical range, its universality and beauty became apparent exactly in the millimeter and submillimeter ranges, what pre-determined the areas of its application. They are mainly radiophysics and electronics of millimeter and submillimeter waves and first of all vacuum electronics, in particular, its new branch – diffractional electronics. Fundamental direction of diffractional electronics is creation of high-coherent sources of millimeter and submillimeter waves, generators of diffractional radiation (GDR), whose operation principle is based on the effect of diffractional radiation. SPE is applied also in antenna technology (in particular, in the design and construction of antennas where the wave analogue of SPE is used), for creating element base, units and radio systems, logical elements of optical computers, appliances carrying out effective input of energy in fiber devices as well as for working out primary devices for measuring the parameters of charged particles *et. al.*.

In the third part we will reveal various capabilities of Smith-Purcell effect together with practically unbounded ways of application of this remarkable natural phenomenon.

CHAPTER SEVEN

Generators of diffractional radiation

In the first two chapters it has been investigated physical properties and physical characteristics of SPE which formed the background for diffractional electronics and, in particular, for the design of GDR.

It is not enough to have only general information about SPE for carrying out experimental studies aimed to the creation of generators (namely, the conditions of radiation, its directions and polarization together with rough estimate of field's amplitude in the beams occurring in real situations) and to limit oneself to known ideas concerning open resonators. However, almost all the experiments which were carried out in the USSR and other countries (USA, Japan) failed because they were based on the most general characteristics of SPR and properties of the simplest quasi-optical devices what enabled the researchers to achieve the intensity sufficient for registration of the radiation itself. OR which were applied in the first generators were formed by two mirrors such that one of them was entire reflection diffractional grating; they were perhaps the least successful type of electrodynamical structure for creating effective generators. The first experiments confirmed the possibility of their use in millimeter (and even in submillimeter) ranges. From the other side, these very experiments promoted false belief about low efficiency of such generators in practical applications.

As it has become clear, the reason of failure of the first experiments in diffractional electronics was strong dependence of SPR on geometrical parameters of diffractional gratings and the open structure which could guarantee effective inverse communication. It was shown in the first part of the book that radiation's energy characteristics had resonance behaviour with respect to varying grating's parameters and the width of these resonances was comparatively small. The probability to create and use periodical structure with satisfactory geometrical parameters is negligible if diffractional gratings are chosen in a random way. Such situation was typical on the initial stages of working out the models of generators and investigating the effect of diffractional radiation.

Resonance properties of SPR are caused by multi-wave character of interaction of electrons with radiation field. There simultaneously exist two qualitatively different types of electromagnetic fields which are closely connected with each other: slow waves propagating along the grating and the spectrum of homogeneous plane waves (fast waves) forming the radiation itself. The work of a generator is determined by the condition of synchronization between the electron flow and one of grating's slow waves. In this very case flow's kinetic energy is transformed to SPR.

Fulfillment of the condition of automatic phase adjustment (AP condition) is necessary for providing generator's operation efficiency. It may seem that one should concentrate most part of the field in slow waves in order to satisfy these two conditions, and, on the contrary, to redistribute the field's amplitude for the benefit of volume waves without taking into account slow waves in order to obtain maximal output of diffractional radiation. In reality, there exists reasonable ratio of the amplitudes of these two different parts of scattered field and corresponding optimal conditions must depend on geometrical parameters of the structures providing required redistribution.

The original mathematical methods presented in the first part of this book correspond to such complicated multi-wave process occurring in the devices of diffractional electronics and enable us to create mathematical models of the considered phenomena.

Experimental studies of open resonance structures which guaranteed the necessary level of inverse communication preceded the creation of operating generators based on the Smith-Purcell effect. In the simplest cases they were made of OR formed by one concave mirror and one plane mirror covered by a narrow strip of diffractional grating whose geometrical parameters were chosen in accord with recommendations of the theory of diffractional radiation. One had to study the spectrum of resonator's eigenfrequencies and corresponding eigenoscillations. Such investigations could be carried out only experimentally since the modern theory of OR had not elaborated theoretical methods for the study of so complicated resonance structures.

These problems formed the essence of the first stage of experiments. It was completed by working out recommendations concerning optimal choice of the configurations of resonator's mirrors and the grating. Those parameters were called optimal which provided excitation of oscillations with required frequency, Q-factor and amplitude. They should have also such field's distribution which is favorable for interaction of electron flow with electromagnetic field and admits excitation in the easiest way by diffractional radiation.

Creation of operating generator of diffractional radiation had been undertaken only after working out the detailed theory of the effect of diffractional radiation and completing "cold" experimental studies of resonance systems. Understanding of the main physical properties of diffractional radiation and its essential features helped to choose the most perspective direction for the practical use of this phenomenon in electronics. Therefore, even the starting attempts of creating the generator were successful: output power of more than 1 Wt in the short-wave part of millimeter range under the condition of continuous generating regime was achieved already by the first model of GDR what was three orders greater than average output power of orothron [121] and of the generator proposed in [115]. Output power of GDR was considerably increased during the short period of time.

"Hot" experimental studies of the generator were carried out in the beginning with the help of dismountable model. The main purpose was to work off optimal conditions of excitation of oscillations in resonators by an electron flow which had been established in the process of "cold" studies without a flow. As a result, it were selected all soldered-off elements of the devices suitable for exploitation. The construction and manufacturing technology of various types of GDR have been thus elaborated.

For the first time in world practice it was proposed and realized several GDR of various modifications differing by excitation regime and functional capabilities. The following types of generators have been investigated in the most exhaustive way: flight GDR (FGDR), reflection GDR (RGDR), impulse GDR (IGDR) and GDR-autodyne. The results of these studies formed the basis for creating the package GDR.

§20. Basic generator of diffractional radiation (continuous flight regime)

Physical mechanism of the work of GDR was subject of discussion from the time of constructing their first samples described in [115, 117] and in [116] (orothron). There were two points of view: a) GDR is a variety of resonance lamp with inverse wave [149, 150]; b) operation principle of GDR is based on the phenomenon of diffractional radiation. Let us discuss in details the second concept.

Physical properties of the Smith-Purcell effect stipulate the operation of GDR. SPE is observed when electron flow is directed along diffractional grating and it may be interpreted as a result of diffraction of electrons' eigenfield on periodical grating's surface. Scattered field may be a superposition of surface (slow) waves localized close to the grating and homogeneous (fast) waves going off the grating. The first or the second type of the field prevails depending on excitation conditions and grating's resonance geometrical parameters.

1. Considerable difference between GDR and other electron devices with

long interaction (LIW, LRW) is in the presence of high Q-factor resonance system in the form of OR, hence, GDR is similar to optical quantum generator. Oscillation system of GDR determines its stability, radiation's spectral composition, range of readjustment, output power, starting current and other parameters. The most general scheme of the first basic samples of GDR [3–5] is presented on Fig. 126. Reflection diffractional grating 1 partially covers one of the mirrors 2; electron flow 3 which is focused by longitudinal magnetic field goes along the grating. Energy is taken out through the second mirror by means of a junction with waveguide 5. Combined readjustment of GDR is achieved by coordinated variations of the distance between mirrors and accelerating voltage.

2. Creating the basic models of effectively operating GDR is connected with detailed study of open electrodynamical structures possessing specific properties.

It is essential that the parameters of OR must be measured without perturbing the vacuum. In this way visualization technique [151, 152] which was originally destined for the measurements in free space should be changed in order to make possible the registering of fields' distribution in vacuum domains. With this purpose the special scanning mechanism was constructed providing the movement in OR of a small absorbing body along the given trajectory. Here we consider the plant similar to the one proposed in [153] for investigating the fields in operating GDR. Optimization of the parameters is closely connected with generator's excitation conditions what, in turn, causes the necessity of describing the fields ' structure of operating GDR. Corresponding measurements are carried out in the four-millimeter range for several GDR.

The scheme of experimental plant is presented on Fig. 127. Oscillation system of GDR is placed to vacuum volume 1. Open resonator of GDR consists of spherical (2) and plane (3) mirrors with reflection diffractional grating 4; electron flow flies by it and excites the resonator. Slide 5 and clips 6 support small absorbing body 7 hung up on quartz thread 8. Scanning mechanism 9 initiates the movement of absorbing body in the given plane of resonator's volume along plane spiral with the step $d \approx 0.7$ mm converging to the axis of OR (or diverging from it).

Batteries B feed potential-meter primary devices 10 connected with scanning mechanism which are put in motion by electric motor 11. These devices allow to present the trajectory of small body on a screen or on a paper with the help of two-coordinate self-writing potential-meter. Appropriate choice of battery's voltage enables one to obtain the trajectory on the screen or on the paper which does not differ from the one of absorbing body moving in the volume of OR and synchronized with it.

Obtaining the visible picture of amplitude field's distribution in the transversal plane of OR is based on the changing intensity due to bringing absorbing body in the area of field's spot. If this body is in the spot of the field of established oscillation, oscillation's amplitude decreases together with output signal. When the body goes out of the area of field's spot, the energy absorbing by the body falls and (modulated) output signal grows. This signal comes to amplifier 16 from generator's output 12 through T-joint 13, attenuator 14 and detector 15. Then the signal is sent from amplifier's output to indicator of signals' level 17.

When absorbing body is in the area of field's spot, it absorbs a lot of energy what yields low output signal, then the level's indicator disconnects feeding chain of the relay supporting potential-meter's pen 18 in "writing" position, the pen stops to write but continues to move synchronously with absorbing body.

When absorbing body leaves the area of field's spot, output signal grows, level's indicator closes the circuit and the relay together with the pen depicts required trajectory of the body. After the body crosses the whole given plane, the pen draws the picture of the oscillation established in generator's OR, where the blank spots correspond to the areas occupied by the field whose level is higher than the given one. The drawing is controlled by attenuator, amplifier and level's indicator. Shape, size and position of blank spots repeat shape, size and position of field's spots in OR owing to synchronous movement of the pen and absorbing body.

One can use electron oscillograph in place of potential-meter in order to obtain visible picture of field's distribution. Then the contrasting picture will appear on the screen. Attenuators 1, 19, wave-meter 20 and detector with indicator on the scheme presented on Fig. 127 control the generator's frequency.

3. Before to consider experimental results, let us arrange with enumeration of oscillations excited in OR by electron flow. Theoretical analysis of the wave pencil of diffractional radiation which is formed by the bounded electron flow moving close to reflection diffractional grating is given in [154]. Electrodynamical size of such pencil may be determined by the formulas from

[94]. The pencil has much greater degree of divergence in the direction perpendicular to flow's movement (the axis Ox) than along the direction of its own movement (the axis Oy) as it is shown on Fig. 128, a. In this way one should expect excitation of oscillations with variations chiefly along the axis Ox.

Transversal fields' distributions shown on Fig. 128, b, d, f, g, h are obtained in OR of the four-millimeter GDR by visualization method with the help of the plant presented on Fig. 127.

One can see that TEM_{20q}-oscillation (Fig. 128, c) is mainly excited what is caused by the field's structure in transversal plane of OR. Other types of oscillations in GDR with semi-spherical OR (most of them are shown on Fig. 128) are excited not so frequently. Comparison of grating's position on the plane reflector (Fig. 128, a) and the field's spot of TEM_{20q}-oscillation (light areas on Fig. 128, c) makes it clear that field's minimums of intensity ("zeros") lie on the boundary between the grating and the plane what yields low ohmical and diffractional losses in OR. Let us note that TEM_{20q}-oscillation has even index m as well as TEM_{210q}- and TEM_{40q}-oscillations (Fig. 128, d, f) whose positions on plane reflector with the grating similar to the location of TEM_{20q}-oscillation. However, TEM_{210q}- and TEM_{40q}-oscillations have more field's variations and, hence, greater ohmical and diffractional losses. Higher order oscillations with even indices m are not excited at all.

The present analysis allows to conclude that TEM_{20q}-oscillation has the highest Q-factor among all oscillations with even indices m which can be excited in OR what well agrees with the data of "cold" measurements. GDR most frequently operates at TEM_{20q}-oscillation; it has the strongest output power and the lowest starting currents. Let us give the values of these parameters for TEM_{20q}- and TEM_{21q}-oscillations:

TEM_{20q}: P_{out}=1.2 Wt; $I_{(work)}$=140 μA; $I_{(strt)}$= 90 μA;
TEM_{21q}: P_{out}=0.03 Wt; $I_{(work)}$=160 μA; $I_{(strt)}$=140 μA;
(in both cases working wavelength $\lambda = 4$ mm).

Open resonators in GDR may be also excited at oscillations with odd indices m: TEM_{10q}- and TEM_{30q}-oscillations (Fig. 128, f, h). Fields' spots of such oscillations lie partially on plane mirror and partially on diffractional grating; boundary between the mirror and the grating goes through the maximum of field's intensity (Fig. 128, a, b, h). Diffractional grating naturally brings in big ohmical and diffractional losses for such oscillations. TEM_{10q}- and TEM_{30q}-oscillations are excited less frequently than TEM_{20q}-oscillations and for greater starting currents.

More complicated oscillation, which is also observed in GDR, is shown on Fig. 128, g. Let us call it $TEM_{(20+11)q}$-oscillation because here at first TEM_{20q}- and then TEM_{11q}-oscillations are excited in OR. The reason for such order of excitation's priority is caused by greater ohmical and diffractional losses occurring for TEM_{11q}-oscillation due to the influence of diffractional grating, while for the excitation of TEM_{20q}-oscillation greater power in OR is required. GDR cannot be excited at pure TEM_{11q}-oscillations. Coincidence of these types of oscillations is natural since they are degenerated with respect to indices $(m+n)$.

4. Let us apply interrelations from [94] describing oscillations' spectrum in OR with smooth mirrors (without electron flow and diffractional grating) in order to estimate the influence of diffractional grating and grouped electron flow on resonance distance of OR of GDR operating in continuous regime. One can estimate the required influence by comparing oscillations' spectrum of OR with smooth mirrors constructed in the form of the function $L = F(f)$, where f is the frequency parameter, with the spectrum of OR containing inhomogeneities which is calculated in [155].

Both these dependencies are shown on Fig. 129. Dot lines indicate resonance distances of OR with smooth mirrors for TEM_{00q}-oscillations which are close to experimentally obtained resonance distances of TEM_{20q}-oscillations; small deviations $\Delta \approx h$ (the same result is valid if to replace TEM_{20q}-oscillation by TEM_{21q}-oscillation) where h is grating's depth, are caused by the influence of diffractional grating. In order to reduce these deviations one should count resonance distances not from the surface of reflection but from the cavities' bottom. Then the distance between reflectors will be $L' = L-\Delta$ where L', L are these distances in OR with or without diffractional grating, respectively.

Dot-point lines indicate resonance distances for TEM_{2011q}-oscillations (other types of oscillations are not presented here). Resonance distances obtained experimentally are marked by bold points. Fields' structure have been also measured in these points.

Finally one may conclude that, if the index n is constant, the change of field's variations along the direction of electron flow does not perturb Δ. Analogous results hold for the measured deviations between TEM_{10q}- and TEM_{11q}-oscillations ($\Delta \approx h$).

Measurement data for the functions $L = F(f)$ give the following results: $\Delta \approx 1.34$ mm for TEM_{30q}-oscillations and $\Delta \approx 1.25$ mm for TEM_{40q}-oscillations. Hence, the deviation for TEM_{40q}-oscillation increased and the difference was + 0.25 mm, while for TEM_{40q}-oscillation it was + 0.35 mm. It is obvious that this shift is connected with reactive load brought into OR by the grouped electron flow, since additions caused by dielectric permittivity of the flow and by lengthening of the wave in the slot of reflection grating are much smaller than the obtained values of deviations.

It is natural to expect that the grouped electron flow will also cause perturbations for GDR operating at TEM_{20q}- and TEM_{10q}-oscillations but of less order, hence, for their estimation it is necessary to measure resonance distance more precisely. If to take into account the additions caused by the grouped electron flow and reflection grating, one gets due to [94]

$$\frac{2(L-\Delta)}{\lambda_0} = q + \frac{1}{2\pi} \arccos\left(1 - \frac{L-\Delta}{R}\right)$$

where Δ =const for the given type of oscillation.

The dependencies $L = F(f)$ constructed in accord with this formula enable one to determine the types of oscillations in OR without measuring their fields' structure.

5. It has been established above by the visualization method that mainly the oscillations with even indices m having field's variations along the axis Ox perpendicular to flow's movement are excited in OR of operating GDR. The only oscillation which is excited independently is the TEM_{21q}-oscillation with even index m. Oscillations with odd indices m having field's variations along the flow are not excited at all. Diffractional grating yields big ohmical and diffractional losses for such oscillations.

Diffractional grating and electron flow do not essentially perturb transversal fields but noticeably influences oscillations' spectrum of OR. Diffractional grating displaces spectral lines and the value of this shift is approximately equal to cavities' depth h, while the grouped electron flow flying over diffractional grating acts like reactive load and causes additional shift of spectral curves. Visualization technique allows to estimate such shifts in vacuum high Q-factor OR formed by arbitrary mirrors.

Analysis of the pictures of fields' distributions enables us to conclude that the most widespread types of oscillations have greater degree of periodicity in the direction perpendicular to the flow due to the wider diffractional far field pattern in this direction (since the flow's width is much less than its length). Hence, one may essentially reduce dimensions of lower mirror in the direction of electron flow and thus to decrease the gap in focusing magnetic field. Diffractional losses are determined by the ratio of the sizes of the mirror and field's spot. Therefore, in order to influence the Q-factor in an equal way, one may choose for TEM_{20q}-oscillation the mirror placed along the flow whose dimensions are in 1.8 times less than its transversal size. These properties of OR formed the basis for creation of unsoldered GDR with reduced size along the direction of flow's movement and enabled one to apply constant magnets with smaller gaps, i. e., with smaller weight and dimensions of the elements.

6. Initial data for constructing unsoldered GDR suitable for operating devices were obtained in the process of investigation of generators' models under the conditions of continuous pumping. The first basic version of such GDR with waveguide output of energy described in [156] (see Fig. 130, where GDR has the form of circular cylinder 1 with external diameter 80 mm and the length 200 mm, strip reflection diffractional grating 3 is drifted on the plane mirror 2 occupying about 15 per cent of its area). This version of GDR contains oscillation system in the form of semi-spherical OR whose spherical and plane mirrors have the diameters $2a_s = 60$ mm, $2a_p = 50$ mm, respectively, radius of the sphere is $R_s = 110$ mm, width and length of reflection diffractional grating are $B = 10$ mm and $l = 46$ mm, respectively.

Anode with a cannon (which is adjusted with the accuracy ± 0.01 mm) is fixed on one side of the plane mirror and collector's slot 4 together with isolated collector 5 are fixed on the other side. Plane mirror and collector are cooled by a flow of running water 13. Spherical mirror 10 and its shifting mechanism 9 (providing variation of distance between the mirrors within the interval 19–29 mm) are connected with the upper lid of GDR by a leg 7. Communication slot 6 in the mirror's center gradually transfers to the standard four-millimeter waveguide 8 with the cross-section 1.8×3.6 mm^2. In this type of GDR it is usually applied the electromagnet of the Russian standard SP-55A (-55). Speed of the focused electron flow 11 is controlled by high-voltage inputs 12.

It has been worked out the four-millimeter version of GDR with waveguide output of energy to the 63 mm gap in the magnet's poles. Such types of GDR operate with constant magnets made on the basis of materials with directed crystallization which provide the field in this gap exceeding 4 kE [151]. It is known that in order to support equal magnetic field the magnet's weight must vary proportional to the cube of distance between poles. Special modeling of fields' behaviour in operating GDR allowed to reduce the gap between poles from the value of 80 mm to 63 mm (Q-factor and starting currents remained unchangeable). Design of this device enables one to apply spherical mirrors with different apertures and cirvature radii. The only demand is the presence of cuts made on spherical and plane mirrors on the collector's side perpendicular to flow's movement. Hence, spherical and plane mirrors have the sizes of not more than 54 mm and 41 mm, respectively, along the line anode-collector.

Stronger focusing magnetic field is required for shorter waves. At the same time size of the spot and length of reflection grating decreases proportional to $\sqrt{\lambda}$ and λ, respectively. It is natural that with shortening λ size of the device becomes smaller what leads to decrease of the gap of focusing magnet. Hence, one can increase magnetic field keeping its energy capacity on the same level. It has been constructed GDR operating in 3 mm – 1 mm range of wavelengths with poles' gap 40 mm [157]. Here constant magnetic field may reach 6.5 kE. Spherical mirrors having the apertures with the length up to 44 mm and plane mirrors with the diameter not greater than 40 mm are applied in such GDR. Reducing the field's spot due to shortening wavelength together with the decrease of curvature radius of concave mirror 1 enables one to use narrow reflection grating 3 with the width up to 7 mm. Width of diffractional far field pattern generated by electron flow 2 is in four times less than dimensions of the spot in the direction of flow's movement. This interrelation together with specific configuration of electromagnetic field yield the size of spherical and plane mirrors (see Fig. 131) to be equal to 36 mm and 25 mm, respectively, from the side of cannon 8 and collector having the form of 2–mm copper plate fixed to the mirror (or opposite side of the device). Digits on Fig. 131 denote the following objects: 9 – communication slot; 10 – shifting mechanism; 13 – SHF-output; 12 – mica window; 11 – lateral wall of GDR; 5 – observation window; 4 – water cooling system. This construction is the basic for submillimeter devices; here it is applied reflection diffractional gratings with the period l =0.5–0.1 mm.

7. Various unsoldered samples of GDR work in the range of wavelengths λ =8 mm – 0.96 mm. Reflection diffractional gratings with the period l =0.8–0.1 mm made of rectangular bars having the thickness 0.5λ– 0.6λ and height 0.25λ are used in such generators. Combined frequency readjustment in one device reaches one octave; accelerating voltage changes within the limits of 1 kV – 5 kV and output power takes the values beginning from hundreds of

μW up to tens of W while anode current reaches the level 280 μA.

OR may be excited at different oscillations; hence, in general, the dependence of frequency on accelerating voltage is ambiguous and the same frequency may be excited at the voltages which have up to 10 per cent difference between each other.

GDR, as well as any resonance device, may operate under the conditions of different degrees of synchronization between the velocity of electron flow and the velocity of space harmonic interacting with the flow. Accelerating voltage then has variation reaching several per cent what inevitably yields frequency readjustment of hundredth parts of per cent which is called the electron frequency readjustment. Its level is determined by the strip of transmission of OR. Width of the strip of transmission depends on magnetic field and anode current; it varies in the interval 10 MGz – 20 MGz for the four-millimeter GDR when output power has 1 dB fall. Broadening of the strip of electron readjustment together with increasing power versus (growing) working current and magnetic field are shown on Fig. 132. Steepness S of electron readjustment increases with growing frequency f and does not depend on anode current and magnetic field: $S = 0.26$ MGz V^{-1} for f =76.4 Ghz while $S = 7.85$ MGz V^{-1} for f =63 Ghz, for the generator working in the regime of LIW, i. e., the steepness increases in 30 times.

One can vary the frequency of GDR for constant accelerating voltage by shifting the concave mirror, i. e., by changing resonance frequency of oscillation contour. Changing frequency yields variation of velocity of an established oscillation what, in turn, perturbs synchronization. The work of GDR at the same type of oscillation leads to gradual variation of frequency and it has a sharp break when the type of working oscillation changes. Raised starting currents are required, as a rule, if OR works at higher order oscillations, when output power falls. That is why it is of considerable practical importance to investigate readjustment of the device working at one type of oscillation. Such readjustment for constant voltage reaches the strip 1.5 Ghz and it is accompanied by power overflows up to 10 dB (sometimes even by the break of oscillations as on Fig. 133, a). It is easy to find the domain with readjustment close to 400 MGz in the case of small distances between the mirrors, where power overflows do not exceed 3 dB. Steepness of mechanical readjustment, which is equal to 3 Ghz mm^{-1}– 4 Ghz mm^{-1}, depends on the frequency and distance between the mirrors (see Fig. 133, b; with shortening the wavelength the range of mechanical readjustment grows).

In order to obtain maximal power and operating stability of the device, it is usually applied combined frequency readjustment, namely, simultaneous variation of resonance frequency and velocity of electron flow what enables one to obtain optimal degree of synchronization (it corresponds to the maximum of curves on Fig. 132). In the case of combined frequency readjustment the same change of accelerating voltage causes much stronger frequency variation (in 50 times and more) than in the case of electron readjustment. If to assume that the field in OR of GDR has the form of plane wave which falls perpendicular to infinite reflection grating and to neglect the absence of synchronization, then the following simple interrelation holds for the grating's period l, generated wavelength λ and accelerating voltage U:

$$\lambda = \frac{cl}{p\sqrt{2lUm^{-1}}} \approx 505\frac{l}{p\sqrt{U(b)}},$$

where p is the number of incident surface harmonic equal to the number of periods of high-frequency oscillation required for the electron to cover the distance between centers of two slots of reflection grating. Difference between experimental and idealized data (obtained by the theoretical formula presented above) for the frequency versus accelerating voltage (Fig. 134) does not exceed 10 per cent, if the device works at different oscillations. When GDR works at TEM_{20q}-oscillation this difference is about 2–3 per cent.

Self-excitation of GDR occurs when the power of SPR spent for feeding the resonator exceeds the power of losses. The latter are determined by the loaded Q-factor which is a variable quantity for the given OR and continuously varying frequency. The power of diffractional radiation, as well as the coefficient of feeding the resonator by this radiation, is a complicated function of accelerating voltage, frequency, type of oscillation, magnetic field, parameters of electron flow, etc. Theoretical calculation of the trustworthy values of starting current in the whole range of GDR readjustment is rather difficult, but the analysis of experimental data enables us to establish some appropriatenesses. In particular, minimal value of starting current is observed in the domain where $\lambda = 4h$, i. e., when there exist optimal conditions for interaction of electron flow with high-frequency field. This value may be estimated by the methods of linear theory of GDR. If to draw the rounding curve through the minimums of starting currents, it would be smooth because of decreasing size of field's spot and growing degree of communication with the load due to the shortening of generated wavelength. Splashes of starting currents may be connected with the influence of slow wave resonance on the Q-factor. Decreasing the depth and the period of reflection grating cause the minimums of starting current at arbitrary frequency and these minimal values turn out to be much lower than the ones obtained for big depths. However, the minimum of starting current usually grows with shortening wavelength due to increasing relative amplitude of pulsations and ohmical losses.

Output power as a function of accelerating current in the range of readjustment may be calculated with the help of nonlinear theory of GDR. Corresponding experimental data are evidence of vanishing the output power when the difference between working current and starting current tends to zero; when it becomes equal to zero the generation breaks and vice versa, greater this difference, sharper the growth of output power.

Interaction of electron flow with the field of OR improves when the depth of reflection grating $h \approx 0.25\lambda$; then output power increases and starting current falls. Shortening the wavelength also promotes the growth of the degree of communication what may shift, in turn, power's maximum and all these factors considerably influence characteristics of operating GDR. Output power in the range of readjustment considerably depends on the power spent on electron flow (raise of power in the short-wave range).

Interaction of every electron with the field of OR in GDR depends on the distance to the reflection grating because electron flow interacts with surface harmonics of the field of fast waves, while this field itself is used for inverse communication. Electron flows formed by diode cannons have the lateral pulsating layer whose thickness is determined by cannon's parameters and focusing magnetic field. With growing magnetic field this layer decreases and at a certain moment its thickness becomes less than inhomogeneities of the grating. Here one observes lowering of starting currents and raising output power. Further growth of focusing magnetic field sometimes worsens electron's interaction with the field, because electron flow slightly deviates from rectilinear direction in the vicinity of field's spot what facilitates the adjustment and improves interaction (strong magnetic field may perturb electrons' trajectory worsening thus their interaction with the field; this effect causes the necessity of additional adjustment). With falling magnetic field starting current grows at first gradually (inversely proportional to the level of magnetic field) and then more rapidly.

§21. Reflection, impulse and autodyne operation regimes of the generators of diffractional radiation

Operation mechanism of reflection GDR (RGDR) is based on SPR excited by electron flow repeatedly interacting with the field of OR. In the impulse regime oscillations are generated by SPR when packages of electrons fly near the grating. Autodyne effect is in their reaction on the own signal reflected from an object which becomes apparent when internal generator's characteristics are observed, namely, field's amplitude in OR, generation frequency and pencil's current.

1. Let us consider operation principles of RGDR [27–29].

The basic feature of RGDR is the repeated movement of electron flow near diffractional grating (under the condition of generation of diffractional radiation) due to specific electrons' reflection close to the collector and cannon's cathode.

Model of the generator and distribution of the potential in the domain cathode – space of interaction – electrons' reflector are given on Fig. 135.

At the initial moment of time $t = t_1$ non-modulated strip electron flow 3 generated by diode cannon comes to the volume of OR formed my mirrors 1, 2; diffractional grating is drifted on one of them. Axis Oy coincides with flow's axis; Ox and Oz are parallel to grating's lamellae and resonator's axis, respectively; $v_0 = \sqrt{2\eta_e U_0}$, $i_1 = k_1 i_0$ are the constant components of electrons' velocity and pencil's current in the resonator's input ($L = 0$); k_1 is the current coefficient characterizing pencil's current coming to the space of interction at the initial moment of time $t = t_1$; i_0 is the cathode's current; l, L are the length and the period of diffractional grating; D_{C-A} is the distance between cathode and anode, D_0 is the distance between the grating and reflector; $U_R \leq 0$ is the reflector's voltage, $U_0 = |U_C|$ is accelerating voltage and U_C is the cathode's voltage.

After the primary flight through the space of interaction ($0 < y < L$) at the moment of time $t = t_{1L}$ electron flow comes to the braking electrostatic field of the reflector placed behind resonator's volume in the plane $y = L + D_0$ and then returns to OR for $t = t_2$ being modulated with respect to both the velocity and charge's density. After the second flight the constant components of pencil's current in the resonator's input takes the form $i_2 = k_2 i_0$; here it is taken into account current subsidence of the flow on diffractional grating and reflector's electrodes and coefficient $k_2 : 0 \leq k_2 \leq 1$ determines the fall of the current of reflected pencil in comparison with starting current.

At the moment of time $t = t_{2L}$ and in the beginning of interaction interval, where $y = 0$, electrons of reflected flow again comes to the braking field of the cannon's cathode. Then after additional modulation they return to the space of interaction for the third time supported by the cathode's field.

The number of such flights are limited, firstly, by the electrons' phase condition (repeated movement should be continued until the electrons "leave" the braking field), secondly, by current subsidence of the flow on diffractional grating and reflector's electrodes, thirdly, by the space charge.

When electrons fly in the direction cathode–reflector, the flow mainly interacts with direct wave of electromagnetic field and when they fly in opposite direction, the flow interacts with inverse wave. It allows to reduce the solution to the problem of excitation of GDR in the case of repeated flights of electrons through the same space of interaction to the problem of excitation of successively connected GDR separated by the phases (time) of electrons' flight through the braking static fields of the reflector and cathode of electron cannon [27]. Here one can calculate all necessary characteristics of GDR.

In the general case, as well as for FGDR, starting current of RGDR grows with increasing thickness of electron flow and falls with increasing resonator's Q-factor ($\sim Q^{-1}$) and the length of the space of interaction ($\sim L^{-3}$). The most effective interaction occurs for the flow with cross-section $S_e = 2z_g\lambda\beta_0\pi^{-1}$. Electrons' layer having the thickness $\lambda\beta_0\pi^{-1}$ is the most effective for "thick" pencil. If $M = 1$ and $k_{m\geq 2} = 0$ (electrons' reflector is earthed), the starting current is equal to the one of FGDR.

Multiplicity (number) of electrons' flights through the space of interaction exerts strong influence on the processes of energy exchange and starting parameters.

One of the features of RGDR is the possibility to control the processes of energy exchange by variation of electrons' flight angles in the space of interaction (Φ_s) and in the field of reflector ($\bar{\theta}_R$). In accord with these factors one can separate three excitation regimes of RGDR:

a) the regime A – generator's characteristics are controlled with the help of parameter Φ_s (or accelarating voltage of diode electron cannon) for constant value of $\bar{\theta}_R$ (or constant reflector's voltage); in the particular case of "earthed" reflector these characteristics corresponds to the ones of FGDR;

b) the regime R – generator's characteristics are controlled with the help of parameter $\bar{\theta}_R$ for constant Φ_s; here it is provided "non-power" control;

c) the regime of "pure" synchronization when $\Phi_s = 0$ and $\bar{\theta}_R$ varies;

FGDR cannot operate in this regime while in the case of RGDR discrete generation zones are mainly excited analogous to classical zones of reflection clystron.

Excitation regime may be chosen in accord with specific practical task.

Resulting power of interaction for multi-flight GDR is determined by partial powers P_1, P_{kR} connected with different ways of forming the convectional curent. The contribution of each partial power depends on the parameters $\Phi_s = 0$, $\bar{\theta}_R$ and k_m.

Fig 136, a illustrates the process of energy exchange for RGDR-2 in the regime A where it is given partial powers of interaction versus $\Phi_s = 0$ for different values of k_2 and flight angles $\bar{\theta}_R$. Curve 1 corresponds to the power $P_1 P_0^{-1}$ of FGDR, curve 2 – to $P_{1R} P_0^{-1}$, curve 3 – to $P_\Sigma P_0^{-1}$.

One can see that for definite values of Φ_s, $\bar{\theta}_R$ the power given by electron flow to SHF field of OR (shaded areas) considerably exceeds the value of P_1. The main contribution to the total power P_Σ is made by P_{1R}; the power P_{2R} is negligible since $\bar{\theta}_R << \Phi_s = 0$; for big flight angles ($\bar{\theta}_R >> 2\pi$) P_{2R} grows and may reach the level P_{1R} (when $\bar{\theta}_R \sim \Phi_s = 0$); Unlike FGDR, the "working" domain of the parameters of synchronization in RGDR essentially depends on $\bar{\theta}_R$ and k_m: in the "principal" area of FGDR $\Phi_s = 0 - 2\pi$ for certain values of $\bar{\theta}_R$ it is excited one zone, either narrow or wide with non-uniform variation of power versus Φ_s ($\bar{\theta}_R \approx 1.6\pi$, $k_2 = 0.3$; such excitation may occur only for $k_2 < 1$ when the constant component of the current of reflected pencil in the input of OR is less than initial current i_0). Several generation zones may be also excited (for example, two zones for RGDR-2, as it is shown on Fig. 136, a, where $\bar{\theta}_R = 2\pi p_1$, $p_1 = 0, 1, 2, \ldots$).

It is of practical interest to study interaction power versus parameter $\bar{\theta}_R$, i. e., in the regime R. The curves for partial powers of RGDR-2 for this case are shown on Fig. 136, c, d. Here two cases are possible: 1) RGDR is excited only in discrete zones with varying $\bar{\theta}_R$ (Fig. 136, d); 2) for the given Φ_s interaction power is an oscillating function of $\bar{\theta}_R$. The first case takes place when there are no losses of reflected electrons caused by current subsidence ($k_m = 1$); if $k_m < 1$, excitation occurs in this case only if the power P_1 of the primary flight exceeds interaction power of reflected electrons P_R. If these conditions are violated, P_Σ has the form of an oscillating function with respect to $\bar{\theta}_R$.

Further analysis of Fig. 136, a allows us to conclude that in the case of "pure" synchronization of FGDR ($\Phi_s = 0$, curve 1) active component of electron's interaction power is equal to zero. However, active component of electron's interaction power in this regime does not vanish due to additional phase shift between resonator's SHF field and convectional current of reflected electrons (see Fig. 136. b). Discrete generation zones are excited in RGDR when $\bar{\theta}_R$ varies.

The simple expression holds for the starting current of RGDR-2 with double interaction:

$$J_{2n_{\min}} = J_{1S}\left(1 + k_2 + 2k_2 \frac{A_L|M|}{\Phi_s^2|F_a|}\right)^{-1}.$$

It follows from this formula that starting current of RGDR may be much les than J_{1S} of FGDR. It will be shown below that this difference essentially depends on the current of reflected electrons (k_m).

Calculation and experimental data presented above enables one to establish qualitatively new properties of RGDR, in particular, two ways of effecive control of output characteristics (the regimes A and R). Non-power control of signal's frequency and amplitude within the limits of generation zone by variation of U_R, when the required generation frequency is chosen with the help of accelarating voltage (like in the reflection clystron) is preferable for the heterodyne-type devices. One should apply the regime A in high-voltage impulse RGDR where generation power may exceed the one in the regime R (in the case of earthed reflector generator works as FGDR, i. e., as GDR with raised power).

Starting current of RGDR in the case of minimal losses of electrons due to current subsidence ($k_m = 1$) decreases like the cube of the number of electrons' flights through the space of interaction. If $k_m < 1$, the degree of the fall of J_S is determined by the current of reflected electrons in the space of interaction and the minimum of J_S in RGDR is always less than the starting current of FGDR.

Calculations show that when $k_m = 1$ in the regime A for the given angle of electrons' flight in the field of reflector increasing number of the flights leads to excitation of additional generation zones, where J_S reaches its minimum in the domain $\Phi_s = -\pi n^{-1}$ and which are not as wide as in FGDR (with respect to parameter Φ_s).

It becomes possible to control the frequency generation in RGDR because the range of electron frequency readjustment is maximal when generator is excited in discrete generation zones, where the frequency has tangent-type variation, and in the regime R it is bounded by the value of oscillation's amplitude of the frequency as a function of $\bar{\theta}_R$.

Experimental studies of RGDR have been carried out with the use of its model (Fig. 137) in the wave range λ =3.8 mm–5.7 mm. Spherical or spherical-cylindrical OR in this generator is formed by spherical mirror 2 with energy output and plane (cylindrical) mirror 5 with diffractional grating over which the electrons fly. Electron flow of the cross-section 5×2 mm^2, which is focused by magnetic field, is formed by diode cannon 1 and comes to the domain of reflector 3 at the resonator's output; there electrons are braked by electrostatic field and returns to the space of interaction. Depending on the trajectory, reflected electrons either uniformly subside on the grating (double flight) or comes to electrostatic field of diode cannon and again returns to the space of interaction (repeated flight). Additional electrode 4 allows to estimate the losses of inverse electrons at the entry to the space of interaction comparing pencil's currents for single flight ($k_2 = 0$) with the ones in the reflection regime ($k_2 > 0$).

The model operates under the conditions of constant pumping out ($p = 5\cdot10^{-7}$) and is placed on the special adjusting table which makes possible to change orientation of the axis of direct and inverse electron flows with respect to lines of force of magnetic focusing field. Electrons' current subsidence on periodical structure is controlled with the help of pencil's current for $k_2 = 0$ and visually through the window in the frame of vacuum cover.

Construction of the model admits the replacement of all its main units: electron cannon, electrons' reflector, mirrors of OR and diffractional grating. The scheme of experimental plant and the methods of investigations are described in [4]. Here we will limit ourselves to the analysis of the most valuable experimental results having practical importance.

Let us consider at first starting characteristics of RGDR-2 operating in the generation regime A. It follows from theoretical studies presented above that for the given angles of electrons' flight in the field of reflector ($\bar{\theta}_R$) either one (but not as wide as in FGDR) or two generation zones must be excited for varying synchronization parameter Φ_S (or acceleratimg voltage U_0 of electron cannon). One of these two zones is situated out of FGDR generation zone from the side of its long-wave part and we will call it the long-wave zone, while the other, which is closer to the domain of the parameters Φ_S of FGDR, will be called the short-wave zone. Excitation of either one or two generation zones depends on flight angle $\bar{\theta}_R$ and current of reflected pencil (current coefficient k_2), determining the phase of convectional current and power of intercation of electron flow with resonator's field. Gneeration zones shift with varying U_R and two zones may transform to one which is wider than in FGDR.

These appropriatenesses are confirmed experimentally. The curves on Fig. 138 give experimental data for the starting current of RGDR-2 versus accelerating voltage for different values of U_R and k_2 (generation frequency $f_0 \approx 75$ GGz). In order to find interrelations for the parameters $\bar{\theta}_R$, Φ_S and voltages U_0, U_R we will use the formulas

$$\Phi_S = \frac{2\pi L}{l}\Big(\frac{505l}{\lambda\sqrt{U_0}}\Big); \qquad U_R = U_0\Big(1 - \frac{8\pi D_0}{\beta_0 \lambda \bar{\theta}_R}\Big),$$

where the axis U_0 corresponds to the values of Φ_S.

One can see that the starting current of RGDR-2 is in five-six times less than the one of FGDR; two (Fig. 138, a) or one (Fig. 138, b, curve 3) generation zones are excited in RGDR-2 and the starting current of long-wave zone is less than J_S of the short-wave oscillation; the shift of generation zones is observed for changing U_R and they transform to one zone for small currents of reflected electrons when J_{2S} varies non-uniformly with respect to U_0 (Fig. 138, b, curve 2).

Let us summarize the results of investigations of starting current for RGDR operating in the regime R when it is provided non-power control of amplitude, frequency and spectrum of HF signal by variation of reflector's voltage U_R while the voltage U_0 of electron cannon remains constant. In this case it is excited either discrete generation zones or generation zones with continuous (oscillating) character of the behaviour of starting current. These conclusions are confirmed experimentally: the curves for J_S are shown on Fig. 139, a (for RGDR-2, $k_2 = 0.5$, $f_0 = 75.1$ GGz). Analysis of such data (obtained also for other types of RDGR and not presented here) allows to assert that discrete generation zones are always excited in RGDR with parabolic variation of J_S (theory states this property only for $k_2 = 1$). If $k_2 < 0.4$, discrete zones transfer to generation zone with oscillating starting current. When theoretical curves (bold lines) are compared with experimental curves (dot lines) on Fig. 139, b, the initial data for the former (in order to determine the constant coefficient) are chosen in experimental points A_1, A_2.

Growing number of electrons' flights leads, as it is predicted by theory, to the fall of starting current, increase of the number of generation zones (with respect to RGDR-2) or emerging of oscillations in starting characteristics, what is caused by the changing phase of convectional current of successive flights in the domain of the reflector and cathode of electron cannon.

Let us point out that in RGDR with non-symmetrical channel of electrons' flight (when reflection grating is placed on a lane mirror, [4]) the number of flights is limited by high losses of the flow due to current subsidence occurring when electrons return to the space of interaction. In some models of GDR it is realized up to four flights what guarantees the reduction of J_S in twenty times. Even greater reduction takes place in RGDR with symmetrical channel (see Fig. 88, a) where the starting current decreases approximately in 1000 times in comparison with FGDR [30].

Hence, excitation conditions of SHF oscillations in RGDR essentially depend on the phases of electrons entering the space of interaction and their losses caused by current subsidence. Variation of corresponding parameters enables one to create RGDR with qualitatively new properties (broaded generation zone, possibility of the non-power control of output characteristics, considerable reduction of starting current, etc.) which are perspective for application of RGDR in the millimeter and submillimeter ranges.

The following assumptions form the basis of theoretical model of RGDR. Non-modulated electron flow enters the space of interaction ($y = D_1$) which generates diffractional radiation in the direction

$$\cos\gamma(\xi) = \frac{c}{v_e(\xi)} + \frac{\lambda}{l}, \tag{7.1}$$

if the condition (1.8) holds. Here $\gamma(\xi)$ is the angle between vector of electrons' velocity and wave vector of radiating wave; $\xi = yl^{-1}$. Initial electrons' velocity (anode's voltage U_0) is found from the relation $\cos\gamma(\xi) = 0$ (when diffractional radiation is directed along resonator's axis, $\kappa = \beta_0$, $\gamma = 90°$).

In the space of interaction electron flow interacts with the space harmonic of SHF field having synchronous velocity. In the case of FGDR ($m = 1$) electron efficiency is determined by the variation of electrons' kinetic energy

$$\eta_e = 1 - \frac{1}{2\pi}\int_0^{2\pi}\left[\frac{v_e(\xi)}{v_0}\right]^2 d\phi_0, \tag{7.2}$$

where $\phi_0 = \omega t_0$ is the phase of electrons entering the space of interaction. If repeated interaction takes place, this interrelation has the form

$$\eta_m = \sum_{m=1}^{M} k_m \eta_m; \qquad \eta_m = \frac{1}{2\pi} \int_0^{2\pi} \frac{v_{m-1}^2(\phi_0) - v_m^2(\phi_0)}{v_0^2} d\phi_0, \tag{7.3}$$

where η_m is the electron efficiency; v_{m-1}, v_m are the velocities of electrons entering (input) or leaving (output) the space of interaction; $k_m = J_{0m} J_0$ is the current coefficient; J_0 is constant current at the entry to the space of interaction (of the m-th flight).

In the process of energy exchange average electrons' velocity changes what yields variation of SPR voltage for the fixed wavelength in accord with expression (7.1) and may lead to "lighting-off" losses of diffractional radiation (from OR to free space), if the radiation angle lies in the interval

$$\arccos \frac{\xi}{\sqrt{\xi^2 + R_1^2}} < \gamma(\xi) < \arccos \frac{D - \xi}{\sqrt{(D-\xi)^2 + R_1^2}}. \tag{7.4}$$

From the space of interaction electrons come to the domain of drift (having the length (D_1, D_2), like on Fig. 135) and then – to the vicinity of reflector, where, as well as in the previous domain, the clots of electrons regroup. Time of the flight through the reflector's domain is characterized by the flight angle $\bar{\theta}_R$.

Interrelations (7.1)–(7.4) give the possibility to obtain self-coordinated system of nonlinear stationary equations for RGDR-M (in the case when fields have time dependence $e^{-i\omega t}$).

After interaction with the field of OR, electrons enter the domain of drift where they are influenced on ly by one force of space charge. Unlike the clystron [158], electrons enter the domain of drift being modulated with respect to velocity and charge's density. One can also write down the system of equations describing the movement of electrons in this domain. Integro-differential equations modeling the behaviour of electron flow near the reflector together with the previous system are transformed to the system of ordinary differential equations and the latter is solved numerically.

Equation of electrons' movement in the domain of drift may be represented in the form

$$\frac{d^2\theta_m^n}{d\xi^2} = -\left(1+\frac{1}{\psi_0}\frac{d\theta_m^n}{d\xi}\right)^3 \frac{2i_{0m}p^2}{N_m} \sum_{n'=0}^{N_m-1}\sum_{n=1}^{\infty} \frac{\Gamma_{0n}}{n\pi} \sin\left[n(\phi_{n'}-\phi_n)+n(\theta_m^{n'}-\theta_m^n)\right], \tag{7.5}$$

where $p = \frac{\omega_p}{\omega_r}\psi_0\sqrt{\Gamma_{01}}$ is the parameter of space charge; $\omega_r = \omega_r' - i\omega_r'(2Q)^{-1}$ is the fundamental frequency of OR; Γ_{01} is the depression coefficient of the forces of space charge taking into account shielding effect of OR and finite size of electron flow; i_{0m} is the density of the flow of the m-th flight. In the domain of reflector dynamical angles of electrons' flight are determined by interrelation

$$\theta_R^n = \frac{v_n}{v_0}\frac{4\psi_0}{1-U_RU_0^{-1}},$$

where v_n $(n = 0, 1, 2, \ldots, N_m - 1)$ is the velocity of n-th electron at the entry to the domain of reflector.

Initial conditions in the case of the first flight of electrons are determined at the entry to the space of interaction by the formula $\theta_1^n(\phi_0, \xi) = 2\pi n(N_1 - 1)^{-1}$, where $\eta_1^n = 0$ for $\xi = D_1$ $(n = 1, 2, \ldots, N - 1)$.

The trajectory analysis of electron' flight phase and composition of harmonics of convectional current of RGDR-2 shows that during the first flight in the braking domain it is formed the clot of electrons which gives its kinetic energy to the field of OR. Later on this clot disintegrates. Already in the first flight electrons' movement is essentially nonlinear, since their trajectories intersect. The new clot of electrons is formed during the second flight where all electrons are separated into the ones with closed trajectories captured by the field of OR and flight particles with unclosed trajectories. Velocity of flight particles has less variation than of the captured electrons, which form the clot disintegrating under the influence of static forces. After the satiation of oscillation's amplitude and with further growth of synchronization parameter b the clot is destroyed and a part of electrons transfers to accelerating phase, what causes the fall of energy output and breakdown of generation. But if to increase the number of flights simultaneously with growing b, the new clots of electrons may emerge in the field's braking phase. At the same time, appearing of new clots yields "indented" output power and non-uniform electron frequency readjustment. In order to eliminate the influence of this effect one should reduce G_i (the working current) what is acceptable, since the starting current falls with growing number of electrons' flights; generation zones in RGDR with repeated flight will have minimal distortions.

Let us consider the work of RGDR in the regime R. For $k_2 \geq 0.5$ the presence of discrete generation zones is characteristic for the whole interval of variation of U_0. Maximal power of generation zones for $N = 7, 8$ ($D = 0.8$ mm) are achieved when the values of accelerating voltages correspond to the high-voltage domain of the considered oscillation excited in OR. The angle of electrons' flight in the field of reflector varies from 15.2π to 13.3π in the origing of generation zones ($U_R = -120, -560$ V). Under such conditions (when $\bar{\theta}_R << \phi_L$; $\phi_L = 100\pi$) the main energy effect in RGDR is caused by grouping of electrons in the space of interaction and intensive changes of their phases what yields comparable levels of generation power in the zones having different indices N. With varying phase of reflected pencil generator is effectively readjusted with respect to frequency and in the considered generation zones the frequency f is a linear function of P_0 at the level 3 dB; zones with big values of N have maximal redjustment steepness $S_R = \partial f(\partial U_R)^{-1}$: for $N = 8$ $S_R = 0.08$ MGz$\cdot$V^{-1} and for $N = 7$ $S_R = 0.4$ MGz$\cdot$V^{-1} what is in several times less than in the case of generator's redajustment with respect to accelerating voltage. But due to essential broadening of generation zones in the regime R one can obtain the range of readjustment which is equal to the one of the regime A in FGDR.

Discrete generation zones may be also excited when the current of reflected electrons' pencil is decreased at the generator's entry. Such generation regime is characteristic for small values of synchronization parameter of the considered oscillation.

For the given parameters of electrodynamical system of RGDR one should either reduce D_0 or increase U_R (or, in other words, to decrease electrons' flight angles in the domain of reflector) in order to excite generation zones with small indices N. Corresponding experimental investigations carried out with the help of the model of RGDR and two separate feeding sources for the voltages U_0 and U_R varying within the limits 0–4 kV confirmed the possibility of excitation of such generation zones.

The characterictic distinction of these zones is the broader interval of oscillations' excitation with respect to U_R, decrease of the steepness of non-power frequency readjustment and growth of limiting current's density. Let us present some particular excitation parameters for RGDR-2. Generation zones having the numbers $N = 1, 2$ (the first and the second) were excited for the oscillation with $f_0 = 62$ GGz ($U_0 = 1680$ V). Center of the second zone was situated in the domain $U_R = -168$ V ($\bar{\theta}_R \approx 4.6\pi$) and center of the first zone – in the domain $U_R = -900$ V ($\bar{\theta}_R \approx 3\pi$); minimal values of starting currents increased up to 30 μA for $J_{1S\min} = 60\,\mu$A. Variation of power versus pencil's current for the second zone in the case of single interaction of electrons with SHF field of OR ($M = 1$) is given in Table 4.

As it follows from the data presented in this Table, satiation of generation power is observed for $J_0 = 90 - 100\,\mu$A ($i_{\text{inp}} = 9 - 10$ A$\cdot$cm^{-2}). For the first generation zone maximal level of power in the domain of limiting values of current's density increases up to 360 μWt. When low-voltage oscillations of RGDR are excited ($U_0 = 1000 - 1500$ V), limiting current's density in both generation zones decreases up to 5–6 μWt what is comparable with the level of i_{inp} in the higher order generation zones when high-voltage oscillations are excited.

In RGDR, as well as in FGDR, variation of electron current yields electron frequency shift. In the case of single interaction generation frequency gradually falls with growing J_0 because of changing ratio of reactive and active components of interaction power characterizing the losses of average electrons' velocity (when they give energy back to SHF field of OR). In RGDR this ratio depends also on the angles $\bar{\theta}_R$, $\theta_{\text{C-A}}$ which change with growing electron current due to the influence of constant component of space charge on the electrons' phases in the domain of reflector and cannon's cathode.

Longitudinal magnetic field causes cyclotron pulsations of electrons' trajectories perturbing thus energy, frequency and starting parameters of RGDR.

For example, generation zones of RGDR has the displacement of 300 V with respect to U_R for the oscillation with U_0 =1500 V when induction of magnetic field varies from 0.4 Tl to 0.5 Tl. Here the amplitude and width of generation zones have negligible perturbations.

The power and width of generation zones change more essentially in the domain of small values of magnetic fields ($B \leq 0.2$ Tl). However, variation of power versus pencil's current remains almost the same and there exists the value $i_{\lim}$ corresponding to satiation of P. Decreasing of magnetic field up to starting level (generation breakdown occurs for $B = 0.1$ Tl) leads mainly to the growth of starting current and fall of generation power.

Let us analyze briefly spectral characteristics of RGDR operating in single-frequency excitation regime and compare them with the parameters of FGDR, reflection clystron and some other devices.

It has been established experimentally that minimal spectrum of the SHF signal of RDGR is distorted in the case of double electrons' flight. If the current coefficient $k_2 \geq 0.4$ and current's density $i_0 < i_{\text{lim}}$, electron frequency readjustment is carried out in discrete generation zones analogous to the ones of FGDR, reflection clystron and resonance lamp of reflected wave. The values of ΔF (at the level 0.1 with respect to the main spectrum's amplitude) and $\partial f(\partial U)^{-1}$ in the centers of generation zones of the considered devices are presented in Table 5 for the following operation characteristics: $\lambda \sim 4$ mm, $J_0 \approx 3J_S$; $p \approx 10^{-7} - 10^{-8}$ mm rt st; stability of feeding sources is $10^{-3} - 10^{-4}$ and pulsations do not exceed 15 μV; index of generation zone is $N = 8$.

It follows from experimental results that the quality of the spectrums of considered generators essentially depends on the steepness of frequency electron readjustment. FGDR has minimal value of ΔF and the width of spectral curve of RGDR grows in 1.5 times due to additional parasitic modulation of the phase of electron flow in the domain of reflector (owing to pulsations of constant voltage on the reflector).

Steepness of frequency electron readjustment of the lamp of reflected wave and reflection clystron has the growth of more than one order increasing thus the distortions in the spectrums of generation's signals. Parasitis modulation of HF oscillations is caused by the presence in the flow of chaotic (solid spectrum) and periodical (discrete spectrum) fluctuations. In the general case parasitis fluctuations are caused by hit effects, instability and pulsations of feeding voltages, secondary electrons, transversal oscillations of electron flow being focused by magnetic field, etc. This phenomenon weaker influences the quality of generation signals in the devices of diffractional electronics, if to compare them with the lamp of inverse wave and clystron, because in GDR there exitsts high Q-factor inverse communication of electron flow with the field of open quasi-optical resonator.

Results of experimental studies are evidence of practical possibility to obtain monochromatic and stable oscillations' spectrums in RGDR. Essential broadening of the width of spectral line of generation signals is observed for repeated electrons' flight in the regime of excitation of stochastic oscillations.

One should also point out that in RGDR with discrete generation zones it is realized qualitatively new phenomenon (in comparison with FGDR) of effective frequency and amplitude modulations of SHF signals for the reflection phase of inverse electrons' pencil. Analogous way of modulation is used in reflection clystron but in RDGR operating in the millimeter range it has higher output power and raised stability of generation frequency what is especially important for various practical applications. RGDR is characterized by the wide range of frequency mechanical readjustment (having the order 0.6–0.8 octave) and therefore it may replace five-six reflection clystrons.

6. Let us summarize the results of investigations of low-voltage (up to 15 kV) impluse GDR (IGDR) with particular emphasis of forming the rounding and bearing frequencies of both minimal (bounded by the time of oscillations' establishment) and long (tens of microseconds) duration of impulses.

Establishment of oscillations in IGDR are studied theoretically by solving nonlinear non-stationary equations of electronics [24, 33, 158–160]. Within the frame of assumption of weak nonlinearity the field's amplitude in single-mode GDR is determined by the equation

$$\frac{dF}{d\tau} - \alpha F + \alpha \frac{F^3}{F_{\text{st}}} = 0, \tag{7.6}$$

where $\alpha = \frac{\omega_0}{2Q}(\frac{J}{J_S} - 1)$ is the increment of oscillation's growth.

If to assume that field's amplitude grows up to the value F_{st} beginning from a certain initial value F_0 (which is determined in general case by fluctuation characteristics of electron flow), one obtains

$$F(\tau) = \nu F_{\text{st}}[1 - \nu^2(1 - e^{2\alpha\tau})]^{-0.5} e^{\alpha\tau} \tag{7.7}$$

where the coefficient ν determines the amplitude $F_0 = \nu F_{\text{st}}$.

The time of oscilations' establishment is given by formula [31]:

$$t_e = \frac{1}{\alpha} \ln \frac{\nu}{\nu_1}[(1 - \nu_1^2)(1 - \nu^2)^{-1}]^{0.5}, \tag{7.8}$$

it is assumed that in the process of establishment the field's amplitude grows from the minimal value $A_1 = \nu_1 A_0$ to the maximal value $A_2 = \nu A_0$ where A_0 is the amplitude of stationary oscillations.

If to neglect nonlinear effect in (7.8), this formula coincides with the one obtained in [31] for $\nu_1 = 0.1$, $\nu_2 = 0.9$.

It is confirmed numerically that nonlinear effects increase the time of establishment t_e approximately in 1.4 times. Duration of the process of establishment in GDR falls and steepness of the growth of field's amplitude of generation signal increases with growing pencil's current $(J \cdot J_S^{-1})$. For the what, in turn, yields the growth of t_e.

One can obtain generation impulse with minimal time of establishment (10–15 nsec in the four-millimeter range) and minimal vibration of the front side, if to vary pencil's current, accelerating voltage (synchronization parameter), type of excited oscillation and length of the space of interaction (size of field's spot on the grating). These parameters may be also controlled by appropriate variations of the shape of modulator's voltage impulse when competing oscillation is excited.

Let us consider formation of generation impulses in IGDR. From one side, the high Q-factor of generator's OR improves the quality of generation impulse (the monochromatic property) but from the other side it increases the duration of transfer processes of oscillations' establishment and back side of impulse signal limiting thus the possibilities to obtain short impulses (with $t_e < t_i$). Duration of rectangular impulse $t_i = (1-2)(\Delta f)^{-1}$ (where Δf is spectrum's interval containing most part of impulse's energy); if for OR $\Delta f = f_0 Q^{-1}$, then

$$t_i = (1-2)\frac{Q}{f_0}; \quad Q \approx \frac{1}{2} t_i f_0. \tag{7.9}$$

Hence, OR with high loaded Q-factor "makes way" with minimal distortions for only definite generation impulse (having the duration t_i with bearing frequency f_0). This assertion is idealized because in real GDR the parameters of HF impulse essentially depend on the way of interaction of electron flow with the field of OR.

Formation of rounding impulse of HF signal excited in OR by current's impulse is shown on Fig. 140 (here and in what follows lower impulse corresponds to the form of high-voltage one of the modulator's voltage U_I at the anode of diode electron cannon and upper impulse corresponds to generation signal). It is assumed that generation impulse is excited by the "top" of U_I-impulse having minimal inclination $\Delta U = 0$ and duration t_U. Then the duration of the front side of generation impulse is determined by the time t_e given by (7.8) and the back side – by the duration t_b of the process of relaxation oscillations (when generator is switched off).

In practice the front side of HF signal is usually identified with the growth of amplitude of rounding magnitude beginning from $0.1A_{\max}$ up to $0.9A_{\max}$ (we will call it the time of establishment of oscillation's impulse t_e). Hence, the total duration of the process of establishing the amplitude of generation impulse is determined by the sum $t_0 + t_e$ (where t_0 is the time of delay of generation impulse related to the U_I-impulse) which may be easily found from interrelation (7.6) if to assume that formation of generation impulse begins from the level $\nu = 0.01$:

$$t_0 = \frac{4.6Q}{\omega_0}\left(\frac{J}{J_S} - 1\right)^{-1}; \quad t_e = \frac{6Q}{\omega_0}\left(\frac{J}{J_S} - 1\right)^{-1}; \quad t_b = \frac{4.6Q}{\omega_0}. \tag{7.10}$$

One can find by means of (7.10) the top t_T of the impulse and the duration of HF signal t_I at the level $0.1A_{\max}$

$$t_T = t_U - \frac{10.6Q}{\omega_0}\left(\frac{J}{J_S} - 1\right)^{-1}; \quad t_I = t_U + \frac{4.6Q}{\omega_0}\left(2 - \frac{J}{J_S}\right)^{-1}\left(1 - \frac{J}{J_S}\right)^{-1}. \tag{7.11}$$

It follows from the obtained formulas that for the given resonator's Q-factor and bearing frequency f_0 the duration of generation impulse essentially depends on the ratio of working and starting currents: if $J_S < J < 2J_S$, the duration $t_I < T_U$; for $J = 2J_S$ $t_I \approx t_U$ and when $J > 2J_S$ the difference between the duration of generation impulse and t_U is approximately equal to the time t_b of establishment of relaxation oscillations. The shape of signal's rounding curve also changes when the value of $J \cdot J_S^{-1}$ varies what is connected with variation of the time of oscillations' establishment. Minimal duration of impulses is determined by t_e and t_b (when $t_T = 0$, the rounding curve of HF signal has the form of a triangle); for the long generation impulses ($t_T >> t_e$, $t_T >> t_b$) the period of time $t_I \approx t_U$. Hence, appropriate choice of the parameters $Q, \omega_0, J \cdot J_S^{-1}$ always allows to produce generation impulse with satisfactory shape of signal's rounding curve and it is confirmed by a number of experiments [31–33, 161]. However, excitation of high-quality impulse signal in GDR is possible only under definite additional conditions which may be stated experimentally.

It has been proved in the process of experimental studies that high-coherent impulses may be excited in OR of GDR when the duration $t_I > t_e$. The quality of generation impulses having either given or limiting duration depends on external factors connected with operation conditions of IGDR and internal physical processes prceeding in a real device. In particular, it is required definite shape of modulator's voltage impulses as well as focusing magnetic field, special feeding sources for the cathode of electron cannon bringing minimal distortions in the shape of U_I-impulses, definite value of pencil's current and other limitations.

Duration of the front side of generation impulses decreases and its steepness grows with increasing impulse current. If $J \leq 1.2J_S$, strong vibration of the front side is observed together with its instability in the millimeter range. For $J > 1.2J_S$ quality of the front side essentially improves. If generation impulses have the duration $t_I \geq$20 nsec ($Q = 1500 - 2000$, $f \approx 75$ GGz, $B = 0.4$ Tl, $J = 2J_S$), then $t_e \approx 15 - 20$ nsec what well agrees with calculation data.

Experimental investigations show that deviations of the shape of U_I-voltage impulse from the rectangular form may considerably distort SHF signal. In order to achieve minimal distortions one should provide fulfilment of the following requirements: fronts' steepness of modulator's voltage impulse is not greater than $U_0 t_e^{-1}$. Analysis of the oscillograms for the rounding curve of IGDR generation impulses [31, 161] enables one to state the conditions of minimal distortion for the 10 nsec signal, to follow its destruction in the case of "bad" shape of modulator's voltage impulse and excitation of the "package" of generation impulses (when modulator's impulse has the special form). Generation impulses in the "package" may have different duration and bearing frequency while minimal duration of partial impulse is bounded by the sum $t_e + t_b$. Entering oscillation of the package generation impulses is determined by the "slit" property of modulator's voltage impulses; the number of impulses in the package, their duration and bearing frequency may be chosen in accord with specific practical task.

Special models of IGDR with continuous pumping out of residual gases and different types od open electrodynamical system (OR with diffractional grating) were used in laboratory experiments. Let us describe operation stages of the typical model. Impulse strip electron flow with cross-section 5×0.2 mm^2 is generated by diode cannon fed by impulse modulator with external initiation which forms rectangular voltage impulses having the amplitude 1–15 kV and duration $t_U = 0.05 - 10\,\mu$sec. The time of growth and the period of fall U_I is chosen in such a way that generator is initiated only by the "top" of voltage impulse. Varying inclination of the upper part of U_I-impulse within the limits $\Delta U = 3 - 40V$ is stipulated in the interval t_U. These variations correspond to 0.03-0.8 per cent deviations from the amplitude value $\bar{U}_0 = 0.5(U_1 + U_2)$ or to $0.3 - 4$ V$\cdot\mu$sec^{-1} shifts with respect to the time of duration t_U. Here $\Delta U = U_1 - U_2$, U_1, U_2 are the voltages in the beginning and near the end of impulse's top U_I (Fig. 140). Oscillation of the movement of impulses U_I is changed by the feeding generator within the limits 0.5 – 10 KGz. IGDR has spherical-cylindrical OR with homogeneous diffractional grating. Loaded Q-factor of OR is 1500-3000.

Low-voltage (up to 3 kV) and high-voltage (up to 15 kV) IGDR have been studied experimentally. Here we will mainly analyze the properties of IGDR with high-voltage diffractional grating since such generators have much stronger generation power (up tp 1 kWt) and efficiency (about 6 per cent) what allows to investigate their characteristics in wider range of variation. The qualitative features of impulse parameters are typical also for IGDR with low-voltage periodical structure [33].

Generation impulses with duration $t_I = 0.02 - 10\,\mu$sec, high spectrum's quality, raised stability of bearing frequency and increased generation power have been obtained experimentally. For example, if $t_I \leq 0.5\,\mu$sec, the spectrum of HF signal has minimal distortions (caused by parasitic amplitude and frequency modulation of bearing frequency) and its width is approximately equal to the theoretical value $\Delta F \approx (1-2)t_I^{-1}$. At the same time, for $t_I \leq 0.5\,\mu$sec the duration of front and back sides of generation impulse noticeably influences its rounding curve.

If $t_I \leq 10\,\mu$sec, the output impulse power P_I and efficiency η of low-voltage generators are approximately in 1.5 times higher than in continuous regime, what is connected with decreased influence of ionization effects on impulse electron current. However, the levels of P_I (several tens of watt) and η (about 1 per cent) in IGDR with high-voltage diffractional grating are rather low nowadays.

Efficiency of diffractional radiation and, hence, generation power increase in high-voltage IGDR [32]: in IGDR operating at accelerating voltage $U_I = 7$ kV the output power for 0.1 μsec-impilses reaches 100 Wt and the efficiency – 2.5 per cent while for $U_I = 11$ kV the output power for the impulses with duration 0.5–5 μsec reaches 0.7 kWt and the efficiency – 8 per cent.

Let us consider nonlinear excitation of oscillations og IGDR. Unlike GDR operating in stationary regime, IGDR works within the period of time equal to the duration of impulse U_I. Therefore, the efficiency of interaction of electron flow with the field of OR as well as formation of impulse signal and the quality of its spectrum depend on many factors which are not distinctive for continuously operating GDR: duration and quality of the impulse U_I (initial conditions), time of establishment and attenuation of oscillations in OR, interrelations between duration of SHF impulse and its bearing frequency, "slit" property of impulses and other parameters.

One of the main features of IGDR is the shift of its output characteristics under the influence of dynamical (connected with variation of flow's parameters in the process of interaction with HF field during the time t_e) and statical (caused by the deviation of impulse U_I from the rectangular form) electrons' displacements in the space of interaction within the limits of the interval of time t_U when the impulse U_I, determining electrons' initial velocity, generates the anode voltage.

It has been established experimentally that both single-mode and multimode oscillations may be excited in IGDR. If the impulse U_I has approximately rectangular shape (inclination ΔU is minimal), the behaviour of P_I versus U_I in the generation zone qualitatively coincides with analogous characteristics of GDR operating in continuous regime. Rounding curve of long impulses and their spectrum have minimal distortions; the spectrum of short impulses broadens what points out to the presence of parasitic modulation in bearing frequency.

One can see on Fig. 141 that the growth of amplitude U_I leads, as well as in continuous regime, to the linear increase of the power and efficiency and then to "satiation" of P_I, η accompanied by oscillation's breakdown.

Let us consider the spectrum of generation impulses in GDR and state the influence of deterministic harmonic processes connected with changing parameters of electron flow on SHF oscillations [33, 161].

In accord with linear theory describing excitation of GDR, deviation of generation frequency for the principal space harmonic occurring beacause of changing velocity of flow's electrons may be found by the formula

$$\frac{\Delta\omega}{\omega_0} = \frac{L\omega_0}{8Q\beta c}\frac{\Delta U}{U_0}.$$

If to assume that changing electrons' velocity in the space of interaction causes linear variation of the amplitude and frequency during the time interval $t_I = \tau$ of rectangular impulse (the time of oscillations' establishment and the time of resonator's "ringing" are much less than the duration of the top of rounding impulse for SHF oscillations), the spectrum of such a signal may be registered in the form [33] which enables one to analyze the signal as a function of various generator's parameters. If generation impulse has only amplitude modulation, which yields non-symmetrical spectrum, the width of the latter approximately corresponds to theoretical value

$$\Delta F \approx (1 - 2)\frac{1}{\tau}.$$

In the presence of frequency modulation the spectrum of generation impulse has negligible distortions and its width is close to this value δF. If the condition providing the frequency modulation does not hold, growth of the inclination of $\Delta U(U_0\tau)^{-1}$ yields spectrum's distortions, it broadens and periodical oscillations emerge in the rounding curve.

With growing resonator's Q-factor the value of $\Delta U(U_0\tau)^{-1}$, providing minimal spectrum's distortions, increases. It is important for generating long impulses in IGDR. Such impulses considerably influence spectrum's quality and variation of electrons' velocity due to neutralization of volume charge by the ions of residue gases.

Spectrum's distortions were repeatedly observed in experimental investigations of the quality of impulse signals in IGDR. Analysis of oscillograms of the rounding curves and spectrums of signals excited in IGDR of the millimeter range ($\lambda = 3.8 - 5.6$ mm) for different values of m_a and ΔU allows to assert that for generation impulses with duration $t_I = 10\mu$sec when $m_a = (A_2 - A_1)(A_0\tau)^{-1} = 0$, $\delta U = 2$ V (A_1, A_2 are the initial and ultimate values of impulse's amplitude) oscillations in the spectrum are negligible but its width 0.6–0.8 MGz exceeds theoretical value ($\delta F = 0.1$–0.2 MGz). Growth of inclination of modulator's voltage impulse ($\Delta U \approx 20$ V) leads to noticeable spectrum's distortions: parasitic oscillations emerge and the spectrum broadens up to the width 3–5 MGz. Comparing the signals one can show that the growth of modulation coefficient m_a yields greater violation of spectrum's symmetry and its partial broadening. The spectrum of generation impulse essentially improves and its width becomes approximately equal to theoretical value $(1 - 2)\tau^{-1}$ when $\Delta U \approx 0.2$ V.

It has been established that in the case of appropriate choice of the parameters and excitation conditions impulses of SHF oscillations with duration $t_I = 0.15 - 10\,\mu$sec have negligible distortions, minimal amplitude modulation of the rounding curve ($m_a \approx 0$) and satisfactory spectrum's quality (without parasitic modulation); width of the spectrum is approximately determined by the formula for ΔF.

We may conclude that the dependence of impulse power and generation frequency versus impulse voltage has the approximately the same form as in continuous regime: generation power is non-symmetrical; with growing pencil's current the zone of frequency readjustment broadens and for the majority of excited oscillations it is observed sharp power's breakdowns close to zone's edges with displaying of hysteresis phenomena; frequency generation behaves like a growing linear function of (increasing) $\bar{U}_0$ with the steepness 0.1–0.2 MGz·V^{-1}.

However, it has been established in the process of experimental investigations that for some oscillations hysteresis phenomena are negligible and there exist the domains with small steepness ($f'_U \approx 0$) what is important for generating the impulses of SHF oscillations with minimal spectrum's distortions. In addition to this, for the majority of excited oscillations the spectrum of short generation impulses ($t_I \leq 1\,\mu$sec) has much less distortions than the spectrum of long impulses ($t_I > 1\,\mu$sec).

It is known that focusing magnetic field determines static form of the trajectory of electron flow in the space of interaction. Quasi-static flow's pulsations in the devices with long interaction may considerably influence spectral characteristics of generation signals if to take into account ionization processes.

The spectrum of generation impulses in IGDR has negligible variation at the level $P \geq 0.5P_{\max}$ and essential spectrum's distortions are observed in the domain of starting values.

The degree of influence of ionization processes on spectrum's quality in the devices with long interaction increases with growing duration of the impulse of electron current and the pressure of residual gases. Experimental investigations show that for generation impulses with duration $t_I \leq 10\,\mu$sec there are no noticeable spectrum's distortions when the pressure in the device grows from 10^{-7} mm rt st to 10^{-4} mm rt st.

3. Let us analyze autodyne operation regimes of GDR. For registering the variations of generation amplitude and frequency in this regime it is applied either external detecting by semi-conductor detectors [162] or internal detecting by internal electron detectors [163] (for measuring the increments of pencil's currents caused by the received signal).

Theoretical analysis of aytodyne regimes may be carried out with the help of equation describing the dynamics of single-mode generator influenced by the signal having the frequency close to generation frequency. Here it is assumed that oscillations in OR are excited by the electron flow and external signal through the slot connecting OR with the waveguide. This equation verified on the basis of solution to the problem of excitation of waveguides and resonators by given currents [164] with taking into account the influence of communication element has been obtained in [46]:

$$\frac{C_s(t)}{dt} - [i(\omega - \omega_s) - \tilde{\omega}_s]C_s(t) + \hat{\omega}_s C_s(t) = -\bar{J}_e(t), \tag{7.12}$$

where $C_s(t)$ is the field's amplitude of s-th resonator's mode; ω is the frequency of radiated signal; $\omega_s = \omega_s' - i\omega_s'' = \omega_s'(1 - 0.5iQ_0^{-1})$ is the complex eigenfrequency of the s-th mode. Correction to the imaginary part of eigenfrequency is caused by the influence of communication element, complex correction $\tilde{\omega}_s$ takes into account influence of reflected signal:

$$\bar{J}_e(t) = \frac{1}{2\pi}\frac{1}{N_s}\int_{\omega t}^{\omega t+2\pi}\int_{V_e} j(z,t)E_s(z)e^{i\omega t}dVd(\omega t),$$

where $j(z,t)$ is the current's density of electron flow; N_s is the norm of the s-th mode in OR; V_e is the volume occupied by electron flow; $E_s(z)$ is the function describing field's distribution of the s-th mode along the direction of propagation of electron flow.

Equation (7.12) coincides with interrelation obtained in [165] where it is assumed that reflected signal yields only the change of loaded Q-factor (Q_L^*) and generation eigenfrequency (ω_s^*) of the autodyne generator. Rigorous analysis of (7.12) confirms these assumptions and enables one to write down explicit formulas for the required parameters in the presence of reflected signal in GDR:

$$\omega_s^* = \omega_s\left(1 - i\frac{\tilde{\omega}_s + \hat{\omega}_s}{\omega_s}\right); \tag{7.13}$$

$$\frac{1}{Q_E^*} = \frac{1}{Q_E}\Big[1 + i\Gamma\cos\delta(t)\Big]; \tag{7.14}$$

$$\frac{1}{Q_L^*} = \frac{1}{Q_0}\Big\{1 + \sigma_0\Big[1 + i\Gamma\cos\delta(t)\Big]\Big\}, \tag{7.15}$$

where Q_E^* is the external Q-factor of GDR; $\delta(t) = \delta_1(t) + \phi_\Gamma$ is the total phase shift of reflected and generated signals; $\Gamma = |K_\Sigma||\Gamma_0|\sigma_0^{-1}$ is the relative reflection coefficient determining relative variation of external Q-factor caused by reflected signal; K_Σ is the coefficient taking into account total power losses of a signal on its way from OR to an object and back; Γ_0 is the coefficient of reflection from an object; σ_0 is the communication coefficient between GDR and the load.

Solution of equation (7.12) is sought in the vicinity of stationary state $C_s = A_0 + a$, $\omega = \omega_0 - \delta\omega$ when there is no reflected signal and has the following form in the regime of established oscillations

$$a(t) = -\frac{1}{2}K_a A_0 \cos(\delta - \psi_\tau); \tag{7.16}$$

$$\delta\omega(t) = \Gamma\left[\frac{\omega_s'}{2Q_E}\sin\delta - \frac{1}{2}K_a q\cos(\delta - \psi_\tau),\right], \tag{7.17}$$

where $\psi_\tau = \omega_D t_a$;

$$K_a = \omega_s' \sigma_0 Q_0^{-1}(p^2 + \omega_D^2)^{-0.5}. \tag{7.18}$$

is the internal coefficient of autodyne amplifying which gives the ratio of the amplitude of autodyne response a to the amplitude ΓA_0 of the reflected signal.

In the process of external detecting it is measured [165] the variation of generation power ΔP_L coming to the load which is caused by reflected signal

$$\Delta P_L \approx -P_{L0}\Gamma K_a^E \cos\delta, \tag{7.19}$$

where P_{L0} is unperturbed power coming to the load when there is no reflected signal.

Coefficient of autodyne amplifying in the case of external detecting has the form

$$K_a^E = \left(\frac{\omega_s'}{2Q_0} - p\right)\left(\frac{\omega_s'}{2Q_0}\right)^{-1} K_a = \frac{Q_E}{Q_0 + Q_E}. \tag{7.20}$$

Internal detecting presupposes registering the variation of total pencil's current caused by reflection signal. Current's increment at the output of the space of interaction may be represented in the form

$$\Delta J(t) = -\frac{1}{2}\Gamma K_j \cos\delta, \tag{7.21}$$

where

$$K_j = \left[\frac{\partial J}{\partial A}\left(\frac{J}{A}\right)^{-1}\right]\Big|_{A=A_0}$$

and K_a is the coefficient of autodyne acceleration related to current.

Sensibility and acceleration coefficient of GDR-autodyne with internal detecting may be increased if to choose operation regime providing the value of current's derivative with respect to amplitude to be much greater than unit.

Let us present the results of theoretical and experimental studies of the coefficient of autodyne acceleration and amplitude-frequency characteristics as functions of the parameters of GDR which allow to determine optimal operation regimes.

Relative generation amplitude A, durabiluty of limiting cycle p and starting current $k = J_S \cdot J_{S\min}$ versus velocity difference

$$b = \frac{v_0 - v_\Phi}{v_\Phi} = \frac{U_0 - U_s}{U_s}, \quad U_s = \frac{m}{2e} v_\Phi^2 \tag{7.22}$$

(corresponding to readjustment by accelarating voltage) are shown on Fig. 142, a.

Behaviour of the durabiluty of limiting cycle p is typical for GDR with small increase of energy per oscillation period, i. e., for Thompson generators [166]; it sharply falls near zone's edges and reaches maximal value in the point corresponding to the minimum of starting current in generation zone when $k \approx 1$. In addition to this, durabiluty of limiting cycle always falls with decreasing k and tends to zero when $k \to 1$.

Variation of the external coefficient of autodyne amplifying within the limits of generation zone when $\sigma_0 = 0.6$ for different values of k is shown on Fig. 142, b; it reaches maximal values at the edges of generation zone and in the vicinities of bifurcation points. Besides, this coefficient decreases in the center of generation zone with growing k and remains almost constant near the edges.

One may conclude that maximal sensibility to external factors together with maximal autodyne amplifying of GDR is provided when k is close to unit. Corresponding values of p are realized close to the edges of generation zone for arbitrary k and in the whole zone when $k \approx 1$.

With increasing communication coefficient σ_0 the loaded Q-factor falls what yields the growth of starting current. Hence, it is of practical interest to study the variation of the coefficient of autodyne amplifying K_a^E versus σ_0 when the ratio of starting and working currents are fixed. Corresponding curves are presented on Fig. 143 for different points of generation zone. As it has been pointed out above coefficient of autodyne amplifying K_a^E reaches its maximal values at the edges of generation zone (curves a, e); for $\sigma_0 > 0.2$ these values are much greater than at the right edge. The curve c corresponds to the minimum of starting current and the curve d – to the maximum of amplitude (and efficiency). In these points K_a^E gradually increases with growing σ_0 without satiation. The curve b describes variation of K_a^E in the low-voltage part of generation zone where $U_0 \geq U_s$ (to the left of the point of minimum of starting current). Here one observes satiation of K_a^E when $\sigma_0 \geq 1.0$ (closer to the point of minimum of starting current greater the value of satiation point) and its further decrease accompanying by growing degree of communication. The domain $\sigma_0 \leq 1.5$ corresponds to the satiation in high-voltage part of generation zone. Further increase of σ_0 is inexpedient in both cases.

As it follows from interrelations (7.16), (7.17) (7.19), amplitude and frequency characteristics of GDR-autodyne are periodical functions with respect to the phase of reflected signal and essentially depend on internal generator's parameters $(K_a^E, q, Q_0, \sigma_0)$ as well as coefficient of reflection from an object. Investigations of autodyne characteristics were carried out in the 8-millimeter range in the regime of external detecting. It turned out that the shape of amplitude and frequency characteristics essentially depended on the parameter Γ: being periodical for small Γ it became almost periodical with big amplitude. Such measurements enable one to determine the boundaries of applicability for theoretically calculated autodyne correction of the generation frequency. The level of working current, namely, the value JJ_S^{-1} also considerably influences the shape of amplitude and frequency characteristics.

One has to point out that owing to the lowest steepness of electron readjustment GDR is characterized by high operation stability close to the points of generation's breakdown what allows to reach essential levels of autodyne amplifying even when the standard feeding sources are used. The values of ΔP_S, P_{S0} measured in different points of generation zone enable one to calculate the coefficients of autodyne amplifying by means of the formula (7.19) for the chosen values k and σ_0. Behaviour of experimental curves for K_a^E coincide with theoretical data and essentially depending on k and the position of working point in generation zone; the ratio of its levels at the edges and in the center of generation zone reaches one-two orders depending on the working current of autodyne. With decreasing k the coefficient of autodyne amplifying grows and becomes maximal when k is close to unit.

Generation regimes in RGDR occur at such levels of starting currents which are in one-two orders less than in FGDR. Readjustment in FGDR may be carried out by means either of variation of reflector's potential (U_R) or of accelerating voltage U_0 (electron readjustment of frequency and power). It becomes possible to reduce the length of the space of interaction and the volume occupied by the generator in magnetic field, if to provide the conditions for repeated flight of electrons from the reflector to cathode and back. In spite of the fact that RGDR have less limiting powers in comparison with FGDR, the characteristics of RGDR stated above are of practical interest for their applications in autodyne operation regimes.

Experimental studies of RGDR-autodyne in the regime of external detecting were carried out with the help of specially constructed model having enlarged communication hole. It made possible to vary the level of communication with a load and to investigate its influence on generator's autodyne characteristics.

In RGDR every value of accelerating voltage U_0 of the zone of flight regime corresponds to the specific partial generation zone of reflector's potential. Stating discrete (increasing) values of U_0 and varying U_R one obtains a set of generation zones where the maximal power at first grows and then falls. Measurements of the amplitudes of autodyne responses in partial zones show that their behaviour considerably differs from the one observed in analogous regimes of RGDR. Namely, maximums of autodyne responses are situated not in the lateral zones (where the amplitudes remain constant for all values of U_R) but in the central zone (where two opposite maximums are observed) corresponding to the center of the zone of flight regime.

Characteristics of RGDR-autodyne in the case of external detecting differ from the ones of FGDR: with growing current the coefficient of autodyne amplifying in RGDR at first grows (while it falls in FGDR) and then falls. Variation of K_a^E as a function of current is much slower than in FGDR.

There are two more essential differences between RGDR and FGDR: 1) higher sensibility of the former to reflected signal at the voltages corresponding to the center of flight zone; 2) change of the phase of response to the opposite one within the limits of partial zone when reflector's potential varies.

RGDR are characterized by smaller coefficients of autodyne amplifying but, at the same time, by low starting currents, symmetrical generation zones and the possibility of non-current control of generation frequency and power. These features are very attractive for practical applications of RGDR in various radio systems.

Autodyne devices with external detecting being sensible and simple in manufacturing posess a number of limitations connected with the presence of semi-conductor detector in the scheme. One can avoid these difficulties and at the same time to provide wide dynamical range, possibilities of parameters' readjustment and high reliability of electro-vacuum SHF measurement devices if to use electron detectors [167, 168]. Their operation principle is based on changing characteristics of electron flow under the influence of received signal. Detector signal is separated out of the pencil's current either in detector's feeding chains or, for example, in the diode gap of electron cannon with the help of potential barrier of space charge.

Diode gaps with the potential barrier of space charge, namely, the modified diodes [169], provide high-quality detecting and exceed the sensibility of semi-conductor diodes. They find application as independent electron detectors of centimeter range in the O-type devices of millimeter range where they are placed at the pencil's exit out of the space of interaction.

One may conclude that internal detecting or auto-detecting should be considered as the more natural and universal regime for auto-dyne SHF-devices than external detecting.

There is one more advantage of electron detecting: if SHF electro-vacuum device operates as an element of radio system, the autodyne device may be transferred to the regime of auto-detecting by reducing the pencil's current up to the values less than the level of starting current; from the other hand, it may work in the regime of generation both as a mixer and a heterodyne. In what follows we will present the results of investigations of GDR with auto-detecting operating as an elecron detector and in autodyne regime.

If to analyze experimental data presented in this book, one may conclude that RGDR is the most suitable for the use in a system of interaction of electron detector and it is natural to choose diode construction with potential barrier of space charge as a registering element because it has high sensibility with respect to perturbations of the velocity in electron flows caused by external SHF-signals. They are characterized also by manufacturing simplicity, small size, heat and electrical reliability.

Let us discuss experimental data obtained for GDR of electron detector operating in before-generation and generation regimes when receiving the signal of external source.

In order to determine volt-ampere characteristics (VAC) of the diode gap, one can vary the collector's potential within the limits of 20 per cent of pencil's potential. VAC are measured by registering the current in collector's chain versus its potential for fixed J_a and U_0 (see Fig. 144, b). VAC of GDR-detector contains three parts which are typical for diode gaps with potential barrier of space charge: regime of current's delay, regime of current's limitation by space charge and regime of total current's transmission. Hence, the structure of VAC of the considered device is analogous to the characteristics of modified diodes and other types of electron detectors applied in electro-vacuum SHF devices where it is used similar diode gaps for separating the signals out of electron flows. They differ from the idealized VAC, which have higher steepness, by the "washing out" of initial electrons' velocities at the exit of electron-optical system forming the pencil.

Current sensibility of such detectors are usually defined as the difference between the gap's current registered in the presence of SHF signal and without the signal. However, such a method is time consuming especially in the series of many experiments.

Here we use slightly different approach when it is registered not the current of generator's response but the fall of its voltage on the load's resistance R_L in the collector's chain. It enables one both to measure voltage detector's sensibility, which may be used for re-calculation of current sensibility when internal resistance of the detector is determined, and to control the shape of detected signal.

One of important characteristics of all detectors is their dynamical range, i. e., the dependence of sensibility on the power of received signal. Behaviour of the amplitudes of responses versus signal's power is different in different points of VAC: they reach maximal values when collector's potential approaches U_0 and remain practically constant in the wide range of readjustment of P_c-signal close to the point corresponding to current's delay. Hence, one can carry out wide range electron control of detector's output signal, i. e., to realize automatic regulation of sensibility.

It is known that GDR is a device with magnetic focusing of electron flow and flow's movement along diffractional grating essentially depends on the level and degree of homogeneity of focusing magnetic field.

The considered device is placed in electromagnet which can change the magnetic field in the gap within broad limits. The voltage U_K^B of detector signal versus induction of magnetic field B in the center of the gap occupied by the device is measured by GDR-detector in the before-generation regime when it is adjusted to the maximal sensibility of external signal. Two-times growth of field's induction (from 0.3 Tl to 0.6 Tl) yields the same growth of U_K^B. This effect may be explained, if to take into account better transition of pencil's current in the channel of diffractional grating (it is confirmed by the increase of collector's current J_C).

The steepness of VAC of the diode gap increases with growing B because the structure of electron pencil improves due to reduced pulsations caused by transversal components of velocity at the exit of electron-optical system. The use of optimal constructions of electron cannons may noticeably raise the efficiency of such electron detectors.

The use of high Q-factor resonance system yields certain restrictions for the range characteristics of GDR recieving external signal and it is natural to assume that corresponding detector also posesses resonance properties.

Hence, the analyzed detector may serve as a basic element for creating narrow-band and sensible input receiving chains in the millimeter and submillimeter ranges with mechanical and electron readjustment.

GDR-detector is used in the cases when the source's signal is received (non-coherent receiving). Now let us consider the characteristics of GDR which receives its own signal (coherent receiving).

Let us describe the operation of such device in autodyne regime. When generation regime is established, a part of the power of GDR is radiated to free space. The signal reflected by an object causes the change of field's amplitude in OR and perturbs the spectrum of velocities of electron flow. Electrons are separated with respect to their velocities in the domain of virtual cathode (see Fig. 144) and the current produced by "information" electrons going to the collector is, in fact, the detector current containing information about the object.

It has been established that the maximum of voltage sensibility of GDR-electron detector is situated in the initial part of VAC where the generation of GDR emerges with maximal coefficient of autodyne amplifying.

High sensibility of GDR with signal's autodetecting is provided by this very phenomena, when the domain in the zone of generation having maximal amplifying coefficient corresponds to the interval of maximal sensibility of generator's electron detector. The potential of the considered GDR-autodyne varies in accord with the behaviour of sensibility reaching maximal values in the points of its growth and falling while sensibility decreases. It is known that the potential of a receiver-transmitter determines its operation distance. Analysis of the influence of regimes of the considered autodyne transmitter-receiver on its potential enables us to assert that within certain limits the operation distance of autodyne does not depend on generation power.

After comparing sensibility and potential with analogous characteristics of other devices, in particular, of reflection clystrons, which are the most sensible among all the investigated SHF generators, we will point out that in the regime of electron detecting clystron receives the signal measured in the regime of modulated oscillations under the condition of double exceeding the noises which is 58 dB lower than 1 MWt [170]. GDR-autodyne with 60-dB attenuation in the section registers the 1 μWt signal which is in 1000 times less than the one registered by the clystron; limiting values of the potentials of autodyne receivers-transmitters (10^6–10^8) given in [171] are exceeded by GDR almost in two orders reaching the level $A \approx 8 \cdot 10^9$.

Experimental investigations of GDR are evidence of the fact that such a device may work in autodyne regime both with external detecting and with autodetecting of reflected signals. However, direct comparison of these regimes is not easy since their characteristics are described in different units: in the first case they are given in the units of amplifying coefficient K_a^E and in the second case the are expressed via sensibility ($P_{\mathrm{T,min}}$, [Wt·Gz$^{-0.5}$]). In order to avoid this difficulty the comparison was carried on one device and by measuring the same parameters, namely, the voltages of autodyne responses of external U^E and internal U_K^E detectors as well as voltage sensibilities calculated with the help of these parameters.

Efficiency of autodyne operation regimes may be also compared by determination of their voltage sensibility using the methods for measuring the limiting sensibility of GDR in the regime of electron detecting and registering the power of external signal together with voltage of the response at sorresponding exit of GDR-autodyne.

§22. Package generators of diffractional radiation

Physical processes occurring in GDR together with different operation regimes and the ways of excitation are considered in §20, 21. They form the basis for working out the package generators, when the main purpose is to minimize their size and mass. Hence, the problems arise of providing operation stability and reliability of GDR as well as their compability and term of service under various conditions of exploiting. Such questions were not considered in the studies of dismountable models, where electron slow was usually focused by an electromagnet with big storage of induction of magnetic field and the space of interaction.

Manufacturing technology for constructing the diffractional structures and mirrors of OR is of special importance for creating the package GDR of the millimeter range and especially of its short -wave part. The number one problem here is the construction of reliable electron-optical systems being able to generate thin strip electron flows with the thickness less than 0.1 mm and high current's density [172] (greater than 10 A·cm^2). The material of magnetic system and its type considerably influence the output parameters of GDR [173].

Elaboration of the package GDR posessing the given properties (parameters) is completely based on the previous physical investigations and it becoms possible only if one uses simple explicit methods for calculating the main parameters. Success of the work is determined by a series of specific recommendations concerning the choice of optimal interrelation between varios characteristics of GDR and by availiability of simple methods suitable for general estimation of the efficiency of measurements of the main GDR output characteristics.

The behaviour of generator's parameters with respect to working current, wavelength, accelerating voltage and communication coefficient with a load has been established.

Numerical optimization of package GDR does not always give the desired results for some of its parameters because it is connected with the use of simplified assumptions concerning physical processes in GDR and inaccurate initial data. Very seldom the reason for the failure of numerical optimization is in unproper of the applied technique. Hence, the of principle computational problem here is elaborating truthworthy requirements for optimal interrelations between the main external and internal parameters of package GDR. Such requirements are necessary for replacing compliciated numerical simulations by simple computational procedures providing guaranteed accuracy. It is natural that the obtained simplified explicit analytical expressions should be treated very carefully. They must be permanently controlled by rigorous theoretical methods and by the possibilities to apply them for the concrete device.

1. In the package GDR applied in the millimeter range diffractional grating is often placed along the direction of propagation of electron flow non-symmetrically with respect to field's caustics. Sometimes such situation is arranged artificially, for example, in order to increase generator's efficiency by shortening the grating's length from the side of collector. Random shift of diffractional gratingwith respect to the axis of OR may also occur, for example, when the mirrors are warped .

It is purposeful to consider the efficiency of energy exchange between $F_0^{\max}$ and $F_1^{\min}$ in GDR operating under the following condition: amplitude's distribution of SHF field along the direction of electron flow corresponding to the principal and the first higher order oscillations are given by the Gauss-Hermit function with respect to the distance (shift) between diffractional grating and the center of field's caustics. Here we assume that equivalent reflection surface of diffractional grating coincides with smooth surface of resonator's mirror, i.e., phase inhomogeneity is compensated and therefore amplitude's distribution of resonance field is not perturbed when the grating changes its position in OR.

Variation of the amplitude of resonace field in OR along the electron's movement is described by the functions of Gaussian wave pencils.

Explicit expressions for the power of flow's interaction with synchronous field of diffractional grating versus its y-coordinate are obtained in [174] for the first and second types of amplitude's distribution of resonance field. These expressions have the form $P = \zeta F_n(\alpha, \phi, \phi_p)\tau r^{-1}$ where ζ is the coefficient depending on the parameters of electron flow and OR; F_n is the efficiency function of energy exchange whose maximal value $F_n^{\max}$ depends on τ - distance between the grating and origin of resonator's caustics, $\alpha = L(2r)^{-1}$, L is the grating's length, r is the radius of the field's spot of the principal oscillation.

The efficiency of energy exchange reaches its maximum when diffractional grating is placed symmetrically in the field of Gaussian wave pencil. Shift of the grating to the left or to the right yields the fall of efficiency and the growth of the optimal value of synchronization parameter.

One can find the optimal value of starting current by the known function $F_0^{\max}(\alpha, \phi\mathrm{opt}, \phi_p, \tau r^{-1})$. The growth of starting current for the grating which is shifted in the caustics in the case of Gaussian field's distribution, when $\phi_p > \phi$, is determined by the formula

$$J_S = \frac{J_S^{(0)}}{\nu(\alpha, \tau)},$$

where

$$\nu(\alpha, \tau) = 6.25 \cdot 10^{-2}[\mathrm{erf}(\alpha + \tau r^{-1}) + \mathrm{erf}(\alpha - \tau r^{-1})]^4$$

$J_S^{(0)}$ is the starting current of GDR with diffractional grating for which $L \geq 4r$ and $\tau = 0$. This factor may be naturally replaced by the experimentally measured magnitude.

Let us consider the behaviour of $F_1^{\max}$ choosing $\alpha = 1$. When $\tau r^{-1} \leq 0.6$ the function $F_1(\alpha = 1, \phi, \phi_p)\tau r^{-1}$ has two intervals of variation of ϕ where it is possible effective energy transmission to SHF field ($F_1 > 0$). The first zone corresponds to the negative values of synchronization parameter ϕ and the second zone – to its positive values, what is characteristic for the generator operating at higher order oscillations when the phase shift of SHF field lies in the space of interaction. If $\tau r^{-1} > 0.6$, there are no field's variations on the grating and the function $F_1(\alpha, \phi, \phi_p)\tau r^{-1}$ changes its usual form. The function $F_1^{\max}(\alpha = 1, \phi_o pt, \phi_p)\tau r^{-1}$ is symmetrical with respect to τ, since in the linear regime active power of interaction of electron flow with synchronous field of diffractional grating is proportional to the difference of squared values of the moduli of field's spectral density determined for those spectral wavenumbers which correspond to the cases of exact synchronization with either SWSC or FWSC in electron flow. When the peremeter of space change $\phi_p > 2\pi$, the efficiency of energy exchange may be increased by choosing the value of τ which provides displacement of the phase shift of SHF field out of interaction domain.

Experimental data obtained for the devices constructed on the basis of GDR with diffractional grating placed non-symmetrically in resonator's caustics well agree with calculation results. It opens the way apply these theoretical formulas for the design of various modefications of the package GDR operating in millimeter and submillimeter ranges.

In such generator starting current's density $i_s \approx 1 - 10$ A·cm^2 grows proportional to generation frequency. In order to provide effective work, the density of working current in GDR must be in several times greater than the density of starting current what may be achieved by decreasing the distance $d_{\mathrm{C-A}}$ between cathode and anode up to the level $d_{\mathrm{C-A}} < 1$ while the working voltage varies within the limits $U_0 \approx 2 - 4.5$ kV.

However, transveral components of electrons' velocities grow with decreasing $d_{\mathrm{C-A}}$ and, hence, one has to raise induction of focusing magnetic field in order to reduce radius of electrons' pulsations. Induction may reach the level $B \approx 0.3 - 0.6$ Tl and the average cyclotron wavelength $\bar{\lambda}_a$ becomes less than $d_{\mathrm{C-A}}$.

It turns out that starting current has oscillation behaviour with respect to the induction of magnetic field. If to assume that electrons move with uniform acceleration, the calculated pulsations' wavelength in accelerating gap will be equal to $\bar{\lambda}_c \approx 1.51$ mm. The first minimum of starting current is observed when induction of magnetic field in the gap's center reaches the level $B \approx 0.37$ Tl and it is developed when pulsations' wavelength approximately coincides with the length of gap, i. e., when $d_{\mathrm{C-A}} \approx \bar{\lambda}_c$. Optimal length of the gap anode-cathode may be determined by the formula

$$d_{\mathrm{C-A}} \geq \frac{1}{2}\frac{4\pi}{B}\sqrt{\frac{U_a}{2em^{-1}}} \approx 0.0106\,\frac{\sqrt{U_0(B)}}{B},$$

where $d_{\mathrm{C-A}}$, [mm]; B, [Tl]; e, m are the charge and the mass of electron, when $\bar{\lambda}_c$ is equal to one half of cyclotron wavelength in the drift space.

The formula

$$B = B_{\mathrm{opt}} \approx (2\lambda)^{-0.5}$$

where B, [Tl]; λ, [mm], gives interrelation between the induction of magnetic field and minimal wavelength which is optimal for creating package devices with minimal mass.

In order to obtain raised output power and required range of readjustment one uses nagnetic systems with 10 per cent higher field's induction: $B_p \approx 1.1 B_{\mathrm{opt}}$ what yields 35 per cent growth of the mass of nagnetic system.

Experimental data presented in [175, 176] are evidence of considerable fall of resonator's Q-factor and essential perturbations of the field's amplitude-phase distribution due to the ohmical losses of diffractional grating placed in the caustics of OR in GDR. In fact, ohmical losses and amplitudes of the space field's harmonics in OR determine the majority of generator's characteristics.

Reflection coefficient and amplitudes of space harmonics (with taking into account the losses) have been obtained in [177, 178] on the basis of rigorous solution to the problem of diffraction of H-polarized wave on the grating made of rectangular bars.

It is of special interest to study the behaviour of phase of the wave reflected by diffractional grating since the character of its variation determines coordination of grating's structure with the smooth mirror of OR (covered by this grating). This problem is treated in [134] but without considering the influence of ohmical losses.

Calculation and analysis of the phase of reflected wave with the help of obtained complex reflection coefficient, which takes into account ohmical losses, allows to specify the coordinates of equivalent reflection surface and, consequently, to determine grating's position on the surface of smooth mirror which guarantees uniform level of the phase of reflected wave over the whole mirror's surface. These results enable one to reduce diffractional losses in the oscillation system of GDR. Maximal correction occurs in the case of a grating with small filling coefficient having not very deep lamellae (depth is close to the quarter of wavelength).

The density of power's flow in the center of field's spot for the principal oscillation of OR is inversely proportional to the squared value of spot's radius (for the fixed value of power's flow going through transversal cross-section). Hence, if the communication hole of GDR with a load is placed in resonator's origin, the absolute value of the density of power's flow falling on the hole will grow with shortening the wavelength in addition to the increase of relative power passed through the hole. This effect yields considerable change of the communication coefficient (proportional to λ^{-2}) within the rangc of combined readjustment.

If to place communication hole in the lateral zones of OR, one can compensate the "central" growth of power's flow in a certain point situated at the distance y_0 from resonator's origin by variation of the flow using its lateral fall and by the decrease of the radius r of field's spot. Optimal position of the communication hole determined experimentally is given by the value $y_0 = r$.

5. It has been established already for the first models of GDR [22] that in order to provide constant communication of the generator with the load the output power must be proportional to the difference between working and starting currents: $P_{\text{out}} \approx k(J_w - J_S)$.

If working currents are not very strong, $J_w \leq 3J_S$, the total power of diffractional radiation and, hence, its electron efficiency is proportional to the energy stored in OR (or to the ratio of loaded Q-factor and resonator's Q-factor without a load $Q_L Q_0^{-1}$) and one can write down the formula for output power as the function of communication coefficient of loaded generator

$$P_{\text{out}} = 2\eta_e U \frac{\sigma}{\sigma+1} \frac{Q_L}{Q_0} \Big(J_w - J_S \Big) = 2\eta_e U_{\text{opt}} \frac{\sigma}{\sigma+1} \Big(\frac{J_w}{\sigma+1} - J_{S\min} \Big), \quad (7.23)$$

where η_e is the coefficient which does not depend on σ and which coincides with maximal value of electron efficiency; $\frac{\sigma}{\sigma+1}$ is the resonator's contour efficiency; $Q_L Q_0^{-1} = \frac{1}{\sigma+1}$; $J_{S\min} = J_S(\sigma+1)^{-1}$ is the starting current for $\sigma = 0$; U_{opt} is the accelerating voltage for the maximal output power $P_{\text{out}}(U)$. Then optimal value of communication coefficient providing maximal output power under the condition of given working current is determined by the expression

$$\sigma_{\text{opt}} = \frac{J_w - J_{S\min}}{J_w + J_{S\min}}, \quad (7.24)$$

which repeats experimental interrelation obtained earlier.

Optimal value of communication coefficient should be increased if the working current grows. For arbitrary constant value of this coefficient the dependence of output power versus working current is almost linear and with growing communication coefficient the inclination angle of straight lines, i. e., the value $k_1 = 2\eta_e U\sigma(\sigma+1)^{-2}$ such that $P_{\text{out}} = k_1(J_w - J_S)$, at first rapidly grows and then gradually reaches its maximum for $\sigma = 1$.

In accord with theoretical results, the steepness of frequency electron readjustment measured in the maximum of output power is inversely proportional to resonator's loaded Q-factor:

$$\frac{\Delta f}{\Delta U} = \frac{1+\sigma}{2Q_0} \pi k_m p N_0 \frac{f_0}{U_{\text{opt}}}, \quad (7.25)$$

where k_m is the coefficient depending on the density of space charge determined by the type of oscillation; p is the number of space harmonic, usually $p = 1$; N_0 is the number of periods in the field's spot of principal oscillation; f_0 and U_{opt} are the working frequency and accelerating voltage in the maximum of output power.

It follows from the results of our experimental studies that with growing working current J_w the range of electron readjustment of GDR asymptotically approaches the strip of transmission of OR. Assuming that this dependence is exponential, we will represent resonator's Q-factor for arbitrary values of communication coefficent and working current as a function of the range of electron readjustment at the level 0.5 P_{max}:

$$Q_0 = \frac{f_0}{\Delta f_{0.5P_{max}}}(\sigma + 1)\left[1 - e^{(1 - J_w J_S^{-1})}\right]. \qquad (7.26)$$

There is usually no interaction between different types of oscillations in GDR, since the oscillations with low transversal indices ($m + n < 5$) are mainly excited in OR. Neighboring oscillations are excited at the frequencies spaced at the intervals of several hundreds MGz and the generation zone of each oscillation is tens of MGz [3]. Covering of one of the mirrors by strip diffractional grating placed along electrons' movement yields both additional spectrum's rarefying and destroying of degeneration for oscillations with equal sums of transversal indices ($m + n = const$) [3].

However, the presence of two levels of reflection surfaces on the composed mirror may cause reduction of the intervals between frequencies when different type of oscillations are excited in OR. Here we will investigate the influense of intertype osillations' interaction having different sums of transversal indices ($m + n \neq const$) on excitation of package GDR (where OR with diffractional grating is used as an oscillating contour) operating in the short-wave part of millimeter range.

Degeneration of oscillation frequently occurs in such GDR. One has to point out that it is difficult to separate degenerated types of oscillations having equal distribution along periodical structure, i. e., when $n_1 = n_2$ (for example, TEM_{00q} and TEM_{20q}). In this way oscillations with different longitudinal indices (like TEM_{21q} and TEM_{40q}) are more convenient for the study of intertype communication and degeneration and the latter may be either established or destroyed for them by means of appropriate generator's readjustment in magnetic field.

The steepness of electron readjustment falls in the domain of interaction for oscillations excited in generator's OR. This effect may be used for creating the generators with raised frequency's stability. Interaction between resonator's oscillations may have the form of auto-modulation when GDR is automatically switched from one oscillation to another; such gradual transference without the breaks of power and frequency considerably broadens the zone of electron frequency readjustment.

2. The following problems must be solved in the process of design and construction of package GDR:

– optimization of resonator's parameters and appropriate choice of OR and diffractional grating in order to provide maximal intensity of HF field in the domain of movement of electron flow;

– working out reliable electron-optical system providing required current's density and quality of electrons' focusing in a thin flow moving close to diffractional grating;

– possibility to apply modern manufacturing technology for diffractional grating as well as high quality mirrors' processing which guarantee the accuracy of mounting generator's units.

Creation of package devices having minimal mass may be conditionally divided into three independent stages: construction of magnetic focusing system with minimal mass providing the given value of induction of magnetic field in the air gap; determination (in the given frequency range) of the optimal level and distribution of the induction of magnetic field in the gap where the device will be placed; establishing optimal dimensions of the device in the direction of propagation of electron flow (the latter is usually reduced to the calculation of mirrors' apertures and the length of effective space of interaction).

Optimization of package GDR usually begins from the determination of size L_{ef} of the effective space of interaction. The influence of the dimensions of this space on mass and size of package devices has of principle character (they are proportional to L_{ef}^3).

At the same time, optimal length of this space is almost constant with respect to the wavelength. For example, its registered value is L_{ef} =10–12 mm for the electron flows having equal energy parameters (accelerating voltage 2–4 kV, working current ~0.12 A, current's density ~30 A·cm^2). It allows to apply these results in wider frequency ranges as well as to use them for the design of package devices with magnetic focusing systems having equal gap between poles.

Since optimal value of the induction of focusing magnetic field is proportional to $\sqrt{\lambda}$, GDR is not very sensible to the increase of magnetic field and there exists additional possibility for optimization by carrying out variation of the distance between cathode and anode, one may use the limited set of magnetic focusing systems in the whole millimeter range.

Indication of the optimal level of starting current $J_{S\text{opt}} = (1 - \sigma_{opt})J_w$

where σ_{opt} is the optimal value of the coefficient of communication between GDR and the load gives simple and reliable relationship for determining L_{ef}.

One uses all possible ways for the reduction of starting current. However, it is necessary to optimize L_{ef} and σ in order to reach maximal output power, for example, to increase σ and to reduce the length of the space of intercation if $J_S << J_w$, since efficiency of the device is inversely proportional to L_{ef}.

For the principal oscillation L_{ef} is approximately equal to the size of field's spot: $L_{ef} \approx 2r_y$. Assuming that L_{ef} is given one can find the curvature of spherical mirror along the space of intercation

$$R_{sp} \approx L_{ef}\left[\left(\frac{\pi}{4}\frac{L_{ef}}{\lambda}\right)^2 + 1\right].$$

Here we use quasioptical approximation for the radius of field's spot r_y assuming that distance between the mirrors $D \approx L_{ef}$ (one should also satisfy contradictory requirements: D must be maximal to provide the high Q-factor and, at the same time, this value must be small enough to sufficy the condition $k_{fd} \sim 1$ where k_{fd} is the coefficient describing the level of diffractional radiation in OR) and the time of inverse communication $t_{ic} \approx 2Dc^{-1}$ (t_{ic} must be much less than the time of electrons' flight through the space of interaction, $Dc^{-1} << L_{ef}v_0^{-1}$).

In order to reduce resonator's diffractional losses to the level which is in two times lower than its ohmical losses, the mirrors' aperture must have the dimensions $2a_y \approx 2L_{ef}$. Quantitative estimate of diffractional losses may be approximately found with the help of graphs presented in [182].

Aperture and curvature of the mirrors in transversal direction may be determined (as well as the starting current) if to take into account real field's distribution in OR with one of the mirrors partially covered by diffractional grating [25]. Field's distribution which is closest to experimental data may be obtained, if to take the normalized polynomials given in [183] instead of normal Gauss and Hermit polynomials used in the known expressions for this magnitude.

Since TEM_{20q} oscillation with two transversal field's spots (two zeros) is the most wide-spread type of oscillation in OR of GDR, the size of central spot should be reduced in order to decrease diffractional losses, i. e., field's zeros must be shifted to grating's boundary. If the field's intensity is given by normal Hermit polynomial and Gauss function, the width of central spot of

TEM_{20q}-oscillation is equal to the radius of field's spot (r_{0x}) of the principal (TEM_{00q}) oscillation measured in the transversal direction.

If normalized Hermit polynomials are taken instead of normal ones, the following relationship should be fulfilled: $b = \sqrt{2}r_{0x}$ where b is the width of diffractional grating. Then the curvature of astigmatic mirror in the direction perpendicular to the movement of electron flow is given by the formula

$$R_{sp} \approx D\left[\left(\frac{\pi}{\lambda D}\frac{b^2}{2}\right)^2 + 1\right],$$

and hence $R_{cx} = R_{cy}$ if $2b^2 = L_{sp}^2$.

Electron flow must give back corresponding amount of energy when flying through the space of interaction to provide the required efficiency of GDR. If to assume that average electrons' velocity at the exit of the space of interaction is equal to phase velocity of synchronous harmonic of diffractional grating with which they interact in the thin layer of electron flow situated close to the grating, one may determine the electron efficiency η_e

$$\eta_e = \frac{\Delta U}{U_a} \approx 2\frac{v_0 - v_\Phi}{v_0},$$

where $v_0 = \sqrt{2eU_a m^{-1}}$ is the velocity of electrons at the entry to the space of interaction; e and m are the charge and the mass of electron.

General efficiency of GDR is given by the formula

$$\eta = \eta_e \eta_c \eta_q,$$

where $\eta_c = \frac{\sigma}{1+\sigma}$ is the contour efficiency; σ is the communication coefficient between GDR and the load; η_q is the coefficient which determines feeding of OR by diffractional radiation. Thus, for $\eta_q = 1$

$$\eta_e = \frac{\Delta v}{v_0}\frac{2\sigma}{\sigma + 1}.$$

When current's density i =20–30 A·cm^{-2}, the length of plasma wave becomes less than the length L_{ef} of the effective space of interaction. Therefore, maximal output power is observed in the case of phase raid of grouped electron flow and phase-synchronized harmonic along the whole space of interaction for $\phi = 2.5\pi$ what yields relationship

$$l = 0.4 L_{ef} \eta \frac{1+\sigma}{\sigma}$$

for the grating's period l versus L_{sp}.

For the given wavelength λ accelerating voltage is given by the formula

$$U = 2.56 \cdot 10^5 \left(\frac{l}{\lambda}\right)^2 \left(1 + \eta_e \frac{1+\sigma}{\sigma}\right), [\mathrm{V}]$$

One may conclude that the coefficient determining the use of electron flow is assumed to be close to unit since the given degree of the brake of electrons must be provided in the thin layer close to diffractional grating. However, it is necessary to take into account this coefficient with respect to both width and thickness of the flow in order to calculate the starting current.

The optimal value of L_{ef} may be tested if to determine starting current for the given parameters of OR, diffractional pencil and electron flow. Experimental investigations give more precise value. If L_{sp} is chosen correctly, then $J_S \approx (1-\sigma)J_w$.

Mastering of the millimeter range is connected with creating small-structure electron optics together with generating strong magnetic fields having required distribution of the tensity in the working volume.

Each constant magnet is a source of magnetic field with distributed parameters. Hence, rigorous computations of the field generated by magnetic elements is possible only with the help of the equations of electromagnetic field theory. Application of such rigorous approach for modeling magnetic focusing systems of complicated structure is connected with considerable difficulties. That is why they often use either simplified methods giving low accuracy of the results or computer simulation.

During the last decade essential decrease of mass and size of magnetic focusing systems has been achieved owing to the use of energy consuming rare earth materials with constant magnetic property [173]. It opened the way for practical realization of the required distributions for the vector of magnetizing of magnetic material and, hence, for their optimization.

The new family of magnetic focusing systems with minimal mass and size has been created on the basis of the method [185, 186] of synthesis of magnetic focusing systems for electro-vacuum SHF devices, in particular, for package GDR. Such magnetic focusing system is made of rare earth magnets (samarij-cobalt) having maximal energy product $(\mathrm{BH})_{max} \approx (130 - 140)$ kDj$\cdot$m^{-3} amd are intended for focusing long electron flows in the regime of pencil's limitation by a high-tensity magnetic field on the cathode. Each type of magnetic focusing systems applied in GDR has optimal configuration for the directions of vectors of magnetizing related to the centers of working gaps and, consequently, the maximal efficiency.

Main characteristics of some experimental samples of magnetic focusing systems for the package GDR which have been created in the Institute of Radiophysics and Electronics of the Ukrainian Academy of Sciences are given in Table 6.

Small-size magnetic focusing systems are applied in the package GDR operating in the millimeter and submillimeter ranges for the focusing of electron flows with high current's density. They provide 2 per cent limitation for the radial inhomogeneity of the longitudinal component of magnetic field in the 4-millimeter working gap. Their efficiency is in 1.6–2 times greater than the efficiency of the systems consisting of either axially or radially magnetized blocks. Magnetic focusing systems with more than five optimal directions of magnetizing vector may have the efficiency which is even in five-eight times greater for definite dimensions of the gap. Small size and magnetic compability of magnetic focusing systems provided by their weak scattering fields enable one to use GDR packed in them as elements of radio apparatus.

Creation of package GDR in the whole millimeter range having low accelerating voltages ($U_0 = 2 - 4$ kV) is inevitably connected with the solution to a number of problems. Shortening the range of working wavelenghts (for the given voltage) causes decrease of the period of diffractional grating what, in turn, yields the growth of ohmical losses and focusing magnetic field, simplification of manufacturing technology, lowering of the coefficient determining the use of electron flow, and, hence, the necessity to apply thiner and denser flows.

Certain difficulties occur in the design and construction of package GDR in the long-wave part of the millimeter range where the wavelength is greater than grating's width and comparable with L_{ef}; here diffractional losses strongly increase and resonators' Q-factor does not exceed 1000. In addition to this, calculations of open electrodynamical structures with the help of quasioptical approximation become impossible.

One of the main problem arising in the modeling and design of package GDR is the creation of effective oscillation system and optimization of its parameters.

Increasing of the coefficient describing transformation of SPR to resonator's eigenoscillations and field's intensity in the domain of the movement of electron flow is carried out by means of astigmatic mirrors when pressing the field of OR in the direction perpendicular to the flow is accompanied by keeping the given length of the effective space of interaction.

Composed astigmatic mirrors are applied in GDR operating in the short-wave part of the millimeter and submillimeter ranges. They provide increase of the length of the effective space of interaction up to the value $L_{ef} = \frac{3}{4}L_{tot}$, where $2L_{tot}$ is the total length of this space, when the mirror is made of three surfaces. The set of package GDR has been worked out on the basis of single construction (Fig. 131) having the size of 40 mm in the direction of force lines of focusing magnetic field which completely cover the range of wavelengths 0.8–8 mm with the help of 11 devices (see Table 7).

Typical behaviour of output power and working current versus generation frequency (accelerating voltage) in the range of combined readjustment is shown on Fig. 145. It has up to 7 dB power falls in more than 20 per cent range of readjustment due to reflections of waves lighting out of OR from the walls of generator's vacuum cover and external influence of resonances at slow waves to which a part of energy stored in OR is transformed in the vicinities of inhomogeneities. The most powerful oscllations are excited at the frequencies whose wavelengths are at least in five times greater than grating's depth (Fig. 145, a). The 2-millimeter GDR has less power fall than the 4-millimeter one in the whole range of readjustment [187] owing to the use of communication device with wider band in the form of open waveguide placed in the lateral zone of OR (instead of wedge-type waveguide placed near the mirror's origin in GDR-4 and GDR-6) and greater number of excited oscillations in the equal interval of variation of the distance D_m between mirrors.

Oscillations at longer waves are excited for smaller D_m; with shortening the wavelength the most powerful oscillations are excited for greater D_m (see Fig. 146) while the radius of field's spot remains almost constant both in transversal and longitudinal directions. The depth h of grating's cavities (in the case of a comb) varies in the interval $0.17\lambda - -0.21\lambda$. Lowering of working current (Fig. 145, b) leads to the decrease of the range of electron frequency readjustment and the level of output signal; working frequency grows but its steepness has no noticable changes.

Let us consider in details the dependence of output power and frequency on working current (Fig. 145, c): for the power it is almost linear when the voltage is optimal for every value of the current; frequency gradually grows with increasing working current. If accelerating voltage is constant, the frequency decreases with growing working current and the power noticeably deviates from the linear function due to the shifts of chosen voltage from its optimal value for the given working current (when working point moves along the zone of electron readjustment).

3. Application of OR with fixed distance between the mirrors allows to provide short-time friquency stability in GDR with taking the advantage of reduction of the diffractional losses by partial "closing" the resonators what is especially important in the long-wave part of millimeter range. Besides, one can carry out stable control of the distribution of field's amplitude along the space of interaction. With this purpose GDR containing various modifications of OR have been investigated.

Experimental results show that GDR with symmetrical OR is the optimal construction. However, further reduction of the gap occupied by a generator in the magnetic focusing system leads to the growth of starting currents and to the fall of output power (for fixed working current and field's tensity). In this way it is purposeful to study generators with non-symmetrical OR in order to elaborate the conditions for decreasing the gap and total mass of the device as well as for improving generator's parameters: reduction of starting currents and increase of output power without changing generator's mass and size. Strong dependence of these parameters on the character of field's distribution along the direction of movement of electron flow in the spase of flow's interaction with HF field opens the way for optimization of GDR with non-symmetrical OR.

Analysis of experimental data obtained for the 2-millimeter GDR with reduced size and fixed distance between mirrors enables us to make the following conclusions:

– it is possible to use OR made of two identical cylindrical mirrors;

– in order to optimize the field's amplitude one of the mirrors is fixed at a certain angle to resonator's axis; the wedge is soldered into the mirror or the cavity is cut in place of the wedge to provide the required field's fall towards the collector and shift of the field's spot along resonator's axis;

– characteristics of GDR containing OR made of cylindrical and spheroidal mirrors may be considerably improved due to optimization of the amplitude of field's distribution along the space of interaction (with respect to efficiency) by placing spheroidal mirror at the appropriate angle to resonator's axis.

Raise of efficiency enables one to reduce the size of the device without violating its main characteristics. At the same time, if size, mass and output power are constant, the increase of generator's efficiency due to optimization of field's distribution allows to reduce scattered power along the space of interaction and, consequently, to lower the requirements for the cooling system.

4. The main factors causing long-time instability of frequency and power in GDR are the drift of the voltage and current of electron flow, pulsations of feeding voltage and the dependence of distance between the mirrors on the temperature of a device [188]. Variation of frequency is very sensible to the distance between mirrors while the power - to the working current.

Output power of GDR grows almost linearly with growing working current:

$$\Delta P = P \frac{\Delta J}{J_w - J_S}$$

and $\Delta P \cdot P^{-1} = 2\Delta J \cdot J^{-1}$ for $J_w = 2J_S$.

Hence, one must provide 0.5 per cent accuracy of the level of working current (for example, equal to 100 μA) in order to support the output power with 1 per cent accuracy.

Relative variation of frequency caused by changing temperature may be estimated by the expression

$$\frac{\Delta f}{f} \approx -\frac{\alpha \Delta T}{1 + \alpha \Delta T} \approx -\alpha \Delta T,$$

where α is the coefficient of linear broadening of resonator's materical (for the copper $\alpha = 16.5 \cdot 10^{-6}$ grad^{-1}); ΔT is the variation of temperature of the device. In the 2-millimeter range 1° temperature's variation yields variation of frequency $\Delta f = 2.5$ MHz. Changes of temperature do not exceed 0.5^o, if the water cooling is applied and the device operates in a stable regime. Therefore, one can provide long frequency instability $\Delta f \cdot f^{-1} \approx 10^{-5}$ without thermo-compensation and thermo-stabilization. Using the balanced thermo-compensation of resonator's heat broadening, one can essentially decrease frequency deviations (in more than 10 times) in readjusting GDR within the wide frequency range. In the case of fixed distance between mirrors such deviations may be practically excluded.

Changing temperature may cause variation of grating's period what, in turn, influences optimal level of the parameter determining synchronization of the velocity of electron flow with phase velocity of the space harmonic of diffractional grating and, hence, generation frequency. These effects are especially noticeable being the result of heat variation of the distance between resonator's mirrors. In some samples of GDR operating in the short-wave part of millimeter range it is observed bright luminescence of separate grating's lamellae in the moment of generator's adjustment to a certain type of oscillations. It is generated by SHF heating of drifted inhomogeneities on the lamellae in the places of maximal field's concentration.

But it has been established in the process of long use of such devices that this effect does not lead to noticeable worsening of the stability.

Drift of the difference frequency separated after crystallis mixer mixing the signals of two similar GDR are measured in order to estimate long frequency instability.

Frequency instability caused by the drift of the voltage and current of electron flow is determined by the steepness of electron shift of generator's frequency. Calculations of this shift with taking into account the parameter of space charge well agree with experimental data.

Conclusion.

The progress in science and engineering is connected not only with new ideas but with something which is may be even more important, namely, with their realization in the form of new devices, technical tools and technologies. Working out the lasers gives a bright example. Nowadays laser is a commonly used device but what a revolution it has made in sciense, technology and in everyday life!

Laser is a radiophysical unit, generator of coherent electromagnetic oscillations being able to generate them in the widest range beginning from the X-rays up to submillimeter waves.

Its main feature is the presence of oscillating system in the form of high Q-factor OR using connected electrons of crystall's atoms as an active medium. Atoms transfer to excited state under the influence of special type of pumping and their electrons occupy corresponding energy levels generating thus quantums of energy which feed OR of a laser. It produces high- coherent, stable and powerful radiation, because the quantum's selection is very high due to the properties of OR.

GDR have many features which brings it together with a laser. The main difference between them is the way of resonator's feeding. Free electrons form active medium in GDR. They are produced by specially created active surfaces - cathodes. Free electrons form electron pencils supported by electron optical systems and constant magnets. Pencils move close to diffractional grating generating SPR which feeds OR. In this connection GDR is often called [191-193] laser on free electrons (LFE).

GDR also initiated definite revolutionary transformations in science and technology. In this book we have considered only few problems concerning applications of GDR in spectroscopy and polarization of atomic nuclei [48], diagnostics of thermo-nuclear plasma, biophysics, remote sensing, etc. The results of investigation carried out in the millimeter and submillimeter ranges enable us to distinguish new directions in radio engineering: GDR-spectroscopy, GDR-radio location, etc. We have not touched application of GDR in quantum electronics, communication, semi-conductor technology, high-temperature super-conductivity *et.al.* Probably, these problems will be treated in other generalizing publications (see detailed discussion on this topic in [5]).

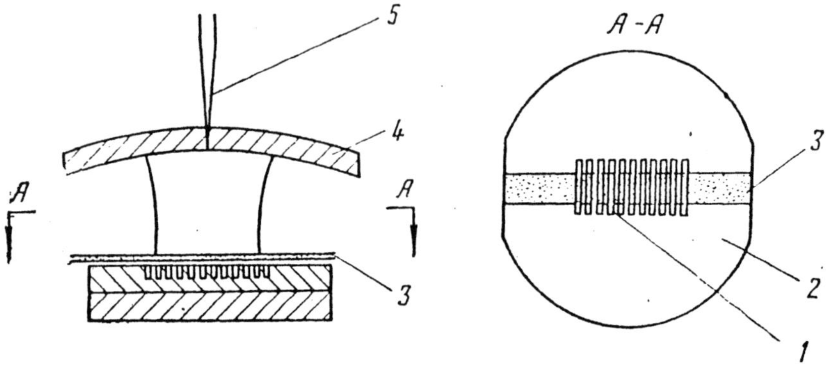

Fig. 126. First basic samples of GDR.

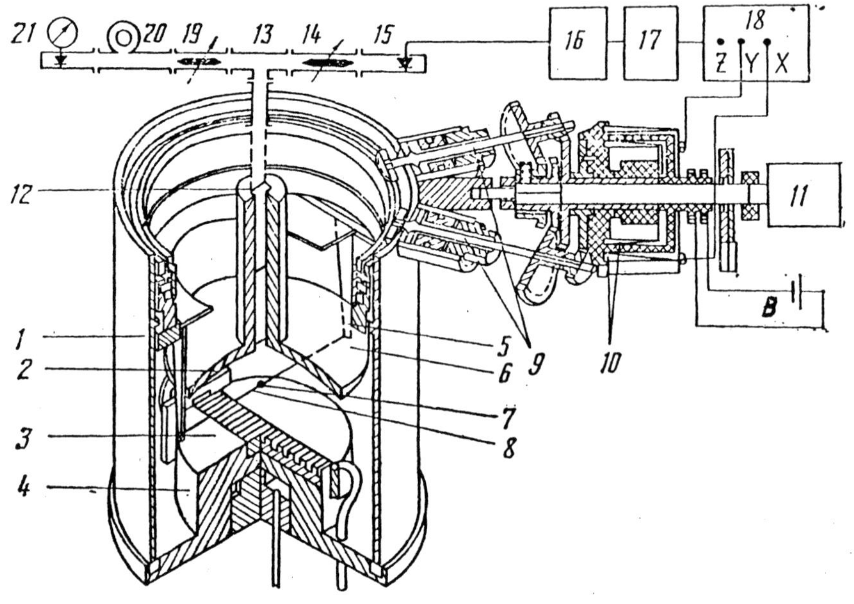

Fig. 127. Experimental plant for the visualization of fields of operating GDR.

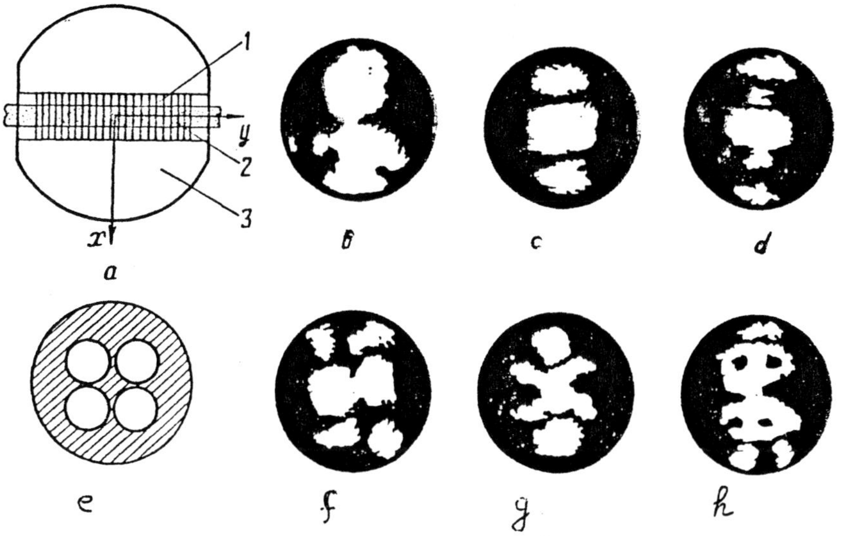

Fig. 128. Reflection diffractional grating (1), electron flow (2), plane mirror (3) in the semi-spherical open resonator (a) and fields' distribution in operating GDR obtained by the method of vidualization for the following types of oscillations: b) – TEM_{10q}; c) – TEM_{20q}; d) – TEM_{40q}; e) – TEM_{21q}; f) – $TEM_{(11+02)0q}$; g) – TEM_{30q}; h) – the assumed TEM_{11q} oscillation.

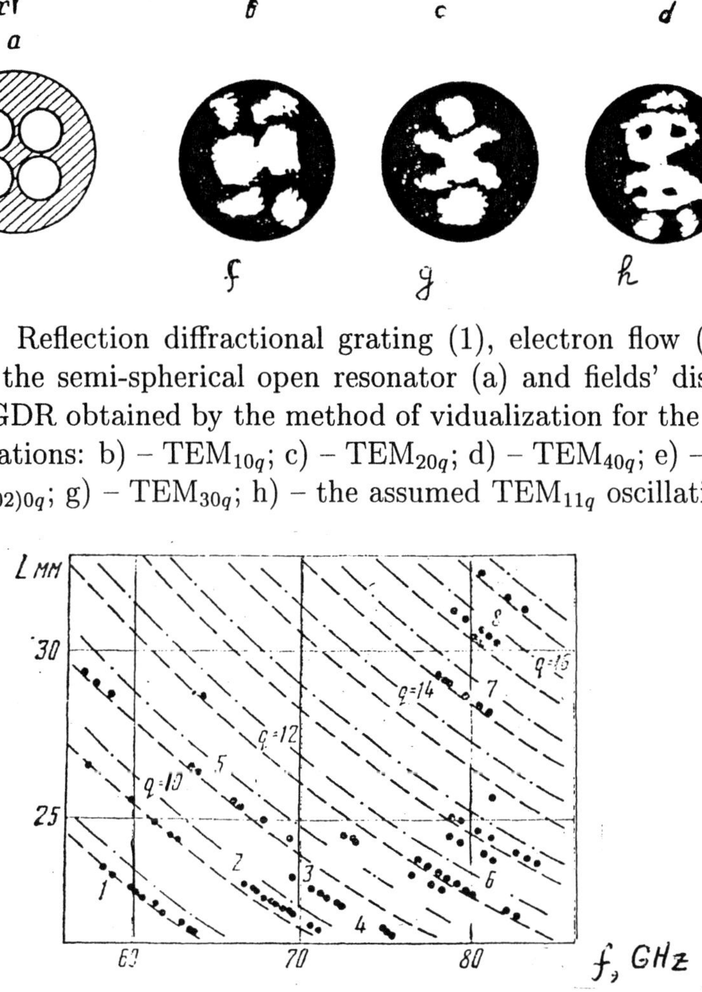

Fig. 129. Spectrum of oscillations of the semi-spherical open resonator with smooth mirrors and the spectrum of the resonator whose plane mirror is covered by the grating with electron flow flying over it for the following types of oscillations: 1 – TEM_{20q}; 2 – TEM_{40q}; 3 – TEM_{10q}; 4 – TEM_{30q}; 5 – TEM_{20q}; 6 – $TEM_{(11+20)q}$; 7 – TEM_{20q}; 8 – TEM_{21q}.

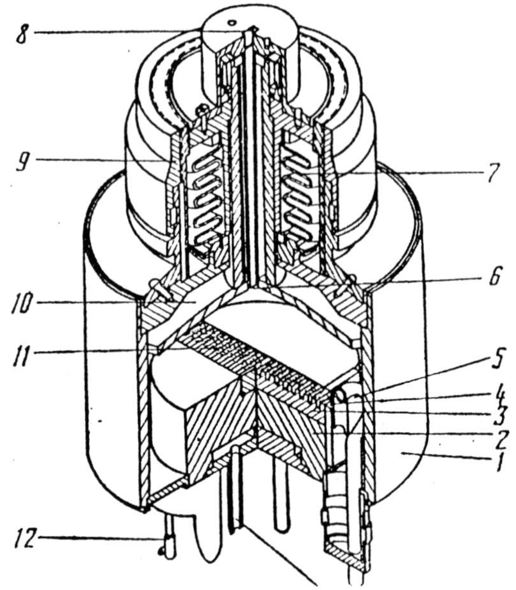

Fig. 130. Main construction units of the first sample of basic GDR with waveguide output of energy for $\lambda = 5.5 - 2.5$ mm.

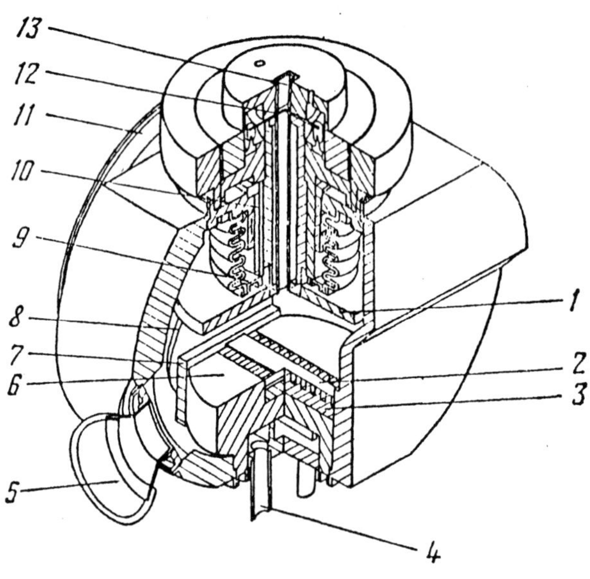

Fig. 131. Basic two-millimeter GDR with waveguide output of energy.

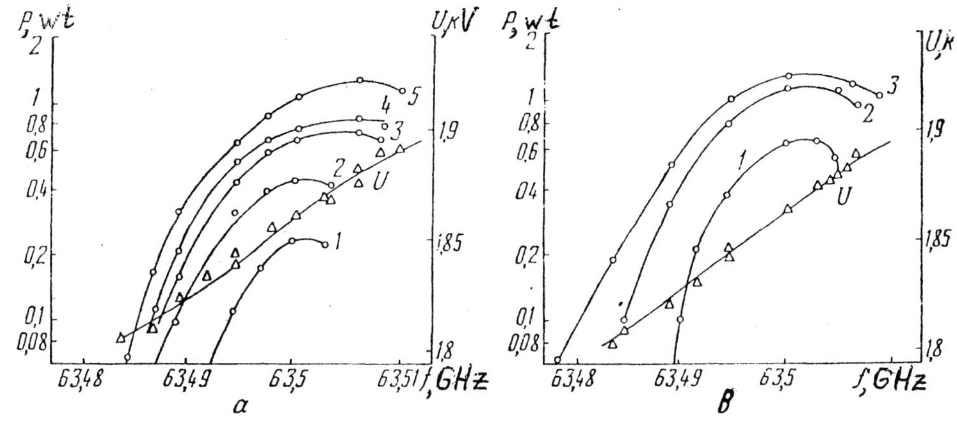

Fig. 132. Broadening of the band of electron readjustment and power of GDR with growing working current and magnetic field: a) $I_S = 72\mu$A; $H = 4$ kE; $L = 24$ mm; q=10 (1 – $I_S = 72\mu$A; 2 – $I_S = 100\mu$A; 3 – $I_S = 115\mu$A; 4 – $I_S = 122\mu$A; 5 – $I_S = 137\mu$A);

b) $I_S = 120\mu$A; $H = 4$ kE; $L = 24$ mm; q=10; steepness is 0.28MHz·V^{-1}; (1 – $H = 3.4$ kE; 2 – $H = 5.25$ kE; 3 – $H = 7.20$ kE; $H_{cr} = 2.50$ kE).

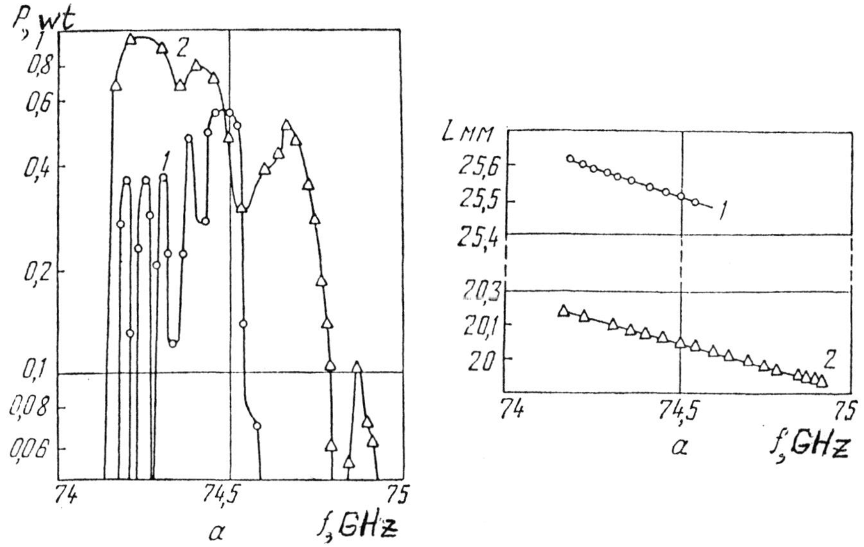

Fig. 133. Mechanical frequency readjustment of oscillations in GDR at the principal type of oscillations of open resonator: 1 – TEM_{20q}; 2 – TEM_{10q} oscillations.

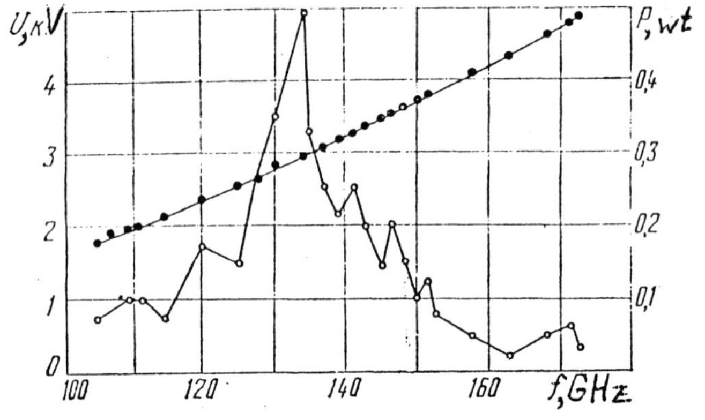

Fig. 134. Accelerating voltage and radiation power of GDR versus frequency: $l = 0.24$ mm; $d = 0.10$ mm; $h = 0.55$ mm; $\mu_p = 1.2\mu\text{A}\cdot\text{V}^{-1.5}$.

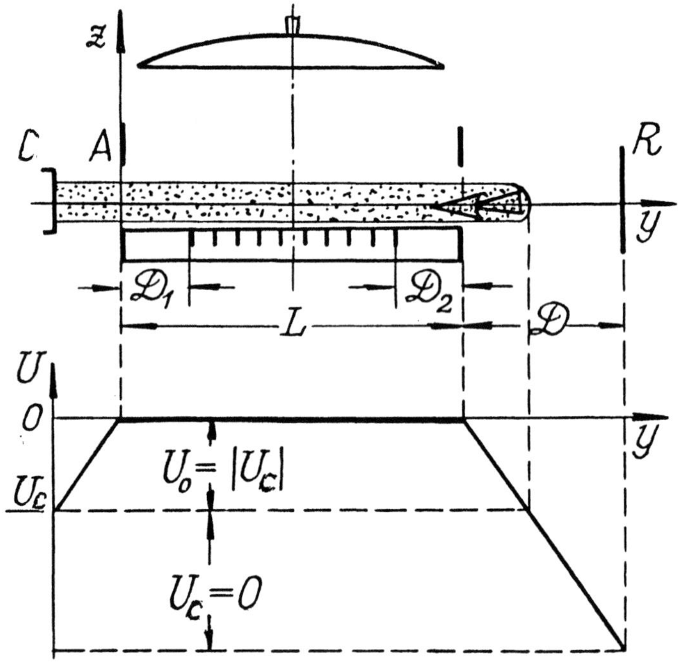

Fig. 135. Theoretical model of reflection GDR: 1 – upper mirror of open resonator; 2 – lower mirror with diffractional grating; 3 – electron flow; A – anode; K – cathode; R – reflector; V_0 – accelerating potential; V_R – reflector's potential.

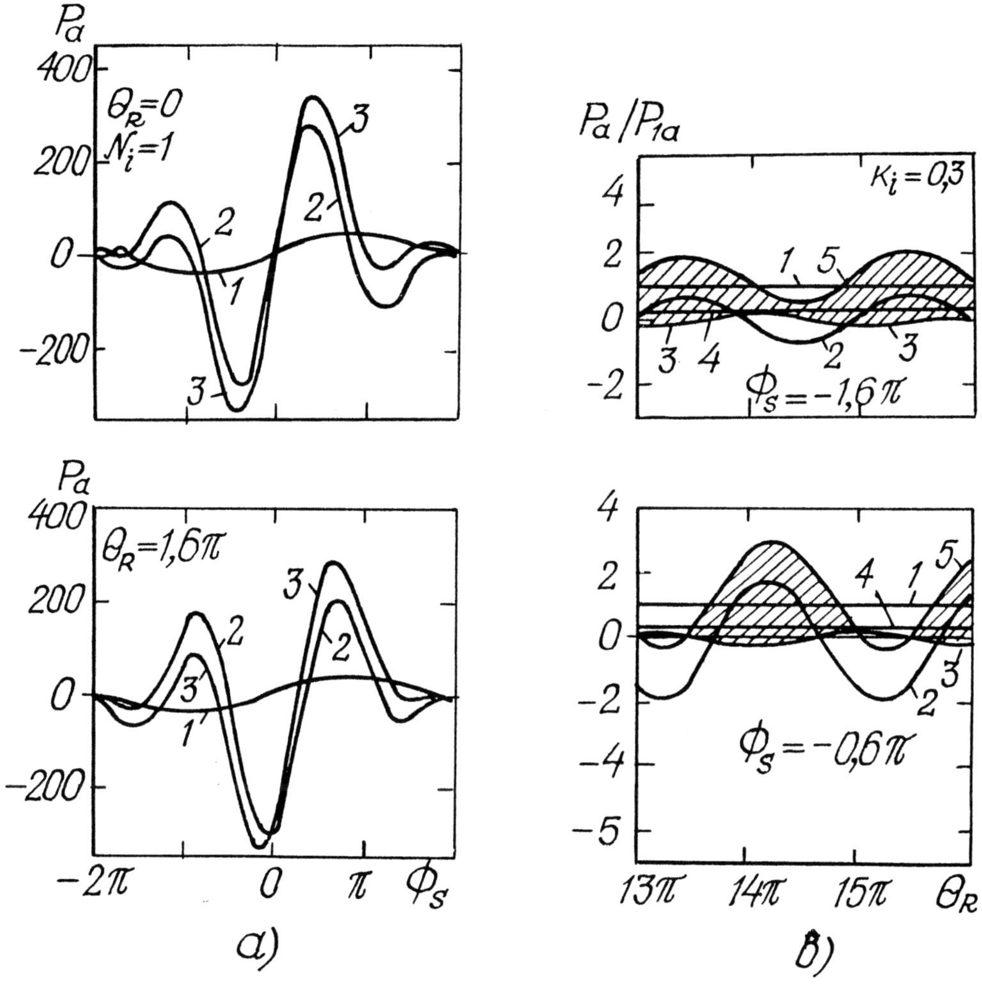

Fig. 136. Interaction power in reflection GDR: a,b – regime A: 1 – P_1; 2 – P_R; 3 – P_Σ; c,d – regime R: 1 – P_1; 2 – P_R; 3 – P_{2R}; 4 – P_{3R}; 5 – P_Σ; a) – $\bar{\theta}_R = 2\pi p_1$, $p_1 = 0, 1, 2, \ldots$; $k_2 = 1$; b) – $\bar{\theta}_R = 1.6\pi$; c) – $k_2 = 0.3$; $\Phi_s = -1.6\pi$; d) – $\Phi_s = -0.6\pi$.

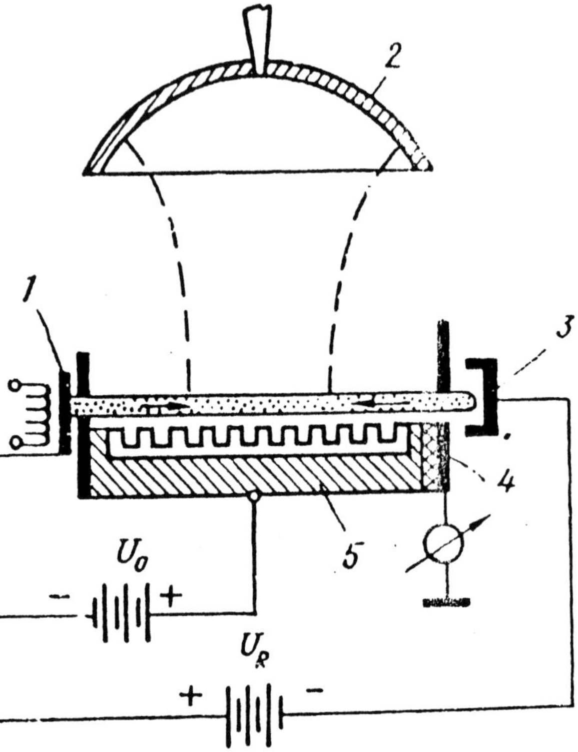

Fig. 137. Electrical scheme of reflection GDR.

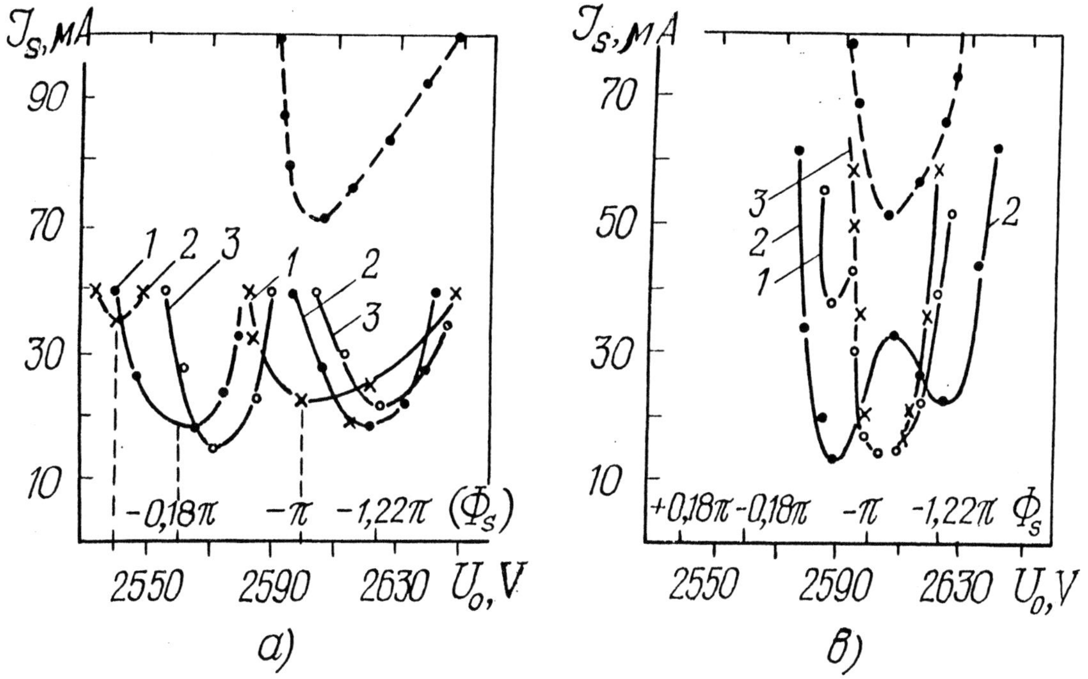

Fig. 138. Experimental data for the starting current of reflection GDR-2 in the generation regime A: a) – $k_2 = 0.5$; 1 – $U_R = -100$ V; 2 – $U_R = 220$ V; 3 – $U_R = 460$ V;

b) – $k_2 = 0.2 - 0.3$; 1 – $U_R = -130$ V; 2 – $U_R = 280$ V; 3 – $U_R = 380$ V; dot line indicates the starting current of flight GDR.

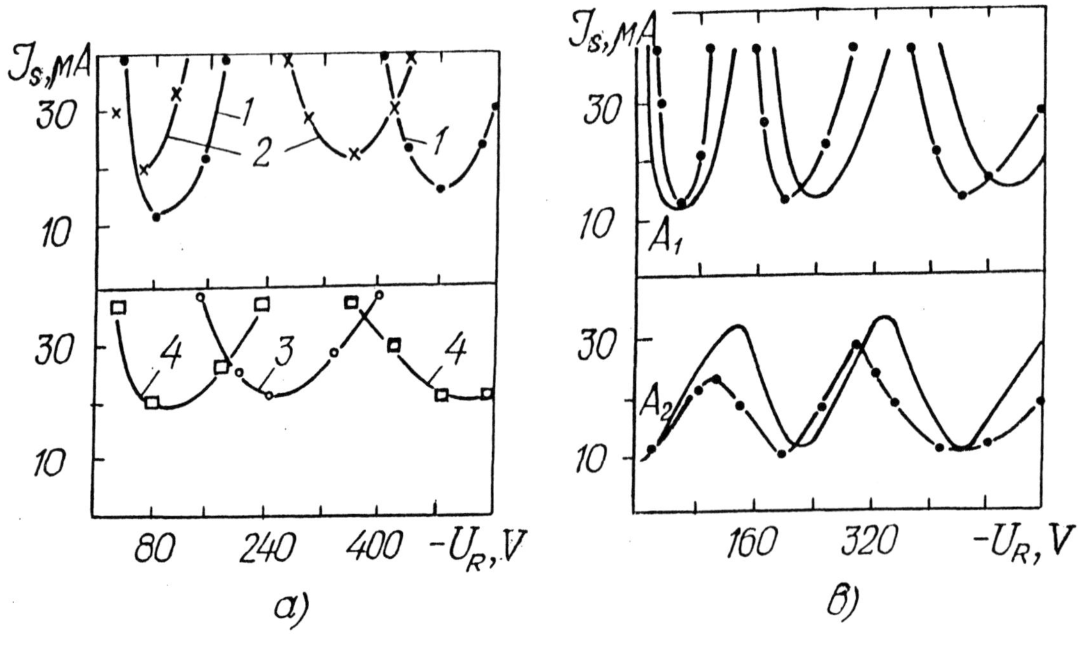

Fig. 139. Starting current of reflection GDR-2 in the regime R: a) discrete generation zones; 1 – $U_0 = 2568$ V; 2 – $U_0 = 2581$ V; 3 – $U_0 = 2602$ V; 4 – $U_0 = 2629$ V;

b) comparison of theoretical and experimental data; at the top – discrete generation zones; ($k_2 = 0.5$; $U_0 = 2585$ V; $\Phi_s = -0.5\pi$) at the bottom – oscillating generation zones; ($k_2 = 0.2$; $U_0 = 2601$ V; $\Phi_s = -\pi$).

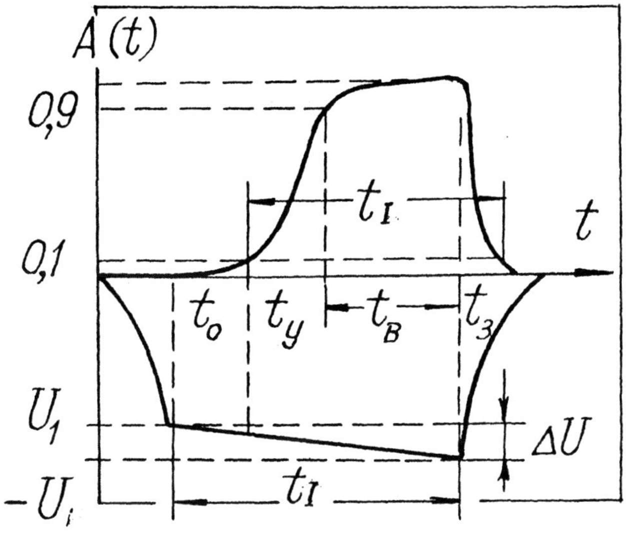

Fig. 140. Theoretical model of the formation of generation impulse in GDR: 1 – rounding curve of HF impulse; 2 – modulator's voltage.

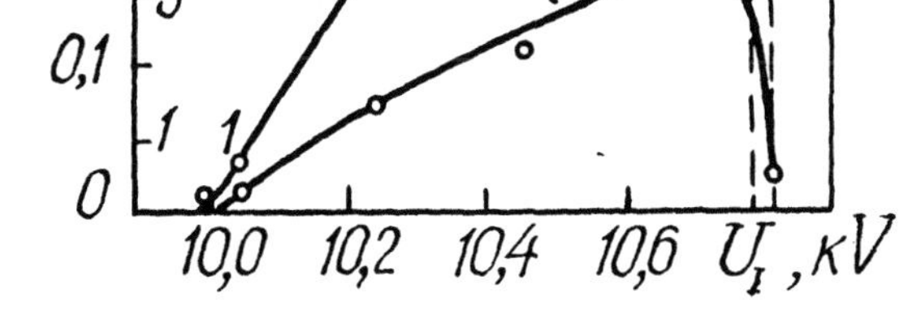

Fig. 141. Single-mode generation regime in impulse GDR ($t_I = 10\mu$sec; $f_0 = 71.78$ GHz; $\Delta U = 40$ V).

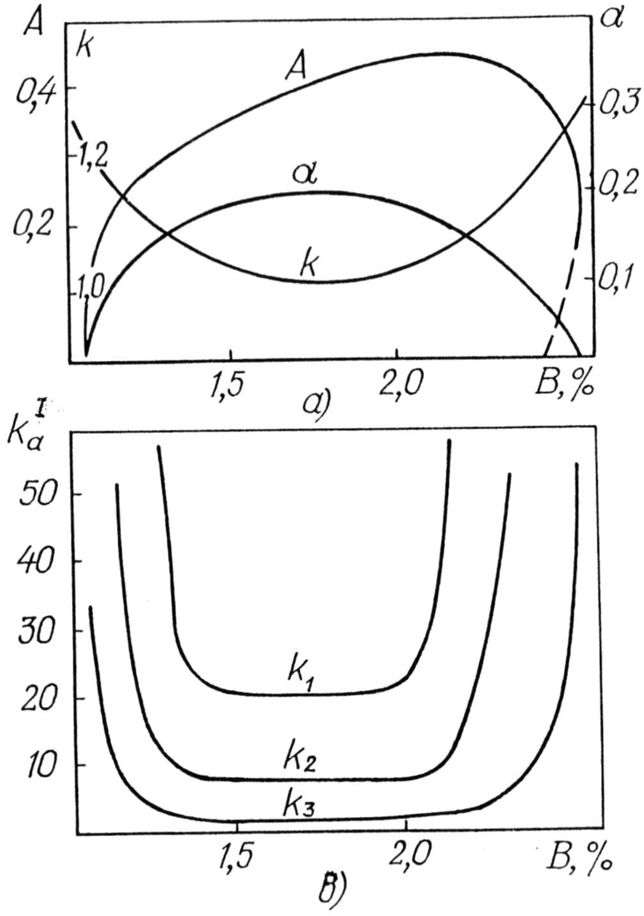

Fig. 142. Generation amplitude A, solidity of limiting cycle p, starting current J_S (a) and coefficient of autodyne amplifying K_a^I ($\sigma = 0.6$) within the limits of generation zone.

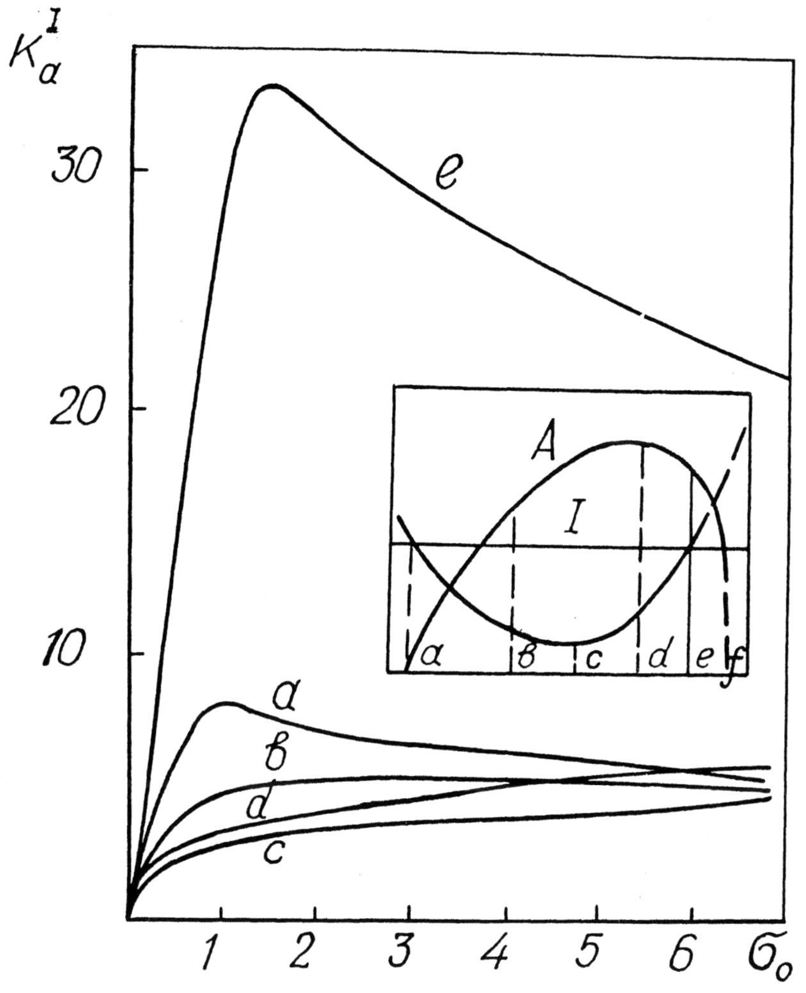

Fig. 143. Coefficient of autodyne amplifying K_a^I versus the value characterizing the degree of communication between GDR-autodyne and the load in the different points of generation zone A for $k = 1, 2$.

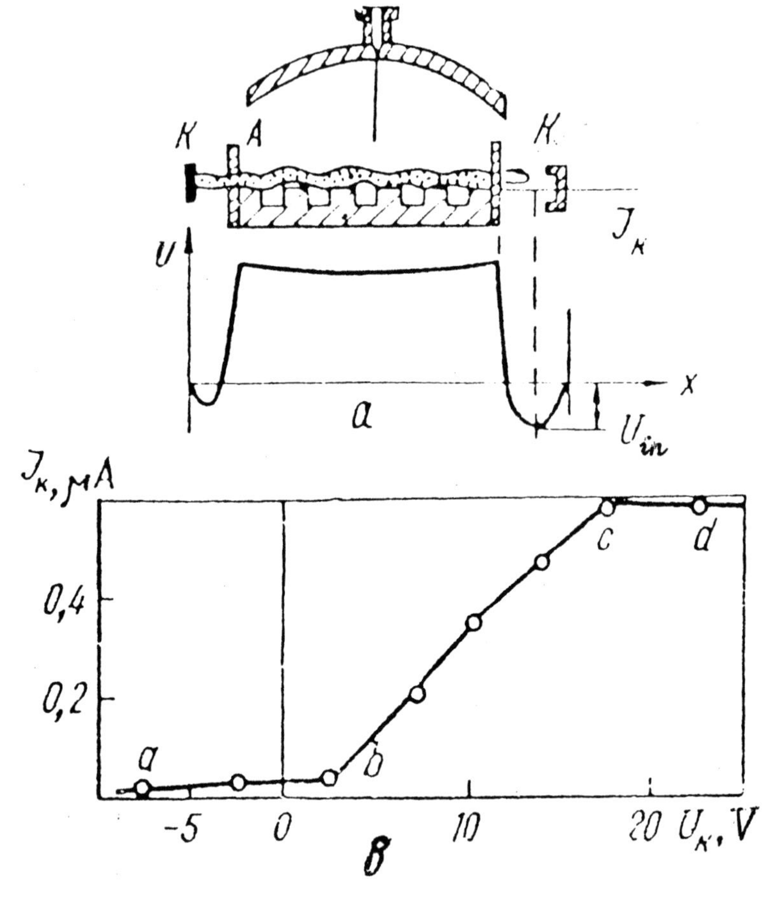

Fig. 144. GDR-electron detector (a) and distribution of statical potential of VAC (b).

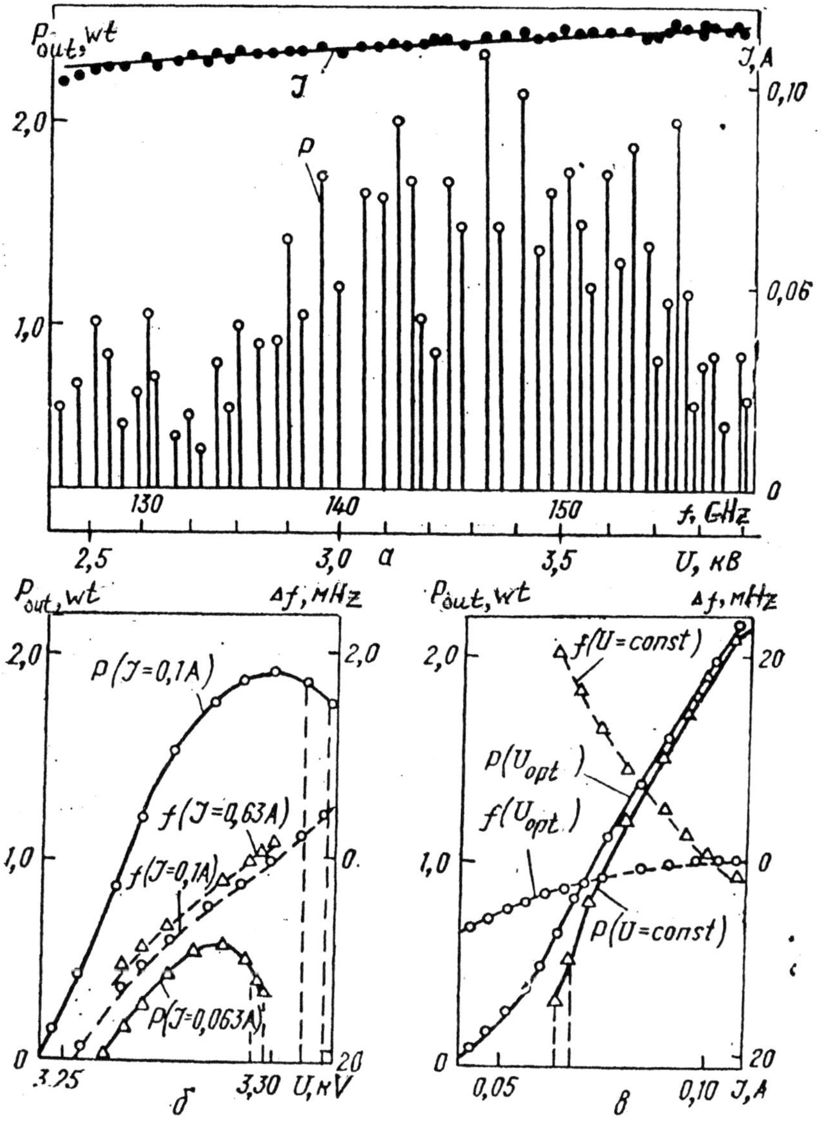

Fig. 145. Main characteristics of GDR: a) output power in the range of combined frequency readjustment; b) variation of output power and frequency within the limits of generation zone; c) output power versus frequency and pencil's current.

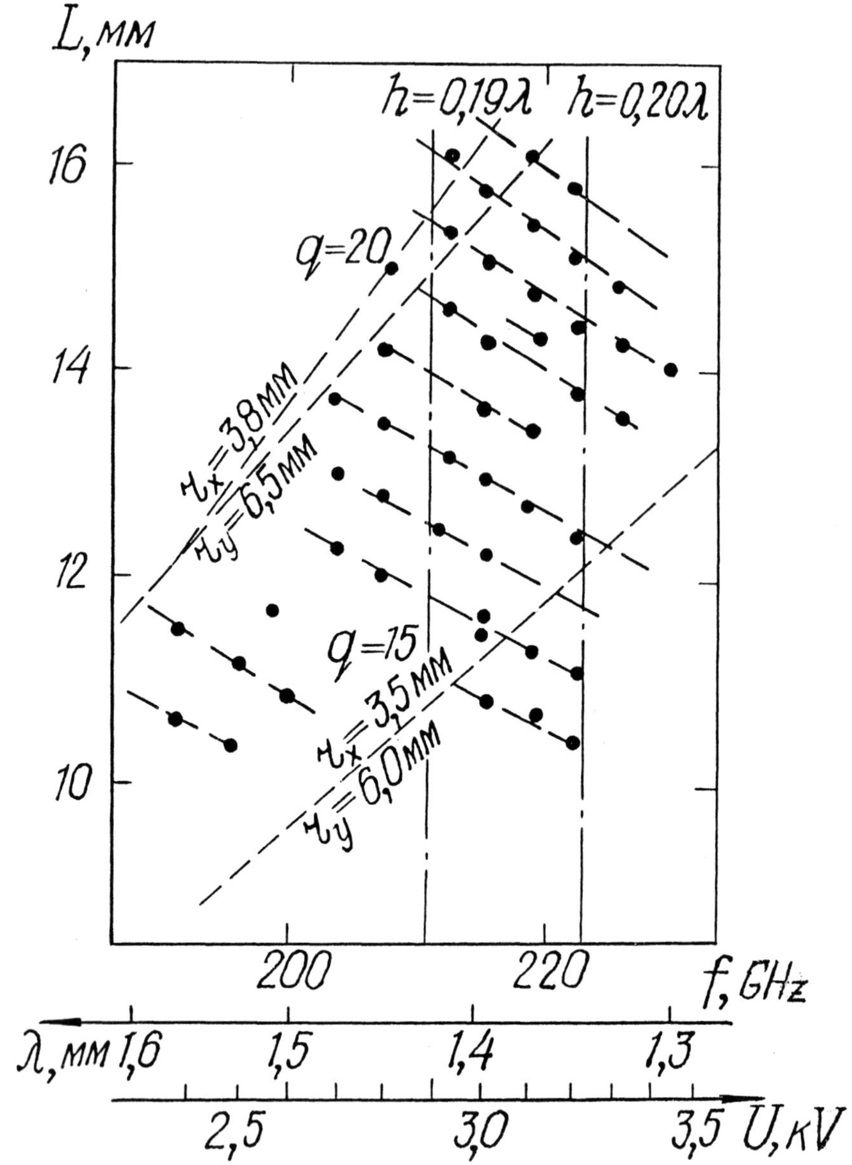

Fig. 146. Spectral characteristics of GDR.

Table 3

Type of the grating	Parameters of the grating [mm]			
	2l	h	d	
Reflection grating made of rectangular bars	2.78	1.0	1.26	
	2.0	9.0	0.45	
Semi-transparent grating made of bars	2.04		0.68	
Echelette, symmetrical	2.345			With rectangular tooth
Echelette, non-symmetrical	8.00			

Table 4 Generation powers of the flight GDR and RGDR-2

T_0,[μA]	P,[μWt](μ=2)	P,[μWt] (μ=1)	T_0,[μA]	P,[μWt] (μ=2)	P,[μWt] (μ=1)
30	0	0	80	240	80
40	80	0	90	250	120
50	130	0	100	260	160
60	170	0	110	240	200
70	240	40	120	80	240

Table 5 Spectral characteristics of signals

Type of generator	$\delta f / \delta U_0$,[MHz*V^{-1}]	$\delta f / \delta U_R$,[MHz*V^{-1}]	δF,[KHz]
GDR	0.2	-	32
RDGR	0.2	0.08	48
LIW	2.3	-	100
Clystron	-	5.2	240

Table 6. Main characteristics of magnetic focusing systems (MFS)

Types of MFS	Tensity of magnetic field in the center of working gap, [Tl]	Type of working gap	Length of working gap, [mm]	Mass, [kg]	Size, [mm]
1	0.42	Planar	42	2	ϕ84 * 114
2	0.61		42	9.5	ϕ130 * 231
3	0.5	Two-gap	33	1.9	ϕ82 * 97
4	0.83		30	5.75	ϕ120 * 135
5	1.0		33	14	ϕ146 * 194

Table 7. Package GDR, covering the millimeter range ($T_w = 0.10 \div 0.15$A; $U_0 = 2.4 \div 3.9$; L = 41 mm)

Type of GDR	Working range λ, [mm]	Output power P_{out}, [Wt]	Electron frequency readjust-ment [MHz]	Steepness of readjustment, kV		Mass of MFS		Mass of optional MFS made of rare earth materials, [kg]	Magnetic field [Tl]
				Versus voltage [MHz V^{-1}]	Versus current [MHz μA^{-1}]	Made of Alnico [kg]	Made of rare earth materials, [kg]		
GDR-1	0.8 – 1.1	0.1 – 0.5	25	0.5 – 0.7	0.0 – 0.7	96	12	8.0	0.7
GDR-1,5			25	0.5 – 0.7	0.0 – 0.6	96	12	8.0	0.7
A	1.1 – 1.4	0.1 – 0.7							
B	1.4 – 1.8	0.2 – 1.0							
GDR-2			30	0.4 – 0.6	0.1 – 0.5	36	7.0	3.2	0.5
A	1.7 – 2.1	0.3 – 1.5							
B	2.1 – 2.6	0.4 – 2.0							
GDR-3			35	0.3 – 0.5	0.1 – 0.3	36	7.0	3.2	0.5
A	2.5 – 3.2	0.6 – 3.0							
B	3.1 – 3.9	1.0 – 5.0							
GDR-4			40	0.2 – 0.4	0.2 – 0.4	18	4.5	2.0	0.4
A	3.8 – 4.8	1.5 – 7.0							
B	4.7 – 6.0	2.0 – 1.0							
GDR-6			60	0.1 – 0.3	0.1 – 0.3	18	4.5	2.0	0.4
A	5.9 – 6.9	3.0 – 18							
B	6.8 – 8.0	5.0 - 30							

CHAPTER 8 Applications of Smith-Purcell effect in other areas of science and technology.

In the previous chapter application of SPE in the vacuum diffractional electronics has been studied in details. We justified the idea of creating high-coherent sources of millimeter and submillimeter waves which itself was one more evidence confirming the existence of SPE.

Here we will continue analysis of further possible applications of SPE. Among them are working out the methods of remote sensing of electron flows; creation of antennas where SPE is used for generating directed radiation of electromagnetic waves; design of the units and element base of millimeter and submillimeter technology, in particular, the devices for effective input (and output) of energy to dielectric waveguides (and optical fibers); prospective design of logical elements for optical computers; methods of solution to the problems of generation and amplifying of electromagnetic waves by semi-conductor devices *et. al.*

§23. Measurement of the dynamical parameters of electron flow (method of remote sensing based on Smith-Purcell radiation).

Information on statical and dynamical characteristics of electron flow has of principle character for the creation of SHF devices as well as for optimizing their parameters. These problems have been investigated in a number of theoretical and experimental works [19,194]. One should separate among experimental approaches the methods where apertures, diaphragms, sensors and volume resonators are applied. Photo-methods have been recently developed. They are based on the effect of luminescence in optical range which emerges when electrons fall on metal surface [195]. In [196] it is shown the possibility of application of transfer radiation for the study of dynamical characteristics of modulated electron flow. Such methods allow to measure flow's characteristics in electro-vacuum devices of the short-wave part of SHF range in the far zone, i.e., without perturbations of the flow. However, they have comparatively low accuracy due to wide far field pattern, weak coherent factor and intensity of the transfer radiation.

Nowadays various new ways for the measurement of flow's dynamical parameters have been proposed and realized [18,20].

1. Let us consider determination of the reduction coefficient of plasma frequency for the modulated electron flow. It has been established in the process of investigation of SRP thin structure that modulation of electron flow by SHF field leads to excitation of space and time spectrum of electron

waves propagating with different phase velocities. If to let this flow pass close to periodical structure, one can generate diffractional radiation in the directions determined by the phase velocities of WSC and structure's parameters stated in the conditions (2.25) providing excitation of diffractional radiation. For the lower order waves, SWSC and FWSC, having maximal amplitude of convectional current, one gets

$$\cos\Gamma_n^{\pm} = \frac{c}{v_0} - \frac{2\pi c|n|}{\omega l} \pm R\frac{\omega_p c}{\omega v_0}, \tag{8.1}$$

where $\Gamma^{\pm}$ are the angels of maximal intensity of SPR far field patterns; $\omega = 2\pi f$ is the flow's modulation frequency; $\omega_p = 1.83 \cdot 10^{10} i0^{-0.5} U_0^{-0.25}$ is the plasma frequency; i_0 and U_0 are the current's density and accelerating voltage, respectively; $v_0 = \sqrt{2eUm^{-1}}$ is the electrons' velocity at the exit of modulator; c, e, m are the velocity of light, charge and mass of electron, respectively; $n = -1, -2, -3, \ldots$ is the number of spase diffractional harmonic.

Different kinds of excitation of the far field patterns of diffractional radiation (when modulated electron flow 1 flies close to periodical structure 2) are given on Fig. 147.

It follows from relationship (8.1) that for small values of current's density

$$i_0 \leq 0.46 \cdot 10^{-24} f^2 U_0^{1.5} \sin^2\delta \tag{8.2}$$

the influence of WSC on the direction of diffractional radiation is negligible and electron flow excites one-petal far field pattern (Fig. 147,a)

$$\Gamma_n^+ \approx \Gamma_n^- \approx \Gamma_n^0 = \gamma_n \pm \delta$$

where $\gamma_n = \arccos(cv_{-1} - 2\pi c|n|(\omega l)^{-1})$; $\delta \approx (0.5 - 2) \cdot 10^{-2}\pi$. Hense, by establishing the current's density which satisfies (8.2), one obtains initial (basis) radiation angle Γ_n^0 which serves as an origin for all further measurements.

If to increase current's density up to the level greater than given in (8.2), two-petal far field patterns is excited (Fig. 147,b) by SWSC and FWSC for which $\Gamma_n^{\pm} = \Gamma_n^0 \mp \alpha^{\pm}$, where $\alpha^{\pm}$ are the angles of inclination of maximal intensity related to the initial angle. It is easy then to find the coefficient of reduction of plasma frequency

$$R = 3.4 \cdot 10^{-13} f U_0^{0.75} i_0^{-0.5} (\cos\Gamma_n^+ - \cos\Gamma_n^{-1}). \tag{8.3}$$

Other way for determination of R, which is based on the measurement of only one of the angles $\Gamma_n^{\pm}$, gives the formulas

$$R = 6.8 \cdot 10^{-13} f U_0^{0.75} i_0^{-0.5} (\cos \Gamma_n^{+} - \cos \Gamma_n^{-1}). \tag{8.4}$$

$$R = 6.8 \cdot 10^{-13} f U_0^{0.75} i_0^{-0.5} (\cos \Gamma_n^{0} - \cos \Gamma_n^{-1}). \tag{8.5}$$

which are convenient when one of the far field patterns 4 or 5 is situated close to grating's surface (Fig. 147, c, d; $\alpha^{\pm} \geq \Gamma_n^0 - \theta^0$ where θ^0 is the width of diffractional far field pattern). In such cases (see Fig. 147, c, d, b, h) one can also measure the angles $\beta^{\pm}$ determining the turn of far field patterns 4, 5 towards the direction Γ_n^0 caused by the decreace of accelerating voltage from U_0 to U_1 or by its increase from U_0 to U_2 and find the coefficient of reduction by modified formulas

$$R = 6.8 \cdot 10^{-13} f U_0^{0.75} i_0^{-0.5} (\cos \Gamma_n^{+} - \beta^{-} - \cos \Gamma_n^{-1}).$$

$$R = 6.8 \cdot 10^{-13} f U_0^{0.75} i_0^{-0.5} (\cos \Gamma_n^{+} + \beta^{+} - \cos \Gamma_n^{-1}). \tag{8.6}$$

Accuracy of the measurements of angles and minimal distance between the first two maximums of intensity of the radiation far field pattern depends on its width $2\theta^0$ which is estimated by the formula [3]

$$2\theta^0 = 65 \frac{c}{fNl}, \tag{8.7}$$

at the one-half power level, where N is the number of grating's periods.

Hence, one can always choose such length of periodical structure for which radiation far field pattern will have small width. Thus, it provides the possibility to measure the angles $\Gamma_n^{+}, \Gamma_n^{0}, \beta^{\pm}$ with high accuracy.

2. The present approach forms the basis for creating the analysers of electron flow in the short-wave part of millimeter range as well as in the centimeter range where experimental results are of considerable practical interest due to the importance of their comparison with the data obtained by the method of transfer radiation [196].

Let us describe this analyser operating in the centimeter range (see Fig. 148). Electron flow 7 produced by electron cannon 5 is modulated by thoroidal resonator 6 and goes along periodical structure 8 being scattered by collector 13. Two types of periodical structures were used: a comb and a plate with rings (see Fig. 148); their periods $l = 5.94 \cdot 10^8 U_0^{0.5} f^{-1}$, [mm]; $U_0, [V]; f,$

[Hz] chosen for the principal space harmonic ($n = -1$) what corresponds to the fulfilment of the condition providing excitation of monochromatic diffractional radiation in the direction perpendicular to grating's surface (Table 8).

Resonators differ by the shape and size of the hole through which electron flow enters their volume. Mechanical adjustment of the resonators (6) enables to vary their eigenfrequency in the interval $f = 8.7 - 10$GHz. Periodical structure 8, electron cannon 5, resonator 6 and collector 13 are placed in vacuum glass cover 12. Reflection clystron 2 operating in the 3-centimeter range is a sourse of supporting signal; it is adjusted with respect to varying resonator's eigenfrequency in the process. Diffractional radiation is received by mouthpiece antenna 9 (80×100mm^2) containing the detector section placed at the distance 900 mm from the structure.

Radiation comes from the detector through amplifier 10 to oscillograph or microampermeter 11. The model of analyzer is placed in the gap between the poles of electromagnet 4 determining the interval of measurements for radiation angles $150° \leq \Gamma^0_{-1} \leq 30°$. Poles' edges and a part of glass ** are covered by absorbing material 1 in order to reduce the influence on radiation far field patterns. Accelerating voltage generated by feeding source 3 varies in the interval 0..3500 V; current's density $i_0 = 0.05 - 4$ A·cm^{-2}.

One can thus gradually synchronize WSC with harmonics of periodical structure and change the direction of SPR far field patterns by means of the control of electrons' velocity for either constant current's density and frequency modulation or for constant voltage and frequency modulation.

It has been established experimentally that one-petal far field pattern 3 with $2\theta \approx 30°$ is formed for the current's densities $i_0 \approx 0.05 - 0.1$ A·cm^{-2} what well agrees with theoretical estimates given by formulas (8.1), (8.7). Maximal intensity is observed at the angle $\Gamma^0_{-1} = 90° \pm \delta$, ($\delta \approx 2.5° - 3°$) for the following values of parameters: for the comb $f = 9.74$ Ghz, $U_0 = 1820$ V; for the plate with rings $f = 8.9$ Ghz, $U_0 = 2000$ V.

Experimental data were obtained for radiation angles $\Gamma^{\pm}_{-1}$ and reduced plasma frequency versus i_0 of the modulated electron flow moving close to the comb and in the channel of periodical structure having the form of plate with rings. Measurements were carried out for low densities of pencil's current and constant values of U_0 satisfying the condition of excitation of diffractional radiation at the angle 90°. When the density of pencil's current varied within the limits $0.5 - 1$ A·cm^{-2} the reduction coefficient $R = 0.79 \pm 0.02$ and the absolute error of determination of plasma frequency was ± 12 MGz (denoted by vertical lines in the points of measurements). One can see on Fig. 149, b that the considered strip and cylindrical flows (flying over the comb and in the channel of the plate with rings, respectively) have close plasma frequencies f_R but these values essentially differ from the plasma frequency f_p of unbounded electron flow (curves 1,2). Fig. 150 shows intensity of diffractional radiation and reduction coefficient versus accelerating voltage for the cylindrical modulated flow moving in the channel of the plate with rings. Behaviour of the curves 1,2 is evidence of the formation of one-petal far field pattern for $i_0 = 0.1$ A·cm^{-2}, $U_0 = 2000$ V which splits for $i_0 > 0.1$ A·cm^{-2} into separate petals corresponding to the radiation of SWSC and FWSC. For $i_0 = 1.2$ A·cm^{-2} (curve 2) maximal intensity is registered at the angles $\Gamma^{\pm}_{-1} = 62°, 118°$ what corresponds to the reduction coefficient $R = 0.78 \pm 0.02$. Let us point out that for the considered parameters the measured values of R for cylindrical flow are close to theoretical results (curve 3) calculated in [194] for electron flow in homogeneous channel and this model may be used in some cases for practical calculations of the reduction coefficient for periodical system in the form of plate with rings.

Experimental investigations are evidence of the efficiency of SPE applications for determining the reduced plasma frequency of modulated electron flows. The proposed method for measuring the reduction coefficient has higher accuracy in comparison with other known methods. The quality of measurements grows with increasing frequency of flow's modulation. In this way it may be considered as a reliable tool for the use in electronics, plasma physics and other fields of the millimeter and submillimeter technology. Here one should apply GDR for modulation of electron flow.

3. Let us consider the measurement of dynamical parameters of electron flow interacting with scattered field. In the devices with long interaction, for example, in GDR, the dynamical parameters of electron flow may essentially differ from the ones of pre-modulated pencil or of the flows in classical O-type devices. At the same time, the known methods of classical electronics cannot be directly applied for such measurements. The new approach [18] has been recently elaborated on the basis of theoretically established properties of the process of energy exchange between electron flow and scattered field (see §13).

It follows from theoretical analysis that if to establish the velocity $v_0 = v_\Phi$ for the real electron flow by accelerating voltage for the given current's density i_0, SWSC and FWSC excite diffractional radiation in the directions (4.11). In order to synchronize SWSC with the space diffractional harmonic one should increase electrons' velocity so that its increment $\Delta v^+ = v_0^+ - v_0$ ($i_0 = const$) and then

$$\cos\gamma^+_{n\,syn} = \frac{c}{\tilde{v}^+ + \Delta v^+} - \frac{m\lambda}{l} = \sin\alpha$$

what gives the expression for phase velocity of SWSC

$$\tilde{v}^+ = v_0\left(2 - \sqrt{\frac{U_0^+}{U_0}}\right) \tag{8.8}$$

and for the velocity of FWSC

$$\tilde{v}^- = v_0\left(2 - \sqrt{\frac{U_0^-}{U_0}}\right), \tag{8.9}$$

where $v_0 \approx 2.93 \cdot 10^7\, U_0^{0.5}$; v_0, [cm·sec^{-1}]; U_0, [V].

One can determine other dynamical parameters of the flow by means of (8.9): reduced plasma frequencies ω_q, reduction coefficients $R^\pm = \bar{R} \pm \tilde{R}$ ($\tilde{R}$ is the dynamical correction to the statical value $\bar{R}$ of the reduction coefficient), optimal synchronization parameters $b^\pm$, relative flight angle of electrons $\Phi_0^\pm$ for SWSC and FWSC, wave numbers of the electron WSC ($h^\pm \approx \omega(v^\pm)^{-1}$)

$$f_q^+ = f(M^+ - 1)(2 - M^+)^{-1};\ b^+ = 1 - M^+;\ M^+ = \sqrt{\frac{U_0^+}{U_0}};$$

$$f_q^- = f(1 - M^-)(2 - M^-)^{-1};\ b^- = 1 - M^-;\ M^- = \sqrt{\frac{U_0^-}{U_0}};$$

$$R^+ = 0.345 f U_0^{\frac{1}{4}} i_0^{-\frac{1}{2}} (M^+ - 1)(2 - M^+)^{-1}; \tag{8.10}$$

$$R^- = 0.345 f U_0^{\frac{1}{4}} i_0^{-\frac{1}{2}} (1 - M^-)(2 - M^-)^{-1};$$

$$\bar{R} = \frac{1}{2}(R^+ + R^-);\ \tilde{R} = \frac{1}{2}(R^+ - R^-);\ \Phi_0^\pm = 33.73\pi N l f(1 - M^\pm)(U_0^\pm)^{-\frac{1}{2}}.$$

Here f, [Ghz]; i_0, [A·cm^{-2}]; Φ_0, [rad]; the voltage U_0 may be either measured experimentally, if to define it as $U_0 = 0.5(U^+_{0min} + U^-_{0min})$ where $U^\pm_{0min}$ is the voltage registered in the case of minimal current's density (i. e., when absorbing and amplifying of SHF signal begins) or calculated by the formula

$$u_0 \approx 284.37 \frac{lf}{1 + 0.033 lf \sin\alpha}.$$

Hence, one can find dynamical parameters of the flow interacting with scattered field by the formulas (8.8), (8.10), if to measure radiation intensity (or oscillation's power in OR fed by an external source) versus accelerating voltage for different currents and to determine then $U_0, U_0^{\pm}$ by the measured data (see the graphs on Fig. 81). This way of measurements is carried out with the help of experimental plant presented on Fig. 81, a (this method is applied in the Institute of Radiophysics and Electronics of the Ukrainian Academy of Sciences for measuring the dynamical parameters of electron flows in the specific constructions of GDR). Some results of the measurements are given on Figs. 151, 152, in particular, dynamical parameters of the flow versus current's density i_0 when incident field is perpendicular to the grating (corresponding data, which is obtained for the same flow, is shown on Fig. 81, c). One can see that phase velocities of SWSC and FWSC essentially varies with increasing current's density for "small" pencil's currents: phase velocity of FWSC grows with increasing i_0 while phase velocity of SWSC falls (Fig. 151, a). However, for a certain value of current's density gradual variation of $v^{\pm}$ is violated due to non-linear effects of energy exchange in the flow of electrons.

A number of important features may be observed on Fig. 151, b. Difference between the reduction coefficient of the plasma frequency of electron flow interacting with scattered field and its statical component $\bar{R}$ is equal to dynamical correction $\tilde{R}$ what leads to different values of R for SWSC (R^+) and FWSC (R^-). With growing current's density the coefficients $\bar{R}, R^{\pm}$ decrease and absolute value of the dynamical correction $\tilde{R}$ grows; it may be negative (when $i_0 > 4.8$ A·cm^{-2} for the considered flow) or positive depending on the relationship between the reduction coefficient R^+ and the statical component $\bar{R}$. Non-uniform variation of R versus i_0 for minimal values of $\bar{R}, R^{\pm}$ (when $i_0 \approx 11.6$ A·cm^{-2}) has been discovered.

Unlike the reduction coefficients, effective plasma frequencies $f_q^{\pm}$ as well as non-reduced frequency f_p of unbounded electron flow (depicted by dot line on Fig. 152, a) grow with increasing i_0 and the situations may occur when either $f_q^+ > f_q^-$ or $f_q^+ < f_q^-$.

Relation of desynchronization parameters $b^{\pm}$, relative flight angles $\Phi_0^{\pm}$ and wavenumbers $h^{\pm}$ to current density i_0 given in Fig. 152, b confirm the theoretical inferences, particularly those relating to the increase of $b^{\pm}$, $\Phi_q^{\pm}$ (by module) with growing current density.

Thus, the characteristics of diffraction radiation, the effects of absorption and intensification of diffraction field by the electron flow provide for highly accurate determination and studying of the dynamic parameters of electron flow in millimeter and submillimeter wave ranges. Simplicity of the considered methods points to the promising nature of their employment in diffraction electronics, plasma physics, relativistic electronics and other fields of science and engineering.

§24. Antenna properties of Smith-Purcell radiation

The antenna system based on surface-to-volume wave conversion (transformation), i.e. the one utilizing the wave analog of Smith-Purcell effect comprises two elements, viz. a source of inhomogeneous plane wave and a periodic scatterer disposed in the surface wave field (the field (where the region of surface wave interaction with the array elements is considerably larger than the wave length and is, essentially, a radiating aperture). The source of surface waves utilizes various dielectric waveguides. It is known that, should the thickness of a flat waveguide be smaller than the wavelength, a considerable proportion of waveguide energy is carried in the space adjoining its surface; the field intensity diminishes exponentially from its boundaries. The characteristics of dielectric waveguides of certain cross-sections (circular, elliptical, rectangular) whose linear transverse dimensions are commensurable with the length of waveguide wave are close to those of a flat waveguide at preset conditions. It turns out that in studying the surface-to-volume wave transformation the governing laws stay true if the source of surface waves is constituted by a dielectric circular, rectangular and elliptical waveguide. Selection of these waveguides is also dictated by convenience of experiment in the millimetric and submillimetric wave bands.

The surface-to-volume wave converters are represented by various gratings which can arbitrarily be divided into two classes, viz. transmission and reflection gratings. In the former ones the directional radiation is generated in both half-spaces relative to the grating; in the latter class, only in the above-grating space. The grating elements may be of different geometric shape, such as metal strips, rectangular or round bars, louver or echelette type, etc. The strip grating enables studying the main characteristics of generated volume waves at a minimum number of parameters in case of diffraction of surface waves on this grating. This, in turn, warrants the use of gratings of a more sophisticated geometric shape, e.g. reflection and transmission bar gratings. It is essential that, regardless of the shape of grating elements their fundamental antenna characteristic, i.e. radiation direction of the main peak of the directional pattern, holds true for any one of them

$$\sin\theta = \frac{l_d}{d}\gamma - \frac{\lambda}{d}n, \tag{8.11}$$

where $\gamma = \frac{c}{v_\Phi}$ is deceleration of phase velocity v_Φ in dielectric waveguide; l_d is geometric difference of waveguide lengths in two grating sections; d is the distance between grating elements. The first term in (8.11) is determined by the type of the excitation grating (channeling line) and the second one points to the nature of dispersion properties of the grating proper. Thus, judging from the viewpoint of shaping the radiation direction of the main peak and radiation of the n-th space harmonic of the grating, all periodical structures with the same period and an unchanged source of their excitation are identical. This property is well known for Smith-Purcell radiation.

The amplitude-phase and polarization characteristics of the grating depend on the geometrical shape of individual scatterers, grating-to-waveguide distance (aiming distance) and the weaveguide properties. Experimental determination of these values is associated primarily with measuring the field intensity near an open structure; in this case the first thing is to study the radiating properties of individual grating elements (strip, bars) then those of the entire grating. A comprehensive pattern of the surface-to-volume wave conversion of the waveguide by the grating can be obtained after finding the power and polarization characteristics not only near the structure but also in the distant zone (antenna measurements). It should be noted that experiments are conducted on the basis of the main theoretical prerequisite, i.e. on an analysis of surface-to-volume wave conversion in the immediate proximity of the preset field. Moreover, the real structures have a limited size, final conductivity of the metal elements, and losses in the dielectric. All this has predetermined the method of measurements and the properties of the block diagram of the experimental test setup.

1. A complex experimental stand for determining electromagnetic fields in the distant and near zones in conversion of surface waves into volume ones is comprised of two large units (Fig. 153). Let us consider the one adapted for investigating the fields in the distant zone, i.e. the unit for measuring antenna characteristics of gratings. Its functions include transforming a surface wave to a volume one; measuring the space distribution of the field in the distant zone in H and E planes, in angle sectors from 0° to 180° and with an absolute error of determining the angular coordinate amounting to about ±0.5°; monitoring the working length of the generator wave; determining retardation of the waveguide; matching the grating + waveguide complex with the main waveguide power supply line; varying the degree of waveguide-to-grating coupling; polarization measurements. A test setup has been devised to meet these requirements (Fig. 153). The generator 1 is loaded on the waveguide 11 across the isolating 2 and measuring 3 attenuators, the regular

portion of directional coupler 4, measuring line 5 and horn junction 6. The other end of the waveguide is connected by an identical junction with the regular portion of the directional coupler 9 which is loaded on matched load 10. Part of generator energy is branched off by the directional coupler 4 to the cymometer 7 and the indication and oscillation frequency monitoring system 8. The measuring line 5 is cut into the waveguide line for monitoring the degree of matching. The required position of the waveguide is set by two foam plastic discs whose radius is $R \geq 40\lambda_0$. The grating is disposed on a stage which is moved by a micrometric screw and a special stop; the grating is rigidly pressed against the stop by a catch. These measures provide for strict maintenance of aiming distances. The adjusting unit allows the grating surface to be tilted from the waveguide axis by turning it about the axis passing through the middle point of the grating parallel to the generatrix of its elements. The entire portion of the setup is transmitting and is disposed on a movable platform rotating in two relatively perpendicular planes. The provision of appropriate potentiometer transducers and instruments of reversible motors ensures synchronous recording on two-coordinate continuous transport tapes of automatic recorders. The receiving portion of the setup comprises an antenna 14 with a detector section disposed on a revolving rod at a distance defined by the distant measuring zone. The receiving antenna unit provides for polarization studies too.

The far field pattern has been measured by the movable antenna method. The relative level of side lobes, the width of main radiation peak and radiation direction are determined by a graphically plotted directivity pattern with calibrated amplitude and angular coordinate. The width of the main lobe is set by the length in degrees of the chord connecting the points of the main lobe corresponding to a 3-dB level. The middle of this chord shows the radiation direction of the main petal. To decrease an error of angle coordinates the rotation axis in E-plane running through the radiating aperture and is combined with the axis of the grating turn. The rotation axis in H-plane coincides with the wave guide longitudinal axis on the regular section. It should be noted that in experiments grating apertures were equal to 20-30 λ_0. It allows to conduct measurements on the usual laboratory table anechoie chambers and large open areas. The carried out analysis has shown that effective measurement of required values in conveniens to make using wave guides for which $\frac{V_\Phi}{c} \approx 0.65 - 0.85$ and a shape of a cross section is elliptical. For such waveguides propagating main wave has stable polarization characteristics. Here it is necessary that along the elliptical waveguide there is propagating a wave which has E-component of the field that is analogous to the similar field of a planar waveguide. In our experiment the waveguide is excited by smooth transition – have exciter to which the energy of SHF field is supplied along the standard metallic rectangular waveguide with the H_{10}- wave. In the elliptical waveguide the ratio of small and large axises is equal to 0.5 providing $\epsilon = 2.25$ and $\frac{V_\Phi}{c} = 4.0 - 4.51$ will be equal to 5-5.5 mm. While studying transformation of surface waves into body ones by means of the grating along with the directly characteristics it is required a detail study of energy dependouces. It should be taken into account that energy properties of the flouring wave antennas are closely connected with the directivity characteristics. Measurements of E-vector absolute values of the radiation electromagnetic field are usually made for aperture antennas depending on the angle co-ordinates in E and H-planes of the beam pattern with subsequent integration of the received distributions in the range of the space being investigated. This method is used in super high frequency range with the corresponding measuring devices. There are no such devices for millimeter waves that will be able to measure with low errors the required energy parameters. In this connection a special procedure was developed (17) which did not require absolute measurements. Its principle is to measure the power ratio of the propagating wave in the waveguide with the grating and without it.

2. Experimental study of near fields while waveguide +grating radiation is important due to the fact that all processes connected with transformation of surfaces waves into body ones occur namely in the space occupied by a wave guide and grating. Knowing of near fields in such structures allow to model processes of interaction of electron flows with the grating fields, especially in diffractive electronics [3-5] ,integral optics and electronics optics [197-199]. As it is known, measurement of the field amplitude-phase distribution in the antenna uncover allows to determine the shape of its beam pattern. Classic methods of measurement of near SHF-fields use low size probes:electrical dipoles, magnetic loops. Directed electromotive force along high frequency cables is supplied to the recording device. In millimeter and submillimeter range the method of scale modelling can not practically repeat the corresponding measurements and the probes used in SHF-range. At the some time the known methods may be modified relatively to these range. There we are to exclude the operation of energy transmission from th probe

to the indicator device by means of channelizing SHF raod. A low size antenna in the shape of a dipole of 0.1-0.2 λ_0 size is placed in the controlled area of the space where there are a waveguide and grating. The electromagnetic field overradiated by the dipole is accepted by an auxiliary high-directed antenna (for ex.. a horn). Its position in chosen taken into account area division of the field overradiated by the dipole and the field of the structure main radiation (it is necessary to take into account add) As a probe is in the field of the waveguide with the grating it is natural to expect its effect on the field being investigated. To make clear this problem a number of beam patterns were made providing various probe position along the uncover being considered. It has been established that the probe with the size up to 0.5 λ_0 does not practically deform the near field.

In millimeter and submillimeter ranges while dissipation of surface waves on various obstacles and body waves appearance interferometer measurement methods are advisable to be used. For this purpose the second block of the measuring bed (Fig. 153) is modified. The chanel of supporting signal is formed by special means. As it is clear from Fig.154, the chanel of a supporting signal includes a transmitting 19 and reciting 14 antennas. The transmitting antenna has the shape of a comb excited by the surface wave of a waveguide (the grating parameters: periods number, grooth depth and aimed distance are chosen in such a way that the beam pattern of about 6-8 o width is formed in E-plane). It allows to record by the reciting antenna a signal with practically constant amplitude and phase in the $\pm 2^\circ$ sector relatively to the direction of the main maximum radiation (if $R \approx 100\lambda_0$ the limits of linear transmission are $\pm 4\lambda_0$). Such an interval is quite enough to measure amplitude-phase distribution on the grating ($l \approx \lambda_0$) The whole wave leading measuring part of the test bed (Fig.154) with components 13-16 was assembled on a separate platform with adjustment devices. The initial (zero) position of the probe in the form of an open end of a rectangular metal waveguide 3.6 x 1.8 mm^2 in size, narrowed in E-plane to 3.6 x 0.1 mm^2 (corresponding to resolution of 0.025 λ in 4-mm band) relative to the plane of aperture under test is registered by the presence of electric contact (using ohmmeter) between the probe and the object.

When using a measuring line to determine the phase of the electromagnetic field the minimum image position and the standing wave in it must be determined by registering the minimum coordinate of $vecE$ vector. A change of phase of one interfering wave will change the minimum coordinate proportionally to the phase shift. The relation $\Delta\Psi = \frac{2\pi}{\lambda_g}\Delta z$ (λ_g -length of wave in waveguide) definitely correlates the minimum coordinate Δz (mm) with the phase difference $\Delta\Psi$ (the measured phase is negative when the minimum is displaced from the initial point towards the reference arm).

The setting of a general-purpose experimental setup (Fig. 153) for measuring the near and faraway fields of single and periodically-spaced obstacles located in the field of a surface wave calls for parallel motions, coaxiality of rotation axles of receiving horn antennas with radiating apertures, coordination of the main line with the power take-off units. The platform with a measuring waveguide line must be moved in strict parallelism with the longitudinal axis of the regular part of the waveguide (deflection should not exceed 0.1 mm and the accuracy of the linear motion readings should be monitored by dial-type indicator heads and should be 0.01 mm).

The radiating antenna of the reference signal channel should also be disposed in a strictly defined location since correct matching of the transmitting antenna with the waveguide precludes the development of a reflected wave in the waveguide and distortion of the amplitude-phase distribution of the field. One of the fundamental reasons for mismatching of the antenna with the main line is radiation directed normally to the plane of the volumetric wave array developing in the open waveguide + array structure. Experimentally, this situation may occur at $\frac{l}{\lambda_0} = \frac{V_\phi}{c}$ i.e. in case of Bragg diffraction [146] ($\kappa = m\frac{\beta}{2}$, $m = 1, 2\ldots$ Bragg diffraction order).

Therefore, the array shaping the wave of the measuring stand reference channel takes the form of a comb with $l = 2.78$ mm, $\kappa = 0.66$ while the waveguide has $\beta = 0.78$. This rules out the initiation of Bragg diffraction while the coefficient of energy reflection into the line is merely 0.0475 (standing wave coefficient = 1.1).

The position of the transmitting antenna of the reference channel is selected on the condition that the main lobe of its directivity diagram is perpendicular to the longitudinal axis of motion of the platform mounting the second unit of the installation (tuning is carried out by an auxiliary antenna 17 disposed on a revolving rod).

The all-round checks of measuring stand performance and methods for investigating the phase distribution of the electromagnetic field have been carried out on a dielectric waveguide with well-known theoretical and experimental characteristics [137]. The obtained experimental dependence of the phase run-on on Z distance reveals that the phase front in the central region of the directivity pattern is flat. Hence, creating a reference signal channel by means of a specially-selected array which converts surface waves into volume ones agrees with the conditions of experiment while adjustment of the entire experimental setup based on phase measurements of the waveguide provides for getting reliable data on the amplitude-phase characteristics of fields originated in the open waveguide-array structure.

3. The above-described experimental techniques of measurements in the millimetric waveband have constituted a basis for studying the characteristics of various antennas utilizing the Smith-Purcell effect. These antennas have been widely used in designing passive and active aerospace-based radar systems for remote probing of the Earth and other planets (for example in the "Kosmos-1500" sputnik, etc.).

Let us consider as an example a receiving-transmitting millimetric waveband antenna complex with a high isolation coefficient utilizing the Smith-Purcell effect.

As a rule, modern super-high frequency transmitting and receiving complexes comprise a single antenna. The required isolation of channels is ensured by an antenna changeover switch. At present, vast experience has been gained in developing antenna switches based on various design circuits [200, 201]. Passive antenna switches (hybrid waveguide junction, circulator, turnstile joint, different polarization) are sophisticated waveguide couplings. Therefore, their efficiency, even in the super-high frequency band, is restricted by the precision required in manufacturing their elements. Thus, the magnitude of isolation easily attainable by hybrid waveguide junctions ranges from 20 to 30 dB. An obvious disadvantage of the switch in the form of a hybrid junction is a 50of transmitted and received power. The ferritic isolators are free of this disadvantage. Their isolation is 50 dB. The active antenna switches (balanced and branching types) are sophisticated combinations of waveguide elements and controllable instruments (gas-discharge and semiconductor ones). Total losses being 1-21 dB, such switches ensure isolation of channels and transmission amounting to 60-80 dB. The uses of active antenna switches are limited only by radar complexes operating under pulsating conditions.

The principles of designing antenna switches in the millimetric waveband based on scaled analogs of the super-high frequency band prove to be either ineffective or unacceptable. Making passive switches in the millimetric waveband involves serious technological difficulties associated with the requisite high manufacturing precision of waveguide elements. Designing of active switches is hindered by absence of controllable gas-discharge or semiconductor instruments specially adapted for switching the millimetric waveband channels.

It is widely known [201] that the maximum isolation can be obtained by the use of two spaced antennas. Isolation grows with increased directivity of antennas and the distance therebetween. Adherence to this obvious principle of isolating the receiving and transmitting channels does not always produce efficient isolation even in the millimetric waveband where high directivity of antennas is attained quite easily. Limitation of isolation of spaced antennas is caused by the limitations of their design parameters. When solving the problem of efficient isolation of receiving and transmitting channels in the millimetric waveband by the method of spaced antennas and in case of a short distance between them or their poor directivity, one should turn to new physical principles of shaping radiating (receiving) apertures. Thus, the problem of producing satisfactory isolation of receiving-transmitting lines in the millimetric waveband is confined to providing two apertures (transmitting and receiving ones) connected but slightly with each other.

Let us consider a receiving-transmitting antenna based on the Smith-Purcell effect, i.e on conversion of surface waves into volume ones with efficient isolation of channels in the millimetric waveband [51].

The directional radiation of the antenna based on this effect is realized due to diffraction of an inhomogeneous flat (surface) wave of the waveguide system on elements of periodical structure coupled with the waveguide field.

Let us assume that a flat electromagnetic wave impinges on a periodical structure. Diffraction of the wave impinging upon the array produces a spectrum of both homogeneous and inhomogeneous flat waves localized near the array. The diffraction harmonic may propagate like a flat wave at a certain angle α_n to the array if this angle defined by the relation

$$tg\alpha_n = (n - \kappa sin\alpha_n)[\kappa^2 - (n - \kappa sin\alpha)]^{-1/2}$$

is real. The direction of propagation of this harmonic depends on the incidence angle α of number n, dimensionless parameter $\kappa = \frac{l}{\lambda}$ (l - array period; λ - length of wave) but does not depend on the period filling coefficient. If $\kappa < |n - \kappa sin\alpha|$ then $tg\alpha_n$ - purely imaginary value so that the corresponding harmonic is in this case an inhomogeneous (surface) flat wave decaying exponentially in the direction away from the array. The phase velocity of this wave in the direction perpendicular to the array lines is smaller than the velocity of light and is equal to $V_\Phi = c\kappa(n - \kappa sin\alpha)^{-1}$. Now, if we connect the surface wave of the array with the waveguide system of surface waves whose propagation constant approaches that of one of the inhomogeneous flat waves of the array, such a waveguide will be excited. The function of waveguide exciter for exciting surface waves by the energy of incident waves is fulfilled by a diffraction grating. For effective excitation of the dielectric waveguide by the diffraction method, a homogeneous flat wave shall be directed at the periodical structure. In this case there is a resonant incidence angle defined by the array period, length of wave and waveguide characteristics (dimension of cross section, dielectric permeability). Hence, the system: surface wave waveguide + array can both radiate and receive electromagnetic waves.

Let us consider the functioning of a receiving-transmitting antenna based on the Smith-Purcell effect of surface-to-volume wave conversion for which purpose let us examine an electrodynamic system comprised of an array with two identical dielectric waveguides (Fig. 155). The waveguides are oriented relative to the array so that their axes are perpendicular to the mark lines. Let the surface wave propagate from arm 1 to arm 2. In the region of waveguide-array interaction each array period is excited in succession. If the radiation condition $|n - \frac{\kappa}{\beta}| < \kappa$ where $\beta = \frac{V_\Phi}{c}$ is satisfied, the energy of the surface wave is transformed into volume flat Smith-Purcell waves which break off from the array, shaping a directivity pattern in the free space. If a scatterer be located in the distant zone in the direction of maximum radiation, then in accordance with its dispersion diagram a part of incident energy will be reflected in the opposite direction. At $\frac{L}{R} \ll 1$, where L - size of array, R is the distance from array to scatterer it can be believed that the array receives a flat wave returned from the scatterer. The incidence angle of this wave is equal to the angle at which the wave of waveguide I has left the array. In accordance with the reciprocity theorem in waveguide II the flat wave reflected by the obstacle will excite in resonance [52] an electromagnetic wave propagating towards arm 3. It should be noted that the waveguide I will also produce a waveguide wave (towards arm I). However, intensity of this wave is smaller by several orders of magnitude than that of the wave propagating through the waveguide I from the generator (arm 1) to the array. Thus, the considered system (array with two identical waveguides of surface waves) is a transmitting-receiving antenna. Its inputs are as follows: 1 - generator arm,

3 - receiving arm while matched loads are connected to outputs of arms 2,4. From obvious reasons it follows that waveguides I and II may change places. Let us point out that the receiving generator can be connected to inputs 2,4.

Such an antenna is identical with the above-described one with the sole difference that its radiation (reception) angle is symmetrical to the normal to the array plane.

Characteristics of a laboratory specimen of the described antenna have been investigated at a wavelength $\lambda = 4.30$ mm on an experimental stand (Fig. 153). Dielectric waveguides I, II of elliptical cross section with an axis relation of $\frac{b}{a}$ have been disposed on a reflecting array comprised of rectangular bars with a period $l = 2.80$ mm. Indicators cut into arms 2, 3 have been calibrated by means of a measuring attenuator in the transmission channel. Then isolation of channels and transmission has been determined. Measurements have revealed that at the width of directivity pattern equalling to $2\Delta\theta_E = 2°$ and at a transmission coefficient $\eta = 0.98$ of the transmitting channel the isolation was -70 dB.

Investigations have been carried out not in the echo-free chamber so that reflections of local objects hindered determination of the true isolation. Nevertheless, it can be stated with certainty that isolation was not less than -70 dB. Such a high isolation coefficient in receiving and transmitting channels at a low directivity and short distances between the radiating and receiving apertures should be attributed to one of the properties of the physical phenomenon used in designing the antenna. The case in point is phase synchronism of the surface-to-volume wave conversion effect. The role of this property of the used physical phenomenon in attaining a high isolation coefficient is emphasized clearly by measuring the isolation of a pair of classic optoacoustic antennas. Isolation of two horn antennas with a directivity higher than that of the investigated antenna amounted only to -40 dB.

§25. Phase Measurements for the Millimeter and Submillimeter Waves Based on Smith-Purcell Effect

Automation of phase measurements in millimeter and submillimeter wave ranges calls for setting up radio systems which will enable creation of various integral circuits. With this purpose in view, efforts have been exerted to develop SPE-based frequencies enabling automatic investigations of phase characteristics of electromagnetic fields.

1. Transformation of the frequency of electromagnetic oscillations forms the basis of processing the signals carrying definite information.

Generating the signal whose frequency has small deviation of 1–10 Hz with respect to the initial one is the particular case of signals' transformation; corresponding devices are called frequency shifters. They are applied in the systems for calibration of phase-sensible receivers. Shortening the working wavelength in such devices leads to the necessity of creating appropriate testing apparatus for the use in millimeter and sub-millimeter ranges. Frequency shifters should provide high frequency of the output signal and the possibility to change the shift itself. Frequency shifters applied in laboratory conditions for the phase measurements of high-frequency electromagnetic oscillations transmitting information about initial phase on the frequency equal to the shift must satisfy similar conditions. Here we will consider several types of shifters and specify their parameters.

It is known that formation of the signal with shifted frequency is not difficult in the radio wave range. Such a signal is a result of shifting the main signal with respect to the one of heterodyne on non-linear element and separating either total or difference frequency by means of filtration. Application of balance mixers allows to suppress both the main and heterodyne signals at the exit what is equivalent to suppression of amplitude modulation of the signal with shifted frequency. Generation of such signal at higher frequencies from SHF up to optical range requires application of other methods since effective filtration of closely situated spectrum's components becomes rather complicated.

One of the method of obtaining the shifted signal is based on Doppler's effect. The frequency shifter has been created which contains dielectric parallelepiped and ultrasonic radiator at one of its sides. The latter forms the moving grating in the form of running ultrasonic wave. Light beam, which goes through dielectric parallelepiped, is scattered by the moving grating what causes Doppler's frequency shift for all space harmonics except of zero harmonic.

Big wave dimensions of the device limit its application in the longer wave ranges.

It is known also other types of frequency shifters: transformator of the linear polarization of incident wave to circular polarization, rotating grating and the transformator of circular polarization to linear polarization. Shift of initial phase of electromagnetic wave at the exit varies in a linear way from 0° to 360° when the grating makes the uniform complete circular turn. The shift of output signal turns out to be equal to the frequency of grating's rotation. Unfortunately, such frequency shift contains essential amplitude modulation of output signal due to non-ideal character of the grating acting as a polarizer. Frequency transformation of E-wave by the grating is equivalent to the exit transmission of non-shifted signal. Pulsations of both shifted and non-shifted signals are observed as the amplitude modulations of output signal at the difference frequency. Efficiency of this frequency shifter is bounded by the value 50 per cent since only the wave with linear polarization may go through the grating.

In addition to this, special conditions for suppressing amplitude modulations of output signal and raising the efficiency are created in the considered frequency shifter.

2. Let us consider the system consisting of dielectric waveguide and diffractional grating placed close to its lateral surface. Assume that slow electromagnetic wave with the frequency f_0 propagates along the waveguide and grating's elements move to meet the wave along the waveguide's axis. One can show that it yields Doppler's frequency shift $\Delta f = -nvl^{-1}$ for all space harmonics of the grating except of one harmonic having the index $n = 0$, where v is the velocity of movement of grating's elements, l is the grating's period. Thus, in order to separate the shifted signal one must provide transmission of energy from zero grating's harmonic to an additional transmission line. Here diffraction of the pencil of volume waves on moving grating is replaced by the diffraction of inhomogeneous plane wave (IPW) of dielectric waveguide. As a result, it is observed localization of the domains where simultaneously exist divisible spectrum's components. The choice of appropriate scheme of distributed communication between grating's harmonics and principal wave of the transmission line provides high quality spectrum of output signal [4].

Let us consider some practical realization of the principles stated above. A sample of frequency shifter contains dielectric waveguide together with directional coupler and rotating grating placed successively along it. Two-shoulder directional coupler is made of a section of dielectric waveguide situated at the distance of approximately a quarter of wavelength from the first waveguide. Rotating grating has the form of periodical set of rectangular cavities (with the period $l = 0.5\lambda_1$ where λ_1 is the wavelength of the wave propagating in the first dielectric waveguide) drifted on the lateral surface of metal cylinder (parallel to its axis) and fixed at the distance of not more than one half of the wavelength from the first waveguide. This grating acts as a reflector for incident wave and the relationship between period and wavelength means fulfillment of the condition of the first order Bragg diffraction which provides effective distributed communication between direct and inverse waves of dielectric waveguide by means of the minus first space harmonic of diffractional grating.

Here there are practically no diffractional losses caused by lateral effects in the domain of communication of the waveguide and grating.

Rotation of the grating leads to the Doppler's frequency shift $\Delta f = vl^{-1}$ of reflected wave. Directional coupler separates direct (non-shifted) and inverse (shifted) waves in such a way that a part of the energy of the former goes to direct shoulder while a part of the energy of the latter goes to inverse shoulder.

Efficiency of the device may reach 99 per cent when directional coupler has relaxation factor 0.458 dB; then 90 per cent of the power goes to direct shoulder, 9 per cent goes to inverse shoulder and 1 per cent is reflected to the source. Directional coupler created on the basis of two connected identical dielectric waveguides is the best type of such device for practical applications since it has high degree of direction and its transition relaxation is easily regulated (here the Y directional coupler may be replaced by the circulator with given characteristics).

Rotating grating may have the form either of metal cylinder having screw rectangular cavity (with the possibility of rotation around cylinder's axis parallel to the axis of dielectric waveguide) or of a set of radial metal strips on dielectric disk (with the possibility to rotate around its axis; axis of dielectric waveguide must be parallel to the plane of the disk and the distance between the axes of disk and waveguide must be several wavelengths greater than disk's radius).

The second type of frequency shifter contains two identical dielectric waveguide, the input one and the output one, which are parallel to each other and the distance between them is equal to several wavelengths. Rotating and additional (immovable) gratings are placed on both sides of the waveguide at the distance less than one half of the wavelength facing each other. Their periods are equal (to l) and satisfy inequality $\lambda l^{-1} + cv_{\Phi}^{-1} < l < \lambda v_{\Phi} c^{-1}$ where cv_{Φ}^{-1} is the deceleration coefficient of dielectric waveguides. The first grating is the periodical system of rectangular cavities drifted on the lateral surface of metal cylinder parallel to its axis and the second grating has the form of periodical cavities situated on the larger side of metal parallelepiped. Axes of the gratings are parallel to each other and perpendicular to the plane where the waveguides' axes lie.

This shifter operates in accord with the scheme given below. Surface wave propagating in the first dielectric waveguide, being scattered by the first grating, radiates to the space between waveguides. It follows from the relationship for the period l and from the condition $\cos\psi_n = cv_{\Phi}^{-1} + n\kappa^{-1}$; $(n = -1, -2, \ldots)$ that only one harmonic of the grating's space spectrum radiates and its index $n = -1$. Rotation of the first grating causes Doppler's frequency shift $\Delta f = vl^{-1}$ for the radiated wave. Diffraction of this wave on the system consisting of the second grating and dielectric waveguide leads to excitation of the surface wave in the second dielectric waveguide. Angle of incidence of volume wave falling on the second grating is equal to radiation angle ψ_{-1} and this angle is optimal for the receiving since both the waveguides and gratings' periods are identical. This effect provides wide-band property of the device.

Efficiency of this frequency shifter may be close to 100 per cent, since transformations of inhomogeneous plane waves of the gratings to homogeneous volume waves and vice versa are very effective. Amplitude modulation of output signal is suppressed with respect to arbitrary given value by increasing the distance between waveguides and excluding the direct transmission of the signal at main frequency to the second waveguide. Interference of volume waves with different frequencies and parasitic amplitude modulation cannot take place here because only the signal with shifted frequency may have the form of volume waves in the space between waveguides (as well as in other domains of the space).

3. Experimental testing of the second type of frequency shifter has been carried out on the model containing rectangular polyethylene $2\times3\text{mm}^2$ dielectric waveguides ($\epsilon = 2.25$) whose wider walls are directed to the grating.

Principal HE-wave is the working mode of the waveguides. Grating with the period $l = 3$ mm has the form of a comb with narrow slots cut on the lateral surface of metal cylinder having the diameter 137.5 mm .

Experimental sample has the following parameters: Doppler's frequency shift $\Delta f = 0.1 - 5$ Khz for the case when grating's rotation velocity is 0.9–45 revolutions per second; amplitude modulation of output signal related to the amplitude of bearing frequency is -60 dB; efficiency of the frequency transformation unit (dielectric waveguide 1 – grating 3) is not less than 80 per cent. The model keeps the capacity for work within the whole range of wavelengths $\lambda = 3.8 - 5.4$ mm.

4. Frequency shifters of the millimeter range may be elaborated only with applications of specific phase rotators. Phase rotator created on the basis of in dielectric waveguide may have as an active element the layer of segnetoelectric whose dielectric permittivity is changed by the control voltage. Initial phase at the exit of dielectric strip waveguide is varied by the change of phase velocity at the definite part of the waveguide which takes place when a metal plate approaches its surface.

However, it is difficult to obtain linear dependence of the phase shift on control parameter (voltage or displacement of the metal plate). in such phase rotators.

Phase shifters considered above may be also used as phase rotators. The grating in the form of screw cavity on the lateral surface of metal cylinder is the most acceptable for high precision phase rotators. In the general case there may be several (m, $m \geq 1$) spiral cavities. Phase rotators with screw gratings provide the shift of electromagnetic wave $\Delta\Phi = m\Delta\alpha$ where $\delta\alpha$ is the angle of cylinder's turn around its axis. For $m = 1$ the phase shift coincides with turn angle.

The quality of comparative characteristics of the considered frequency shifters, especially high Doppler's frequency shift (up to 50–100 Khz in the millimeter range) and the frequency of output signal, allow to assert that they have better parameters than other known types of shifters. In addition to this, it is possible to combine such shifters successively in one chain; the total shift then is equal to the sum of partial shifts in each cascade.

§26. Interferometer for the integrated schemes of millimeter range

Phenomenon of interference is widely used in different areas of science and technology. In the optical range it is applied in metrology and for testing the quality of smooth surfaces [202]; in integrated optics – for creating the logical elements of optical computers [54]. In the millimeter range interferometers form the basis for constructing various bridges, power divisors, frequency filters *et. al.*

It is well known the type of interferometer applied in the millimeter and submillimeter wave ranges which consists of a grating and dielectric waveguide. The edge of dielectric waveguide serves as a channel of total signal; wave pencils in free space falling on the system dielectric waveguide–grating at the radiation angles of corresponding grating's harmonics are the channels of interfering signals.

In this interferometer there exist, however, undesirable scattering of energy in free space. Variation of phase difference between incident pencil leads to addition or subtraction of the waves in dielectric waveguide. In the first case pencils' energy is transformed to the energy of these waves and in the second case it is reflected from the system dielectric waveguide–grating at the angles of mirror reflection what causes scattering of energy in free space and emergence of background radiation. Interference is explained in [52] from another point of view as a result of displacement of the domains of constructive and destructive interference with respect to grating's elements.

3. Working out of interferometers on the basis of integrated schemes of the millimeter range mainly requires the use of operation regimes where surface waves are applied with minimal radiation to the free space. Let us consider the system dielectric waveguide–grating working as interferometer assuming that the condition of the second order Bragg diffraction is fulfilled ($l = \lambda_g$, Fig. 156). Both edges of dielectric waveguide serve as channels of interfering signals and the result of interference is registered in free space. Angle of maximal radiation in the case of diffraction of surface wave on diffractional grating (SPE) is determined by the equality $\cos\psi_n = cv_\Phi^{-1} + n\lambda l^{-1}$; $n = -1, -2, \ldots$. The condition $l = \lambda_g$ corresponds to radiation angle $\psi_{-1} = 90°$ of the minus first grating's harmonic, hence, interference will be observed in the parallel pencils. There exists practically one pencil with varying amplitude which depends on phase difference of the waves propagating in dielectric waveguide. This wave pencil is the Smith-Purcell radiation.

Experimental testing of the proposed model is carried out for the grating with period $l = 3.2$ mm and polyethylene elliptic waveguide for the HE working wave ($cv_{\Phi}^{-1} = 1.18$; $\lambda n = 3.78$). Big axis of elliptic cross-section is parallel to grating's surface.

Explanation of the result of interference in parallel pencils is always accompanied by the question [202]: how does the energy of interfering waves go off in the process of their opposite-phase addition? Limiting transition from the small but finite interference angles to the infinitely small ones leads to the effect called localization of the strips of interference in infinity. In real devices resulting pencils are always have finite width, therefore, one must determine all possible channels of energy losses for every concrete structure. In the case of opposite-phase addition presented on Fig. 156, a there is practically no radiation to free space and the losses are caused only by lateral effects of the grating; feeding edges of dielectric waveguides, where the regime of standing wave is established (with the coefficient $\rho = 22$ dB for the comb having narrow 0.2–0.4 mm slots), serve as leaky channels. It is easy to justify this result if to consider field's distribution in dielectric waveguide (see Fig. 156, a): field's amplitude on the slot is minimal and the vector of tensity of electric field changes the direction after crossing the slot's center. The structure of exciting field corresponds to optimal conditions of excitation of the second type of wave in the waveguide between two metal planes and of the second type of oscillation in the slot resonator formed by a cavity in metal plane. The latter operates at principal resonance in the long-wave range, hence, higher order oscillations are practically not excited and the grating acts like a metal plane. One cannot use here the comb with wide slots (for example, 2.8 mm) since the second type of oscillation excited in the slot resonator leads to big ohmical and non-directional radiation losses.

2. Syn-phase addition of interfering waves is shown on Fig. 156, b. Structure of the field in dielectric waveguide provides excitation of the principal oscillation in slot resonator. Inhomogeneous wave of the dielectric waveguide is transformed to the radiation field – SPR; the coefficient of standing wave in the waveguide is $\rho = 2.5$ dB. If one channel of interferometer is switched off, coefficient of standing wave in the second channel is $\rho = 9.5$ dB under the condition that the signal is sent only to this entry. Such a comparatively high value of this coefficient is due to resonance reflection of surface wave from the grating and the latter holds because system's parameters satisfy the condition of the second order Bragg diffraction. If this condition is violated ($\psi_{-1} \neq 90°$), the waveguide is well coordinated with the grating transfor-

mator of surface waves to volume waves. If $\psi_{-1} \neq 90°$ but $\psi_{-1} \approx 90°$, the strips of constructive and destructive interference are observed within the limits of total far field pattern (when the signal is sent to both edges of the waveguide).

The considered interferometer with one channel of volume wave (SPR) may be used as an element of measuring plants operating in the millimeter range and in receiving sections of radio location stations. It may be applied for testing the quality of dielectric waveguides (constancy of deceleration coefficient) based on the analysis of the shape of far field pattern of the system dielectric waveguide-grating in the regime of normal Smith-Purcell radiation.

§27. The device for excitation of planar dielectric waveguide and other elements of millimeter technology based on Smith-Purcell effect

1. The devices for input of energy to planar dielectric waveguide, for example, prism and grating elements carrying out the communication between volume and surface waves on a definite area of waveguide's surface, are used in integrated optics [197]. These units are very effective but the necessity to work with volume fields often impedes their application.

Planar technology of the units of integrated optics requires working out the elements of communication between the lines of surface waves and planar dielectric waveguides [197]. The simplest known type of such elements is the gradual transition from the waveguide having the cross-section of about one half of the wavelength to the waveguide whose cross-section has the size of several tens of wavelengths [111]. Its longitudinal dimensions are much greater than the cross-section's diameter. Another type of transition has the form of linear and planar dielectric waveguides connected through over-critical gap between linear waveguide and beveled butt-end of planar dielectric waveguide [111]. Mouthpiece-parabolic transition [147] has the smallest size but its main disadvantage is in the presence of non-coinciding axes of of linear and planar waveguides what is also characteristic for the previous type of transition.

Let us consider the way of communication between linear and planar dielectric waveguides carried out by means of the grating having the form of beveled comb. Diffraction of inhomogeneous plane wave of linear dielectric waveguide on such grating yields formation of a pencil (Fig. 157) where the lines of equal phases in the plane of comb are parallel to the lines of equal phases in exciting waveguide for the regime of phase synchronization and phase field's structures are completely identical (Fig. 158, a). Hence, if to provide communication in the domain of the pencil, planar dielectric waveguide will be effectively excited. Since vector of the group velocity of comb's surface wave and vector of the phase velocity have different directions (see Fig. 158, a), amplitude distribution in the cross-section of planar waveguide is uniform. Unlike pencils' transformators where the phenomenon of Bragg diffraction is applied [202], communication in the given scheme is carried out by means of the comb's principal harmonic what forms the conditions for existence of modified SPE.

2. Experimental testing of excitation unit in accord with proposed proce-

given value of $J \cdot J_S^{-1}\ t_e$ is minimal within the limits of generation zone in the domain of maximal increment α.

Processes of oscillations' establishment were studied experimentally in the millimeter range ($\lambda \approx 2 - 4$ mm). One can single out a number of features characteristic not only for the initial stage but as a whole for the processes of excitation of impulse autogenerator with high Q-factor OR.

The presence of high Q-factor (in operating GDR of the millimeter range resonator's Q-factor has the order 1500–5000) limits obtaining short generation impulses, since t_e increases. In fact, if to define Q-factor as $Q = f_0(\Delta f)^{-1}$, minimal duration of rectangular impulse $t_{\min} \approx (1 - 2)Qf_0^{-1}$, hence, for the given Q-factor it is possible to generate shorter impulses at higher frequencies and for the given Q and bearing frequency f_0 one can guarantee only definite minimal duration of SHF impulse. In the experiments oscillations' establishment were considered only for generation impulses having the duration exceeding $t_{\min}$.

Initial conditions stated by the shape of modulator's voltage impulse (anode voltage of diode electron cannon) essentially influence the quality of excited impulse signal (and its fore-part characterizing the processes of oscillations' establishment). In the theory initial signal is taken as the idealized rectangular current's impulse. in the real experiment voltage's impulse has front and back sides and its top has a small "rake". These deviations may cause variations of output impulses' parameters and it is taken into account when t_e is measured (modulator's voltage impulse has sufficiently steep front side providing generation of autooscillation only in the beginning of its top having minimal inclination).

The necessity of "thin" measurements of the processes of oscillations' establishment in OR make great demands of experimental generator's model which, in particular, must contain special low-capacity high-voltage inputs for feeding voltage and high incandescence of cannon's cathode. It is necessary also to elaborate the methods for measuring the front side of generation impulse, since in the experiment one uses two-beam wide-range impulse oscillograph providing simultaneous observation of modulator's voltage impulses (or impulses of electron current on the collector) together with scanning in time of the rounding high-frequency signal (with the accuracy up to several nano-seconds).

In the presence of competing oscillation (in multi-particle regime) the duration of nonlinear stage of transfer process for the main signal increases dure shows that dielectric waveguide of the width 40 mm is effectively excited by the beveled comb and irregularity of transversal amplitude distribution in the planar waveguide does not exceed ± 1.2 dB (Fig. 159). Length and width of excitation unit are almost equal and comparable with the length of mouthpiece-parabolic transition; continuations of waveguides' axes coincide. Direction of pencil's propagation in dielectric waveguide may be changed within small interval ($\pm 10°$) by the variation of grating's turn angle α. Inclination of the lines of equal phases in the grating's plane, whose orientation is always similar to the phase distribution in planar waveguide (Fig. 160), corresponds to varying direction of wave's propagation in planar waveguide (Fig. 158). The possibility to control the direction of propagation is the basic advantage of this scheme.

3. Combination of slotted cylinder with planar dielectric waveguide is also effective used in a number of devices. For example, mouthpiece-parabolic excitator [147] is a combination of an excitator of cylindrical wave spreading in the plane of dielectric waveguide and of focusing phase corrector in the form of parabolic reflector. The latter is made as the metal-coated butt-end of planar waveguide having parabolic shape. This reflector may also have the form of a grating made of slotted cylinders. Being put inside dielectric planar waveguide, such grating completely reflects the incident H-wave.

The grating of cylinders whose slots face each other in pairs, completely lets through the incident H-wave at resonance frequency and may serve as non-reflecting waveguide's support; besides, in this case there are practically no radiation losses.

Grating made of slotted cylinders may also replace the grating of solid cylinders in the "fence" waveguide formed by two parallel lines of cylinders in a dielectric layer. It will essentially reduce losses in the layer when the H-wave is excited there.

Hence, one can apply different ways of pencils' formation in planar dielectric waveguides. Devices for excitation of such waveguides are used for creating small-size antennas of the millimeter range whose operation principle is based on the effect of transformation of inhomogeneous plane waves to volume waves – SPE [4] and which have narrow far field pattern in two reciprocally perpendicular planes. Focusing devices, filters and non-reflecting supports are applied for constructing functional units in the integrated schemes of millimeter wave range.

§28. Diffractional input of energy to dielectric waveguide

Methods for carrying out the input of energy to dielectric waveguide through its lateral surface are well known in the optical wave range. Dielectric prisms and diffractional gratings connected with dielectric waveguide (whose corrugated surface having rectangular or sinusoidal shape, dielectric permittivity being a periodical function versus longitudinal coordinate, etc.) are mainly used in the optical range as elements of communication of homogeneous plane wave and waveguide mode. Such communication elements are usually manufactured by the methods and technology based on lithography, ion implantation, optical holograpny *et. al.* [203, 204].

Principles of manufacturing the gratings inside the volume of dielectric waveguide, which are applied in the optical range, are unacceptable in the millimeter range and, consequently, the input of energy to dielectric waveguide of the millimeter and submillimeter ranges through its lateral surface should be carried out in accord with specific requirements which enable one to take into account distinctive features of this part of waves' spectrum.

1. Effective ways of communication between electron flow and excited periodical structures have been established in different problems of diffractional electronics. It is essential here to determine conditions providing the transformation of surface waves of the flow, which under certain conditions may be treated as dielectric waveguide, to the volume waves. Such results form the basis for series of experiments carried out in order to find optimal conditions for the input (output) of energy to (regular) dielectric waveguide with the use of SPE. A part of waveguide's surface is occupied by a grating [53]; the accuracy of manufacturing allows to apply it in the millimeter and submillimeter ranges.

Such a construction of the element of energy input may be used also in the optical range when external grating may be drifted by the methods of holography.

Let us investigate experimentally [53] the energy characteristics of the input of energy to dielectric waveguide carried out by means of the system dielectric waveguide-grating where the latter is placed at a certain (aimed) distance from waveguide's surface.

In particular, we will study the behaviour of generated power versus aimed distance.

Model of the plant is presented on Fig. 161. It consists of the generator fixed on rotating bar, reflection grating made of rectangular bars and dielectric waveguide connected with the grating and adjusted to the indicator of incident power in the form of detector and amplifier with measuring attenuator. Typical experimental curve (1) of the power generated in the waveguide versus aimed distance d has extreme character. Hence, intensity of the wave excited in the waveguide has resonance behaviour with respect to varying degree of communication because transmission coefficient of the considered system is mainly determined by two factors: the value of waveguide's feeding coefficient k_1 in the domain of communication, which exponentially decreases with growing d due to similar character of the fall of slow grating's exciting harmonic, and the coefficient $k_2 = 1 - P_{rad}P_0^{-1}$ (P_0 is the power brought into the waveguide and P_{rad} is the power of radiation losses) describing radiation losses of the wave propagating along the grating. One may conclude that maximal power brought into the waveguide corresponds to zero aimed distance. However in such regime (of strong communication) the losses of energy radiated out of the waveguide to free space (coefficient k_2) are maximal. It is shown in [52] that with increasing d the ratio $P_{rad}P_0^{-1}$ falls almost exponentially and it is the reason for the growth of k_2 with decreasing level of communication in the system waveguide-grating. The total system's transmission coefficient has the form $k = k_1k_2$ what yields its extreme character.

Hence, anomalous decrease of the power generated in the waveguide for small aimed distances is connected with radiation losses of waveguide wave in the domain of interaction. It is confirmed by curve 2 (Fig. 161) which has been obtained in such a way that incident angle of the exciting plane wave is adjusted with maximal intensity of waveguide wave for every value of aimed distance. Identical behaviour of the curves 1,2 on Fig. 161 are evidence of the fact that extreme properties of the power of waveguide mode versus aimed distance is determined by radiation losses of waveguide's energy.

Decrease of the transmission coefficient occurring for the element of diffractional input caused by diffractional losses may be obviously followed, if to compare two types of systems waveguide-grating (having equal period $l = 2.78$ mm, slotwidth $d = 1.26$ mm and slotdepth $h = 1.0$ mm): 1) the grating is external with respect to the waveguide; 2)the grating is a single whole with the waveguide. The output ratio of generated powers for both types of systems was 12–13 dB for the wavelength $\lambda = 5.78$ mm.

Efficiency of the diffractional input of energy to dielectric waveguide through its lateral surface were estimated by comparing output and incident powers in accord with calculation formula

$$\eta = \frac{\int_0^{\theta_0} f^2(\theta)d\theta}{\int_0^{2\pi} f^2 \sin(\theta)d\theta}, \tag{8.11}$$

where θ is the polar angle in spherical coordinate system; $f(\theta)$ is the far field pattern of irradiating device; $2\theta_0$ is the angle of observing the aperture from the point of irradiation. Experimental data together with calculations carried out by (8.11) show that the behaviour of input efficiency versus frequency has resonance character reaching the level of 75 per cent in the central part. The value of resonance frequency is determined by grating's parameters (l, h, d).

The considered method of diffractional input has high efficiency and is characterized by the possibility to choose optimal aimed distance what opens the way for its applications as elements of integrated schemes in the millimeter, submillimeter and optical ranges.

§29. Diffractional logical elements of optical computer

The charged particle, electron, or the flow of electrons moving in a conductor in the given direction of the definite electric chain serve as carriers of information in modern computers. Operation speed depends on the duration of signal's transmission which is carried out by means of logical elements. Raise of heating due to increasing degree of integration leads to premature destruction of these elements. In this connection it has been pointed out in [205] to the necessity of creating new types of logical elements consuming low power.

One of the trends here is applications of the methods and technologies used in fibre optics [205]. Devices manufactured on the basis of optical fibres may serve both as active elements and transmission channels. They promote noise protection, provide stability with respect to external influences and efficiency in non-stationary conditions, enable to lower size and weight. Two parallel (coupled) fibres (light waveguides) may be considered as the principal standard block of the logical optics.

Work of the known thin-film laser logical elements realizing the "no" operation is based on extinguishing the laser's generation. Application of electro-optical effect in the system of two fibers, where one acts as a detector of light energy and modulates the wave of the other, provides the fulfillment of "and", "or" and "no" operations.

The schemes of logical computers are also described where it is used the dependence of light transmission coefficient on the voltage applied to dielectric and the phenomenon of interference of wave beams.

Characteristic feature of the logical elements both made of connected fibers and the ones whose work is based on the passage of volume light pencils through dielectric plates is the same nature of waves, namely, surface waves of either fibers or light pencils. It is used here the phenomenon of transformation of surface waves to volume waves and vice versa, the modified SPE, what once more confirms its broad capabilities.

1. Let us consider the of use of plane dielectric waveguide connected with diffractional grating [54] as a logical element. Assume that the fiber is is placed in the closest vicinity of the grating perpendicular to its strokes and well-focused light pencils having plane phase front within the spot of incident field falls on the grating. Then the scattered field may be found as a result of solution to the problem of diffraction of homogeneous plane wave on periodical structure [62]. It consists of two qualitatively different parts: spectrum of homogeneous plane waves going off the grating to the free space and inhomogeneous plane waves localized close to the grating. The following relationship holds:

$$\tan \alpha_n = \frac{n - k \sin \alpha}{\sqrt{k^2 - (n - k \sin \alpha)^2}}, \tag{8.12}$$

where α is the incident angle of light pencil, α_n is the angle at which the n-th harmonic goes off the grating, l is the grating's period, λ is the wavelength of incident field, $k = l\lambda^{-1}$.

Under certain conditions ($k < |n - k \sin \alpha|$) the left-hand side in (8.12) is purely imaginary magnitude and corresponding harmonic is exponentially decreasing inhomogeneous plane wave whose phase velocity

$$v_\Phi = c \frac{k}{n - k \sin \alpha} \tag{8.13}$$

is less than the velocity of light c.

Every waveguide of surface waves (a fiber in the considered case), being connected with periodical structure, is excited in a resonance way and it is the key moment in the solution to the problem of input of radiation to the waveguide (fiber). Resonance angle, for which the energy generated in the fiber will be maximal, is determined by (8.12), (8.13). If the incident angle differs from its resonance value, the level of generated power is in several orders less than in the case of resonance.

Note that incident light pencil and fiber's exit are called input and output of the element, respectively. Hence, output values of zero and unit will correspond to the same input values in the case of space resonance.

The capabilities of this scheme (see Fig. 162, where the principal (a) and equivalent (b) schemes of the element are presented) essentially broaden, if one uses two coherent wave pencils with different incident angles. Assume that orientation angle of the first pencil yields resonance excitation of the waveguide at another harmonic. Then emergence of output signals corresponding to 0 and 1 will be determined also by the phase difference of two interfering waves and one obtains the element with two inputs X_1, X_2 and one output Y_1. Let us note that equivalent schemes of the considered logical element and of every other one coincide.

2. It is known that logical element realizes logical function, if the output Boolean vector corresponds to the values of this function. In addition to this, logical elements are characterized by definite direction of operations, flexibility and efficiency of the mounting of logical schemes.

Let us consider as an example the logical addition ("or") which is described by the following output table of values

X_1	0	1	0	1
X_2	0	0	1	1
Y	0	1	1	1

Let us go back to the scheme shown on Fig. 162 operating in such regime that phase difference of the channels X_1, X_2 yields synchronous addition of powers at the output Y. Assume 0 and 1 corresponds to such input and output signal which may be definitely separated by the registering device. Hence, in order to apply the considered element for realization of this logical function it is sufficient to consider signal's behaviour at the output Y for different values of X_1, X_2.

Corresponding data presented on Fig. 163 are evidence of the identity of input and output information and one may conclude that the considered electrodynamical system consisting of diffractional grating and of the waveguide connected with it realizes logical addition and may be called optical logical element. The same system operating with other set of parameters (incident angles of light pencils, phase difference between them) enables one to realize other logical functions. Their capabilities become clear after the analysis of Table 9.

Let us note that the proposed system fiber-diffractional grating may serve as a logical element, if to take fibers with arbitrary cross-sections and types of gratings.

§30. Formation of holograms with the help of Smith-Parcell radiation.

1. Obtaining of holograms is connected with registering of interference picture formed by superposition of supporting wave and subject wave scattered by radiating object and coherent with the former. Photo-plate in the domain of waves' interaction becomes identical to diffractional grating after the processing and exposition. In order to reconstruct the subject wave photo-plate is placed at the same area and with the same orientation as in the process of exposition and the wave identical to the supporting one is directed to it.

Holograms differ by the ways of processing of photo-sensitive layer (phase or amplitude), depth of the fixed interference picture (plane or volume), combination of phase fronts (plane or spherical), types of subject and supporting waves and holography angle (axis or non-axis). Diffraction of volume (plane or spherical) waves of reconstructing field on a hologram acting as diffractional grating is applied in all types of holograms at the stage of reconstruction of the image.

In the case of two homogeneous plane waves having the wavelength λ the hologram-grating is formed; it has the period l and interference angle 2ϕ such that $l = \lambda(2\sin\phi)^{-1}$, i.e., $l \approx 0.5\lambda$ when $\phi \sim 90°$ (the minimal period of Denisuk's hologram [206]). For small holography angles $l >> \lambda$ what yields emergence of several images after the reconstruction of holograms and, consequently, lowers the quality and diffractional efficiency of the main image. Hence, the nature of physical phenomenon of the diffraction of volume waves on periodical structures leads to the presence of background flow which is observed together with reconstruction of real and imaginary pictures after irradiating the hologram by the light from reconstructing source. In order to raise diffractional efficiency one should apply the Smith-Parcell effect, namely, diffraction of surface (inhomogeneous) waves on interferogram playing the role of diffractional grating [62].

2. Let us consider formation of hologram as a result of interference of volume (subject) and surface (supporting) waves [208]. Assume that planar dielectric waveguide is excited by a homogeneous plane light wave (the case of normal incidence, Fig. 164) while the slow wave, which is coherent with incident wave, propagates along the axis Oz close to waveguide's surface. The lines of equal phases of slow (bold vertical lines on Fig. 164) are perpendicular to the surface and the lines of constant amplitude are parallel to the surface.

Let us describe interference picture emerging after interaction of running inhomogeneous plane wave and incident plane wave (Fig. 164). There it is taken into account that interference maximums (shown by bold lines inclined at the angle $\alpha = \arctan \beta^{-1}$ to Oz, $\beta = v_\Phi c^{-1}$, v_Φ is the phase velocity of slow wave, c is the velocity of light in vacuum) correspond to the points in the space where phase difference of interfering waves vanishes [209] and form diffractional grating with the periods $l_1 = \beta\lambda$ along the axis Oz and $l_2 = \lambda\sqrt{\beta^2(1+\beta^2)^{-1}}$ along the direction perpendicular to the lines of maximums. Hence, one can change hologram's period in broad limits for fixed holography angle varying the deceleration of waveguide's supporting wave. Lines of constant values of dimensionless grating's period in the direction of the axis Oz are shown on Fig. 165 ($k = l\lambda^{-1}$, 2ϕ is the interference angle, β is the deceleration of surface wave). It is essential that the considered method where surface wave is applied as the supporting wave enables one to get the hologram whose period is smaller than of the one obtained by the method of homogeneous plane waves. This factor creates preconditions for reducing the number of images when reconstructing the hologram.

If to place now the photo-sensitive layer in the domain adjoining the fiber (see Fig. 164), one will get the hologram with characteristics given by the formulas presented above after exposition and development of the photo-layer.

3. Advantages of the analyzed method of holography become clear after comparison with other existing approaches. Let us consider specific example. Assume that the hologram is formed by two volume waves one of which (supporting) propagates along the axis Oz and the other (subject) – along the axis Ox. The hologram-grating with the period $l = \lambda$ counted along Oz is formed as a result of waves' interaction. After the reconstruction of

such hologram with the help of homogeneous plane wave one obtains three space harmonics in reconstructed light and, consequently, three object's images. For example, if to use the surface wave with deceleration $\beta = 0.7$, the obtained hologram-grating will have the period $l = 0.7\lambda$. Irradiation of this hologram by homogeneous plane wave yields emergence of two harmonics in reconstructed spectrum, i. e., two images. If to reconstruct the hologram by the method based on SPE, it is sufficient to choose the reconstructing surface wave having relative phase velocity $\beta = 0.7$ and finally one will get the *single* image in reconstructed hologram.

The possibility to get single image in reconstructed hologram is not limited by the holography angle $2\phi = 90°$. Shaded domain on Fig. 165 corresponds to the single-image zone, which exists for practically all holography angles.

We may conclude that application of Smith-Parcell effect for combined creation and reconstruction of holograms allows to obtain single image in reconstructed hologram. It forms the basis for new direction in the holography – integrated holography.

§31. Application of Smith-Parcell effect for generation and amplifying of millimeter and submillimeter waves in semi-conductors

Transformation of surface waves to volume waves and vice versa – the Smith-Parcell effect – are the distinctive features of active semi-conductor structures [210–212].

Let us consider interaction of drifting charges with the field of active system semi-conductor+grating within the frame of quasi-hydrodynamical model [210, 211]. Assume that semi-conductor sample in the form of infinite plate of the width h with mono-polar (electron) conductivity is placed in constant longitudinal (along the axis Oz) electric and magnetic fields at the distance a from the strip metal grating (having period l and gap between strips d) whose axis is directed along the axis Ox.

Perturbation of density of charge's carriers emerges in semi-conductor, which has the form of oscillating component of current's density, under the influence of random electromagnetic field generated in a flow of drifting carriers. Fluctuations of space charge emerging in semi-conductor due to non-balanced character of the system propagate as waves of space charge. After being scattered by the grating they are transformed to electromagnetic

Smith-Parcell radiation. SPR is either localized near the grating or propagates to the free space.

Self-coordinated system of equations for determination of SPR evaluated within the frame of quasi-hydrodynamical model consists of Maxwell equations, continuity equations and equations of movement. Looking for the solution in the form of monochromatic waves, we find the fields in semi-conductor which are described by Bloch functions. In the free space fields are determined as the solutions to homogeneous wave equation. Applying operators of boundary value conditions, one obtains equivalent system of functional equations which is reduced by the method [61] to infinite system of linear algebraic equations of the second kind. Zeros of its determinant or, in other words, roots of characteristic (dispersion) equation of the considered electrodynamical problem determine surface and volume waves of the structure semi-conductor+grating. In particular, one has the following representation for complex propagation constant of the waves of space charge in the case of excitation of one space harmonic

$$z_p = \frac{\kappa}{\beta} + \mu' + i\mu'' = \frac{\kappa}{\beta} + \begin{cases} P\omega_p\omega^{-1}T_1 + 0.5i\nu\omega^{-1}T_2, & \omega^2 >> \nu^2; \\ P\omega_R\omega^{-1}T_3, & \omega^2 << \nu^2; \end{cases} \tag{8.14}$$

which takes into account the losses caused by generation of SPR within the frame of plasma and relaxation mechanism of swinging the charge's oscillations; here ω_p, ω_R are the frequencies of plasma and relaxation oscillations; P are the roots of dispersion equation;

$$T_1^{-1} = \frac{[a_1(3+u)(1-u)^2 + 8](1-u)a_2 + a_3}{(1+a_1)a_3 + a_2\{16 - (1-u)[a_1(3+u)(1-u)^2 + 8(1-a_1)]\}};$$

$$a_1 = \frac{(\epsilon_0^2 - 1)\tanh H}{2\epsilon_0 - (\epsilon_0^2 - 1)\tanh H}; \quad a_2 = 1 + i\kappa a_4 \ln\frac{1+u}{2}$$

$$a_3 = 4i\kappa a_4(1-u)^2; \quad a_4 = 1 - i\frac{1 - \epsilon_0 \tanh \bar{H}}{2\sqrt{\epsilon_0} + i(\epsilon_0 - 1)\tanh \bar{H}};$$

$$H = \frac{2\pi h}{l}; \qquad \bar{H} = kh\sqrt{\epsilon_0}; \qquad u = \cos\frac{\pi d}{l};$$

T_2, T_3 are defined in a similar way; ϵ_0 is the static permittivity of semi-conductor. It is clear that in the case of absence of collisions z_p is also a complex magnitude; its imaginary part characterizes increment (decrement) of the growth (fall) of the waves of space charge.

2. SPR excited by the structure semi-conductor+grating consists of the set of $n\,(n = 0, \pm 1, \pm 2, \ldots)$ space harmonics. Excitation of one harmonic may occur for definite combination of the parameters of semi-conductor, external medium, voltage and dimensions of the grating. Infinite sequence of electromagnetic waves numbered by the index p (with propagation constants z_p given by (8.14)) and connected with the waves of space charge having radiation losses corresponds to every n-th space harmonic. We will assume that phase velocity of fast (slow) wave is greater (smaller) than unperturbed velocity of the drift of charge's carriers in semi-conductor. Slow waves are characterized by the values $\mu'_p > 0$ and fast waves – by the values $\mu'_p < 0$.

Waves may grow or fall while propagating in the semi-conductor. The waves of space charge increasing in the volume of semi-conductor lead to formation of the packages of drifting charge's carriers and, consequently, to generation of coherent SPR. Modulation of carriers' flow increases together with SPR due to inverse influence of radiation to active medium (semi-conductor).

Maximums of the increment characterizing the growth of the waves of space charge was registered for the structures whose period (covered by the metal) was equal to 60–80 per cent. Increment grows with increasing concentration of carriers and falls with increasing collisions' frequency. Small dissipation losses caused by rare collisions do not perturb field's distribution as well as dependence of increment versus structure's parameters; however, they yield increment's reduction, in particular, it falls with decreasing effective mass m^* of the carrier in the domain of semi-conductor's conductivity. Increment essentially grows in the case of resonance distance between reflecting screens.

These results may be generalized for semi-conductors described within the frame of two-zone model when $\tilde{\epsilon} = [0.5\epsilon_g (m^*)^{-1} (0.5\epsilon_g m^* + p^2)]^{0.5}$, where $\tilde{\epsilon}$ is the energy of electron in the domain of semi-conductor's conductivity counted from the middle of prohibited zone; p is the quasi-impulse; ϵ_g is the width of prohibited zone. If to replace $0.5\epsilon_g (m^*)^{-1}$ by c^2, one obtains relativistic relationship for the energy and impulse [213]. It allows by analogy with this technique to use "relativistic" equation of movement for electron gas in the semi-conductor with non-parabolic zone [214] and finally to obtain similar results within the frame of such "relativistic" approach assuming that the frequency of plasma oscillations is normalized in accord with the formula

$\omega_p^2 \leftrightarrow \omega_{rel}^2 = \omega_p^2(1 - 2m^* v^2 \epsilon_g^{-1})^{1.5}$.

Taking into account non-quadratic dispersion leads to the fall of increment due to the growing effective mass and simultaneous increase of carriers' energy caused by "relativistic" effect. Non-linear dispersion in semi-conductors may be the reason of essentially non-linear phenomena like generation of harmonics and combination effects. In case of weak non-linearities initial system splits into linearized equations modeling interaction at every frequency harmonic belonging to spectral field's expansion. One can show that if phase velocities v_Φ^ω and $v_\Phi^{2\omega}$ of the waves of space charge (producing the components of SPR at the frequencies ω and 2ω) are synchronized, corresponding propagation constants are equal.

Hence, coherent SPR or induced radiation, which emerges in the considered structure semi-conductor+grating, may amplify electromagnetic waves. Let us note that it is preferable to apply plasma mechanism of swinging the charge's oscillations ($\omega_p^2 >> \nu^2$). The condition $\omega^2 >> \nu^2$ determines the frequencies providing the most effective excitation of oscillations in the considered structure. As an example let us take semi-conductors (InSb)+grating with the parameters $\epsilon_3 = 18$; $m^* \approx 0.013 m_0$; $n_0 \approx 10^{14}$ cm^{-3}; $2\pi h l^{-1} = 60$; $\nu \approx 10^{11} c^{-1}$; $\epsilon_1 = \epsilon_2 = 1$; $l = 2 \cdot 10^{-3}\lambda$ is the grating's period; 80 per cent of the grating on the period is metalized what yields considerable amplifying of the signal.

The structure semi-conductor+grating placed in OR has been also analyzed [212]. Here the generated SPR feeds OR and one obtains semi-conductor analogue of GDR. Investigation of this generator within the frame of the model of given field enables one to assert that existing level of manufacturing technology of semi-conductor materials ($\nu \approx 10^{11} c^{-1}$) together with high drift velocities of the carriers and small grating's periods guarantee the creation of semi-conductor GDR with high characteristics for applications in the millimeter range.

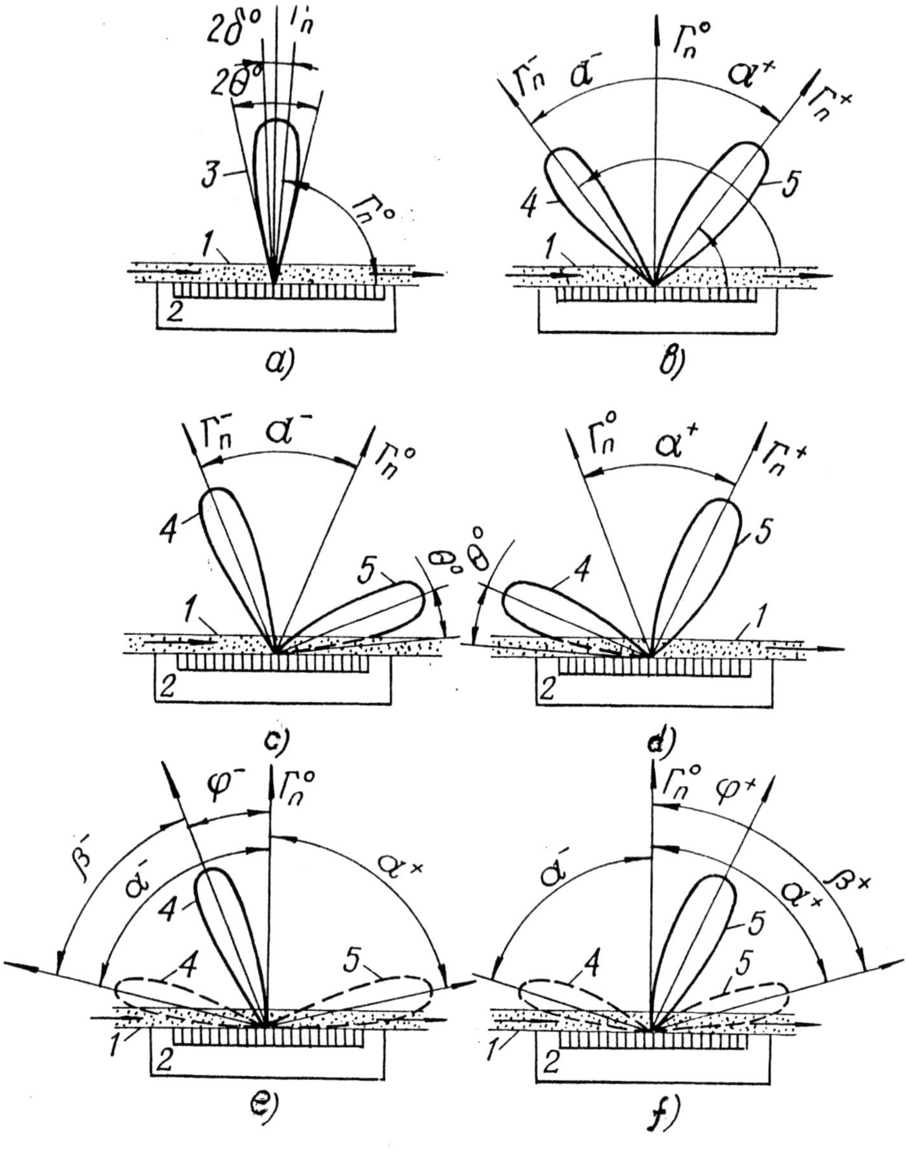

Fig. 147. Variants of excitation of the radiation far field pattern of Smith-Purcell radiation of electron WSC in the case of determining the reduction coefficient of plasma frequency (a–f).

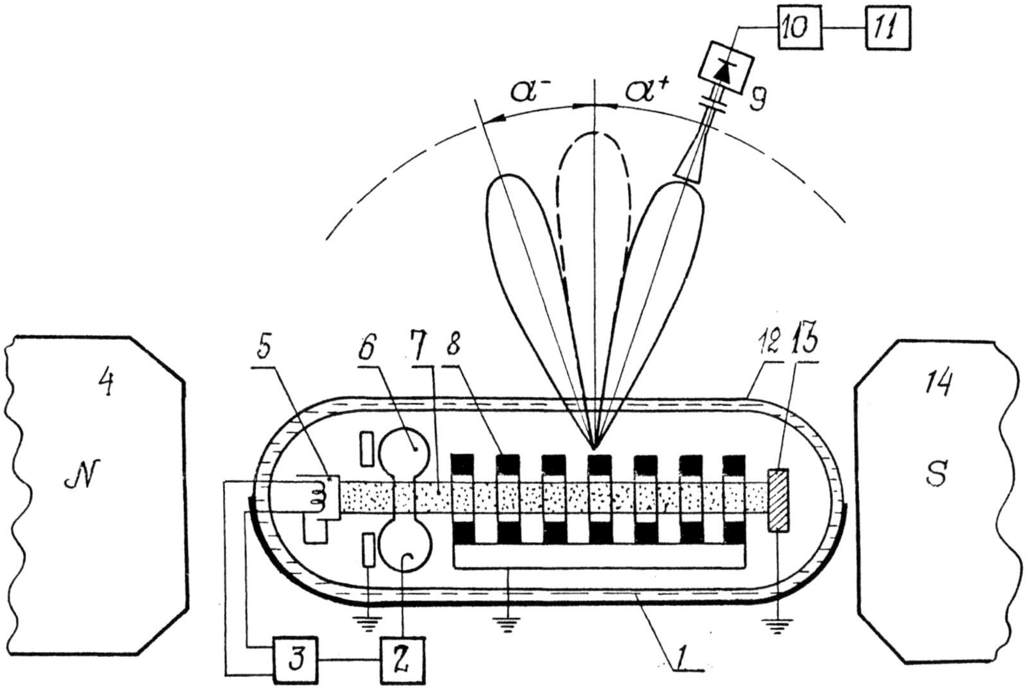

Fig. 148. Analyzer of the modulated electron flow in cm wave range.

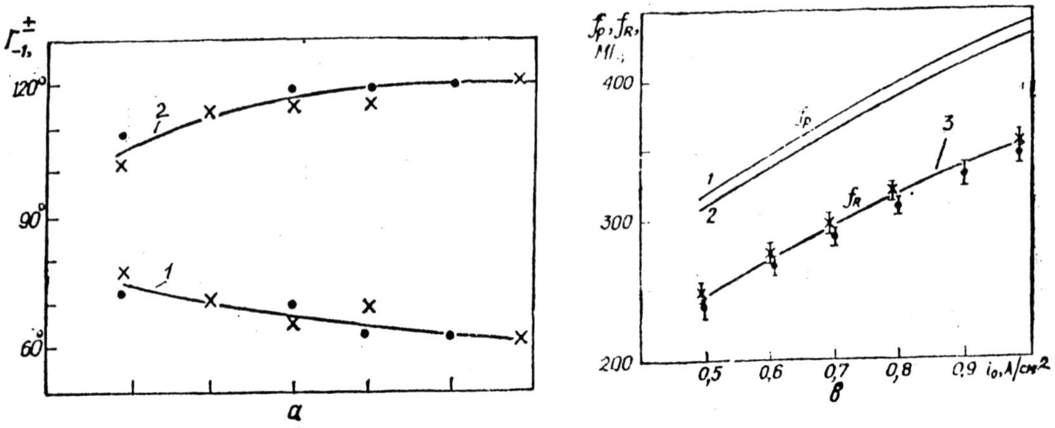

Fig. 149. Measured angles of diffractional radiation (a) and effective plasma frequency (b) versus current's density: a) 1 – radiation angles Γ^{+}_{-1} of SWSC; b) theoretical data for the non-reduced plasma frequency (1,2); 1 – the comb, $U_0 = 1820$ V; 2 – plane with the ring, $U_0 = 2000$ V; 3 – experimental data.

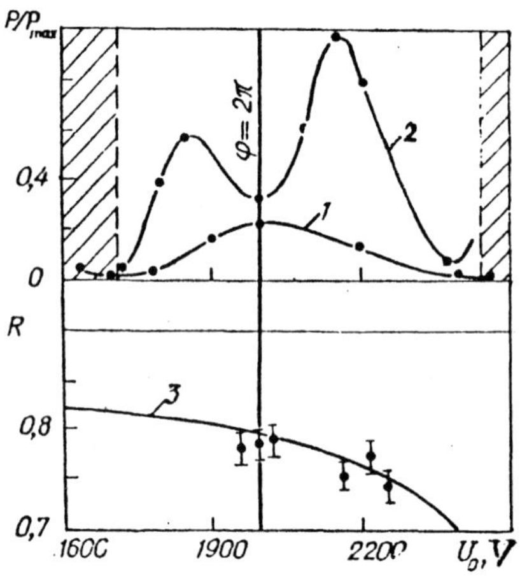

Fig. 150. Intensity of Smith-Purcell radiation and reduction coefficient versus the voltage U_0 of cylindrical flow in the channel of the plane with rings ($f \sim 8.9$ GGz): 1 – $i_0 = 0.1$ A·cm^{-2}; 2 – $i_0 = 1.2$ A·cm^{-2}; 3 – theoretical curve for the electron flow in the tube of drift; vertical dot lines divide the domains of excitation of surface waves (shaded domain) and the waves tearing off the grating; the line of phases' shift at the grating's period $\phi = 2\pi$ corresponds to the value $U_0 = 2000$ V where it is observed radiation at the angle $\Gamma^0_{-1} = 90°$.

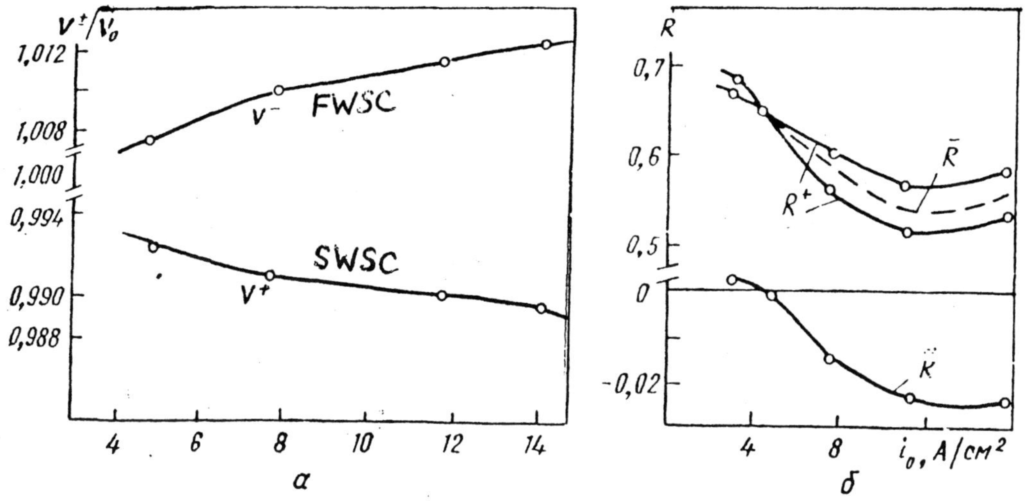

Fig. 151. Influence of current's density i_0 on the phase velocities of electron waves (a) and dynamical reduction coefficient of plasma frequency in the electron flow interacting with the scattered field of the millimeter wave range (b).

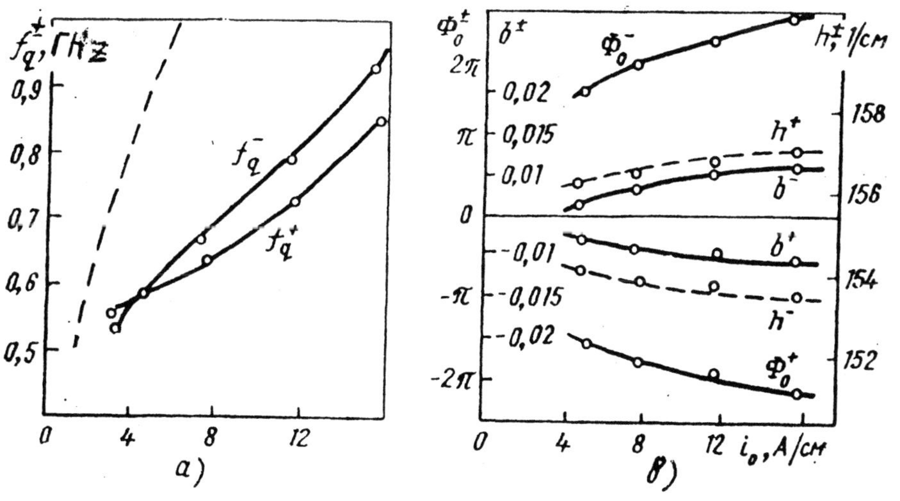

Fig. 152. Experimental data of the dynamical parameters of electron flow interacting with the scattered field in the four-millimeter range.

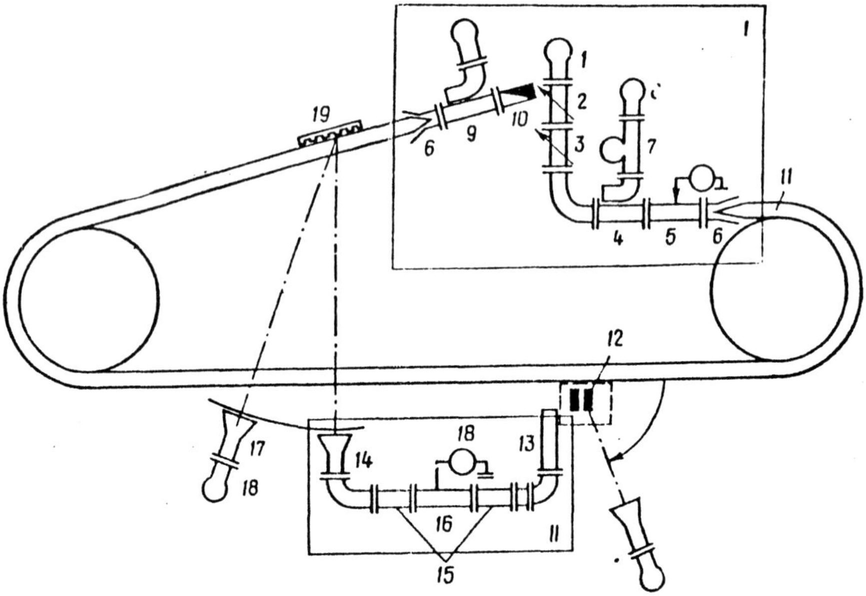

Fig. 153. Scheme of the plant for the antenna measurements of Smith-Purcell radiation.

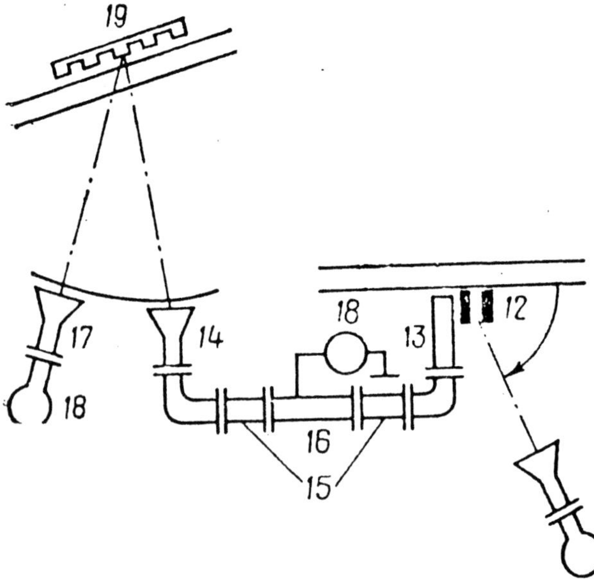

Fig. 154. Interferometer measuring plant for the study of wave Smith-Purcell effect.

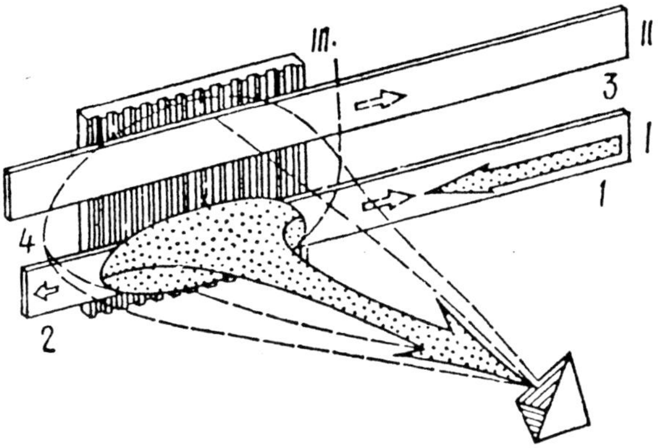

Fig. 155. Receiving-transmitting antenna on the basis of Smith-Purcell effect.

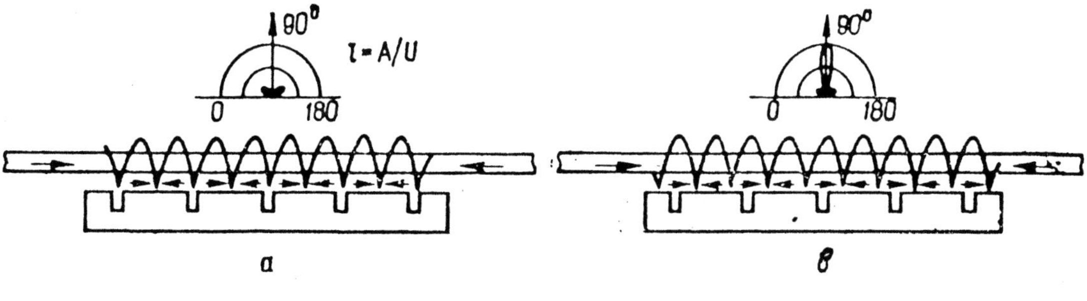

Fig. 156. Operation principle of the interferometer created on the basis of application of Smith-Purcell effect.

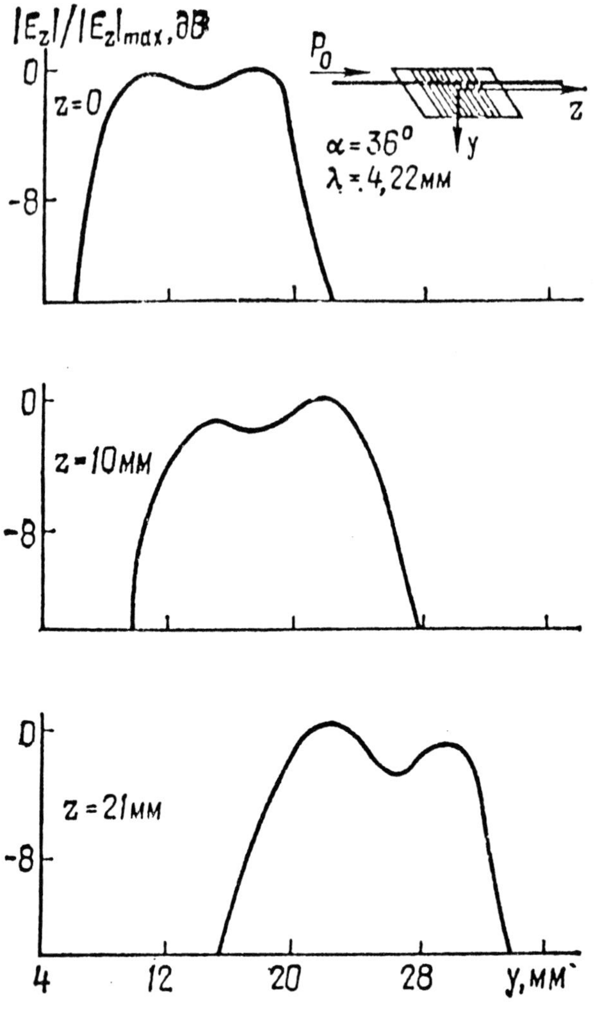

Fig. 157. Formation of field by the beveled grating.

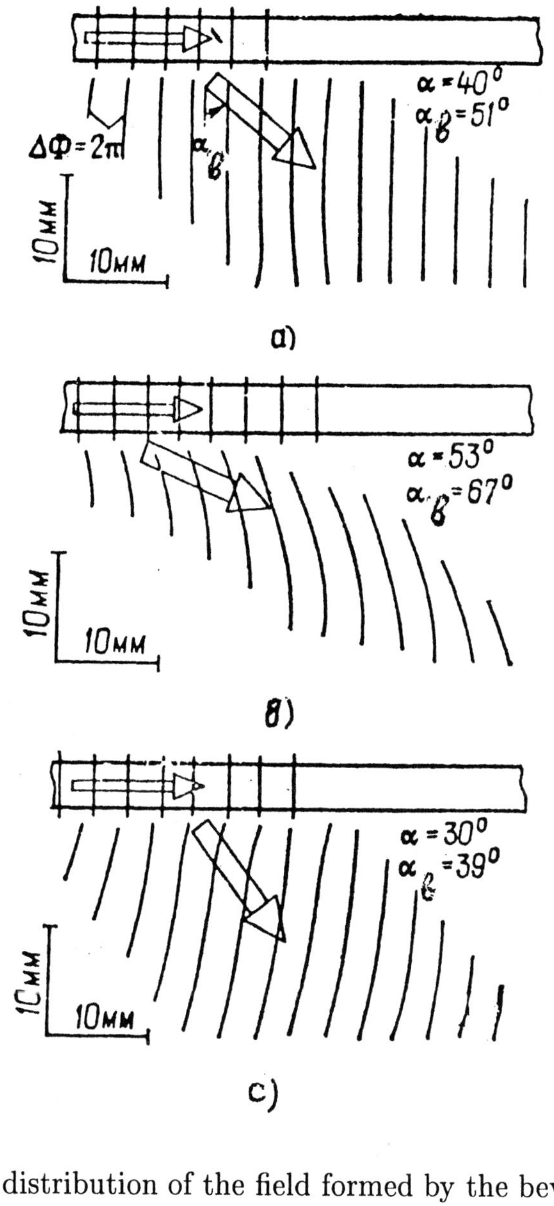

Fig. 158. Phase distribution of the field formed by the beveled grating.

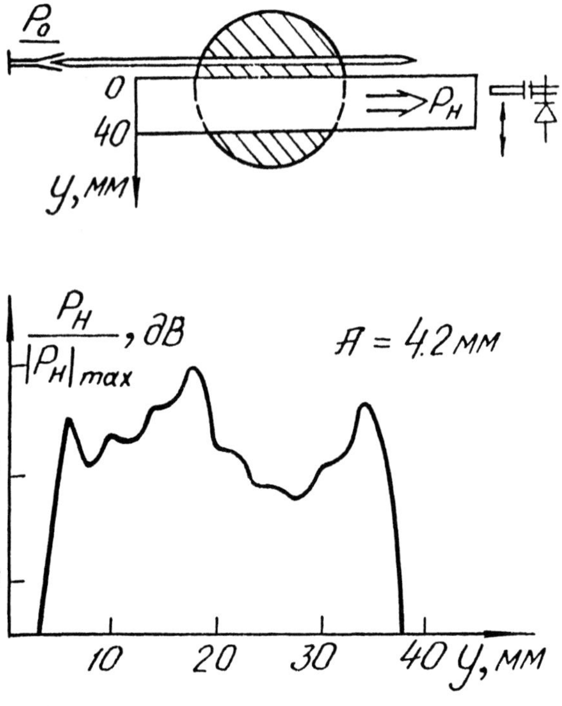

Fig. 159. Excitation of planar dielectric waveguide by the beveled grating.

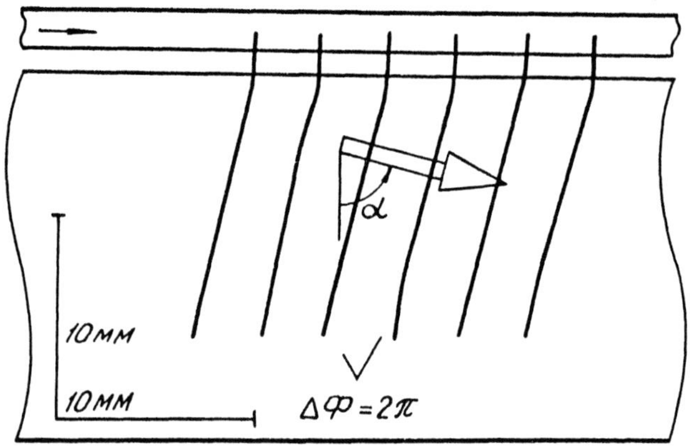

Fig. 160. Lines of equal phases of the planar dielectric waveguide excited by beveled grating.

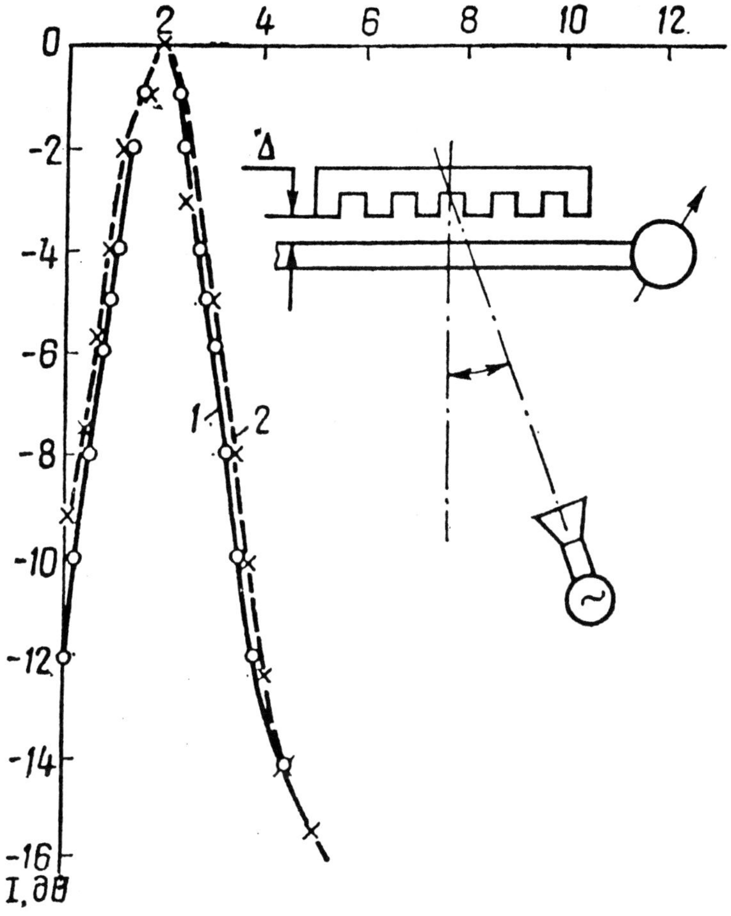

Fig. 161. Input of energy to the dielectric waveguide.

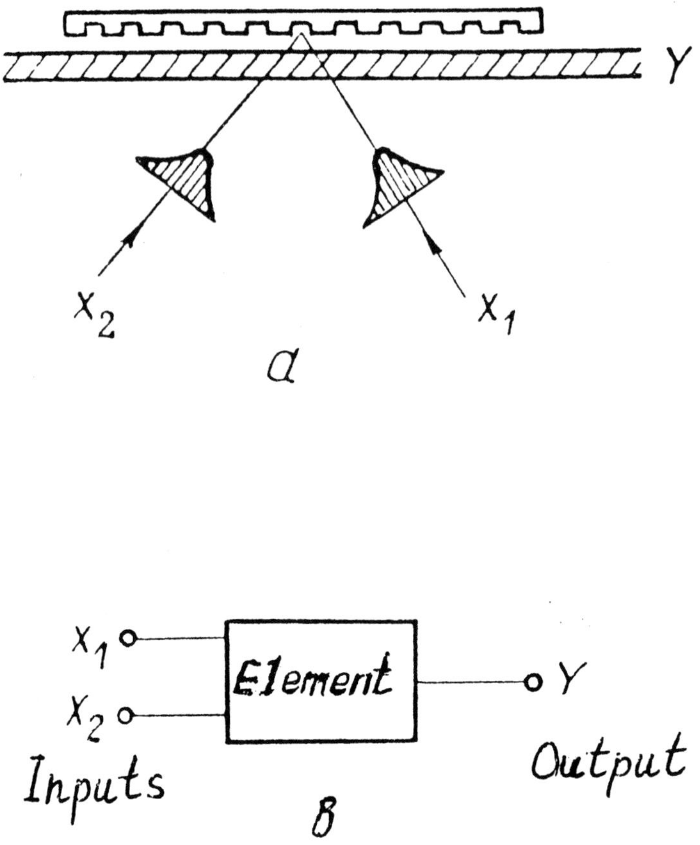

Fig. 162. Scheme of the logical element based on the Smith-Purcell effect.

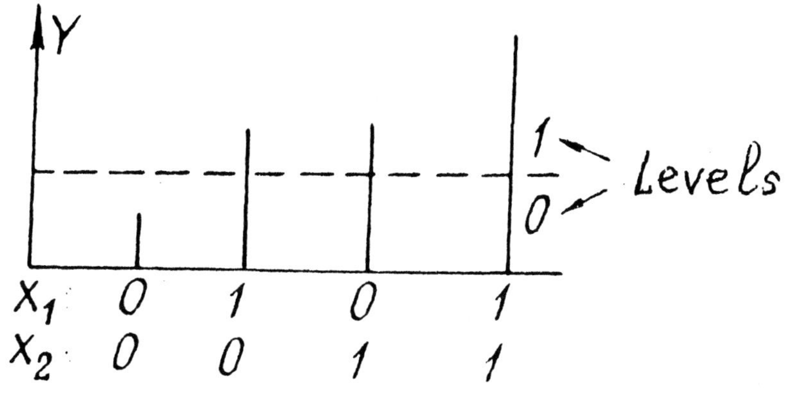

Fig. 163. Scheme of the carrying out of logical function – disjunction.

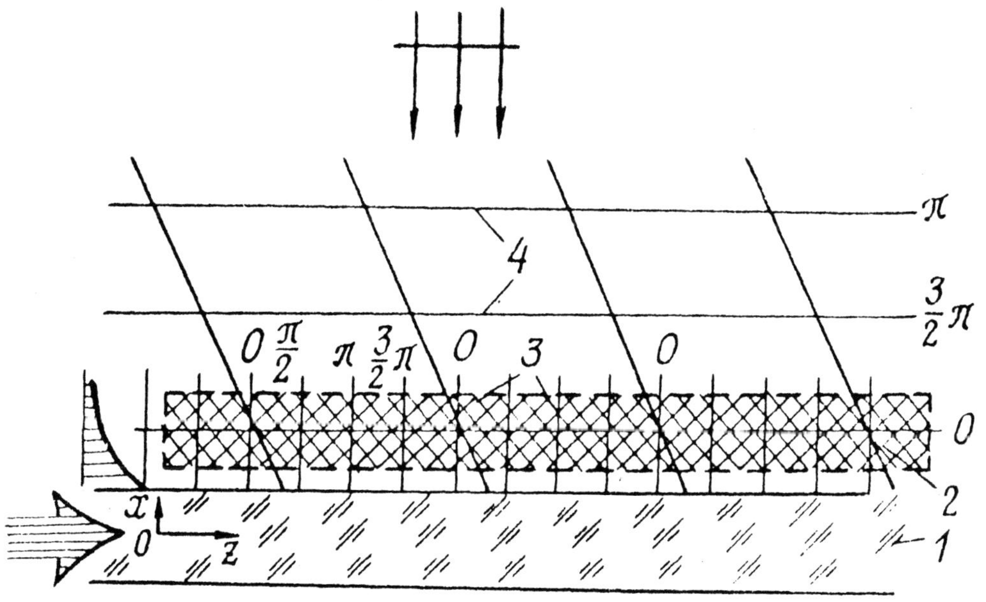

Fig. 164. Scheme of the creation of hologram: 1 – fiber; 2 – photo-sensitive layer; 3 – planes of constant phase of surface wave; 4 – planes of constant phase of homogeneous plane wave.

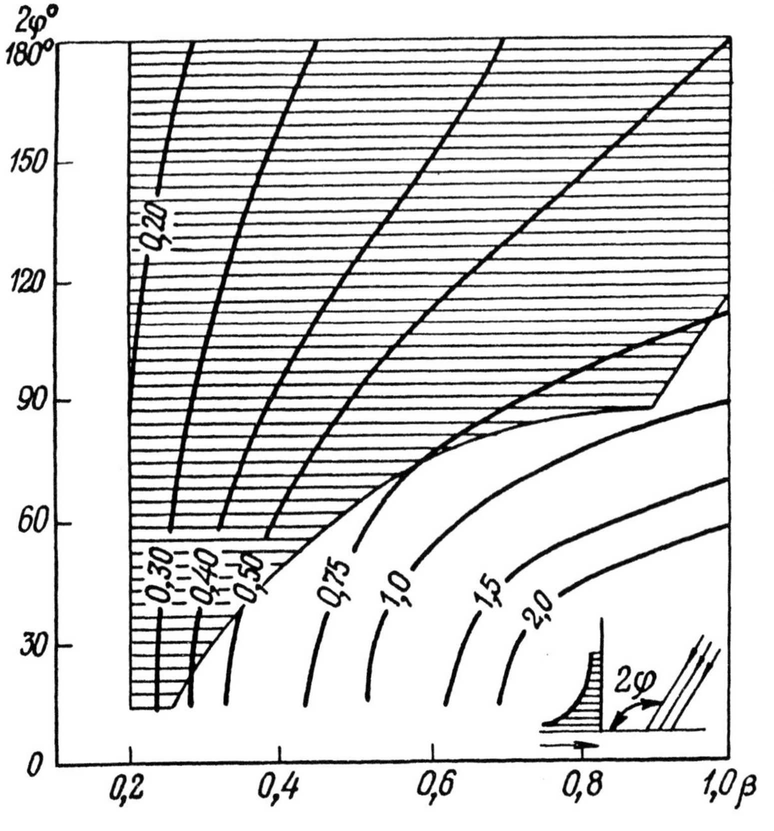

Fig. 165. Lines of constant values of hologram's period. Shaded domain corresponds to the zone of unique reconstructed image.

Table 8. Parameters of the flow and periodical structures

Structure	Length [mm]	Period [mm]	Slotwidth [mm]	Transversal size, [mm]	Flow	Diameter of flow's cross section, [mm]
Comb	65	2.6	1	5	Strip	3 * 0.4
Plate with rings	60	3.0	1.5	3 * 6	Cylindrical	2.6

Table 9. Level of output signal

Operation		Level of output power				Scheme's parameters
	X_1	0	1	0	1	
	X_2	0	0	1	1	
Constant (zero)		0	0	0	0	$\alpha^{(1)}$, $\alpha^{(2)}$, - non-resonance
Non-equivalence		0	1	1	0	$\alpha^{(1)}$, $\alpha^{(2)}$, - resonance
Repetition		0	1	0	1	$\psi_2 - \psi_4 = \pi$ $\alpha^{(1)}$ - resonance - $\alpha^{(2)}$ - non-resonance
Repetition		0	0	1	1	$\alpha^{(1)}$ - non-resonance $\alpha^{(2)}$ - resonance
Addition (or)		0	1	1	1	$\psi_1 - \psi_2 = 0$ $\alpha^{(1)}$, $\alpha^{(2)}$, - resonance

CONCLUSION

The results presented in the monograph reflects to a marked degree international state of the studies of SPE and especially in the countries of the former Soviet Union. The Kharkov Institute of Radiophysics and Electronics of the Ukrainian Academy of Sciences is the universally recognized leader in this area. Fundamental data concerning basic properties of SPR, its display in various branches of physics and possibilities for applications in science and technology have been obtained here on the basis of specially created theoretical and experimental methods.

Legitimate questions arise. Have all the problems been solved and have we achieved the complete physical understanding of SPE (it equally concerns the Cherenkov effect)? Does SPE require elaborating new theoretical and experimental approaches for its investigations? Have all the ways of its applications been exhausted?

These problems as well as many other ones will stir researchers in the future since SPE and the Cherenkov effect are those fundamental phenomena of the nature whose cognition forms the basis for explanation of more complicated effects which arise in the process of movement of charged particles in the presence of various obstacles.

Essence of the first halting place is in the studies of non-linear development of SPE for independently existing positive and negative dispersions. It allows to determine initial stage of the formation of SPR for the flow of charged particles moving close to diffractional grating, to follow this process up to its establishment and finally to determine the efficiency of SPR. Dispersion and non-linearity may cause emergence of non-linear waves whose evolution strongly influences SPR.

The second of principle new stage is devoted to the cardinal problem of investigating SPE in the vicinity of non-linear evolution catastrophic situations. Let us note first of all that until recently non-linear evolution processes have been studied only for independently existing positive and negative dispersions. It is known, in addition to this, that the dispersion near Morse critical point (MCP) may occur for linear processes and its characteristic feature is in sharp transition from positive to negative dispersion and vice versa [215]. In MCP itself the catastrophic situation is observed when small deviation of non-spectral parameter yields strong changes of dispersion and other properties of the system and, hence, essential perturbations of evolution process.

One may propose the following scheme for the study of SPE in the vicinity of MCP. At first the dispersion equation is verified for the given open electrodynamical system. Then the relationships for spectral and non-spectral parameter are obtained in the form of complex-valued functions of several complex variables. These functions are represented in the form of Taylor series in the vicinity of MCP and the cut dispersion equation is treated as quadratic equation. By means of the latter one can write down the linear evolution Klein-Gordon equation. Corresponding partial differential equations have local character and the solution to dynamical system should be investigated for such values of time parameter which provide variations close to MCP. Since the system is characterized by catastrophic changes near MCP, the same situation occurs for (linear) evolution process.

In order to introduce non-linearity (for example, in the case of weak non-linear dependence), the coefficients of linear evolution equation are expanded in the series with respect to the sought for field produced in SPE; then one obtains non-linear evolution equation (NEE) in the vicinity of MCP taking only the first terms. NEE in the vicinity of MCP is the hybrid of Klein-Gordon and wave equations [216]. Such equations have universal character, since the presence of MCP in linear boundary value problems of electrodynamics as well as weak non-linearities in electronics are connected with the natural state of physical system. The next step is in looking for effective ways of solution to NEE. One of the possible approaches is in obtaining "auto-model solutions " which enable one to describe SPE "at the entry", "at the exit" and in the turning point (MCP) itself where the catastrophe variation of transformation of the energy of moving charged particles to SPR takes place. It is precisely the evolution of the catastrophe variation of physical system that is of the interest for deep understanding of physical process, since the "regular" evolution (determined by the negative or positive dispersion) yields the known behaviour of considered system in time.

This program of theoretical studies of SPE is inevitably connected with working out new experimental techniques, in particular, creating new tools for "thinner" measurements, since the display of SPE itself and its evolution near MCP are rather sensible to "movements" of non-spectral parameters. Experience in carrying out corresponding experiments reflected in [217-219] is evidence of the possibility to solve such problems.

The present approach is perspective for future applications in various domains of science and technology. Let us point out only to some of them. If SPR emerges in GDR when non-spectral parameters of its open electrodynamical system vary close to MCP, then there may exist such their values which will provide noticeable change of the steepness of electron readjustment and, consequently, the raise of generator's operation stability. In antenna technology small "movements" of the parameters enables one to control far field patterns in accord with given program. The study of SPE is of special interest in spectroscopy. One can increase the accuracy of spectrometers operating in the millimeter and submillimeter ranges with the help of appropriate choice of the parameters of open electrodynamical systems.

If to sum up, one should mention that in the given monograph some problems have been solved which arise in theoretical and experimental modeling of Smith-Purcell effect. At the same time, a number of original problems have been stated, whose solution would open the new ways for the cognition of this remarkable phenomenon of nature.

REFERENCES

1. S. J. Smith, E. M. Purcell, Visible light from localized surface. Charges moving across a grating, *Phys. Rev.*, vol. 92 , pp.1069-1073 (1953).

2. J. Jelly, *Cherenkov radiation and its application*, Izd. Inostr. Lit., Moscow, (1960).

3. V. P. Shestopalov, *Diffractional electronics*, Izd. Kharkov. Univers., Kharkov (1976).

4. V. P. Shestopalov, *Physical foundations of millimeter and submillimeter technology*, Vol. 1,2, Naukova Dumka, Kiev (1985).

5. V. P. Shestopalov, A. A. Vertij, G. P. Ermak, B. K. Skrinnik, G. I. Khlopov, A. I. Tsvik *Generators of diffractional radiation*, Naukova Dumka, Kiev (1991).

6. I. Paloch, A. Oliner, Self-coordinated Smith-Parcell and Cherenkov theory, In: *Selected talks at the Intern. Symposium on Diffraction*, Mir, Moscow, pp. 167-168 (1966).

7. A. A. Vertij, A. I. Tsvik, V. P. Shestopalov, Experimental observations of the effect of diffractional radiation in the millimeter range, *Dokl. AN SSSR*, vol. 9, pp.35-39 (1984).

8. A. I. Tsvik, L. I. Tsvik, Excitation of electromagnetic oscillations by an electron flow moving along the wave trajectory over diffractional grating, *Radiotekhnika*, No 20, pp. 24-32 (1972).

9. A. I. Tsvik, Influence of the waves of space charge of electron flow on diffractional radiation, In: *Propagation and diffraction of radio waves in the millimeter and submillimeter ranges,* , Naukova Dumka, Kiev, pp. 267-274, (1984).

10. A. A. Vertij, I. V. Ivanchenko, A. V. Nesterenko *et. al.*, Investigation of the thin structure of diffractional radiation, *Izv. Vuzov, Radiofizika*, vol. 28, pp. 812-823 (1985).

11. A. A. Vertij, I. V. Ivanchenko, A. V. Nesterenko *et. al.*, Investigation of the thin structure of diffractional radiation in the millimeter range, *Ibid*, pp. 1443-1449.

12. A. A. Vertij, I. V. Ivanchenko, I. V. Lopatin *et. al.*, Quasioptical complex for the study of radiation effects in the millimeter range, *Ukr. Fiz. J.*, vol. 31, pp. 347-350 (1986).

13. A. I. Tsvik, Electrone-wave theory of diffractional radiation, *Dokl.*

AN USSR, Ser. A, No 12, pp. 53-85 (1985).

14. A. I. Tsvik, Electrone-wave mechanisn of transformation of scattered field to radiation, *Izv. Vuzov, Radioelektronika*, vol. 29, pp. 58-64 (1986).

15. P. P. Maximov, A. I. Tsvik, V. P. Shestopalov, Effect of Doppler's lighting of energy of decelerated electrons in GDR, *Dokl. AN SSSR*, vol. 280, pp. 1353-1356 (1985).

16. A. A. Vertij, N. A. Popenko, A. V. Nesterenko *et. al.*, Investigation of polarization properties of electromagnetic field in a generator of diffractional radiation, In: *Proc. of the X All-Union Conf. on SHF Electronics*, Minsk, pp. 106-107 (1983).

17. A. A. Vertij, I. V. Ivanchenko, N. A. Popenko, A. I. Tsvik, Investigation of polarization properties of radiation of a generator of diffractional radiation, *Izv. Vuzov, Radiofizika*, vol. 30, pp. 104-112 (1987).

18. G. S. Vorobjev, A. V. Nesterenko, A. I. Tsvik *et. al.*, Investigation of physical processes of interaction of electron flow with scattered field, *Izv. Vuzov, Radiofizika*, vol. 31, pp. 805-812 (1988).

19. G. I. Aleksandrov, B. M. Zamorozkov, Y. A. Kalinin *et. al.*, Experimental studies of the structures of electron pencils in O- and M-type devices, *Elektronika*, No 8(108), p. 205 (1973).

20. G. S. Vorobjev, A. Y. Kirichenko, A. I. Tsvik *et. al.*, A method of determination of the reduction coefficient of plasma frequency in SHF devices (USSR 1077501, 01 1 9\42), *Otkr., Izobr.*, No 31, p. 248 (1985).

21. A. V. Nesterenko, A. I. Tsvik, V. P. Shestopalov, Miniaturization of the generator of diffractional radiation *Dokl. AN SSSR*, vol. 277, pp. 84-88 (1984).

22. I. M. Balaklitskij, B. K. Skrinnik, O. A. Tretjakov *et. al.*, Generator of diffractional radiation of the waves in the millimeter and submillimeter ranges, *Ukr. Fiz. J.*, vol. 14, pp. 539-552 (1969).

23. A. A. Shmat'ko, O. A. Tretjakov, V. P. Shestopalov, Non-stationary non-linear theory of the generators based on the effect of diffractional radiation, *Dokl. AN USSR, Ser. A*, No 1, pp. 69-72 (1976).

24. A. S. Bakai, K. A. Lukin, V. P. Shestopalov, Non-stationary non-linear theory of the generators of diffractional radiation, *Dokl. AN USSR, Ser. A*, No 4, pp. 335-338 (1978).

25. I. M. Balaklitskij, V. G. Kurin, B. K. Skrinnik, *et. al.*, Generator of diffractional radiation, (USSR 334605, 01 23\16), *Otkr., Izobr.*, No 12, p. 147 (1972).

26. I. M. Balaklitskij, G. S. Vorobjev, A. I. Tsvik *et. al.*, Experimental investigations of spectral characteristics of the generators of diffractional radiation, *Izv. Vuzov, Radiolelektronika*, vol. 21, pp. 105-109 (1978).

27. I. M. Balaklitskij, G. S. Vorobjev, A. I. Tsvik V. P. Shestopalov, Generator of diffractional radiation with the reflector of electron flow, *Dokl. AN USSR, Ser. A*, No 9, pp. 822-824 (1976).

28. G. S. Vorobjev, A. V. Nesterenko, A. I. Tsvik V. P. Shestopalov, Characteristics of the generator of diffractional radiation with repeated flight of electrons, *Radiotekhnika i Elektronika*, vol. 28, pp. 1611-1616 (1983).

29. G. S. Vorobjev, A. I. Tsvik V. P. Shestopalov, Experimental investigations of the non-linear processes of excitation of SHF oscillations in the reflection GDR, *Preprint No 298, IRE Ukr. Acad. Sci.*, Kharkov (1988).

30. V. K. Korneenkov, V. S. Miroshnichenko, A. I. Tsvik, V. P. Shestopalov, On excitation of stochastic oscillations in the generator of diffractional radiation – laser on free electrons, *Dokl. AN USSR, Ser. A*, No 5, pp. 59-61 (1982).

31. I. M. Balaklitskij, B. K. Skrinnik, A. I. Tsvik V. P. Shestopalov, On the possibility of creating impulse generator of diffractional radiation, *Izv. Vuzov, Radiofizika*, vol. 18, pp. 1528-1537 (1975).

32. I. M. Balaklitskij, G. S. Vorobjev, A. I. Tsvik V. P. Shestopalov, Investigations of spectral characteristics of the impulses of diffractional radiation, *Izv. Vuzov, Radiofizika*, vol. 21, pp. 1853-1861 (1975).

33. A. I. Tsvik V. P. Shestopalov, Experimental investigations of nonstationary non-linear proceses in impulse GDR, *Izv. Vuzov, Radiofizika*, vol. 29, pp. 524-530 (1986).

34. A. A. Vertij, G. S. Vorobjev, I. V. Ivanchenko, *et. al.*, Experimental investigations of transformation of surface waves to volume waves in an open waveguide, *Izv. Vuzov, Radiofizika*, vol. 31, pp. 1242-1254 (1988).

35. V. G. Kurin, B. K. Skrinnik, V. P. Shestopalov, Optimization of communication between generator of diffractional radiation and a load, *Izv. Vuzov, Radiofizika*, vol. 19, pp. 128-134 (1976).

36. I. D. Revin, B. K. Skrinnik, A. S. Sysoev *et. al.*, Linear theory of the generator of diffractional radiation, *Dokl. AN USSR, Ser. A*, No 3, pp. 259-264 (1976).

37. I. D. Revin, B. K. Skrinnik, A. S. Sysoev *et. al.*, On a linear theory of the generator of diffractional radiation, *Izv. Vuzov, Radiofizika*, vol. 20, pp. 764-776 (1977).

38. B. K. Skrinnik, Optimization of the parameters of the generator of diffractional radiation, In: *Proc. of the IX All-Union Conf. on SHF Electronics*, Kiev, p. 263 (1989).

39. C. A. Spindt, Triservice cathode workshop, In: *Proc. of Appl. Surface Sci. Progr. Cathode Res. Workshop*, Boston, USA, pp. 268-276 (1983).

40. V. N. Voroncov, G. P. Ermak, V. G. Kurin *et. al.*, Investigations of frequency instability of the generator of diffractional radiation – laser on free electrons in the short-wave part of millimeter range, In: *Proc. of the Conf. on Metrological Problems of the Measurements of Frequency and Spectral Characteristics of Laser Radiation*, Kharkov, NPO Metrology, p. 97-98 (1982).

41. I. M. Balaklitskij, A. M. Kasumov, Y. V. Maistrenko, V. P. Shestopalov, On fluctuation characteristics of the continuous electromagnetic radiation in electro-vacuum generators of the short-wave part of millimeter range, *Preprint No 64, IRE Ukr. Acad. Sci.*, Kharkov (1976).

42. G. P. Ermak, V. S. Korostelev, V. F. Minaev *et. al.*, Experimental investigation of the frequency fluctuations of "O"-type generatos in the 2-millimeter range, In: *Proc. of the X All-Union Conf. on SHF Electronics*, Minsk, pp. 254-255 (1983).

43. V. K. Korneenkov, A. A. Petrushin, B. K. Skrinnik *et. al.*, Generator of diffractional radiation with spherical-cylindrical open resonator, *Izv. Vuzov, Radiofizika*, vol. 19, pp. 812-821 (1976).

44. V. P. Shestopalov, I. M. Balaklitskij, B. K. Skrinnik *et. al.*, Effect of diffractional radiation and its application in electronics, *Electron. Tekhnika. Ser. 1. Elektronika SVC*, No 12, pp. 50-55 (1972).

45. V. K. Korneenkov, B. K. Skrinnik, V. P. Shestopalov, On determination of efficiency of the generator of diffractional radiation, *Dokl. AN USSR, Ser. A*, No 3, pp. 256-258 (1977).

46. G. P. Ermak, A. B. Lebedev, K. A. Lukin *et. al.*, Autodyne effect in the generator of diffractional radiation, *Izv. Vuzov, Radiofizika*, vol. 29, pp. 957-966 (1986).

47. A. A. Vertij, Y. P. Popkov, V. P. Shestopalov, Spectral relaxation-time measurements in HMBA Cr^{V}-radical at 150 GHz, *Int. J. Infrared and Millimeter Waves*, vol. 5, pp. 191-206 (1984).

48. A. A. Vertij, A. A. Karnaukhov, V. P. Shestopalov, *Polarization of atomic nuclei by millimeter waves*, Naukova Dumka, Kiev (1990).

49. G. P. Ermak, A. B. Lebedev, K. A. Lukin *et. al.*, Theoretical and

experimental investigations of GDR-autodyne, *Preprint No 262, IRE Ukr. Acad. Sci.*, Kharkov (1984).

50. K. A. Lukin, V. P. Shestopalov, Theory of non-linear and turbulent processes in physics, In: *Proc. of the II Int. Workshop. Part 1*, Kiev, pp. 383-388 (1985).

51. S. D. Andrenko, N. D. Devyatkov, V. P. Shestopalov, Antenna arrays of the millimeter wave range, *Dokl. AN SSSR*, vol. 240, pp. 1340-1343 (1978).

52. S. D. Andrenko, V. P. Shestopalov, Experimental investigations of the transformation of surface waves to volume waves in the millimeter wave range, *Preprint No 98, IRE Ukr. Acad. Sci.*, Kharkov (1975).

53. S. D. Andrenko, Y. B. Sidorenko, V. P. Shestopalov, On a transformation of surface waves to volume waves *Dokl. AN USSR, Ser. 4*, No 2, pp. 156-159 (1976).

54. S. D. Andrenko, V. P. Shestopalov, Diffraction logical elements of an optical computer, *Dokl. AN SSSR*, vol. 248, pp. 1070-1073 (1979).

55. K. Ishigaro, T. Tako, An estimation of Smith-Parcell effect as a light source in the infrared region, *Opt. Acta*, vol. 8, pp. 25-32 (1961).

56. S. J. Smith, Emission of light by a charged particle moving near diffraction grating, *Doct. Thesis*, Harvard Univ., Cambridge, MA, USA (1953).

57. V. P. Shestopalov, *Spectral theory and excitation of open structures*, Naukova Dumka, Kiev (1985).

58. V. B. Braginskij, Radiation of electromagnetic waves caused by uniform movement of electric charges near inhomogeneities, *Radiotekhnika i Elektronika*, vol. 1, pp. 225-232 (1956).

59. V. N. Parigin, Radiation of electric clots over diffractional grating, *Izv. Vuzov, Radiofizika*, vol. 1, pp. 139-147 (1958).

60. G. Toraldo di Francia, On the theory of Cherenkov effect, *Nuovo Cimento*, vol. 16, pp. 61-71 (1960).

61. V. P. Shestopalov, *Method of the Riemann-Hilbert problem in the theory of diffraction and propagation of electromagnetic waves*, Izd. Kharkov. Univers., Kharkov (1971).

62. V. P. Shestopalov, Y. K. Sirenko, *Dynamical theory of gratings*, Naukova Dumka, Kiev (1989).

63. O. A. Tretjakov, S. S. Tretjakova, V. P. Shestopalov, Radiation of electromagnetic waves by an electron flow moving over periodical structures,

In: *Proc. of the III All-Union Symp. on Diffraction*, Nauka, Tbilisi, pp. 172-173 (1964).

64. O. A. Tretjakov, S. S. Tretjakova, V. P. Shestopalov, Radiation of electromagnetic waves by an electron flow moving over diffractional grating, *Radiotekhnika i Elektronika*, vol. 10, pp. 1233-1242 (1965).

65. O. A. Tretjakov, S. S. Tretjakova, V. P. Shestopalov, Radiation of electromagnetic waves by an electron flow moving inside the ring waveguide, *Izv. Vuzov, Radiofizika*, vol. 8, pp. 552-562 (1965).

66. B. M. Bolotovskij, G. V. Voskresenskij, Diffractional radiation, *Usp. Fiz. Nauk*, vol. 88, pp. 209-238 (1966).

67. B. M. Bolotovskij, G. V. Voskresenskij, Radiation of charged particles in periodical structures, *Usp. Fiz. Nauk*, vol. 94, pp. 377-402 (1968).

68. R. W. Wood, On a remarkable case of uneven distribution of light in a diffraction grating spectrum, *Phil. Mag.*, vol. 4, pp. 396-403 (1902).

69. J. A. Bradshaw, A millimeter wave multiplier using Parcell radiator, In: *Proc. of the Symposium on Millimeter Waves*, New York, pp. 223-232 (1959).

70. V. G. Sologub, O. A. Tretjakov, V. P. Shestopalov, Excitation of electromagnetic field by an electron flow moving near diffractional grating, *Izv. Vuzov, Radiofizika*, vol. 11, pp. 588-602 (1968).

71. V. G. Sologub, V. P. Shestopalov, Resonance phenomena occurring in the process of diffraction of plane the H-polarized wave by the grating made of metal bars, *Jurn. Tekhn. Fiz.*, vol. 38, pp. 1505-1521 (1968).

72. V. E. Budanov, A. A. Kirilenko, V. P. Shestopalov, Characteristics of diffractional radiation of various diffractional gratings, *Radiotekhnika i Elektronika*, vol. 21, pp. 671-680 (1976).

73. V. P. Shestopalov, A. A. Kirilenko, S. A. Masalov, Reciprocity principle and some physical appropriatenesses of waves' scattering on diffractional gratings, *Vistn. AN USSR*, No 3, pp. 8-22 (1975).

74. A. S. Sysoev, Wave pencils of diffractional radiation, *Doct. Thesis*, Kharkov Univ., Khrakov, (1973).

75. D. S. Kuznecov, *Special functions*, 2-nd Ed., Vishaya Shkola, Moscow, 1965.

76. A. Erdelyi, *Asymptotic expansions*, Dover Publ., Inc. (1959).

77. M. A. Evgrafov, *Asymptotic estimates and entire functions*, 2-nd Ed., Fizmatgiz, Moscow (1962).

78. J. Dechan, P. Mast, Transformation of a pencil propagating in the system of quadratic lenses, In: *Quasioptica*, Mir, Moscow, pp. 189-202 (1966).

79. O. A. Tretjakov, E. I. Chernyakov, V. P. Shestopalov, Radiation of electromagnetic waves by an electron flow moving over diffractional grating, *Jurn. Tekhn. Fiz.*, vol. 36, pp. 33-40 (1966).

80. O. A. Tretjakov, E. I. Chernyakov, V. P. Shestopalov, On the theory of Smith-Parcell effect, *Izv. Vuzov, Radiofizika*, vol. 9, pp. 341-352 (1966).

81. A. Arfken, *Mathematical methods in physics*, Atomizdat, Moscow (1970).

82. H. Kogel'nik, T. Lee, Resonators and lasers' light beams, In: *Zarubezhn. Radioelektronika*, No 3, pp. 102-117 (1967).

83. D. B. Kanareikin, N. F. Pavlov, V. A. Potekhin, *Polarization of radio location signals*, Sov. Radio, Moscow (1966).

84. V. P. Shestopalov, Diffraction of electromagnetic waves by a beveled grating, *Jurn. Tekhn. Fiz.*, vol. 37, pp. 1442-1453 (1967).

85. S. N. Karp, Diffraction by finite and infinite gratings, *Phys. Rev.*, vol. 86, pp. 586-601 (1952).

86. Y. N. Fel'd, Diffraction by of electromagnetic wave by a semi-infinite grating, *Radiotekhnika i Elektronika*, vol. 3, pp. 882-900 (1958).

87. R. F. Millar, Plane wave spectra of grating theory. 1. Scattering by an infinite grating of identical cylinders, *Canad. J. Phys.*, vol. 41, pp. 2106-2115 (1963).

88. N. L. Hills, S. N. Karp, Semi-infinite diffraction gratings, *Comm. Pure and Appl. Math.*, vol. 18, pp. 203-215 (1965).

89. V. G. Sologub, On a solution to a convolution-type equation with finite interval of integration, *Jurn. Vych. Matem. i Matem. Fiz.*, vol. 11, pp. 837-850 (1971).

90. A. S. Borzenkov, V. G. Sologub, On a numerical solution to the problem of diffraction by a strip, *Vest. Kharkov. Univers., Ser. Matem., mekh.*, No 113, pp. 73-85 (1974).

91. V. G. Sologub, On a nethod of investigation of the problem of diffraction by a finite number of strips, *Dokl. AN USSR, Ser. 3*, No 6, pp. 550-568 (1975).

92. O. A. Tretjakov, Theory of the effect of diffractional radiation and its applications in electronics, *Doct. Thesis*, Khrakov Univ., Kharkov, (1972).

93. N. I. Muskhelishvili, *Singular integral equations*, Fizmatgiz, Moscow (1963).

94. L. A. Weinstien, *Open resonators and open waveguides*, Sov. Radio, Moscow (1966).

95. V. P. Shestopalov, V. A. Slusarskij, Experimental study of the diffraction of electromagnetic waves by plane diffractional gratings, *Jurn. Tekhn. Fiz.*, vol. 34, pp. 1649-1655 (1964).

96. O. A. Tretjakov, E. I. Chernyakov, V. P. Shestopalov, Radiation of electromagnetic waves by an electron flow moving over periodical structures, *Jurn. Tekhn. Fiz.*, vol. 37, pp. 233-240 (1967).

97. E. I. Chernyakov, Some problems of the radiation of electromagnetic waves by an electron flow moving over diffractional gratings, *Doct. Thesis*, Khrakov Univ., Kharkov, (1966).

98. A. I. Tsvik, E. I. Chernyakov, Diffractional radiation of shielded grating, *Radiotekhnika*, No 10, pp. 110-114 (1969).

99. L. I. Nikolaenko, A. I. Tsvik, Influence of dielectric layer on the radiation excited in periodical structures with losses, *Radiotekhnika*, No 19, pp. 101-107 (1969).

100. M . Danos, Cherenkov radiation by a long flow, In: *Millimeter and Submillimeter Waves*, Izd. Inostr. Lit., Moscow, pp. 81-90 (1959).

101. A. V. Nesterenko, V. P. Shestopalov, Excitation of Cherenkov radiation of the millimeter range by non-relativistic electron flow in an open dielectric resonator, In: *Proc. of the All-Union Conf. on the Design of Radio-Electronic Devices on Dielectric Waveguides and Resonators*, Tbilisi, pp. 216-217 (1988).

102. L. I. Nikolaenko, A. I. Tsvik, On a problem of excitation of diffractional radiation in periodical structures, *Radiotekhnika*, No 11, pp. 91-97 (1969).

103. O. A. Tretjakov, A. I. Tsvik, E. I. Chernyakov, Excitation of diffractional radiation in periodical structures with losses, *Radiotekhnika*, No 7, pp. 28-31 (1968).

104. V. M. Lopukhin, A. S. Roshal', *Electron-beam parameter amplifiers*, Sov. Radio, Moscow (1968).

105. O. A. Tretjakov, Radiation of charged particle moving over diffractional grating, *Radiotekhnika i Elektronika*, vol. 10, pp. 1343-1352 (1965).

106. E. V. Avdeev, G. V. Voskresenskij, Radiation of point charge uniformly moving near periodical system of perfectly conducting planes, *Radiotekhnika i Elektronika*, vol. 11, pp. 1560-1571 (1966).

107. V. P. Shestopalov, Problems of diffractional electronics, *Vistn. AN USSR*, No 10, pp. 28-40 (1975).

108. I. V. Alyamovskij, *Electron pencils and electron cannons*, Sov. Radio, Moscow (1966).

109. A. V. Gaponov-Grekhov, *Relativistic high-frequency electronics*, Gor'kij (1979).

110. A. Kratzer, V. Franz, *Transcendental functions*, Izd. Inostr. Lit., Moscow, (1963).

111. V. F. Vzyatishev, *Dielectric waveguides*, Nauka, Moscow (1969).

112. Y. B. Sidorenko, Radiation of the system dielectric waveguide-grating, In: *Proc. of the Conf. Inst. Radiophys. Electr.*, Kharkov, IRE, pp. 136-138 (1976).

113. W. W. Salisbury, Emission of visual radiation from free electrons, *J. Opt. Soc. Am.*, vol. 52, pp. 1315-1327 (1962).

114. G. R. Bird, Proposal: the generation of coherent infrared radiation from a diffraction grating, *IEEE Trans. Microwave Theory and Techn.*, vol. MTT-11, pp. 432-441 (1963).

115. N. Tagushi, S. Ono, A proposal of the new type of electron tube with a Fabry-Perot resonator, In: *Proc. of Research Inst. of Electron. Comm.*, Sendai, Japan, pp. 1-25 (1964).

116. F. S. Rusin, G. D. Bogomolov, Electron device for generation and amplifying of oscillations in the millimeter and submillimeter ranges, (USSR 195557, 16.2.1965), *Otkr., Izobr.*, No 10, p. 49 (1967).

117. K. Mizuno, S. Ono, Y. Shibata, Comments on "Orothron" – an electronic oscillator with an open resonator and reflecting grating, *Proc. IEEE*, vol. 57, pp. 2054-2055 (1969).

118. F. S. Rusin, Orothron, the generator of electromagnetic oscillations with an open resonator and reflecting grating, *Doct. Thesis*, Moscow, (1967).

119. F. S. Rusin, G. D. Bogomolov, Orothron as an electronic device containing open resonator and reflecting grating, *Izv. Vuzov, Radiofizika*, vol. 11, pp. 756-770 (1968).

120. F. S. Rusin, G. D. Bogomolov, Orothron as the generator of the millimeter range, In: *Powerful Electronics*, No 5, pp. 45-51 (1968).

121. A. A. Petrushin, I. M. Balaklitskij, V. P. Shestopalov, The plant for investigation of electromagnetic fields in open resonators of the millimeter range, *Pribori i Tekhn. Experim.*, No 2, pp. 147-151 (1970).

122. A. A. Petrushin, I. M. Balaklitskij, Investigation of open resonators applied in diffractional electronics, *Radiotekhnika*, No 7, pp. 111-123 (1968).
123. I. M. Balaklitskij, A. A. Petrushin, O. A. Tretjakov *et. al.*, Generators of diffractional radiation of the millimeter and submillimeter wave ranges, *Ukr. Fiz. J.*, vol. 13, pp. 429-440 (1968).
124. I. M. Balaklitskij, A. A. Petrushin, O. A. Tretjakov *et. al.*, Experimental studies of open resonators with diffractional reflection grating. P. I, *Ukr. Fiz. J.*, vol. 15, pp. 724-741 (1970).
125. I. M. Balaklitskij, A. A. Petrushin, O. A. Tretjakov *et. al.*, Experimental studies of open resonators with diffractional reflection grating. P.II, *Ukr. Fiz. J.*, vol. 15, pp. 893-900 (1970).
126. A. A. Vertij, N. A. Popenko, B. K. Skrinnik *et. al.*, Investigation of influence of phase inhomogeneity in the form of reflection diffractional grating on the characteristics of GDR, *Izv. Vuzov, Radiofizika*, vol. 18, pp. 109-121 (1975).
127. A. P. Koretskij, V. K. Korneenkov, A. A. Petrushin *et. al.*, Experimental studies of open spherical-cylindrical resonators in the millimeter wave range, *Izv. Vuzov, Radiofizika*, vol. 18, pp. 109-121 (1975).
128. V. B. Shteishleger, *The phenomenon of waves' interaction in electromagnetic resonators*, Oborongiz, Moscow (1955).
129. A. A. Vertij, N. A. Popenko, V. P. Shestopalov *et. al.*, Phase inhomogeneities in GDR having the form of reflection diffractional gratings, *Izv. Vuzov, Radiofizika*, vol. 19, pp. 129-138 (1976).
130. V. P. Shestopalov, A. A. Kirilenko, L. A. Rud', *Resonance scattering of waves. Vol. 2. Waveguide inhomogeneities.*, Naukova Dumka, Kiev (1988).
131. A. A. Vertij, Y. P. Popkov, V. P. Shestopalov, Investigations of main the characteristics of quasioptical resonance cell of radiospectrometer in the millimeter range, *Preprint No 208, IRE Ukr. Acad. Sci.*, Kharkov (1983).
132. V. P. Androsov, A. A. Vertij, Y. P. Popkov, V. P. Shestopalov, Application of 150 GHz quasioptical EPR and ENSOR systems to examinations of materials for polarized nuclear targets, *Int. J. Infrared and Millimeter Waves*, vol. 3, pp. 597-606 (1982).
133. A. A. Vertij, Y. P. Popkov, V. P. Shestopalov, Properties of open resonator with inhimogeneity in the form of localized diffractional grating, *Radiotekhnika i Elektronika*, vol. 25, pp. 412-425 (1980).
134. A. A. Vertij, V. P. Shestopalov, Polarization phenomena in the

generators of diffractional radiation – lasers on free electrons, *Dokl. AN SSSR*, vol. 262, pp. 1124-1127 (1982).

135. V. V. Shevchenko, *Gradual transitions in open waveguides*, Nauka, Moscow (1969).

136. Y. S. Komissarov, V. A. Pavluk, G. G. Polovnikov, Experimental justification of the measurement techniques applied for investigations of diffractional phenomena, *Radiotekhnika*, No 7, pp. 94-106 (1968).

137. S. D. Andrenko, A. A. Vertij, V. P. Shestopalov, On a method of measurements of the characteristics of transmission lines of surface waves in the millimeter and submillimeter ranges, *Proc. of the All-Union Symp. on the Propagation of Millimeter and Submillimeter Waves in Atmosphere of Earth and Planets*, Gor'kij, pp. 133-136 (1974).

138. C. W. Yeh, Elliptic dielectric waveguides, *J. Appl. Phys.*, vol. 33, pp. 3235-3247 (1962).

139. K. Walter, *Antennas of running wave*, Energija, Moscow (1970).

140. Y. V. Shubarin, *Antennas of super high frequencies*, Izd. Kharkov. Univers., Kharkov (1960).

141. S. D. Andrenko, V. P. Shestopalov, Electrodynamical modeling of the density modulated electron flow, *Proc. of the III Ukrainian Conf. on the Electron Optics and Applications*, Kharkov, pp. 59-60 (1974).

142. A. Y. Kirichenko, A. I. Tsvik, Transformation of waves of a dielectric waveguide to volume waves, *Izv. Vuzov, Radiofizika*, vol. 29, pp. 128-139 (1986).

143. S. D. Andrenko, Y. B. Sidorenko, V. P. Shestopalov, On a transformation of surface waves to volume waves, *Dokl. AN USSR, Ser. A*, No 2, pp. 156-159 (1976).

144. K. P. Yatsuk, A. I. Sapelkin, V. Y. Kogan, Experimental investigations of the waveguides of the millimeter range, *Radiotekhnika i Elektronika*, vol. 14, pp. 1314-1316 (1969).

145. V. E. Budanov, N. N. Suslov, V. P. Shestopalov, Experimental investigation of the radiation from an open generator of diffractional radiation with the help of "cold" model, *Ibid*, pp. 675-679.

146. S. D. Andrenko, A. P. Evdokimov, Y. B. Sidorenko, V. P. Shestopalov, Scattering of inhomogeneous waves by a beveled grating, *Preprint No 244, IRE Ukr. Acad. Sci.*, Kharkov (1984).

147. S. D. Andrenko, A. P. Evdokimov, Y. B. Sidorenko, V. P. Shestopalov, Control of the characteristics of scattered fields in the systems with trans-

formators of surface waves, *Preprint No 246, IRE Ukr. Acad. Sci.*, Kharkov (1984).

148. R. King, *Scattering and diffraction of electromagnetic waves*, Izd. Inostr. Lit., Moscow, (1962).

149. F. S. Rusin, Linear theory of orothron, In: In: *Powerful Electronics*, No 5, pp. 129-137 (1968).

150. F. S. Rusin, G. D. Bogomolov, Oscillation system of an orothron, In: In: *Powerful Electronics*, No 5, pp. 38-45 (1968).

151. A. A. Petrushin, V. P. Shestopalov, I. M. Balaklitskij, Device for visualization of electronagmetic field in open resonators, (USSR 286004, 06.1.1969), *Otkr., Izobr.*, No 30, p. 43 (1970).

152. I. M. Balaklitskij, V. G. Kurin, B. K. Skrinnik, GDR operating in the regime of LIW, *Ukr. Fiz. J.*, vol. 15, pp. 717-724 (1970).

153. I. M. Balaklitskij, V. G. Kurin, B. K. Skrinnik *et. al.*, The plant for examination of fields' distribution in the millimeter ranges in vaccum open resonators, *Pribori i Tekhn. Experim.*, No 5, pp. 107-111 (1974).

154. I. M. Balaklitskij, I. D. Revin, B. K. Skrinnik *et. al.*, Quasioptical generator of diffractional radiation, *Izv. Vuzov, Radiofizika*, vol. 16, pp. 235-243 (1973).

155. B. K. Skrinnik, Investigation of the spectra of the generator of diffractional radiation, *Radiotekhnika*, No 20, pp. 216-222 (1972).

156. I. M. Balaklitskij, V. G. Kurin, B. K. Skrinnik *et. al.*, Generator of diffractional radiation, of the millimeter wave range, *Electron. Tekhnika. Ser. 1. Elektronika SVC*, No 6, pp. 117-125 (1971).

157. V. P. Shestopalov, I. M. Balaklitskij, O. A. Tretjakov *et. al.*, Effect of diffractional radiation and its application in electronics, *Electron. Tekhnika. Ser. 1. Elektronika SVC*, No 12, pp. 50-63 (1972).

158. V. I. Gaiduk, K. I. Palatov, D. M. Petrov, *Physical foundations of super high frequency electronics*, Sov. Radio, Moscow (1971).

159. Y. I. Evdokimenko, K. A. Lukin, V. P. Shestopalov, On the two-dimensional non-linear stationary theory of GDR, *Electron. Tekhnika. Ser. 1. Elektronika SVC*, No 10, pp. 35-40 (1981).

160. V. I. Isaev, Transition processes in orothron, In: *Vopr. Elektron. Sverkhvis. Chast.*, Saratov, No 10, pp. 65-88 (1977).

161. I. M. Balaklitskij, A. I. Tsvik, Influence of amplitude and frequency modulation on the spectrum of impulse of SHF oscillations in GDR, *Izv. Vuzov, Radiofizika*, vol. 23, pp. 49-53 (1980).

162. V. S. Korostelev, B. K. Skrinnik, G. I. Khlopov *et. al.*, Synchronous measurements of short-time frequency instability of GDR in the two-millimeter range, In: *Proc. of the All-Union Symp. on Millimeter and Sub-millimeter Radio Waves. Vol. 1.*, Kharkov, pp. 19-20 (1984).

163. V. P. Shestopalov, B. K. Skrinnik, G. P. Ermak, Experimental examination of the limiting sensibility of GDR operating in autodyne regime, In: *Physics and Technology of Millimeter and Sub-millimeter Waves*, Naukova Dumka, Kiev, pp. 3-6 (1983).

164. L. A. Weinstien, V. A. Solntsev, *Lectures on super high frequency electronics*, Sov. Radio, Moscow (1973).

165. E. M. Gershenson, B. N. Tumanov, V. T. Buzikin *et. al.*, General characteristics and features of autodyne effect in autogenerators, *Radiotekhnika i Elektronika*, vol. 27, pp. 104-112 (1982).

166. A. N. Malakhov, *Fluctuations in auto-oscillating systems*, Nauka, Moscow (1967).

167. R. V. Lebedev, Y. A. Pirogov, Autodyne detecting in reflecting clystron with registering the response in the reflector's chain, *Vest. Mosk. Univers., Ser. 3*, vol. 21, pp. 75-79 (1980).

168. M. N. Devyatkov, A. I. Kostienko, E. Y. Myasoedov, Lamps of running waves as SHF detectors and mixers, *Radiotekhnika i Elektronika*, vol. 7, pp. 836-843 (1962).

169. A. I. Kostienko, M. N. Devyatkov, Y. A. Pirogov, Influence of millimeter radiation on the gap between electrodes in virtual cathode, *Electron. Tekhnika. Ser. 1. Elektronika SVC*, No 2, pp. 146-150 (1967).

170. Y. A. Pirogov, On a mechanism of detecting SHF signals by means of reflecting clystron, *Vest. Mosk. Univers., Ser. 3*, vol. 12, pp. 617-618 (1971).

171. B. N. Tumakov, V. G. Buzikin, Auto-oscillations in autodyne SHF generators, *Electron. Tekhnika. Ser. 1. Elektronika SVC*, No 2, pp. 3-9 (1983).

172. V. E. Kapitonov, I. V. Lopatin, A. S. Tishchenko *et. al.*, Small-size electron-optical systems for electro-vacuum devices of the millimeter and sub-millimeter ranges, In: *Physics and Technology of Millimeter and Sub-millimeter Waves*, Naukova Dumka, Kiev, pp. 168-176 (1983).

173. A. V. Deryagin, Rare earth magnetic-rigid materials, *Usp. Fiz. Nauk*, vol. 120, pp. 393-432 (1976).

174. T. I. Vasiljeva, I. D. Revin., V. P. Shestopalov, The features of energy

exchange in GDR containing diffractional grating placed non-symmetrcally in the caustics of open resonator, *Dokl. AN USSR, Ser. A*, No 7, pp. 53-85 (1984).

175. I. M. Balaklitskij, V. G. Kurin, B. K. Skrinnik, The plant for investigation of millimeter fields' distribution in vacuum open resonators, *Pribori i Tekhn. Experim.*, No 5, pp. 107-109 (1974).

176. V. K. Korneenkov, V. S. Miroshnichenko, V. P. Shestopalov, The study of fields' distribution in an open resonator of operating GDR, *Izv. Vuzov, Radiofizika*, vol. 28, pp. 1576-1581 (1985).

177. N. V. Kvach, V. G. Sologub, V. P. Shestopalov, Diffraction of plane H-polarized wave by the grating made of conducting rectangular bars, *Radiotekhnika*, No 34, pp. 3-11 (1975).

178. T. I. Vasiljeva, I. D. Revin., Characteristics of reflection grating with losses, In: *Radiophysics and Electronics of Millimeter and Submillimeter Waves*, Kharkov, pp. 42-50 (1988).

179. A. E. Poyedinchuk, Y. A. Tuchkin, V. P. Shestopalov, On the analytical nature of the phenomenon of intertype interaction of eigenoscillations, *Dokl. AN SSSR*, vol. 300, pp. 1356-1359 (1988).

180. V. P. Shestopalov, Resonance and intertype oscillations, *Dokl. AN SSSR*, vol. 303, pp. 1131-1134 (1988).

181. J. Birnbaum, *Optical quantum generators*, Sov. Radio, Moscow (1967).

182. R. A. Valitov, s. F. Dubko, V. V. Kamishan *et. al.*, Submillimeter waves technology, Sov. Radio, Moscow (1969).

183. G. Korn, T. Korn, *Mathematical reference book*, Nauka, Moscow (1970).

184. V. E. Kapitonov, A method for calculating magnetic focusing systems with minimal mass and size, In: *Proc. of the VIII All-Union Sem. on the Methods of Calculation of Electron-Optical Systems*, Leningrad, p. 24 (1986).

185. V. E. Kapitonov, A method for synthesis of optimal magnetic systems, *Preprint No 256, IRE Ukr. Acad. Sci.*, Kharkov (1984).

186. V. E. Kapitonov, Optimization of magnetic focusing systems by the method of synthesis, In: *Electronics of Millimeter and Sub-millimeter Ranges*, Naukova Dumka, Kiev, pp. 172-178 (1988).

187. I. D. Revin, B. K. Skrinnik, N. T. Cherpak, Generator of diffractional radiation as a source for pumping of quantum amplifiers of running waves in

the millimeter range, *Ibid*, pp. 3-13.

188. I. D. Revin, B. K. Skrinnik, N. T. Cherpak *et. al.*, Generator of diffractional radiation in quantum radiophysics of the millimeter waves range, *Preprint No 267, IRE Ukr. Acad. Sci.*, Kharkov (1985).

189. B. M. Alendev, M. H. Varshavskij, A. A. Veshchipov, *Measurements of spectral-pure and correlation parameters and characteristics of laser radiation*, Radio i Svyaz', Moscow (1982).

190. V. P. Shestopalov, On non-linear processes in the generators of diffractional radiation, *Dokl. AN SSSR*, vol. 261, pp. 1116-1118 (1981).

191. R. P. Leavitt, D. F. Wortman, C. A. Morrison, The orothron – a free-electron laser using the Smith-Parcell effects, *Appl. Phys. Let.*, vol. 35, pp. 363-365 (1979).

192. K. Muzuno, S. Ono, Y. Shibata, A new electron tube with a Fabry-Perot resonator for the generation of waves, In: *Proc. of the Symp. on Submillimeter Waves*, Polytechnic Press, New York, pp. 115-134 (1970).

193. L. Shengang, C. Jiayu, Y. Shaufu *et. al.*, The study of diffraction radiation oscillations, *Int. J. Infrared and Millimeter Waves*, vol. 8, pp. 885-890 (1987).

194. V. N. Shevchik, G. N. Shvedov, A. V. Soboleva, *Wave and oscillation phenomena in electron flows at super high frequencies*, Izd. Saratov. Univers., Saratov (1962).

195. I. M. Balaklitskij, E. V. Belousov, V. G. Korzh, Photo-method for investigation of electron pencils with high specific power, *Izv. Vuzov, Radioelektronika*, vol. 25, pp. 38-42 (1982).

196. V. D. Eremka, A. Y. Kirichenko, L. A. Lisova, Application of transition radiation for the study of dynamical characteristics of electron pencils, *Izv. Vuzov, Radioelektronika*, vol. 22, pp.153-158 (1977).

197. M. Barnoski, *Introduction to integrated optics*, Mir, Moscow (1977).

198. A. M. Goncharenko, V. P. Red'ko, *Foundations of integrated optics*, Nauka i Tekhnika, Minsk (1975).

199. Y. R. Nosov, *Optoelectronics*, Sov. Radio, Moscow (1977).

200. A. F. Harway, *Super high frequency technology*, Sov. Radio, Moscow (1965).

201. M. Skolnik, *Introduction to the radio location technology*, Mir, Moscow (1965).

202. M. Born, E. Wolf, *Foundations of optics*, Nauka, Moscow (1970).

203. E. M. Zolotov, V. A. Kiselev, V. A. Sychugov, Optical phenomena in thin-film waveguides, *Usp. Fiz. Nauk*, vol. 112, pp. 231-251 (1974).

204. T. Kogel'nik, Introduction to integrated optics, *Usp. Fiz. Nauk*, vol. 121, pp. 695-712 (1977).

205. A. M. Prokhorov, Perspectives of fiber optics, *Vestnik AN SSSR*, vol. 2, pp. 18-25 (1976).

206. Y. N. Denisuk, New methods of holography, *Dokl. AN SSSR*, vol. 144, pp. 1275-1278 (1962).

207. S. D. Andrenko, A. Y. Usikov, V. P. Shestopalov, On a new method of establishing the wave front, *Dokl. AN SSSR*, vol. 244, pp. 83-86 (1979).

208. S. D. Andrenko, V. P. Shestopalov, On a method of forming the holograms, *Dokl. AN SSSR*, vol. 246, pp. 69-72 (1979).

209. L. M. Soroko, *Foundations of holography and coherent optics*, Nauka, Moscow (1971).

210. V. A. Abdulkadirov, Interaction of drifting carriers of charge with the field of periodical structure in a semi-conductor, *Dokl. AN USSR, Ser. A*, No 12, pp. 50-53 (1980).

211. V. A. Abdulkadirov, V. P. Shestopalov, Of principle possibility to amplify, generate and transform high frequency oscillations by space-periodical semi-conductor system, *Dokl. AN USSR, Ser. A*, No 10, pp. 35-38 (1987).

212. V. A. Abdulkadirov, I. D. Revin, V. P. Shestopalov, Interaction of drifting carriers of charge in a semi-conductor, with the field of open resonator, *Ukr. Fiz. J.*, vol. 38, pp. 548-554 (1993).

213. L. V. Keldysh, Deep levels in in semi-conductors, *J. Exp. Teor. Fiz.*, vol. 45, pp. 364-375 (1963).

214. F. G. Bass, V. A. Pogrebnyak, On the analogy between relativistic plasma and electron gas in semi-conductors with non-quadratic dispersion, *Fiz. Tverdogo Tela*, vol. 14, pp. 1766-1799 (1972).

215. V. P. Shestopalov, *Morse critical points of dispersion equations*, Naukova Dumka, Kiev (1992).

216. F. G. Bass, I. E. Pochanina, V. P. Shestopalov, Non-linear evolution equations in the vicinity of Morse critical points, *Dokl. AN Rossii*, vol. 334, pp. 32-33 (1993).

217. V. G. Kurin, B. K. Skrinnik, V. P. Shestopalov, Experimental studies of intertype oscillations, *Dokl. AN SSSR*, vol. 317, pp. 93-96 (1991).

218. V. G. Kurin, B. K. Skrinnik, V. P. Shestopalov, Intertype oscillations (waves) of the generator of diffractional radiation, *Dokl. AN Rossii*, vol. 324, pp. 997-999 (1992).

219. V. G. Kurin, B. K. Skrinnik, V. P. Shestopalov, Phenomenon of frequency auto-stabilization in the generators of diffractional radiation, *Pisma v JTF*, vol. 18, pp. 33-37 (1992).

INDEX

analysis, Smith-Purcell experiments and their, 201-204
antenna properties of Smith-Purcell radiation for beveled gratings, 288-312
antenna properties of Smith-Purcell radiation, 401-410
application of Smith-Purcell effect for generation and amplifying of millimeter and submillimeter waves in semi-conductors, 428-450
application of Smith-Purcell effect in other areas of science and technology, 393-450
autodyne operation regimes of the generators of diffractional radiation, reflection, impulse and, 328-358
basic generator of diffractional radiation, 317-327
beveled gratings, antenna properties of Smith-Purcell radiation for, 288-312
beveled strip grating, theoretical study of the wave Smith-Purcell radiation on the dielectic waveguide and, 183-199
charged clots, Smith-Purcell radiation of the point charge, single clot and finite sequence of, 47-57
Cherenkov radiation, Smith-Purcell and, 109-154
Cherenkov radiations, experimental modeling of Smith-Purcell and, 281-287
Cherenkov radiations, theory of Smith-Purcell and, 110-122
conducting rectangular bars, Smith-Purcell radiation for the gratings made of perfectly, 15-34
continuous flight regime, 317-327
dielectic waveguide and beveled strip grating, theoretical study of the wave Smith-Purcell radiation on the, 183-199
dielectic waveguide and beveled strip grating, theoretical study of the wave Smith-Purcell radiation on the, 183-199
dielectric waveguide strip diffractional gratings placed perpendicular to the waveguide, scattering of waves of, 159-182
dielectric waveguide, diffractional input of energy to, 421-422
diffractional grating, open resonators with local reflection, 235-243
diffractional gratings placed perpendicular to the waveguide, scattering of waves of dielectric waveguide strip, 159-182
diffractional gratings, justification of the method of wave modeling of transformation of surface waves to volume waves by, 269-280
diffractional input of energy to dielectric waveguide, 421-422
diffractional logical elements of optical computer, 423-425

diffractional radiation, basic generator of, 317-327
diffractional radiation, generators of, 315-392
diffractional radiation, package generators of, 359-392
diffractional radiation, reflection, impulse and autodyne operation regimes of the generators of, 328-358
dynamical parameters of electron flow, measurements of the, 393-400
electron flow over strip grating, unbounded, 5-14
electron flow with the field of Smith-Purcell radiation in open resonators, interaction of, 244-247
electron flow, measurements of the dynamical parameters of, 393-400
electron-wave excitation of Smith-Purcell radiation, 123-154
electron-wave model of Smith-Purcell effect, 109-154
excitation of open resonators by Smith-Purcell radiation, 235-268
excitation of planar dielectric waveguide and other elements of millimeter technology based on Smith-Purcell effect, the device for, 419-420
experimental methods for the study of Smith-Purcell radiation in the millimeter range, 206-214
experimental modeling of Smith-Purcell and Cherenkov radiations, 281-287
experimental wave modeling of Smith-Purcell effect, 269-312
experiments of Ishigaro and Tako, 205-206
finite sequence of charged clots, Smith-Purcell radiation of the point charge, single clot and, 47-57
finite size of gratings on the Smith-Purcell radiation, influence of the, 58-109
formation of holograms with the help of Smith-Purcell radiation, 426-427
generators of diffractional radiation, 315-392
generators of diffractional radiation, reflection, impulse and autodyne operation regimes of the, 328-358
grating, interaction of Smith-Purcell radiation with the field scattered by a, 215-234
gratings made of perfectly conducting rectangular bars, Smith-Purcell radiation for the, 15-34
gratings on the Smith-Purcell radiation, influence of the finite size of, 58-109
holograms with the help of Smith-Purcell radiation, formation of, 426-427
"Hot" experiments with "pure" Smith-Purcell effects, 201-234
impulse and autodyne operation regimes of the generators of diffractional radiation, reflection, 328-358
influence of the finite size of gratings on the Smith-Purcell radiation, 58-109
integrated schemes of millimeter range, interferometer for the, 416-418
interaction of electron flow with the field of Smith-Purcell radiation in open resonators, 244-247
interaction of Smith-Purcell radiation with the field scattered by a grating, 215-234
interferometer for the integrated schemes of millimeter range, 416-418
inverse Smith-Purcell effect, 215-234
Ishigaro and Tako, experiments of 205-206
local reflection diffractional grating, Open resonators with, 235-243

measurements of the dynamical parameters of electron flow, 393-400
millimeter and submillimeter waves in semi-conductors, application of, Smith-Purcell effect for generation and amplifying of, 428-450
millimeter and submillimeter waves on the basis of Smith-Purcell effect, phase measurements for, 411-415
millimeter range, experimental methods for the study of Smith-Purcell radiation in the, 206-214
millimeter range, interferometer for the integrated schemes of, 416-418
millimeter technology based on Smith-Purcell effect, the device for excitation of planar dielectric waveguide and other elements of, 419-420
open resonators by Smith-Purcell radiation, excitation of, 235-268
open resonators with local reflection diffractional grating, 235-243
open resonators, interaction of electron flow with the field of Smith-Purcell radiation in, 244-247
open resonators, polarization effects of Smith-Purcell radiation in, 248-268
optical computer, diffractional logical elements of, 423-425
package generators of diffractional radiation, 359-392
phase measurements for millimeter and submillimeter waves on the basis of Smith-Purcell effect, 411-415
planar dielectric waveguide and other elements of millimeter technology based on Smith-Purcell effect, the device for excitation of, 419-420
Plane waves and quasi-optical wave pencils of Smith-Purcell radiation, 5-108
point charge, single clot and finite sequence of charged clots, Smith-Purcell radiation of the, 47-57
polarization effects of Smith-Purcell radiation in open resonators, 248-268
quasi-optical Smith-Purcell wave pencils, 35-46
quasi-optical wave pencils of Smith-Purcell radiation, plane waves and, 5-108
rectangular bars, Smith-Purcell radiation for the gratings made of perfectly conducting, 15-34
reflection, impulse and autodyne operation regimes of the generators of diffractional radiation, 328-358
remote sensing method based on Smith-Purcell radiation, 393-400
resonators by Smith-Purcell radiation, excitation of open, 235-268
resonators with local reflection diffractional grating, open, 235-243
resonators, interaction of electron flow with the field of Smith-Purcell radiation in, open, 244-247
resonators, polarization effects of Smith-Purcell radiation in open, 248-268
scattering of waves of dielectric waveguide strip diffractional gratings placed perpendicular to the waveguide, 159-182
science and technology, application of Smith-Purcell effect in other areas of, 393-450
semi-conductors, application of Smith-Purcell effect for generation and amplifying of millimeter and submillimeter waves in, 428-450
sensing method based on Smith-Purcell radiation, remote, 393-400

single clot and finite sequence of charged clots, Smith-Purcell radiation of the point charge, 47-57
Smith-Purcell and Cherenkov radiation, 109-154
Smith-Purcell and Cherenkov radiations, experimental modeling of, 281-287
Smith-Purcell effect in other areas of science and technology, application of, 393-450
Smith-Purcell effect, electron-wave model of, 109-154
Smith-Purcell effect, experimental wave modeling of, 269-312
Smith-Purcell effect, inverse, 215-234
Smith-Purcell effect, phase measurements for millimeter and submillimeter waves on the basis of, 411-415
Smith-Purcell effect, the device for excitation of planar dielectric waveguide and other elements of millimeter technology based on, 419-420
Smith-Purcell effect, wave model of, 159-182
Smith-Purcell experiments and their analysis, 201-204
Smith-Purcell radiation for beveled gratings, antenna properties of, 288-312
Smith-Purcell radiation for the gratings made of perfectly conducting rectangular bars, 15-34
Smith-Purcell radiation in open resonators, interaction of electron flow with the field of, 244-247
Smith-Purcell radiation in open resonators, polarization effects of,248-268
Smith-Purcell radiation in the millimeter range, Experimental methods for the study of, 206-214
Smith-Purcell radiation of the point charge, single clot and finite sequence of charged clots, 47-57
Smith-Purcell radiation on the dielectic waveguide and beveled strip grating, theoretical study of the wave, 183-199
Smith-Purcell radiation with the field scattered by a grating, interaction of, 215-234
Smith-Purcell radiation, antenna properties of, 401-410
Smith-Purcell radiation, electron-wave excitation of 123-154
Smith-Purcell radiation, excitation of open resonators by, 235-268
Smith-Purcell radiation, formation of holograms with the help of, 426-427
Smith-Purcell radiation, influence of the finite size of gratings on the, 58-109
Smith-Purcell radiation, plane waves and quasi-optical wave pencils of, 5-108
Smith-Purcell radiation, remote sensing method based on, 393-400
Smith-Purcell radiation, wave modeling of, 155-158
Smith-Purcell wave pencils, quasi-optical, 35-46
strip diffractional gratings placed perpendicular to the waveguide, scattering of waves of dielectric waveguide, 159-182
strip grating, theoretical study of the wave Smith-Purcell radiation on the dielectic waveguide and beveled, 183-199
strip grating, unbounded electron flow over, 5-14
submillimeter waves in semi-conductors, application of Smith-Purcell effect for generation and amplifying of millimeter and, 428-450

submillimeter waves on the basis of Smith-Purcell effect, phase measurements for millimeter and, 411-415
surface waves to volume waves by diffractional gratings, justification of the method of wave modeling of transformation of, 269-280
Tako, experiments of Ishigaro and, 205-206
theoretical study of the wave Smith-Purcell radiation on the dielectic waveguide and beveled strip grating, 183-199
theory of Smith-Purcell and Cherenkov radiations, 110-122
unbounded electron flow over strip grating, 5-14
volume waves by diffractional gratings, justification of the method of wave modeling of transformation of surface waves to, 269-280
wave model of Smith-Purcell effect, 159-182
wave modeling of Smith-Purcell effect, experimental, 269-312
wave modeling of Smith-Purcell radiation, 155-158
wave modeling of transformation of surface waves to volume waves by diffractional gratings, justification of the method of, 269-280
wave pencils of Smith-Purcell radiation, plane waves and quasi-optical, 5-108
wave pencils, quasi-optical Smith-Purcell, 35-46
waveguide and other elements of millimeter technology based on Smith-Purcell effect, the device for excitation of planar, dielectric, 419-420
waveguide strip diffractional gratings placed perpendicular to the waveguide, scattering of waves of dielectric, 159-182
waveguide, diffractional input of energy to dielectric, 421-422
waveguide, scattering of waves of dielectric waveguide strip diffractional gratings placed perpendicular to the, 159-182